Cultural Evolution

Society, Technology, Language, and Religion

Strüngmann Forum Reports

Julia Lupp, series editor

The Ernst Strüngmann Forum is made possible through the generous support of the Ernst Strüngmann Foundation, inaugurated by Dr. Andreas and Dr. Thomas Strüngmann.

This Forum was supported by funds from the Deutsche Forschungsgemeinschaft (German Science Foundation)

Cultural Evolution

Society, Technology, Language, and Religion

Edited by

Peter J. Richerson and Morten H. Christiansen

Program Advisory Committee:
Morten H. Christiansen, Herbert Gintis, Stephen C. Levinson, Peter J. Richerson, Stephen Shennan, and Edward Slingerland

The MIT Press

Cambridge, Massachusetts
London, England

This volume is the result of the 12th Ernst Strüngmann Forum,
held May 27 to June 1, 2012, in Frankfurt am Main, Germany.

Series Editor: J. Lupp
Assistant Editor: M. Turner
Photographs: U. Dettmar
Design and realization: BerlinScienceWorks

The book was set in TimesNewRoman and Arial.

Library of Congress Cataloging-in-Publication Data

Cultural evolution : society, technology, language, and religion / edited by Peter J. Richerson and Morten H. Christiansen.
pages cm. — (Strüngmann forum reports)
Includes bibliographical references and index.
ISBN 978-0-262-01975-0 (hardcover : alk. paper)
ISBN 978-0-262-55190-8 (paperback)
1. Social evolution. 2. Technological innovations—Social aspects. 3. Human evolution—Religious aspects. I. Richerson, Peter J., editor of compilation. II. Christiansen, Morten H., 1963–, editor of compilation.
HM626.C84295 2013
303.4—dc23

2013018994

Contents

The Ernst Strüngmann Forum

Founded on the tenets of scientific independence and the inquisitive nature of the human mind, the Ernst Strüngmann Forum is dedicated to the continual expansion of knowledge. Through its innovative communication process, the Ernst Strüngmann Forum provides a creative environment within which experts scrutinize high-priority issues from multiple vantage points.

This process begins with the identification of themes. By nature, a theme constitutes a problem that transcends classic disciplinary boundaries. It is of high-priority interest and requires concentrated, multidisciplinary input to address the issues involved. Proposals are received from leading scientists active in their field and are selected by an independent Scientific Advisory Board. Once approved, a steering committee is convened to refine the scientific parameters of the proposal and select the participants. Approximately one year later, a central meeting, or Forum, is held to which circa forty experts are invited.

Preliminary discussion for this theme began in 2010, when Morten Christiansen and Pete Richerson brought the initial idea to my attention. As participants from past Forums, they were familiar with our philosophy and approach, and felt that a Forum might assist future enquiry into cultural evolution. The resulting proposal was approved by the Scientific Advisory Board, and a steering committee was convened from June 6–8, 2011. This committee was comprised of Morten H. Christiansen (cognitive scientist), Herbert Gintis (economist), Stephen C. Levinson (linguist), Peter J. Richerson (biologist), Stephen Shennan (archaeologist), and Edward Slingerland (historian). Together, they identified the key issues for debate at the Forum, which was convened in Frankfurt am Main from May 27 to June 1, 2012.

A Forum is a dynamic think tank. The activities and discourse that accompany it begin well before participants arrive in Frankfurt and conclude with the publication of this volume. Throughout each stage, focused dialog is the means by which issues are examined anew. Often, this requires relinquishing long-established ideas and overcoming disciplinary idiosyncrasies, which otherwise would inhibit joint examination. When this is accomplished, however, new insights begin to emerge.

This volume conveys the synergy that arose out of myriad discussions between diverse experts, each of whom assumed an active role. It contains two types of contributions. The first provides background information to key aspects of the overall theme. Originally written before the Forum, these chapters have been extensively reviewed and revised to provide current understanding on these topics. The second (Chapters 6, 11, 16, and 20) summarizes the extensive discussions that transpired at the meeting as well as thereafter. These chapters should not be viewed as consensus documents nor are they proceedings. Their goal is to transfer the essence of these discussions, expose open questions, and highlight areas for future enquiry.

An endeavor of this kind creates its own unique group dynamics and puts demands on everyone who participates. Each invitee contributed not only their time and congenial personality, but a willingness to probe beyond that which is evident. For this, I extend my gratitude to all.

A special word of thanks goes to the steering committee, the authors of the background papers, the reviewers of the papers, and the moderators of the individual working groups: Carel van Schaik, Kevin Laland, Stephen Levinson, and Armin Geertz. To draft a report during the week of the Forum and bring it to its final form in the months thereafter is never a simple matter. For their efforts and tenacity, I am especially grateful to Fiona Jordan, Alex Mesoudi, Dan Dediu, Michael Cysouw, and Joseph Bulbulia—the rapporteurs of the four discussion groups.

Most importantly, I wish to extend my sincere appreciation to Morten Christiansen and Pete Richerson. As equal partners, they worked through each stage, from the development of the proposal to the editing of this volume. Their commitment to this 12th Ernst Strüngmann Forum ensured a most vibrant intellectual gathering.

A communication process of this nature relies on institutional stability and an environment that encourages free thought. The generous support of the Ernst Strüngmann Foundation, established by Dr. Andreas and Dr. Thomas Strüngmann in honor of their father, enables the Ernst Strüngmann Forum to conduct its work in the service of science. In addition, the following valuable partnerships are gratefully acknowledged: the Scientific Advisory Board ensures the scientific independence of the Forum; the German Science Foundation provided financial support for this theme; and the Frankfurt Institute for Advanced Studies shares its vibrant intellectual setting with the Forum.

Long-held views are never easy to put aside. Yet when this happens, when the edges of the unknown begin to appear and gaps in knowledge are able to be discerned, the act of formulating strategies to fill such gaps becomes a most invigorating exercise. On behalf of all involved, I hope that this volume will convey a sense of this and stimulate further enquiry into cultural evolution.

Julia Lupp, Program Director
Ernst Strüngmann Forum
Frankfurt Institute for Advanced Studies (FIAS)
Ruth-Moufang-Str. 1, 60438 Frankfurt am Main, Germany
http://esforum.de

List of Contributors

Quentin D. Atkinson Department of Psychology, University of Auckland, Auckland 1142, New Zealand

Andrea Baronchelli Department of Physics, Northeastern University, Nightingale Hall, Boston, MA 02115, U.S.A.

Robert Boyd School of Human Evolution and Social Change, Arizona State University, Tempe, AZ 85287–2402, U.S.A., and Santa Fe Institute, Santa Fe, NM 87501, U.S.A.

Briggs Buchanan Department of Archaeology, Simon Fraser University, Burnaby, BC V5A 1S6, Canada

Joseph Bulbulia Religious Studies Programme, Faculty of Humanities and Social Sciences, Victoria University, Wellington 6140, New Zealand

Morten H. Christiansen Department of Psychology, Cornell University, Ithaca, NY 14853, U.S.A.

Emma Cohen Max Planck Institute for Psycholinguistics, 6525 XD Nijmegen, The Netherlands

William Croft Department of Linguistics, University of New Mexico, Albuquerque, NM 87131–0001, U.S.A.

Michael Cysouw Forschungszentrum Deutscher Sprachatlas, Philipps University Marburg, 35032 Marburg, Germany

Dan Dediu Language and Genetics, Max Planck Institute for Psycholinguistics, 6525 XD Nijmegen, The Netherlands, and Donders Institute for Brain, Cognition and Behaviour, 6525 HR, Radboud University Nijmegen, The Netherlands

Nicholas Evans School of Culture, History and Language, The Australian National University, HC Coombs, ACT 0200 Canberra, Australia

Emma Flynn Department of Psychology, Durham University, Durham DH1 3LE, U.K.

Pieter François Centre for Anthropology and Mind at the School of Anthropology, Oxford OX2 6PE, U.K.

Simon Garrod Institute of Neuroscience and Psychology, University of Glasgow, Glasgow G12 8QB, U.K.

Armin W. Geertz Department of Culture and Society–Study of Religion, Aarhus University, 8000, Aarhus C, Denmark

Herbert Gintis Santa Fe Institute, Santa Fe, NM 87501, U.S.A. and Central European University, Budapest, Hungary

Russell D. Gray Department of Psychology, University of Auckland, Auckland 1142, New Zealand, and Department of Philosophy, Research School of the Social Sciences, Australian National University, 0200 Canberra, ACT, Australia

Simon J. Greenhill School of Culture, History and Language, College of Asia and the Pacific, Australian National University, ACT 0200, Australia

Daniel B. M. Haun Max Planck Research Group for Comparative Cognitive Anthropology, 04103 Leipzig, Germany

Joseph Henrich Canada Research Chair in Culture, Cognition and Coevolution, University of British Columbia, Vancouver, BC V6T 1Z4, Canada

Daniel J. Hruschka School of Human Evolution and Social Change, Arizona State University, Tempe, AZ 85287, U.S.A.

Marco A. Janssen School of Human Evolution and Social Change, Arizona State University, Tempe, AZ 85287, U.S.A.

Fiona M. Jordan Department of Archaeology and Anthropology, University of Bristol, Bristol BS8 1UU, U.K.

Anne Kandler Santa Fe Institute, Santa Fe, NM 87501, U.S.A.

James A. Kitts Graduate School of Business, Columbia University, New York, NY 10027, U.S.A.

Kevin N. Laland School of Biology, University of St. Andrews, St. Andrews KY16 9TS, U.K.

Laurent Lehmann Department of Ecology and Evolution, University of Lausanne, Quartier UNIL-Sorge, 1015 Lausanne, Switzerland

Stephen C. Levinson Language and Cognition, Max Planck Institute for Psycholinguistics, 6525 XD Nijmegen, The Netherlands, and Radboud University Nijmegen, 6500 HC, The Netherlands

Elena Lieven Max Planck Institute of Evolutionary Anthropology, 04103 Leipzig, Germany, and School of Psychological Sciences, University of Manchester, Manchester M13 9P, U.K.

Sarah Mathew Department of Anthropology, Stony Brook University, Stony Brook, NY 11794–4364, U.S.A.

Robert N. McCauley Center for Mind, Brain, and Culture, Emory University, Atlanta, GA 30322, U.S.A.

Alex Mesoudi Department of Anthropology, Durham University, Durham DH1 3LE, U.K.

Ara Norenzayan Department of Psychology, University of British Columbia, Vancouver, BC V6R IZ3, Canada

Harriet Over Department of Developmental and Comparative Psychology, Max Planck Institute for Evolutionary Anthropology, 04103 Leipzig, Germany

Jürgen Renn Max Planck Institute for the History of Science, 14195 Berlin, Germany

Victoria Reyes-García ICTA, Campus UAB, Universitat Autònoma de Barcelona, 08193 Bellaterra, Barcelona, Spain

Peter J. Richerson Department of Environmental Science and Policy, University of California–Davis, Davis, CA 95616, U.S.A., and Institute of Archaeology, University College, London, U.K.

Stephen Shennan Institute of Archaeology, University College London, London WC1H 0PY, U.K.

Edward Slingerland Asian Center, University of British Columbia, Vancouver, B.C. V6T 1Z2, Canada

Dietrich Stout Department of Anthropology, Emory University, Atlanta, GA 30322, U.S.A.

Claudio Tennie Department of Developmental and Comparative Psychology, Max Planck Institute for Evolutionary Anthropology, 04103 Leipzig, Germany

Peter Turchin Department of Ecology and Evolutionary Biology, and Department of Anthropology, University of Connecticut, Storrs, CT 06269, U.S.A.

Carel van Schaik Anthropological Institute, University of Zürich, 8057 Zürich, Switzerland

Matthijs Van Veelen Department of Economics, University of Amsterdam, 1018 WB Amsterdam, The Netherlands

Harvey Whitehouse School of Anthropology, University of Oxford, Oxford OX2 6PE, U.K.

Thomas Widlok Cultural Anthropology, Radboud University Nijmegen, 6500 HE Nijmegen, The Netherlands

Polly Wiessner Department of Anthropology, University of Utah, Salt Lake City, UT 84108, U.S.A.

David S. Wilson Departments of Biology and Anthropology, Binghamton University, Binghamton, NY 13902, U.S.A.

1

Introduction

Peter J. Richerson and Morten H. Christiansen

Objectives of the Forum

Over the past forty years, the field of cultural evolution has grown from a handful of theorists using concepts and formal tools from population biology to model cultural change to a much larger and more diverse scholarly enterprise. This diversity has contributed valuable insights but it has also generated challenges for integration and comparison. This Ernst Strüngmann Forum provided an opportunity to bring together a cross section of active scholars representing this diversity to consider the present state of the field and to outline outstanding problems and future directions. Given our involvement at past Forums, we felt that this venue would best help us accomplish this task.

Briefly stated, the Forum offers scientists the opportunity to retreat and scrutinize the state of their fields. There are no lectures or presentations; instead, the entire time is spent in discussion. Previously held perspectives are subjected to debate and "gaps in knowledge" emerge; ways of "filling these gaps" are collectively sought, thereby defining possible directions for future research. The essence of these multifaceted discussions is then captured in book form for the purpose of expanding discussion even further.

In this introductory chapter, we wish to provide background to the overarching issues as well as to the specific discussions of this Forum. One of our most important objectives was to assess the extent to which studies of cultural evolution cohere as a common field of investigation. Contributors to this book, and to the field more generally, come from an exceedingly large number of conventional disciplines in the natural sciences, social sciences, and humanities—each of which has its own distinct methodologies and traditional subject matter preoccupations. In an effort to find common ground, Herbert Gintis (2007) has proposed gene–culture coevolution (i.e., cultural evolution plus the biological evolution that affects, or is affected by, cultural evolution) as one of a handful of concepts that can contribute to the unification of the disciplines that study humans, much as biological evolution is one of the major synthetic principles for biology. If this concept is correct, then participants in a highly interdisciplinary group such as the one we assembled—united by little else than an interest in cultural evolution or something that is highly relevant to it—as well as those who will read this volume should find that such differences primarily reflect a comfortable division of labor rather than incommensurable perspectives which form no part of a larger project.

The Forum did give most of the participants a strong sense of being engaged in a common project. This book allows us to share some of the excitement of the Forum with the other members of the cultural evolution community and interested observers. We found the discussions at the Forum to be both positive and productive, a comparative rarity in our experience at interdisciplinary conferences, and hope that they will provoke further discussions throughout the human sciences.

Why the Four Sections

The basic argument in this book is that many aspects of human endeavor can be better understood by adopting a cultural evolutionary perspective, including the topics upon which the Forum concentrated: social systems, technology, language, and religion. We recognize that other broad subject matters, or different ways of dividing the subject matter, could have been used, but limitations in the number of participants precluded further broadening, and many people do think of themselves as specializing in one of these four areas.

Despite the rapid growth of the field of cultural evolution, especially over the last two decades, the number of practitioners and the amount of time they have had to work are still small relative to the size of the problems. The field still feels very young, even if the oldest contributors are quite gray! For example, much of the most sophisticated modeling of culture has been rather generic. The models specify the abstract structure of the inheritance system (vertical, oblique, or horizontal transmission) and a set of forces that drive cultural change (innovation, biased transmission, natural selection on cultural variation, random drift), which are often applied to narrow cases (dairying and adult lactase secretion, the demographic transition). The Forum considered whether intermediate levels of generality hold promise: Do models of the evolution of social organization, technology, language, and religion have interesting similarities but also differences? Can we hope for empirical generalizations at this level of abstraction? These four domains, of course, are not at all isolated from one another in real life; however, the field will benefit from a more systematically organized theoretical and empirical effort, not least because the traditional disciplines are so organized.

In the long run, the cultural evolution project will not fulfill its promise until every student of human behavior feels comfortable using cultural evolution as one of their tools, much as biologists are comfortable with organic evolution. The common problem is change over time in systems where the past influences the present. As Darwin noted, the phrase "descent with modification" fits the evolution of societies, technologies, and languages as well as biological organisms. Still, how do we know that something is the product of an evolutionary process? We expect to observe tree-like phylogenetic relationships across both cultural and biological domains, but culture often produces more heavily reticulated phylogenies. To address such reticulation, methods have been developed (Gray et al., this volume), as have methods to incorporate geographical information and fit explicit evolutionary models to historical and geographical patterns simultaneously (Bouckaert et al. 2012).

Evolutionists in the broadest sense must thus confront problems of the complexity and diversity of the systems they study and of historical contingency. If Gintis's principles of unification are correct, then the history versus science dichotomy is an illusion, as is argued elsewhere (Boyd and Richerson 1992). Biologists have become rather humble about what they are able to know in the face of the complexity, diversity, and historical contingency of the systems they study, even as they exploit every trick they can devise to finesse these problems (e.g., Burnham and Anderson 2002). This change should comfort humanists who have legitimately complained about the arrogant reductionism of many scientists from earlier generations. Indeed, cultural evolutionists have been in the forefront in bringing cultural diversity to the attention of psychologists, who all too often assume that they can tap "human nature" in the laboratory using university undergraduates as research participants (Henrich et al. 2010b).

What Is Cultural Evolution?

If we define culture as the ideas, skills, attitudes, and norms that people acquire by teaching, imitation, and/or other kinds of learning from other people, cultural evolution is fundamentally just the change of culture over time. The authors of this book have a view of cultural change that is based on concepts and methods pioneered by Darwin in the nineteenth century. In this conception, culture constitutes of an inheritance system; variant ideas, skills, and so forth that are transmitted by (usually) more experienced to less experienced individuals. Societies are a population of individuals that we can characterize in terms of the frequency of the cultural variants individuals express in the population at any point in time. As time progresses, many factors impinge upon the population to change the frequency of cultural variants expressed in the population. For example, someone in the population may either invent or acquire from another society a new and better skill of economic importance, such as a new way to make string and rope that is faster than the currently common technique and results in stronger cordage. This new skill will tend to increase in the population, perhaps because (a) users can sell more cordage than competitors and use the resulting proceeds to rear larger families, who perpetuate the new technique, and also because (b) unrelated individuals become aware of the new skill and its success and imitate those who have this skill. To study cultural evolution formally from this perspective means that we must set up an analytical accounting system to keep track of the increase or decrease in the frequency of cultural variants in order to try to establish the causes of the frequency changes (Cavalli-Sforza and Feldman 1981). The concrete reasons for cultural changes in particular populations are almost endlessly complex and diverse. To achieve some generalizable knowledge, we impose a taxonomy that collects the diverse concrete reasons into classes with similar dynamic properties (Henrich and McElreath 2003). The impact of a skill on the size of family one can raise is attributed to "natural selection." The processes of selectively imitating people who display a successful variant are attributed to "biased transmission" or "cultural selection." Biases, in turn, come in many varieties. A new form of speech, for example, might be acquired from someone we consider prestigious or charismatic.

Even good evolutionists sometimes speak of evolutionary "forces," such as natural selection and biased transmission, as if they were similar to gravity. As an analogy, this usage is harmless enough, but it certainly should not be taken literally. The force of gravity is a deep, universal physical law. Evolutionary forces are the outcome of diverse processes which interact to influence survival and reproduction. They have enough in common to permit a relatively small set of mathematical models, with roughly similar structure, to fit the data. Under closer examination, however, evolutionary forces have none of the universality and tidiness of the inverse square law and the universal gravitational constant. The "forces" usage often troubles humanists, who usually want to stick close to the details of particular cases of cultural variation and cultural change. Past attempts to formulate laws of history have had a checkered record, to say the least (Popper 1947). However, thoughtful evolutionists are well aware of the differences between concrete instances of genetic or cultural evolution and the abstraction involved in synthetic analyses based on the estimation of evolutionary forces (Turchin, this volume; Nitecki and Nitecki 1992). Even if reasonably robust findings emerge from our collective efforts, they are unlikely to fit any particular case perfectly.

The Investigation of Cultural Evolution: A Brief History

Humans have almost always had neighbors that spoke different languages or dialects, and many societies were aware that different societies preceded them. Hunter-gatherers living in the Great Basin in the nineteenth century, for example, were aware that the rock art in the area was made by inhabitants who they believed were not their own ancestors. The earliest systematic study of human differences and change was pioneered by historian-ethnographers in Greece (Herodotus, 484–420 BCE) and China (Sima Qian, 145–86 BCE) (Martin 2009). The writing of proper histories and ethnography, using methods designed to produce accurate treatments of other societies and the past, as opposed to myths with negligible attention to veracity, were relatively rare until the last couple of centuries (Brown 1988). Historical scholarship in the West exploded in the late eighteenth century, marked by Edward Gibbon's *History of the Decline and Fall of Rome* (1782). Ethnographic investigations also began to boom as expeditions of discovery became more professionalized, with scientific societies nominating naturalists to serve on them (Sorrenson 1996).

From the late eighteenth century to the middle of the nineteenth century, the field of historical linguistics (comparative philology) flowered into the first truly sophisticated cultural evolutionary research program (Müller 1862/2010; Hock and Joseph 2009). Not only did linguists notice the fairly remote connections between languages, such as those of northern India and Western Europe (Jones 1786/2013), they mapped the pattern of descent with modification of the Indo-European and Semitic languages in some detail, partly using ancient texts from dead languages as historical anchor points. As pointed out by Müller, linguists aspired to develop a theoretically sophisticated causal account of the mechanisms that drive linguistic change and diversification.

Darwin's contributions to the study of evolution were revolutionary. He is remembered primarily as a biologist, but his ideas about biological heredity were very rudimentary. In the preface to the second edition of the *Descent of Man* (Darwin 1874), he insisted that the effects of use and disuse were heritable and spoke of "inherited habits." Further, when discussing the evolution of human societies, he used such terms as "customs," "education," "laws," and "public opinion." In the chapter, "On the Races of Man" (Darwin 1874), he demolished the argument that a race could be considered a species, thus countering the main plank of nineteenth-century "scientific" racism used to justify slavery and other abuses of non-European peoples. He cited Edward Tylor (1871), the pioneering anthropologist who was the first to define "culture" in the way we use it here, to support his argument that differences between the races were due much more to traditions and customs than to organic differences. Darwin made a tolerably good start on a theory of cultural evolution.

It is possible to read Darwin as using cultural transmission as his model of biological inheritance. This would be quite understandable. The process of cultural transmission is partly quite accessible to natural-historical observation, whereas genes must be studied using the careful phenomenological experiments of the Mendelians. Genes only became truly "visible" once DNA was discovered to be the genetic basis of the gene, and in the last decade, gene sequencing has become so inexpensive that biologists can routinely observe genes directly. Darwin's (1877) detailed observations of one of his children's early development made him quite cognizant of the power of imitation and teaching to transmit culture. He might have intuited that an inheritance system which did not conserve acquired variations would waste the efforts that parents put into individual learning and other forms of phenotypic adaptation. Human life, as we know it, would be unimaginable without a cultural inheritance system passing on the knowledge acquired by parents and other adults to children. Wouldn't the inheritance of acquired variation be part of the organic system of inheritance as well? In any event, in 1869 Darwin proposed a theory of organic inheritance called "pangenesis"; this involved all the cells of the body casting off "gemmules," which were collected in the gonads and incorporated into gametes as the hereditary substance responsible for the development of offspring organisms. If an organ had been modified to adapt to the organism's environment, modified gemmules would be produced to reproduce the acquired variation (Darwin 1869:374–405). Twentieth-century biology marked this theory as Darwin's greatest mistake (Ridley 2009). Ironically, in that important respect, Darwin's theory of evolution was a better fit to human culture than to genes, yet Darwin is generally thought of as a biologist whose ideas about human evolution are generally thought to be mistaken.[1]

In the last half of the nineteenth century, Darwin's ideas on cultural evolution had a major impact on important thinkers in psychology and economics, where historical scholarship has been conducted at a high standard (for psychology, see Richards 1987). A considerable number of late nineteenth- and early twentieth-century psychologists were highly evolutionary in their approach to animal and human behavior, including George Romanes, William James, Conwy Lloyd Morgan, Henry F. Osborn, and James Mark Baldwin. Herbert Spencer's influence was large alongside Darwin's, about whose theory of natural selection Spencer was skeptical. Even Alfred Wallace thought that natural selection could not explain the human mind. These scholars were preoccupied with understanding the nature of heredity, the primary processes driving evolution, and the fundamental differences, if any, between human minds and behavior and those of other animals. Prior to the rediscovery of Mendel's principles, everyone's understanding of heredity remained primitive. Darwin's endorsement of the inheritance of acquired variation and Spencer's exclusive dependence on it became controversial with the rise of August Weismann's arguments about the separation of the germ line and the soma early in the embryonic development of most animals. Baldwin particularly struggled to reconcile Darwin's argument for the inheritance of acquired variation with Weismann's doctrine. Eventually he arrived at something like our contemporary understanding of the main issues. He proposed that there were two systems of heredity: organic heredity, which obeyed Weismann's doctrine, and social heredity, particularly important in humans, which does not. He also proposed a form of selection,

1 Darwin may yet be vindicated regarding the inheritance of acquired variation. The development of multicellular organisms depends upon up-regulating and down-regulating genes so as to specialize cell lines for their highly divergent functions. Once specialized, the operational "transcriptome" of each cell type is transmitted to daughter cells in that line by means of various "epigenetic" mechanisms. Recent work on epigenetic inheritance suggests that some modified phenotypes may be transmitted across generations, even in obligate sexually reproducing organisms (Grossniklaus et al. 2013), and may greatly influence evolution (Laland et al. 2011; Jablonka 2013).

"organic selection," to explain the appearance of the inheritance of acquired variation in organic traits. Developed independently by Morgan and Osborn, it became more popularly known as the Baldwin Effect. The idea is that phenotypic adaptation would keep a population from extinction under changed conditions while selection did its work, and phenotypic adaptations would often foreshadow the direction that selection on the Weismannian hereditary material would take. Hence, human cognitive power could influence evolution by a Lamarckian process that is underpinned by social heredity, via the organic selection process operating on germ line heredity that mimicked the inheritance of acquired variation mechanism, without actually depending on it.

Geoffrey Hodgson (2004) provides a thoughtful, detailed analysis of late nineteenth-century ideas about cultural evolution and related topics, centered on the institutional economist Thorstein Veblen, whose creative work took place between 1898 and 1909. Veblen was much influenced by Darwinian psychologists, who are the focus of Richards' (1987) book. His most important contribution was to articulate the concept of institutions—culturally transmitted systems of rules that structure human social life. Like Baldwin, Veblen struggled to understand the relationship between the biological heredity that we share with other organisms and the cultural system that is more or less unique to humans. He insisted that it was important to understand the causal mechanisms, analogous to natural selection, that drive cultural evolution; however, his work on the subject was unsystematic in Hodgson's estimation. Veblen did imagine that innate predispositions, specifically what he called an "instinct for workmanship," might influence technological evolution (Cordes 2005). This concept clearly foreshadows the notion of epigenetic rules, cultural selection, and biased transmission that figure in the late twentieth-century revival of Darwinian theories of cultural evolution. The instinct for workmanship motivated humans to search for elegant functional technological designs that efficiently serve basic human needs. The instinct would motivate the careful production of artifacts, attempts on the part of craftspeople to improve them, and the borrowing of better designs from others.

Given the number, prestige, sophistication, and diversity of Darwin's early twentieth-century followers in the human sciences, one might have thought that the legacy of the *Descent of Man* was secure. Instead, just as Darwinian ideas began to be combined with genetics to form one of the theoretical foundations in biology (Provine 1971), the equally productive ideas of Darwin and his followers regarding cultural evolution, and the link between organic and cultural inheritance in humans, went into a half-century near-total eclipse (Richerson and Boyd 2001). The reasons for this eclipse have not been well told except in the special cases of psychology and institutional economics. Chance may have played a role. Both Baldwin's and Veblen's careers were damaged at their peak by sexual scandals, according to Richards (1987) and Hodgson (2004). Many of the emerging social scientists were keen to distinguish themselves from biology and to downplay the significance of biology for sciences of human behavior. For example, the influential early French sociologist Gabriel Tarde (1903) excluded biological considerations in his pioneering study of the "laws of imitation." Hodgson (1993) described how the greater prestige of physics compared to biology caused economists to look to physics for models of scientific rigor rather than biology, as this discipline began to professionalize around the turn of the twentieth century. The prestige of Darwin's own ideas about evolution reached a minimum around that time, inhibiting the social science pioneers from using him for inspiration, much less authority (Bowler 1983). When Darwinism began to emerge from its eclipse with R. A. Fisher's (1918) paper, which showed how natural selection could be reconciled with the genetic theory of inheritance, it emerged as a contribution to biology (Provine 1971, chapter 5); the contributions of Darwin and late nineteenth-century Darwinians to the study of human behavior were largely forgotten.

The theory of evolution which did inform many of the early twentieth-century social scientists derived from Spencer rather than Darwin. Ideas of progressive evolution stemming from Spencer were popular, often under the misleading label "Social Darwinism." Spencer's main idea was that the same principle of evolution underlay cosmological, geological, biological, and human behavioral change. The principle was that all structures progress from simple, undifferentiated homogeneity to complex, differentiated heterogeneity (Spencer 1862). Physicists will recognize Spencer's principle as the Second Law of Thermodynamics—*backward.* Robert Carneiro (1967) outlined Spencer's impact on twentieth-century social science (see also Freeman et al. 1974). Richard Hofstadter (1945) wrote a famous critique of Social Darwinism which was, in turn, the subject of a sharp countercritique (Bannister 1979).

Perhaps the most sophisticated twentieth-century evolutionist in the Spencerian tradition was Julian Steward (1955), who critiqued the simple unilinear theories derived from Spencer, suggesting that societies progressed lockstep through an invariant succession of stages of complexity. Steward was an ethnographer of very wide experience and even wider reading. He knew that trajectories of change in social complexity and the like were highly variable. He also knew that the correlation between the complexity of such features as technology and social organization was imperfect. Thus he focused his analysis of evolution on what he called the "culture core," which comprised technology and the aspects of other features of culture directly related to the mobilization of technology to provide human subsistence. He described how societies that used hunting and gathering technology varied greatly in the details of their social organization, depending on the exact nature of the resources that are hunted and gathered. Hunting small game and gathering dispersed plant resources favored very simple but highly flexible family-level organization, whereas large herding game typically led to cooperation between many families, and thus more complex social organization. Steward's culture core framework was a sort of commonsense adaptationism overlaying a concept of progressive change. In this it resembled the sociological functionalists (e.g.,

Lenski and Lenski 1982). None of these thinkers were preoccupied with the micro-mechanistic foundations of evolution in the way Darwin and his followers were. There is no doubt, however, that progressivist human evolutionists were onto *something*. The overall trend toward greater complexity of human societies in the Holocene is unmistakable. Paleoanthropologists, especially in the late twentieth century, documented this trend far back in the history of our lineage (Klein 2009). In evolutionary biology, the issue of progress has been vexatious, going right back to Darwin's ambivalence about it (Nitecki 1988).

A revival of a Darwinian approach to cultural evolution began rather modestly in the 1950s when Armen Alchian (1950) suggested that profit-maximizing firms might emerge from natural selection on random variation between competing firms rather than because firm managers consciously chose profit-maximizing strategies. Alchian's paper, in turn, led to the lively field of evolutionary economics, whose single most important classic was Richard Nelson and Sidney Winter's (1982) book: *An Evolutionary Theory of Economic Change*. A few years later, neurophysiologist Ralph Gerard, mathematical psychologist Anatol Rapoport, and anthropologist Clyde Kluckhohn teamed up during a yearlong interdisciplinary meeting at the Center for Advanced Study in the Behavioral Sciences in Stanford to write a rather sophisticated programmatic essay describing how cultural evolution might be studied using the concepts and methods of evolutionary biology (Gerard et al. 1956). This paper influenced several of the contributors to the next wave of cultural evolution work, including Cavalli-Sforza and Feldman (1973), Durham (1982), and Richerson (1977).

In psychology, the key figure in reintroducing Darwinian theory was Donald Campbell (1960, 1965, 1975). In the first of these papers, Campbell argued that creative thought might consist of an intrapsychic process of "blind variation and selective retention" analogous to natural selection, an idea later developed by Gerald Edelman (1987). In his 1965 essay, Campbell developed the concept of "vicarious selectors," genetically evolved mental devices that evolved under natural selection to shape human learning and bias social learning in adaptive directions. This concept, though not Campbell's term, influenced all subsequent Darwinian approaches to cultural evolution. The essay also argued that cultural inheritance would evolve much as genes do, except for the role that vicarious selectors play alongside blind variation and natural selection. The 1975 paper described how genetic and cultural evolution could come into conflict and how the micro-mechanistic Darwinian approach to cultural evolution differed from the neo-Spencerian progressive approaches. These three papers were highly cited and widely influential. Other early contributors to the emerging field included Ruyle (1973), Cloak (1975), and Pulliam and Dunford (1980).

In child development, also in the late 1950s and early 1960s, Albert Bandura began publishing his extremely influential studies of social learning in children (e.g., Bandura and Walters 1963). This work established the critical importance of imitation in the acquisition of human behavior and led, in due course, to a reasonably sophisticated understanding of the capacity for culture acquisition in humans. At the same time, Lev Vygotsky's (1978) neglected cultural-historical approach to child development began to have a major impact on the field. Important modern work in this field includes Tomasello (1999), Whiten and Custance (1996), Carey and Spelke (1994), Bloom (2000), and Harris and Koenig (2006).

Another relevant field is the diffusion of innovations, which traces back to Tarde's work in sociology and to the diffusionist school in anthropology. Because the diffusion of modern innovations is so important to economic growth, the phenomenon attracted the attention of economists and applied economists in the 1940s and 1950s (e.g., Griliches 1957). By the early 1970s, around 1,500 reasonably detailed studies of the phenomenon were known. Everett Rogers and Floyd Shoemaker (1971) did a pioneering meta-analysis of these data and teased out a number of robust strategies which people exposed to innovations used to decide whether or not to adopt them. Robert Boyd and Peter Richerson (1985) derived their taxonomy of bias forces from Rogers and Shoemaker's analysis and studied mathematical models of several of the processes they described. Much subsequent modeling and empirical work has been based on this foundation (e.g., Henrich and McElreath 2003).

Language evolution did not experience the same eclipse in the twentieth century as did other human sciences fields. Explicit theoretical discussions on the evolution of language in the hominid lineage remained largely outside the academic discourse, in part because of the ban on such discussions imposed by the influential Société Linguistique de Paris in 1866. Nonetheless, historical and descriptive linguistics continued to document the histories of language families around the world and their resultant linguistic diversity (for a review, see Evans, this volume). In addition, several innovative research programs in linguistics emerged in the latter part of the twentieth century in parallel to the other fields discussed above. William Labov (1963) initiated a program of detailed micro-mechanistic studies of sound changes (dialect evolution) that eventually produced a rather detailed account of the evolutionary pressures on sound change from within languages and from the external social environment (Labov 1994, 2001). Similarly detailed studies of languages in contact showed how linguistic innovations could flow between speech communities (Thomason and Kaufman 1988). Historical linguists also discovered that function words and morphemes could evolve by the shortening and conventionalizing of constructions using referential words, outlining how grammar evolves by "grammaticalization" (Traugott 1980; Hopper and Traugott 2003).

The early 1990s saw a resurgence of scientific interest in language evolution, following the publication of the landmark paper by Pinker and Bloom (1990) on the role of natural selection in the evolutionary emergence of human language. Theoretical considerations were quickly complemented by formal models of language evolution. Whereas initial computational models focused on the biological evolution of language-specific mechanisms (e.g., Hurford 1989), recent years have seen a shift toward cultural evolution as the primary explanation for the emergence of linguistic structure (Christiansen and Kirby 2003; see also Jäger et al. 2009; Steels 1997). Much of this theorizing came in response to the emerging evidence that social learning plays a much stronger role in language acquisition than the heavily innatist proposals of the early generative grammar period envisioned (contrast Pinker 1994 with Tomasello 2005 and Hurford 2011).

The study of cultural evolution has had a largely conflictual relationship with the most highly visible evolutionary approach to human behavior, human sociobiology. The human sociobiology program was tentatively launched by an important paper by Richard Alexander (1974) and the last chapter in Edward O. Wilson's (1975) treatise on sociobiology, followed shortly by book-length evolutionary treatments of human behavior (Alexander 1979; Lumsden and Wilson 1981). This work was considered a political abomination by many on the left, who (mistakenly) associated evolution with right-wing ideology, as well as by many social scientists who could not imagine how biologists could make any useful contribution to the social sciences (Segerstråle 2000). At the same time, a small number of anthropologists and psychologists embraced the sociobiological turn because they were skeptical of the atheoritical, if not antitheoretical, use of cultural "explanations" in their fields (Chagnon and Irons 1979; Sperber 1984; Tooby and Cosmides 1989). The cultural evolutionists, specifically the dual inheritance theory version of Richerson and Boyd (1976), envisioned from the beginning a much more active role for cultural evolutionary processes than did the original founders of sociobiology or the pioneers of the descendant fields, human behavioral ecology and evolutionary psychology. Indeed, the cultural evolution field owes much more to the other influences described above than to human sociobiology, although it must be said that the temperature of debates with human behavioral ecologists, evolutionary psychologists, and others has diminished in recent years as the empirical importance of cultural evolutionary processes has come to be more widely appreciated, especially by younger scholars.

The history of the last two decades or so in the field of cultural evolution is embodied in the various chapters of this book. To say more at this stage would begin to reiterate their contents.

Common Themes across the Four Areas

One of our major objectives in this book is to explore the commonalities of the evolutionary processes between the four designated areas: the structure of human groups, technology and science, language, and religion. Although each of these areas has attracted the attention of many disciplines, the specific disciplines that have contributed to the study of cultural evolution vary. This is to be expected since the substance of the phenomena which they cover is distinctively different. For example, variation in the details of social organization and technology is perhaps more likely to have consequences for survival and reproduction than variation in the details of language and religion. Much of the variation in the latter, with its highly symbolic phenomena, is adaptively neutral. Different words for "cat" and different rituals for invoking the favor of the gods may matter less for survival and reproduction of users than the species of a tree used to make a bow or the manner in which warriors are recruited, trained, and led. Nevertheless, for the most part, cultural evolution does share important commonalities across different domains.

Gene–Culture Coevolution

In our species, genetic evolution and cultural evolution are inextricably linked. Our bodies are adapted to acquire and use culture, and our cultures are adapted to help our genes perpetuate themselves. The deep entangling of the cultural and genetic evolutionary subsystems, each complex enough in its own right, poses many hard problems in each of the areas we considered (as well as in others). Possibly, the evolution of an innate social psychology, which was predisposed to follow norms and institutions, coevolved with culture-generated social selection (Jordan et al., this volume). Language evolution likely involved gene–culture coevolution—a process that is perhaps still active today (Levinson and Dediu, this volume; Evans, this volume). The issue has not entirely been resolved. Not a few evolutionists adhere to a strongly gene-centric view of even human evolution (Laland et al. 2011). Not a few humanists and humanistically oriented scientists take a dim view of introducing considerations of genes into the study of culture (e.g, Fracchia and Lewontin 2005).

Use of Mathematical Models

Mathematical models have played a key role in the development of our field, as illustrated by the above-mentioned pioneering work in modeling (see also Turchin, this volume), and they will continue to play an important role (e.g., Chater et al. 2009; Henrich and Boyd 2008; Bowles and Gintis 2011). Perhaps the most innovative new use of models is as a data analytic tool. Advances in computing power have made it practical to use maximum likelihood and Bayesian inference methods to fit competing causal models directly to data (e.g., Leonardi et al. 2012; McElreath et al. 2008).

Experimental Methods

Evolution is a population-level process, as evolutionists are wont to say. Experiments that are logistically and ethically feasible may seem too small in scale to be very informative. However, many questions are difficult to answer without some sort of controlled experiment. Experiments are a little like mathematical models: they are simplified caricatures of a real large-scale process, but they can give us nice insights into the workings of components of the process. Robert Jacobs and Donald Campbell (1961) reported the first laboratory-scale experiment on cultural transmission and evolution in a laboratory microsociety. A handful of experiments were subsequently done, including a very ambitious series of experiments on the evolution of leadership by Chester Insko et al. (1983). Recently, laboratory microsociety experiments have been used to study the evolution of in-group favoritism (Efferson et al. 2008a), cumulative cultural evolution of technology (Caldwell and Millen 2008; see also other papers in that issue), the use of social-learning strategies by individuals (Mesoudi 2011b), and the evolution of language (Scott-Phillips and Kirby 2010). Other sorts of experiments are used to test functional hypotheses about the impact of cultural beliefs on behavior such as religious beliefs on helping behavior (e.g., Laurin et al. 2012; reviewed by Norenzayan et al., this volume).

Field Studies of Microevolutionary Processes

In evolutionary biology, field studies aimed at estimating the strength of natural selection and other evolutionary forces are a classic method (Endler 1986) of directly studying microevolutionary processes. Historians, sociolinguists, and students of the diffusion of innovations conduct similar field projects, though they do not use quite the same theory-driven approach to quantification that evolutionary biologists do. Evolutionist anthropologists have pioneered applying the approach of field biologists to human field data, beginning with the work of Soltis et al. (1995) working with extant ethnographic data and that of Aunger (1994) using purpose-collected field data. More ambitious long-term projects have begun to report early results (Bell 2013; Henrich and Henrich 2010). A larger number of field studies address particular evolutionary hypotheses without formally estimating the strength of forces (Mathew and Boyd 2011; Sosis and Bressler 2003; Norenzayan et al., this volume). Field studies by linguists have not only documented the astonishing diversity of linguistic structures, they have also illuminated the mechanisms that drive change (Evans, this volume). Games devised by experimental economists have been used as tools for mapping cultural diversity with respect to prosocial propensities (e.g., Henrich et al. 2006; Herrmann et al. 2008).

Critical Importance of Development

The individual-level process of development is the place where genetic and cultural inheritance systems interact most vigorously. Claims about the developmental process have thus figured importantly in the debates between more gene-centric (e.g., Tooby and Cosmides 1992; Pinker 1994) and more culturally oriented (Tomasello 1999; Richerson and Boyd 2005) conceptions of how the human evolutionary process works. The developmental process is formidably complex, and much has been proven to happen in the first few months of life. Even as the human brain is still quite undeveloped, infants are active observers of other's behavior, capable of exercising attention biases (e.g., Kinzler et al. 2011). Children's tendencies to learn socially from others that are like them culturally are the foundation for the generation of so much between-group cultural variation in our species (Haun and Over, this volume). Fortunately, developmentalists have devised methods to infer what judgments are being made, even by preverbal infants. For example, eye gaze and attention patterns betray an infant's interests and choices (Carey 2009; Boysson-Bardies 1999). Developmentalists study a wide variety of subject domains, including science and religion (e.g., Harris and Koenig 2006), language (e.g., Bloom 2000), social norms (e.g., Chudek and Henrich 2011), and motor skills (e.g., Whiten et al. 2009). Comparative work shows that human children have a powerful imitative system compared to even highly intelligent apes and monkeys (Dean et al. 2012), and human adults are well adapted to support the social learning of children (Csibra and Gergely 2011). The evidence accumulated since the 1990s amounts to a rather devastating refutation of the highly gene-centric cognitive modules view of development (Sterelny 2012), the original inspiration for which was Chomsky's failed principles and parameters approach to language learning. Lieven (this volume) reviews the evidence that at very young ages infants are already highly sensitive to the particularities of the language they are learning. Combined with the development of such devices as shared attention, which also operates in other cultural domains, the powerful imitative capacity of children is sufficient for them to acquire very diverse languages without having an elaborate innate dedicated language-learning system. As discussed further by Lieven et al. in the Appendix of this volume, developmental processes are likely to have a key impact on cultural evolution across a variety of domains.

Accounting for Macroevolutionary Events and Trends

Many of the most interesting evolutionary questions involve large-scale trends and events in human evolution. Gintis and van Schaik (this volume) outline the basic pattern of hominin social evolution over the last few million years. Why did complex cumulative culture evolve so recently, despite the fact that it has made us an extraordinarily successful species (see Boyd et al., this volume)? Why did brain size and cultural sophistication in our lineage increase progressively over the Pleistocene? When and why did our distinctive societies with high rates of cooperation between nonkin arise? Why do we institutionalize cooperation between relatives and long-term partners when the familiar evolutionary mechanisms of inclusive fitness and reciprocity would seem to explain such cooperation without the need to invoke cultural mechanisms (Mathew et al., this volume)? When did something like the modern capacity for language evolve? Why did anatomically modern people disperse out of Africa around 60 KYA when our species is perhaps 100 thousand years older? What was and is the role of religion in the simpler societies of the Pleistocene and transitional societies of the Holocene (Bulbulia et al., this volume; Guthrie 2005)? Why did human populations evolve agriculture, states and their distinctive religions, and industrial production in the Holocene (see chapters by Turchin, Slingerland et al., Norenzayan et al., and Bulbulia et al., this volume)? Why do Holocene societies have boom and bust dynamics?

The concepts and tools of cultural evolution and gene–culture coevolution have devoted substantial attention to these topics (e.g., Steele and Shennan 2009). Innovations in both empirical methods and in modeling and model-fitting data analysis are driving a considerable increase in the sophistication of archaeology and historical reconstructions (e.g., Collard et al. 2010; Turchin and Nefedov 2009). The quest is for synthetic long time span, high-resolution quantitative records constructed from the short qualitative records that are directly available from historians and archaeologists, often using clever proxies for unmeasured variables like population density. Given such time series, we can hope to find informative fits of modestly complex evolutionary models. Gene sequencing techniques are producing a cornucopia of data on human genetic diversity (and some excellent sequences from subfossil DNA). This data produces evidence of past selection and past demography of humans, our parasites, and domesticates. As methods improve, there is hope that genetic data can supplement the sparse conventional paleoanthropological record, especially for things like language and social dispositions which fossilize poorly, by finding evidence for genetic responses to gene–culture coevolution (Pinhasi et al. 2012; Richerson et al. 2010).

Major Ongoing Problems to Solve

For most of the problems reviewed thus far, the field of cultural evolution might be characterized as at the end of the beginning. For these questions we can point to sound methodological approaches and a decent body of findings that are likely to hold up reasonably well to future scrutiny. Here, we want to highlight problems where we are closer to the beginning of the beginning.

Understanding the Epigenetic and Neurobiological Systems that Underpin Culture

Aunger (2002) made a brave attempt to provide a neurobiological foundation for human culture. Since then, Rizzolatti (2005) hypothesized that a mirror neuron system homologous to that detected in macaques using single electrode techniques plus associated regulatory circuits, might produce the human capacity for imitation. Support for this hypothesis is confounded by the number of areas in the human brain that show mirror-like activity in fMRI studies (Molenberghs et al. 2012). The complexity of the human brain circuitry together with limitations of imaging techniques leave us with a very incomplete understanding of the neurobiology of the culture capacity (Stout, this volume). Whitehouse (this volume) proposes a landscape model for the roles of genes, culture, and environment on epigenetic processes. Several quite basic features of the cumulative cultural system are poorly understood (Boyd et al., this volume). While the highly gene-centric cognitive picture of human evolution seems precluded by developmental studies, which clearly identify a powerful early developing capacity or capacities to acquire information by imitation and teaching, the detailed division of labor between innate-cognitive structures and cultural transmission remains quite controversial. Slingerland et al. (this volume) argue that a number of key cognitive structures underpin the phenomenon of religion, whereas Harris and Koenig (2006) imply that simple trust in the testimony of adults can explain many of the mysteries of religious belief. On the other hand, the "core cognition" proposal of Carey (2009), consistent with Harris' proposal, has been criticized as being too innatist by developmental systems enthusiasts (Spencer et al. 2009).

Epigenetics introduces another level of complexity to understanding the mechanistic basis of culture capacities. Provençal et al. (2012) found a large number of changes in the methylome of the prefrontal cortex of macaques reared with mothers present versus only a peer present, and methylation patterns are only one component of the epigenetic system. In humans we might imagine that the epigenetic system is a vehicle for massive cultural influences on gene expression, but it could also be a vehicle for massive contingent epigenetic effects on factors which bias culture acquisition. Further, the possibility that some epigenetic changes can be transmitted to offspring leads to the possibility that transgenerational epigenetic transmission can be confounded with culture and that this represents still another pathway by which genes and culture can influence one another (Jablonka 2013; Daxinger and Whitelaw 2012).

Moving beyond Proof-of-Concept Examples of Gene–Culture Coevolution

Genome-wide scans, which search for genes that have come under strong selection in humans recently enough to leave internal evidence in the genome, have apparently uncovered many such genes (e.g., Sabeti et al. 2002; Hawks et al. 2007). We commented above on the promise of studying gene–culture coevolution by using possible responses to such coevolution in combination with the paleoanthropological record to understand better how our species evolved. However, present evidence for gene–culture coevolution still rests on a few classic cases, such as the evolution of lactase persistence in dairying peoples and the evolution of hemoglobin polymorphisms in malarial areas. So far, the difficulty of discovering the functional significance of the alleles that have apparently been under selection leaves most of the new examples tantalizing but enigmatic. The putative "language gene" *FOXP2* provides a cautionary tale in this regard (Coop et al. 2008; Fisher and Scharff 2009). Similarly, as regards religion, our understanding of the linkages between genetic and cultural components is still primitive (Norenzayan et al., this volume). Without methodological breakthroughs, the promise of genomic studies will remain the prisoner of slow and expensive retail functional biology.

Understanding the Diversity of Micro Processes

Laboratory studies of the strategies individuals use to acquire information from others has revealed a surprising amount of individual variation and much use of suboptimal strategies (Efferson et al. 2008b; McElreath et al. 2008; Mesoudi 2011b). Limited simulation studies conducted thus far suggest that diverse social-learning strategies will persist at equilibrium (Whitehead 2007). There is every reason to think that substantial cross-cultural variation exists in social-learning strategies (Bettinger and Eerkens 1997; Shennan, this volume). For example, just in the last few centuries, the principles of scientific reasoning and the social organization of systemic criticism, which constitute the scientific enterprise, arguably created a novel cultural system institutionalizing new forces that shape an unprecedented form of cumulative cultural evolution (McCauley, this volume). The fact that scientific institutions and the coupling of science to technical innovation are so successful, yet so recent, gives rise to the worry that science as a cultural system may be fragile (Mesoudi et al., this volume). Cross-cultural variation in the use of language as a device for socialization has been documented but not well explored, nor has justice been done to the contribution of peer interactions on the evolution of language during childhood (Lieven, this volume). We are at the very beginning of the effort to understand the diversity within and between cultural systems.

Using History and Living Diversity as a Natural Laboratory for Studies of Cultural Evolution

The use of phylogenetic methods to study cultural evolution is well advanced. However, as mentioned above, empirical methods have advanced to the point where we can use model fitting and model selection methods to try to infer directly the underlying process that drove a particular evolutionary trajectory (Itan et al. 2009; Bouckaert et al. 2012; Turchin and Nefedov 2009). Human documentary history is quite rich, and the human fossil and archaeological records are rather rich. The cornucopia of genetic data that is currently flowing from ever cheaper sequencing technology not only makes this data available, it is pushing developments in bioinformatics which can also be applied to cultural data. Constructing quantitative time series using these data and comparing the fits of alternative models to the data promises a revolution in our understanding of cultural evolution; however, the issues involved are not trivial (Shennan, this volume). Even in linguistics, one of the most sophisticated fields in cultural evolution to use diversity as a natural laboratory, Evans (this volume) identifies no less than seven major challenges. Gray et al. (this volume) highlight three outstanding questions in the evolution of language that can be addressed with computational phylogenetic methods, and Dediu et al. (this volume) formulate a number of challenges facing future research into the cultural evolution of language.

Reducing the Gaps between the Natural Sciences, Social Sciences, and Humanities

In the nineteenth century, the arts and sciences were weakly organized. The great national academies covered all of the sciences. There were few professional positions for scholars; many practitioners were rich gentlemen and enthusiastic amateurs with broad interests. Darwin published on geology, zoology, botany, and anthropology and wrote an account of the Voyage of the Beagle for a popular audience. William Thompson, Lord Kelvin, worked as a theorist on electricity and thermodynamics, on engineering projects, such as the transatlantic telegraph, and on improvements to the mariner's compass. In his attempt to estimate the age of Earth, Kelvin's pioneering geophysical work brought him into conflict with Darwin and many geologists, who inferred a much greater age than the 10–20 million years Kelvin's calculation allowed. Many projects in history and historical linguistics were founded on serious methodological innovations, such as the comparative method. The eighteenth- and nineteenth-century project known as the Quest for the Historical Jesus (Bartley 1984) provides one example. What we normally think of as science-minded anthropologists often make use of such methods today (e.g., Wiessner and Tumu 1998; Currie et al. 2010a). As the history of the study of cultural evolution shows, the professionalization of the sciences and humanities around the turn of the twentieth century resulted in many more active, full-time, paid, specialist scholars who became organized into disciplines that tended to behave in a quasi-tribal fashion (Campbell 1979). The unity of the scholarly enterprise broke down. Even within the social sciences, disciplinary balkanization is a problem (Mesoudi et al., this volume), far more serious than in the much larger field of biology, where subdisciplinary boundaries are not taken all that seriously.

During the political upheavals of the 1960s and 1970s, critical theory and deconstructionism included the natural and social sciences in the analyses of how ethnocentrism, paternalism, and political power distorted the intellectual enterprise. Some natural and social scientists reacted quite defensively to these critiques, penning polemical countercritiques (e.g., Gross and Levitt 1997). Some of the frequently leveled critiques of evolutionary studies by the humanistically inclined—for example, that evolutionary views are connected to conservative political ideologies—are demonstrably false (Tybur et al. 2007; Lyle and Smith 2012). Rants against reductionism and positivism sounded strange to those of us brought up on philosophers of biology arguing that these ideas were purely of historical interest. One of us (PJR) was a participant in a 1981 conference organized by Donald Campbell and Alex Rosenberg to explore what Campbell termed the concept of an "epistemologically relevant internalist sociology of science." He was impressed by the young proponents of the internalist Strong Program in the Sociology of Science (e.g., Bloor 1971), whom he saw as pursuing a valuable, intimately ethnographic look at the micro-scale processes by which science worked. At the same time he had no doubt that the then conventional realist notion that science worked fairly well as an instrument for fallible but real discovery was essentially correct. In fact, he looked forward to the Strong Program contributing to the improvement in the functioning of science as a social system. In effect, Campbell was trying to stop the "Science Wars" before they started. He had no success with either the internalists or realists at the conference. It was clear that the internalists perceived themselves to be young innovators with no use for the "errors" of their elders, whereas the realists saw the internalists as making no useful contribution. At least one paper from each side was subsequently published, thus giving an impression of the passions with which each side pressed its case (Woolgar 1982; Gieryn 1982).

In our view, the "Science Wars" were based on willful ignorance on both sides and have done serious damage to scholarship in the four focal areas of this volume. As some of our participants have argued elsewhere (Slingerland 2008; N. Henrich and Henrich 2007; Boyd and Richerson 1992; Turchin 2008), as have others (Leijonhufvud 1997), "humanistic" and "scientific" methods each make distinctive and vital contributions to understanding the world. In essence, evolving systems are complex and diverse. They cannot be reduced to a single model, and even if they could be so reduced, the model would be far too complex to actually use. Much of our understanding of such systems is bound to remain semantic, qualitative, particularistic, incomplete, and open ended. On the other hand, the discipline of acquiring quantitative data and fitting formal models often yields great insights, albeit fallible insights on a narrow front. Mathematics and quantitative empirical methods are just mental prostheses invented to finesse the unaided mind's weak powers of deduction and inability to estimate quantity accurately. Most scholars do not have serious problems deploying quantitative and qualitative methodologies opportunistically. We have never met a historian or archaeologist, no matter how "humanistic," who objected to using radiometric dating in situations where it would be useful. Evolutionists, even the most "scientific" ones, are usually decent natural historians, historians, or ethnographers whose qualitative command of some segment of the world is essential to their science. We think that you have to don some sort ideological blinders to start a fight over which sort of tools are more valuable. The world we all want to understand is fiendishly hard to comprehend. Why would any sensible scholar reject "on principle" any useful method to advance understanding?

Conclusion

The field of cultural evolution has grown rapidly over the last forty years, particularly as a self-conscious entity. This growth rests on deep foundations in the social sciences and humanities. It also has a solid foundation in behavioral biology, which unfortunately is not covered here in depth. Other animals turn out to have important systems for social learning analogous to human culture, and the last few decades have matured into a veritable golden age of studies of animal culture (Danchin et al. 2004; Whiten et al. 2011; see also Menzel and Fischer 2011). The first two-thirds of the twentieth century were a sharp hiatus in the study of cultural evolution from a Darwinian perspective. Since the mid-1960s, it has taken nearly a half century to make up for the neglect that the field suffered across most of the range of research topics covered in this book.

As evidenced by the chapters contained in each of the four topic areas, understanding human cultural evolution constitutes a similar but not identical problem. The issue of understanding the developmental support for cumulative culture is much the same. The same basic forces which shape evolutionary change work everywhere. For example, borrowing technology and words or grammatical constructions from another culture represent similar processes. Differences, however, are surely important. Few variant words or variant religious beliefs have the same direct impact on well-being as variant subsistence technology. Variant words and religious practices do play important roles in structuring social life and can certainly have an important indirect impact on health and welfare. We do not want to discount the diversity of cultural processes across domains within cultures nor across cultures nor in the historic and especially prehistoric past.

Given the inherent complexities, no publication short of a multivolume treatise could hope to do complete justice to the current field of cultural evolution. Nevertheless, this book provides a broad sample of the work that is ongoing by cultural evolutionists. We hope that you enjoy it as much as we enjoyed the Forum and the resulting editing.

Acknowledgments

We thank the Ernst Strüngmann Foundation and German Science Foundation for its support, and the Forum's delightful and competent staff: Julia Lupp, Marina Turner, and Andrea Schoepski. Program Advisory Committee members Herbert Gintis, Stephen Levinson, Stephen Shennan, and Ted Slingerland were our full partners in devising the program and nominating participants. Moderators, rapporteurs, authors, and each individual group member, of course, made the Forum and this book an intellectual endeavor. As cochairs and coeditors we take responsibility for any remaining deficiencies and rough edges.

Structure of Human Groups

2

Zoon Politicon: The Evolutionary Roots of Human Sociopolitical Systems

Herbert Gintis and Carel van Schaik

Abstract

Our primate ancestors evolved a complex sociopolitical order based on a social dominance hierarchy in multi-male/multi-female groups. The emergence of bipedalism and cooperative breeding in the hominin line, together with environmental developments which made a diet of meat from large animals fitness enhancing, as well as cultural innovation in the form of fire and cooking, created a niche for hominins in which there was a high return to coordinated, cooperative scavenging or hunting of large mammals. This, in turn, led to the use of stones and spears as lethal weapons.

The availability of lethal weapons in early hominin society undermined the standard social dominance hierarchy of multi-male/multi-female primates. The successful sociopolitical structure that replaced the ancestral social dominance hierarchy was a political system in which success depended on the ability of leaders to persuade and motivate. This system persisted until cultural changes in the Holocene fostered the accumulation of material wealth, through which it became possible once again to sustain a social dominance hierarchy, because elites could now surround themselves with male relatives and paid protectors.

This scenario suggests that humans are predisposed to seek dominance when this is not excessively costly, but also to form coalitions to depose pretenders to power. Much of human political history is the working out of these oppositional forces.

Self-Interest and Cultural Hegemony Models of Political Power

For half a century following the end of World War II, the behavioral sciences were dominated by two highly contrasting models of human political behavior. In biology, political science, and economics, a *self-interest* model held sway, wherein individuals are rational self-regarding maximizers. In sociology, social psychology, and anthropology, by contrast, a *cultural hegemony* model was generally accepted. In this model, individuals are the passive internalizers of the culture in which they operate. The dominant culture, in turn, supplies the norms and values associated with role performance, so individual behavior meets the requirements of the various roles individuals are called upon to play in daily life (Durkheim 1933/1902; Parsons 1967; Mead 1963).

Contemporary research has been kind to neither model. There has always been an undercurrent of objection to the cultural hegemony model, which Dennis Wrong (1961) aptly called the "oversocialized conception of man." Behavioral ecology alternatives were offered by Konrad Lorenz (1963), Robert Ardrey (1966/1997) and Desmond Morris (1999/1967), a line of thought that culminated in Edward O. Wilson's *Sociobiology: The New Synthesis* (1975), the resurrection of human nature in Donald Brown's *Human Universals* (1991), and Leda Cosmides and John Tooby's withering attack in *The Adapted Mind* on the so-called "standard social science model" of cultural hegemony (Barkow et al. 1992). Meanwhile, the analytical foundations of an alternative model, that of *gene–culture coevolution* (see below), were laid by C. J. Lumsden and Edward O. Wilson (1981), Luca Cavalli-Sforza and Marcus Feldman (1973, 1981), and Robert Boyd and Peter Richerson (1985).

In opposition to cultural hegemony theory, daily life provides countless examples of the fragility of dominant cultures. African-Americans in the era of the civil rights movement, for instance, rejected a powerful ideology that justified segregation, American women in the 1960s rejected a deep-rooted patriarchal culture, and gay Americans rejected traditional Judeo-Christian treatments of homosexuality. In succeeding years, each of these minority countercultures was largely accepted by the American public. In the Soviet Union, Communist leaders attempted to forge a dominant culture of socialist morality by subjecting two generations of citizens to rigid and intensive indoctrination. This failed to take hold and, following the fall of the USSR, was rejected whole cloth, without the need for extensive counter-indoctrination. Similar examples could be given from the political experience of many other countries, possibly all.

Undermining the self-interest model began with the ultimatum game experiments of Güth et al. (1982), Roth et al. (1991), and many others. These experiments showed that human subjects may reject positive offers in an anonymous one-shot money-sharing situation if they find the split to be unfair. The experiments of Fehr and Gächter (2000, 2002) showed that cooperation could be sustained in a finitely repeated public goods game if the punishing of free riders is permitted, despite the fact that the self-interest model predicts no cooperation. These and related findings have led in recent years to a revision of the received wisdom in biology and economics toward the appreciation of the central importance of other-regarding preferences and character virtues in biological and economic theory (Gintis et al. 2005; Henrich et al. 2005; Okasha and Binmore 2012).

The untenability of the self-interest model of human action is also clear from everyday experience. Political activity in modern societies provides unambiguous evidence. In large democratic elections, the rational self-regarding agent will not vote because the costs of voting are positive and significant, but the probability that one vote will alter the outcome of the election is vanishingly small. Thus the personal gain from voting is vanishingly small. For similar reasons, if one chooses to vote, there is no plausible reason to vote on the basis of the impact of the outcome of the election on one's self-regarding gains. It follows also that the voter, if rational, self-regarding, and incapable of personally influencing the opinions of more than a few others, will not bother to form opinions on political issues, because these opinions cannot affect the outcome of elections. Yet people do vote, and many do expend time and energy in forming political opinions. This behavior does not conform to the self-interest model.

It is a short step from the irrefutable logic of self-regarding political behavior that rational self-regarding individuals will not participate in the sort of collective actions that are responsible for growth in the world of representative and democratic governance, the respect for civil liberties, the rights of minorities and women in public life, and the like. In the self-interest model, only small groups of individuals aspiring to social dominance will act politically. Yet modern egalitarian political institutions are the result of such collective actions (Bowles and Gintis 1986; Giugni et al. 1998). This behavior cannot be explained by the self-interest model.

Apart from professional politicians and socially influential individuals, electoral politics is a vast morality play in which models of the rational self-regarding actor are not only a poor fit, but are conceptually bizarre. It took Mancur Olson's *The Logic of Collective Action* (1965) to make this clear to many behavioral scientists, because virtually all students of social life had assumed, without reflection, the faulty logic that rational self-regarding individuals will vote, and will "vote their interests" (Downs 1957).

Defenders of the self-interest model may respond that voters *believe* their votes make a difference, however untenable this belief might be under logical scrutiny. Indeed, when asked why they vote, voters' common response is that they are trying to help get one or another party elected to office. When apprised of the illogical character of that response, the common reply is that there are in fact close elections, where the balance is tipped in one direction or another by only a few hundred votes. When confronted with the fact that one vote will not affect even such close elections, the common riposte is that "Well, if everyone thought like that, we couldn't run a democracy."

Politically active and informed citizens appear to operate on the principle that voting is both a duty and prerogative of citizenship, an altruistic act that is justified by the categorical imperative: act in conformance with the morally correct behavior for individuals in one's position, without regard to personal costs and benefits. Such mental reasoning, which has been called "shared intentionality," is implicated in many uniquely human cognitive characteristics, including cumulative culture and language (Sugden 2003; Bacharach 2006).

Shared intentionality rests on a fundamentally prosocial disposition (Gilbert 1987; Bratman 1993; Tomasello and Carpenter 2007; Hrdy 2009).

Human beings acting in the public sphere are, then, neither docile internalizers of dominant culture nor sociopathic personal gain maximizers. Rather, they are generally what Aristotle called *zoon politikon*—political beings (Aristotle 350 BC/2002). In this chapter we lay out a rather general framework for understanding this deep property of the human psyche, drawing in various ways on all the behavioral sciences. This framework will be used to elucidate the role of basic human political predispositions in creating and transforming sociopolitical structures.

The Political and Economic Structure of Primate Societies

Humans are one of more than two hundred extant species belonging to the Primate order. All primates have sociopolitical systems for regulating social life within their communities. Understanding human sociopolitical organization involves specifying how and why humans are similar to and different from other primate species. Similarities likely indicate that the trait was already present before humans evolved. For instance, many primate species, including humans, seek to dominate others and are adept at forming coalitions. It is thus likely that their common ancestor also possessed these traits. Dominance seeking and coalition formation in humans, then, are not purely cultural. Rather, humans are endowed with the genetic prerequisites for dominance striving and coalition formation.

On the other hand, although chimpanzees engage in warlike raids where larger parties target and kill much smaller ones, no nonhuman primate species engages in human-style war, with large numbers of individuals on either side of a conflict. Because hunter-gatherer societies do engage in such war, the presumption is that this predisposition is uniquely human and perhaps purely cultural, or derived from more basic genetic predispositions, which themselves may be the response to prior cultural changes, of which insider favoritism may be an example (Otterbein 2004; Bowles and Gintis 2011).

Using this logic, we can examine the social structure of multi-male/multi-female monkey and ape societies (de Waal 1997b; Maestripieri 2007) to identify the elements of human sociopolitical organization that were likely present among the first hominins. The focus here is on males because in human politics, historically, men were the main players. We ask about leadership, dominance, and alliances.

Primates live in groups to reduce the risk of predation (Alexander 1974; van Schaik 1983), to facilitate the exchange of information as to food location (Eisenberg et al. 1972; Clutton-Brock 1974), and to defend food sources against competing groups (Wrangham 1980). However, these benefits largely arise through mutualism or as byproducts of grouping. Thus these groups rarely if ever engage in organized collective action. As a result, the primate form of group living has only limited need for leaders (i.e., individuals instrumental in initiating and coordinating group-level action). Instead, individuals vary in dominance based on pure physical prowess.

In most primate species, both sexes form dominance hierarchies, in which more dominant individuals gain privileged access to food or mates, and tend to have higher fitness as a result (Vigilant et al. 2001; Maestripieri 2007; Majolo et al. 2012). In many primate species, dominant females depend on alliances to maintain their position; for males the same is true in only a handful of primate species, including chimpanzees. Thus dominants rarely perform any group-level beneficial acts. A rare exception includes males displaying toward predators, a behavior seen in a variety of primate species.

Chimpanzees are an archetypical species when it comes to reconstructing the origins of the human political system. Dominant male chimpanzees provide little leadership, and they provide virtually no parenting. In many primate species, dominant males have sufficiently high paternity certainty to induce them to provide protection to infants (Paul et al. 2000), but in chimpanzees, paternity concentration is so low (Boesch et al. 2006; Vigilant et al. 2001), most likely because chimpanzee females are scattered and cannot easily be located at all times, that males tend to ignore rearing the young. The only clear service they provide to the group is that they keep the peace by intervening in disputes (de Waal 1997b; Rudolf von Rohr et al. 2012). In short, the political structure of the chimpanzee society, like that of primates generally, is largely a system for funneling fitness-enhancing resources to the apex of a social dominance hierarchy based on physical prowess and coalition-building talent. This holds basically for the bonobo as well, where monopolization of matings by particular males is even lower.

Chimpanzee males rely on coalitions and alliances more than males in most other primate species. Their coalitions come in two major categories: rank-changing and leveling coalitions (Pandit and van Schaik 2003; van Schaik et al. 2006). At the top of the hierarchy, males often rely on a supporter to acquire and maintain top dominance (Goodall 1964; Nishida and Hosaka 1996; de Waal 1998). Because this implies that the top male does not necessarily have the highest individual fighting ability, he relies on the presence of an ally, and frequently depends on coalitions to protect his position (de Waal 1998; Boesch et al. 1998). In addition, multiple lower-ranking males may form coalitions to keep the top male(s) from taking too big a share of the resources. These coalitions do not change the dominance ranks of the participants, but intimidate the dominants into limiting damaging actions aimed at subordinates. Females similarly form such leveling coalitions to counter the arbitrary power of dominant males, especially in captivity (Goodall 1986). This pattern of political power based on the hierarchical dominance of the physically powerful along with a system of sophisticated political alliances to preserve or to limit the power of the alpha male (Boehm and Flack 2010) is carried over, yet fundamentally transformed, in human society (Knauft 1991; Boehm 1999, 2011).

This data on nonhuman primates, in general, and chimpanzees and other multi-male/multi-female species, in particular, is rather surprising and very important. It is surprising because, Aristotle notwithstanding, political theorists have widely assumed that political structure involves purely cultural evolution, whereas the primate data show roots to political behavior going back millions of years. The result is important because it lays the basis for an evolutionary analysis of human political systems. Such an analysis promises to elucidate the role of basic human political predispositions in reinforcing and undermining distinct sorts of human sociopolitical structures.

The Evolutionary Trajectory of Primate Societies

It would be useful if we could read past social structure from the historical record, but we cannot. The fossil record provides the most concrete answers to our evolutionary history but is highly incomplete. There are, for instance, skeletal records of only about 500 individuals from our hominin past. Moreover, behavior does not fossilize and social structure, up until the past few thousand years, has not left direct marks in the earth. Thus we must investigate the relationship between genetic relatedness and phenotypic social organization from living primate species.

The hominin lineage branched off from the primate main stem some 6.5 million years ago. The watershed event in the hominin line was the emergence of bipedalism. Bipedalism is well developed in *Australopithecus afarensis*, which appeared three million years after the origin of the hominin lineage. *Homo ergaster* (2.0–1.3 MYA) or *H. erectus* (1.9 MYA–143,000 years ago) was the first currently documented obligate biped, having a relatively short arm:leg ratio.

Bipedalism in hominins was critically dependent upon the prior adaptation of the primate upper torso to life in the trees. The Miocene Hominoid apes were not true quadrupeds; they had specialized shoulder and arm muscles for swinging and climbing, as well as a specialized hand structure for grasping branches and manipulating leaves, insects, and fruit. When the hominin line was freed from the exigencies of arborial life, the locomotor function of the upper limbs was reduced so that they could be reorganized for manipulative and projectile control purposes. Both a more efficient form of bipedalism and the further transformation of the arm, hand, and upper torso became possible.

Nonhominin primate species are capable of walking on hind legs, but only with difficulty and for short periods of time. Chimpanzees, for instance, cannot straighten their legs, and require constant muscular exertion to support the body. Moreover, the center of gravity of the chimpanzee body must shift with each step, leading to a pronounced lumbering motion with significant side-to-side momentum shifts (O'Neil 2012). The hominin pelvis was shortened from top to bottom and rendered bowl-shaped to facilitate terrestrial locomotion without sideward movement, the hominin leg bones became sturdy, the leg muscles were strengthened to permit running, and the development of arches in the feet facilitated a low-impact transfer of weight from leg to leg. Thus, bipedality facilitates running efficiently for great distances, although not approaching the speed of many large four-footed mammals.

Today we celebrate obligate bipedality as the basis for human upper-body physical and psychomotor capacities for crafting tools and handicrafts. However, another major contribution of these capacities was for fashioning lethal weapons.

Control of Fire: A Precondition of Social Sharing Norms

The hominin control of fire cannot be accurately dated. We have firm evidence from about 400,000 years ago in Europe (Roebroeks and Villa 2011) and about 800,000 years ago in Israel (Alperson-Afil 2008), but it is likely that this key event happened in Africa much earlier. The control of fire had strong effects on hominin cultural and phylogenetic evolution. First, the transition to obligate bipedality is much easier to understand if the hominins that made it had control of fire (Wrangham and Carmody 2010). Prior to the control of fire, humans almost certainly took to the trees at night, like most other primates, as a defense against predators. Because predators have an instinctive fear of fire, the control of fire permitted hominins to abandon climbing almost completely.

Second, the practice of cooking food was a related cultural innovation with broad gene–culture coevolutionary implications. Cooking presupposes a central location to which the catch is transported, and hence requires abandoning the socially uncoordinated "tolerated theft" distribution of calories typical of food sharing in nonhuman primate species, in favor of a distribution based on widely agreed-upon fairness norms (Isaac 1978b). This major sociopsychological transition was probably made possible by the adoption of some form of cooperative breeding and hunting among hominins and had begun before the origin of *H. erectus* (van Schaik and Burkart 2010). In sum, the control of fire and the practice of cooking were important preconditions for the emergence of a human moral order.

Although the archaeological record does not permit accurate dating for the regular use of fire by hominins, (Sandgathe et al. 2011; Roebroeks and Villa 2011), it is clear that hominins with access to cooked food did not require the large colon characteristic of other primates. This allowed them to reduce the amount of time spent chewing food from the four to seven hours a day (characteristic of the great apes) to about one hour per day. With a smaller gut, less need for chewing, and more rapid digestion, hominins were liberated to develop their aerobic capacity and perfect their running ability (Wrangham and Carmody 2010).

From Gatherer to Scavenger

Beginning around 2.5 MYA there was a major forking in the evolutionary path of our ancestors. The Australopithecines branched in at least two very different evolutionary directions: one led to the robust Australopithecines and a genetic dead end by about 1.4 MYA; the other, eventually, to the first humans.

It is likely that these diverging evolutionary paths were the response to novel environmental challenges. Coinciding with this hominin divergence was a shift in the global climate to frequently fluctuating climatic conditions. Early hominins succeeded by learning to exploit the increased climate instability (Potts 1996; Richerson et al. 2001; O'Connell et al. 2002).[1] The resulting adaptations enhanced hominin cognitive and sociostructural versatility. "Early bipedality, stone transport,...encephalization, and enhanced cognitive and social functioning," Potts (1998:93) argues, "all may reflect adaptations to environmental novelty and highly varying selective contexts." This view is supported by the observation that greater encephalization occurred as well in many mammalian lineages (Jerison 1973).

Eating the meat of large animals provided a niche for emerging hominins quite distinct from that of other primates and thus selected for the traits that most distinguish humans from apes. This much was clear to Darwin in *The Descent of Man* (1871). However, until recently, most paleoanthropologists assumed that meat was acquired through hunting from the australopithecine outset (Dart 1925; Lee and DeVore 1968). In fact, it now appears that early hominins, in the transition from the Pliocene to the Pleistocene, were more likely scavenger-gatherers than hunter-gatherers, of which there is firm evidence dating from 1.6 to 1.8 MYA.

The first proponents of early hominins as scavengers believed that the scavenging was "passive," in that small groups of hominins took possession of carcasses only after other predators, upon being sated, abandoned their prey (Blumenschine et al. 1994). More recent evidence, however, suggests the prevalence of "competitive scavenging," in which organized groups of humans supplied with primitive weapons, chased the killers and appropriated carcasses in relatively intact shape (Dominguez-Rodrigoa and Barba 2006). The implicit argument is that the hominin lethal weapons of the period were sufficient to drive off other predators, and hence presumably to drive off live prey as well. To cripple or kill a large prey item, however, requires considerably more

---

[1] deMenocal (2011) notes that Darwin (1859) long ago speculated on the role of climate change in human evolution, as did Dart (1925), and that modern findings support the importance of climate-based selection pressures (Vrba 1995; Potts 1998), and specifically, climate variability. Examining the environmental records of several hominin localities, Potts (1998) found that habitat-specific hypotheses are disconfirmed by the evidence; however, the variability selection hypothesis, which states that large disparities in environmental conditions were responsible for important episodes of adaptive evolution, was widely supported.

powerful weapons. Thus, before poisoned, stone-tipped spears and arrows, the hunting of large prey was likely unrewarding (but see Liebenberg 2006).

Flaked stone tool making, butchering large animals, and expanded cranial capacity all appear around 2.5 MYA, but there is no evidence that Australopithecines hunted large game. *Australopithecus* and *H. habilis* were in fact quite small: adult males weighed under 100 pounds and females about 75 pounds. Their tools were primitive, consisting of stone scrapers and rough hammerstones. They therefore lacked the sophisticated weapons for hunting large and swift-moving prey. They are unlikely to have hunted effectively, but they could well have scavenged. Modern chimpanzees and baboons are known to scavenge the kills of cheetahs and leopards, so this behavior was likely in the repertoire of the earliest hominins. With highly cooperative and carefully coordinated maneuvers, they could have chased even ferocious predators.

Hunting and scavenging small animals is not cost effective for large primates, while scavenging large animals requires group participation and efficiently coordinated cooperation, both in organizing an attack on predators feeding on a large prey and in protecting against predators while processing and consuming the carcass (Isaac 1978a). Moreover, the only known weapons that might be used to scare off hunters and scavengers and potential predators were stones of the appropriate size and weight to be thrown at high velocity (Isaac 1987). Such stones had to be carefully amassed in strategic sites within a large scavenging area, so that when a scouting party located an appropriate food object, it could call others to haul the stones to the site of the dead animal, as a strategic operation preceding the appropriation of the animal carcass. These were the first lethal projectile weapons.

This scenario is supported by the fact that the fossils of large animals that have bone markings, indicating hominin flaying and scraping with flaked stone tools, are often found with stones that originated several kilometers away. Contemporary chimpanzees carry stones to nut-bearing trees and use them to crack the nuts, so this behavior was likely available to Australopithecenes. Chimpanzees, however, carry stones only several hundred meters at most, while *H. habilis* scavengers carried stones as far as ten kilometers. By contrast, neither the Oldowan tools of the period nor the later and more sophisticated Acheulean tools, found from the early Pleistocene up to about 200,000 years ago, show any sign of being useful as hunting weapons, although besides stones, scavengers of 500,000 years ago probably had sharpened and fire-hardened spears to ward off competitive scavengers and threatening predators, at least after the domestication of fire (Thieme 1997). By contrast, nonhuman primates use tools, but they do not use weapons to battle (McGrew 2004), although chimpanzees have been seen using spears fashioned from nearby tree branches to kill bushbabies that they discovered in tree hollows (Pruetz and Bertolani 2007).

The emergence of lethal weapons, however primitive, was likely key to the evolution of hominid social organization. Bingham (1999) and Boyd et al.

(2010) stress the importance of the superior physical and psychomotor capacities of humans in clubbing and throwing projectiles as compared with other primates, citing Goodall (1964) and Plooij (1978) on the relative advantage of humans. Darlington (1975), Fifer (1987), and Isaac (1987) document the importance of these traits in human evolution. Bingham (1999) stresses that humans developed the ability to carry out collective punishment against norm violators, thus radically lowering the cost of punishing transgressors. Calvin (1983) argues that humans are unique in possessing the neural machinery for rapid manual-brachial movements that both allows for precision stone throwing and lays the basis for the development of language, which like accurate throwing depends on the brain's capacity to orchestrate a series of rapidly changing muscle movements. These changes took place, in all likelihood, more than 700,000 years ago.[2]

Social Hierarchy: Dominance and Reverse Dominance

Hunter-gatherer societies have been classified into *immediate-return* and *delayed-return* systems (Woodburn 1982). In the former, group members obtain direct return from their labor in hunting and gathering, with food lasting at most a few days. The tools and weapons they use are highly portable. In delayed-return societies, individuals hold rights over valuable assets, such as means of production (e.g., boats, nets, beehives), processed and stored food and materials, and herds of animals. In these societies we find forms of social stratification akin to those in modern societies: social dominance hierarchies in the form of lineages, clans, chiefdoms, and the like. The fossil record suggests, however, that the delayed-return human society is a quite recent innovation, appearing some 10,000 years ago, although on ecologically suitable locations, it may have existed earlier—most of these locations are now below sea level. *H. sapiens* thus evolved predominantly in the context of immediate-return systems.

The issue in "delayed return" is not the capacity for delayed gratification or long-range planning, but rather the availability of accumulated wealth. Material wealth allows aspirants to positions of social dominance to control enough allies and resources to offset the capacity of subordinate individuals to disable and kill them. As long as the material gains from a position of social dominance exceed the cost of coalition building and paying guard labor, social dominance of the sort common in other primate societies can be reestablished

---

2 Fossil evidence indicates that hominins developed speech on the order of 1 MYA. The hyoid bone is a key element of speech production in humans. Martinez et al. (2008a) show that hominin hyoid bones from 540,000 years ago are similar, and hence were inherited from their last common ancestor, *Homo rhodesiensis*, around 700,000 to 1,000,000 years ago. Using evidence from the acoustical properties of Middle Pleistocene fossil remains of the hominin inner ear, Martinez et al. (2004) argue that hominins of this period had auditory capacities similar to those of living humans.

in human society.[3] To avoid confusion, we will refer to societies that lack forms of material wealth accumulation as *simple*, rather than delayed-return, hunter-gatherer societies.

Simple hunter-gatherer societies, Woodburn (1982:434) suggests, are "profoundly egalitarian...[they] systematically eliminate distinctions...of wealth, of power and of status." Fried (1967), Service (1975), Knauft (1991), and others likewise comment on the egalitarian character of simple hunter-gatherer societies. What factors are responsible for such unusual egalitarianism? Here, we argue that it is due to the combination of interdependence and ability to punish transgressors.

Cut marks on bones suggest that big-game hunting started only 250,000 years ago, and delegating sharing to a single cutter began 200,000 years ago (Stiner 2002; Stiner et al. 2009). However, cut marks on bones may not be a reliable indicator of how meat is shared (Lupo and O'Connell 2002). Indeed, if Wrangham and Carmody (2010) are correct in dating the control of fire by hominins and the cooking of meat, the problem of the fair distribution of meat among families must have been solved much earlier, and doubtless was a major source of egalitarian sentiment, as well as providing the material substrate for the development of a social morality. Certainly contemporary hunter-gatherer societies are often violent, competitive, and there is considerable political inequality (Potts 1996), but they almost always distribute large game peacefully, based on a commonly accepted set of fairness principles (Kaplan and Hill 1985b; Kelly 1995).

The human ecological niche requires food sharing on a daily basis as well as on a longer-term basis due to the occasional injuries or illnesses to which even the best hunters or gatherers may be subjected (Sugiyama and Chacon 2000; Hill et al. 2011). Thus each individual forager, especially in the immediate-return form of foraging, is utterly dependent on the others in their camp, band, or even wider sharing unit. This strong interdependence dampens the tendency to free ride on others' efforts and favors strong individual tendencies toward egalitarianism, as well as sophisticated fairness norms concerning the division of the spoils (Whallon 1989; Kaplan and Hill 1985a).

Collective hunting in other species does not require a fairness ethic because participants in the kill simply eat what they can secure from the carcass. However, the practice of bringing the kill to a central site for cooking, which became characteristic of hominin societies, is not compatible with uncoordinated sharing and eating. In the words of Winterhalder and Smith (1992:60):

---

3 In fact, the appearance of farming and private property in land led to high levels of political inequality in only a few societies, and states with a monopoly in coercive power emerged only after a millennium of settled agriculture. Nor were early farming societies more economically stratified than hunter-gatherer societies (Borgerhoff Mulder et al. 2009). The accumulation of material wealth is thus merely a precondition for the reestablishment of social dominance hierarchies.

> [O]nly with the evolution of reciprocity or exchange-based food transfers did it become economical for individual hunters to target large game. The effective value of a large mammal to a lone forager...probably was not great enough to justify the cost of attempting to pursue and capture it....However, once effective systems of reciprocity or exchange augment the effective value of very large packages to the hunter, such prey items would be more likely to enter the optimal diet.

Fire and cooking, therefore, are cultural preconditions to the emergence of a normative order and social organization based on normative behavior.

The second element is that egalitarianism is imposed by the community, creating what Boehm (1999) calls a *reverse dominance hierarchy*. Hunter-gatherers share with other primates the striving for hierarchical power, but social dominance aspirations are successfully countered because individuals do not accept being controlled by an alpha male and are extremely sensitive to attempts of group members to accumulate power through coercion. When an individual appears to be stepping out of line by threatening or killing group members, he will be warned and punished. If this behavior continues and he cannot be ostracized, the group will delegate one or more members, usually including at least one close relative of the offender, to kill him. Boehm's message in *Hierarchy in the Forest* is that "egalitarianism involves a very special type of hierarchy, a curious type that is based on *anti*hierarchical feelings" (Boehm 1999:10).

Because of the extremely long period during which humans evolved without the capacity to accumulate wealth, we have become constitutionally predisposed to exhibit these antihierarchical feelings. Of course, in modern societies, there is still enough willingness to bend to authority in humans to ensure that social dominance hierarchy remains a constant threat and often a reality.

Capable leadership in the absence of a social dominance hierarchy in these societies is doubtless of critical importance to their success, and leaders are granted by their superior position, and through the support of their followers, with fitness and material benefits. Leadership, however, is based not on physical prowess, but rather on the capacity to motivate and persuade.[4]

The centrality of reverse dominance hierarchy is assessed in *Moral Origins: the Evolution of Virtue, Altruism, and Shame* (Boehm 2011). Boehm located 339 detailed ethnographic studies of hunter-gatherers, 150 of which are simple hunter-gatherer societies, and coded fifty of these societies from around the world. He calls these simple hunter-gatherer societies "Late Pleistocene Appropriators" (LPAs). Despite the fact that these societies have faced highly variable ecological conditions, Boehm finds that their social organization follows the pattern suggested by Woodburn (1982) and Boehm (1999). Not only

---

4 This account of the growth of intelligence sharply contrasts the Machiavellian intelligence doctrine (Jolly 1972; Humphrey 1976; Byrne and Whiten 1988), according to which encephalization was the product of an arms race in which the gains from intellect were enhanced ability to deceive others and detect deception.

do LPAs exhibit reverse dominance hierarchy, they also subscribe to a common human social morality, operating through internalized norms, so that individuals act prosocially because they value moral behavior for its own sake and would feel guilty behaving otherwise.[5]

How do we explain this unique pattern of sociopolitical organization? Woodburn attributes this to our access to and presence of lethal weapons, which neutralize a social dominance hierarchy based on coercion. "Hunting weapons are lethal," he writes, "not just for game animals but also for people. Effective protection against ambush is impossible...with such lethal weapons." Woodburn adds that under "normal circumstances the possession by all men, however physically weak, cowardly, unskilled or socially inept, of the means to kill secretly anyone perceived as a threat to their own well-being...acts directly as a powerful leveling mechanism. Inequalities of wealth, power and prestige...can be dangerous for holders where means of effective protection are lacking" (Woodburn 1982:436).

Boehm (2011) argues that LPAs inherited from our ancient hunter-gatherer forbears the capacity to control free riders through collective policing, using gossip and informal meetings as the method of collecting information concerning the behavior of group members. Moreover, according to our best evidence, the hunter-gatherer societies that defined human existence until some 10,000 years ago also were involved in widespread communal and cooperative child rearing (Hrdy 2000, 2009) and hunting (Boehm 1999; Bowles and Gintis 2011; Boyd and Silk 2002; Boehm 2011), thus tightening the bonds of sociality in the human group and increasing the social costs of free-riding behavior.

Nonhuman primates never developed weapons capable of controlling a dominant male. Even when sound asleep, an accosted male chimpanzee reacts to hostile onslaughts by awakening and engaging in a physical battle, basically unharmed by surprise attack. In *Demonic Males*, Wrangham and Peterson (1996), recount several instances where even three or four male chimpanzees viciously and relentlessly attack a male for twenty minutes without succeeding in killing him. The ineffectiveness of chimpanzees in this regard is not simply the lack of the appropriate lethal weapon, but the inability to wield effectively potentially dangerous natural objects, for instance stones and rocks. A chimpanzee may throw a rock in anger, but only weakly and rarely will it achieve its target.

The human lifestyle, unlike that of chimpanzees, requires many collective decisions, such as when and where to move camp and which alliances to

[5] The notions of norms and norm internalization are common in sociology and social psychology but are absent from the other social science disciplines. According to the sociopsychological theory of norms, appropriate behavior in a social role is given by a social norm that specifies the duties, privileges, and expected behavior associated with the role. Adequate performance in a social role normally requires the actor to have a *personal commitment* to the role—one that cannot be captured by the self-regarding "public" payoffs associated with the role (Gintis 2009).

sustain or sever. This lifestyle thus requires a complex sociopolitical decision-making structure and a sophisticated normative order. Many researchers incorrectly equate dominance, as found among chimpanzees, with leadership. In some species, such as gorillas, dominants can indeed initiate or influence group progressions, because many rely on the dominant as the main protector and prefer his proximity. In human foragers, there are no such dominants.

Capable leadership, in the absence of a social dominance hierarchy in these societies, is nonetheless of critical importance to their success. However, leaders are granted by their superior position and with the support of their followers, fitness, and material benefits. Leadership, as we have seen, is based not on physical prowess or coercion, but rather on the capacity to motivate and persuade. Eibl-Eibesfeldt (1989) and Wiessner (2006), among many others, have stressed the importance in hominin societies of leadership based on persuasion and coalition building. Wiessner (2009:197–198) remarks: "Unlike nonhuman primates, for whom hierarchy is primarily established through physical dominance, humans achieve inequalities through such prosocial currencies as the ability to mediate or organize defense, ritual, and exchange."

It is important not to confuse reverse dominance hierarchy, which is a predisposition to reject being dominated in an authoritarian manner, with a predisposition for egalitarian outcomes. Rather, persuasion and influence become a new basis for social dominance. The Machiavellian intelligence hypothesis (Byrne and Whiten 1988) is not wrong about the role of hyper-cognition in personal success, but rather about the social basis for this success, which is exhibiting prosocial behavior that enhances the fitness of the group and its members (Clutton-Brock 2009). Wiessner (2006:198) observes that successful small-scale societies "encourage the capable to excel and achieve higher status on the condition that they continue to provide benefits to the group. In no egalitarian institutions can the capable infringe on the autonomy of others, appropriate their labor, or tell them what to do."

Are There Egalitarian Nonhuman Primates?

If there were a multi-male/multi-female primate society without a social dominance hierarchy, and in the absence of lethal weapons, this would cast doubt on the propositions offered herein. Does such a society exist? Here, an important distinction is between egalitarianism that arises due to low intensity of contest competition and egalitarianism, accompanied by high tolerance, that arises due to interdependence or some form of subordinate leverage over dominants (Sterck et al. 1997).

While there are clear behavioral patterns in nonhuman primates that serve as the basis for human reverse dominance hierarchy, all multi-male/multi-female nonhuman primate societies are in fact based on social dominance hierarchy. There may be variation in the degree to which female or male dominance relations are decided and thus their dominance hierarchies are more or less steep, depending on the strength of contest competition for resources (Sterck et al. 1997). It is often argued that bonobos (*Pan paniscus*) are more egalitarian than chimpanzees and more like humans (de Waal 1997a; Hare et al. 2007). However, except for female dominance hierarchy in feeding access for infants, the pattern of dominance in bonobos strongly resembles that of chimpanzees (Furuichi 1987, 1989, 1997). Moreover, differences in the steepness of the dominance hierarchy among males and females are not consistent across studies (Stevens et al. 2007; Jaeggi et al. 2010).

Similarly, reports indicate a rather thoroughgoing egalitarianism among woolly spider monkeys, or muriquis (Strier 1992), which also live in large multi-male/multi-female societies, much like those of bonobos and chimpanzees. They are highly promiscuous and males hardly compete for matings (Milton 1984; Strier 1987). In all the primate examples of egalitarianism in large societies (i.e., not in those forming pairs of polyandrous trios), there is a clear reduction in the intensity of male contest competition as a result of female reproductive physiology that leads to unpredictable ovulation and thus low potential monopolization of matings, and therefore paternity concentration, by top-ranking males (van Schaik et al. 2004b). Thus, egalitarian social relations are the result of scramble-like competition.

In none of these societies do we find the interdependence that we see in human societies. The closest analog are the societies of wild dogs and wolves, which are both cooperative breeders and hunters (Macdonald and Sillero-Zubiri 2004). Even there we mostly, though not always, have a single breeding pair rather than multiple cooperating pairs. We conclude that, on the basis of available evidence, there are no multi-male/multi female egalitarian primate societies except for *H. sapiens*.

Phylogenetic and Cultural Implications of Governance by Consent

Following the development of lethal weapons, successful hominin social bands came to value individuals who could command prestige by virtue of their persuasive capacities. Persuasion depends on clear logic, analytical abilities, a high degree of social cognition (knowing how to form coalitions and motivate others), and linguistic facility (Plourde 2009). Leaders with these traits could be both effective and fearsome, but one intemperate move could lead to their devolution from power. Thus in concert with the evolution of an increasingly complex feeding niche (Kaplan et al. 2000), the social structure of hunter-gatherer life was one contributing factor to the progressive encephalization and evolution of the physical and mental prerequisites of effective linguistic and facial communication. In short, two million years of evolution of hyper-cooperative multifamily groups that deployed lethal weapons gave rise to the particular cognitive and sociopolitical qualities of *H. sapiens*.

The increased cephalization in humans was an extension of a long primate evolutionary history of increased brain size, usually associated with increased cognitive demands required by larger group size (Humphrey 1976; Jolly 1972; Byrne and Whiten 1988).[6] The lethal weapon argument extends this analysis to explain human exceptionalism in the area of cognitive and linguistic development.

The role of lethal weapons in promoting egalitarian multi-male/multi-female hominin groups explains the huge cognitive and linguistic advantage of humans over other species not as some quirk of sexual selection—the favored theory of Darwin (1871), Fisher (1930), Miller (2001), and many others—but rather as directly fitness enhancing, despite the extreme energy costs of maintaining a large brain. Increased cognitive and linguistic ability entailed heightened leadership capacities, which fellow group members were very willing to trade for enhanced mating and provisioning privileges.

In a sense, hominins evolved to fill a *cognitive niche* that was relatively unexploited in the early Pleistocene (Tooby and DeVore 1987). According to Pinker (2010:8993):

> I suggest that the puzzle [of human hyper-cognition] can be resolved with two hypotheses. The first is that humans evolved to fill the "cognitive niche," a mode of survival characterized by manipulating the environment through causal reasoning and social cooperation. The second is that the psychological faculties that evolved to prosper in the cognitive niche can be coopted to abstract domains by processes of metaphorical abstraction and productive combination, both vividly manifested in human language.

Cooperative Mothering and the Evolution of Prosociality

In cooperative breeding, the care and provisioning of offspring is shared among group members. The standard estimate is that some 3% of mammals have some form of allomaternal care; in the Primate order, however, this frequency rises to 20% or more (Hrdy 2009, 2010). In many nonhuman primates and mammals in general, cooperative breeding is accompanied by generally heightened prosociality, as compared with related species with purely maternal care. The most plausible explanation is that cooperative breeding leads to a social structure that rewards prosocial behavior, which in turn leads to changes in neural structure that predisposes individuals to behaving prosocially (Burkart et al. 2009; Burkart and Van Schaik 2010). An alternative possibility is that there is some underlying factor in such species that promotes prosociality in general, of which collective breeding is one aspect.

Human prosociality was strongly heightened beyond that of other primates living in large groups, including cooperative breeders, by virtue of the niche

6 Group size is certainly not the whole story. Multi-male/multi-female monkey groups are often as large or larger than ape groups, although the latter have much larger brains and are considerably more intelligent. The full story concerning cephalization in mammals, in general, and primates, in particular, remains to be told (Navarrete and van Schaik 2011).

hominins occupied, involving coordination in hunting and scavenging, and sophisticated norms for sharing meat. This combination might account for the degree of cooperative breeding in the hominin line. As hominin brain size increased, the duration of immaturity did as well (Barrickman et al. 2008), and immatures had to learn an increasingly large number of foraging and other skills (Kaplan et al. 2000). Hominins evolved a unique system of intergenerational transfers that enabled the evolution of increasingly complex cognitive abilities to support the continual acquisition of complex subsistence skills (Kaplan et al. 2007). Our uniquely prosocial shared intentionality (Tomasello et al. 2005) can be traced back to the psychological changes involved in the evolution of cooperative breeding and hunting (Burkart et al. 2009).

Lethal Weapons and Egalitarian Political Organization from the Holocene to the Present

With the development of settled trade, agriculture, and private property some 10,000 years ago, it became possible for a Big Man to gather a relatively small group of (usually closely related) subordinates and consorts around him that would protect him from the lethal revenge of a dominated populace, whence the slow but inexorable rise of the state, both as an instrument for exploiting direct producers and for protecting them against the exploitation of external states and bands of private and state-sanctioned marauders. The hegemonic aspirations of states peaked in the thirteenth century, only to be driven back by the series of European population-decimating plagues of the fourteenth century. The period of state consolidation resumed in the fifteenth century, based on a new military technology: the use of cannons. In this case, as in some other prominent cases, technology became the handmaiden to establishing a social dominance hierarchy based on force.

In *Politics*, Book VI part vii,[7] Aristotle writes "there are four kinds of military forces—the cavalry, the heavy infantry, the light armed troops, the navy. When the country is adapted for cavalry, then a strong oligarchy is likely to be established [because] only rich men can afford to keep horses. The second form of oligarchy prevails when the country is adapted to heavy infantry; for this service is better suited to the rich than to the poor. But the light-armed and the naval elements are wholly democratic...An oligarchy which raises such a force out of the lower classes raises a power against itself."

The use of cavalry became dominant in Western Europe during the Carolingian period. The history of warfare from the Late Middle Ages to the First World War was the saga of the gradual increase in the strategic military value of infantry armed with longbow, crossbow, hand cannon, and pike, which marked the recurring victories of the English and Swiss over French and Spanish cavalry in the twelfth to fifteenth centuries. Cavalries responded

[7] Available at: http://www.constitution.org/ari/polit_06.htm

by developing dismounting tactics when encountering infantry, using heavy hand-held weapons, such as two-handed swords and poleaxes. These practices extended the viability of cavalry to the sixteenth century in the French and Spanish armies, but gradually through the Renaissance, and with the rise of Atlantic trade, the feudal knightly warlords gave way to the urban landed aristocracy, and warfare turned to the interplay of mercenary armies consisting of easily trained foot soldiers wielding muskets and other weapons based on gunpowder. Cavalry remained important in this era, but even in the eighteenth and nineteenth century, cavalry was used mainly to execute the *coup de grâce* on seriously weakened infantry.

The true hegemony of the foot soldier, and hence the origins of modern democracy, began with the perfection of the hand-held weapon, with its improved accuracy and greater firing rate than the primitive muskets of a previous era. Until that point, infantry was highly vulnerable to attack from heavy artillery. By the early twentieth century, the superiority of unskilled foot soldiers armed with rifles was assured. World War I opened in 1914 with substantial cavalry on all sides, but mounted troops were soundly defeated by men with rifles and machine guns and thus were abandoned in later stages of the war. The strength of the political forces agitating for political democracy in twentieth century Europe was predicated on the strategic role of the foot soldier in waging war and defending the peace (Bowles and Gintis 1986).

Conclusion

It is tempting to focus on the past 70,000 years of human cultural history when theorizing about human sociopolitical organization, because the changes that occurred during this period radically transformed the character of our species (Richerson and Boyd 2005; Pagel 2012). However, the basic genetic predispositions of humans underlying sociopolitical structure were forged over a much longer period of time: the million-plus-year perspective offered in this chapter.

The framework developed here is applicable to many spheres of human culture, although we have applied it only to the evolution of sociopolitical structure. The central tool is *gene–culture coevolution*, which bids us pay close attention long-term to the dynamic interplay between our phylogenetic constitution and our cultural heritage. The second important conceptual tool is the *sociopsychological theory of norms*. Many social scientists reject this theory because it posits a causal social reality above the level of individual actors. This position is sometimes termed methodological individualism. Methodological individualism is not a philosophical, moral, or political principle, but an assertion about reality. As such, it is simply incorrect, because social norms are an emergent property of human society, irreducible to lower-level statements (Gintis 2009). All attempts to explain human culture without this higher-level construct fail.

We have suggested the following scenario for the long history of human sociopolitical dynamics. Our primate ancestors evolved a complex sociopolitical order based on a social dominance hierarchy in multi-male/multi-female groups. Enabled by bipedalism, environmental changes made a diet of meat from large animals fitness enhancing in the hominin line. This, together with cultural innovation in the domestication of fire, the practices of cooking and of collective child rearing created a niche for hominins in which there was a high return to coordinated, cooperative, and competitive scavenging as well as technology-based extractive foraging. This, in turn, led to the use of stones and spears as lethal weapons, and thence to the reorganization of the upper torso, shoulders, arms, and hands to maximize the effectiveness of these weapons, as well as the growth of new neural circuitry allowing the rapid sequencing of bodily movements required for accurate weapon deployment.

The hominin niche increasingly required sophisticated coordination of collective meat procurement, a willingness to provide others with resources, the occasional, but critical reliance on resources produced by others, and procedures for the fair sharing of meat and collective duties. The availability of lethal weapons in early hominin society helped to stabilize this system because it undermined the tendencies of dominants to exploit others in society. Thus two successful sociopolitical structures arose to enhance the flexibility and efficiency of social cooperation in hominins: (a) reverse dominance hierarchy, which replaced social dominance based on physical power with a political system in which success depended on the ability of leaders to persuade and motivate, and (b) cooperative breeding and hunting, which provided a strong psychological predisposition toward prosociality and favored internalized norms of fairness. This system persisted until cultural changes in the Holocene fostered material wealth accumulation, through which it became once again possible to sustain a social dominance hierarchy based on coercion.

This scenario has important implications for political theory and social policy, for it suggests that humans are predisposed to seek dominance when this is not excessively costly, but also to form coalitions to depose pretenders to power. Moreover, humans are much more capable of forming powerful and sustainable coalitions than other primates, due to our enhanced cooperative psychological propensities. This implies that many forms of sociopolitical organization are compatible with the particular human amalgam of hierarchical and antihierarchical predispositions.

This also implies, in particular, that there is no inevitable triumph of liberal democratic over despotic political hierarchies. The open society will always be threatened by the forces of despotism, and a technology could easily arise that irremediably places democracy on the defensive. The future of politics in our species, in the absence of concerted emancipatory collective action, could well be something akin to George Orwell's *1984* or Aldous Huxley's *Brave New World.* Humans appear constitutionally indisposed to accept a social dominance hierarchy based on coercion unless the coercive mechanism and its associated social processes can be culturally legitimated. It is somewhat encouraging that such legitimation is difficult except in a few well-known ways, based on patriarchy, popular religion, or liberal democracy.

Acknowledgments

Thanks to Christopher Boehm, Samuel Bowles, Bernard Chapais, Michael Ghiselin, James O'Connell, Peter Richerson, Joan Silk, Peter Turchin, and Polly Wiessner for help with earlier versions of this paper.

3

Human Cooperation among Kin and Close Associates May Require Enforcement of Norms by Third Parties

Sarah Mathew, Robert Boyd, and Matthijs Van Veelen

Abstract

While our capacity for large-scale cooperation is striking, humans also cooperate with kin and close associates much more than most other vertebrates. Existing theories do not satisfactorily explain this difference. Moreover, mechanisms posited for explaining large-scale human cooperation, like norms, third-party judgments and sanctions, also seem to be essential in regulating interactions among kin and close associates. It is hypothesized that norms and third-party judgments are crucial even for small-scale cooperation, and that kin selection and direct reciprocity alone cannot generate the degree of small-scale cooperation needed to sustain the human life history.

Introduction

Human cooperation clearly differs from that observed in other mammals. Most striking, humans cooperate on much larger scales than other mammals, sometimes through a common cause generated in groups of thousands or even millions of people. To explain the evolution and maintenance of such large-scale cooperation, scholars have studied mechanisms that do not rely on kin selection or direct reciprocity, including indirect reciprocity (Nowak and Sigmund 1998; Panchanathan and Boyd 2004), punishment (Boyd et al. 2003, 2010; Brandt et al. 2006; Hauert et al. 2007), signaling (Smith and Bliege Bird 2000; Gintis et al. 2001; Hawkes and Bliege Bird 2002) as well as genetic (Sober and Wilson 1994; Choi and Bowles 2007) and cultural group selection (Boyd and Richerson 1985, 1990, 2002a; Henrich 2004a). Humans, however, are also exceptional cooperators at smaller scales compared to most other vertebrates.

Division of labor and delayed exchange of valuable commodities, for example, occurs in virtually every human society whereas they are virtually absent in other vertebrates.

The fact that humans engage in more small-scale cooperation than other vertebrates is not satisfactorily explained by existing theories. It is generally accepted that large-scale cooperation, of the kind observed in humans, is not feasible in other vertebrate societies because the mechanisms that maintain large-scale human cooperation hinge on language and culture. However, the mechanisms that are thought to maintain small-scale cooperation—kin selection (Hamilton 1964) and direct reciprocity (Trivers 1971; Axelrod and Hamilton 1981)—should work equally well in many other vertebrates, and while a number of hypotheses have been advanced to explain the greater levels of human small-scale cooperation (Chapais 2008; van Schaik and Burkart 2010; Pinker 2010; Hrdy 2009), no consensus has been reached.

Here, we hypothesize that the enforcement of norms by third parties is crucial for the evolution of both large-scale cooperation *and* small-scale cooperation in humans. Not only do the distinct levels of small-scale cooperation in humans and other animals not follow from the existing theories, large-scale cooperation and small-scale cooperation in humans share much in common. This is particularly surprising if (a) small-scale cooperation evolved due to the effects of genetic relatedness and direct reciprocity and (b) large-scale cooperation evolved through the effects of reputation, sanctions, and cultural group selection. We posit that norms and the sanctioning of norm violators may have emerged to support small-scale cooperation, and that neither kin selection nor direct reciprocity on their own would suffice to produce the high levels of cooperation within families and among friends that is needed to sustain the human life history. This can account for the far greater degree of small-scale cooperation in humans than in most other vertebrates as well as the commonalities between small- and large-scale human cooperation. It also helps explain how, as the rate of cultural evolution accelerated, cultural group selection could have led to norm enforcement at the larger scales of cultural groups.

Insufficient Cooperation among Close Kin and Frequent Associates

There is compelling evidence that the human life history would not be possible without extensive cooperation within the nuclear family and across families (Kaplan et al. 2000). In three well-studied foraging groups, there is substantial net food transfer from husbands to wives, from parents to offspring, and from nonreproductive adults to breeding pairs with dependent offspring (Kaplan and Hill 1985a; Gurven et al. 2000; Kaplan et al. 2000; Hawkes et al. 2001; Gurven 2004; Marlowe 2004; Hill and Hurtado 2009). The high level of small-scale cooperation within foraging bands is essential to sustain the unique human life history traits of large brains, long life, low mortality, prolonged juvenile period, short interbirth intervals, and the utilization of the most nutrient-dense plant and animal resources in their habitats (Kaplan et al. 2000). Kin selection and reciprocity both play some role in supporting this cooperation (Gurven et al. 2000; Gurven 2004; Allen-Arave et al. 2008). However, in other vertebrates, close kin and social partners who repeatedly interact do not achieve similar levels of cooperation. This suggests that close kinship and reciprocity are not sufficient to allow the evolution of the extensive small-scale cooperation in humans.

Observed Levels of Small-Scale Cooperation Are Not Consistent with Kin Selection and Direct Reciprocity

If potential benefits from cooperation are widespread, the theories of kin selection and reciprocity suggest that close kin and frequent associates should cooperate to a much greater extent than they actually do. When we do observe cooperation, it is often among related individuals, consistent with kin selection (e.g., Jennifer et al. 1994; Silk et al. 2004 ; Silk 2006; van Schaik and Kappeler 2006; Langergraber et al. 2007; Hughes et al. 2008). However, a full test of the theory requires also evaluating the absences of cooperation, and not just the occurrences.

Cooperation in other vertebrates mostly involves behaviors that are low cost or even individually beneficial, because interactions are mutualistic or the result of coercion by a dominant individual (Clutton-Brock 2009; Clutton-Brock et al. 2001). Group hunting, alarm calling, joint predator mobbing, and territorial choruses that have been observed in nonhuman mammals and birds may provide direct benefits to the actor (Clutton-Brock 2009; Grinnell et al. 1995).

Reciprocity is rare in nonhuman animals (Hammerstein 2003). One of the few examples is grooming in primates (Frank and Silk 2009), a fairly low-cost activity. Models of kin selection and reciprocity, however, do not predict that cooperation should be restricted to low-cost interactions. It is the ratio of benefits to costs that matters, and so high-stakes forms of cooperation between close kin and regular interactants should be common. For instance, sharing food with a sick kinsman or close associate is costly, but can produce much larger benefits for the recipient. Illness and serious injuries are often fatal when individuals cannot feed themselves. Thus, it would seem that species in which kin provide such aid should be commonplace. Yet, humans are one of the few mammals that do this, and when they do, the benefits are large (Sugiyama 2004). More generally, specialization in food production coupled with food sharing seems to be highly beneficial, leading to lowered mortality, a long juvenile period, and high investment in learning that characterize the niche of *Homo sapiens* (Kaplan et al. 2000). In most animal societies, there is little food sharing even among close relatives.

A few mammals (e.g., mole rats, callitrichid primates, wild dogs, wolves and meerkats) breed cooperatively (Solomon and French 1997; Jennifer et al. 1994; Clutton-Brock 2006), but they are the exception rather than the rule. Moreover, in many cooperatively breeding mammals, reproductive skew leads only to a redistribution of reproductive success in favor of the dominant without net gains in average fitness. Dominants suppress the reproduction of subordinates, who then are left with little choice but to help raise the dominant's offspring. In contrast, cooperative breeding in humans does not lead to increased female reproductive skew and, if anything, decreases the skew. Humans also manage to have exceptionally short interbirth intervals for our otherwise k-selected life history strategy—a clear sign that we actually reap the gains from breeding cooperatively.

Discrepancy Is Unlikely to Be Due to the Lack of Benefits from Cooperation

Our argument depends on the assumption that there are many opportunities for small-scale cooperation in nature that are not utilized in most vertebrates. However, it can be argued that other vertebrates lack extensive small-scale cooperation because cooperation is not beneficial in their ecological niches. While we cannot rule out this possibility, we do not think that it is very plausible. Several lines of evidence suggest that the potential benefits of cooperation—joint efforts, specialization, and trade—are omnipresent.

First, lineages that evolved division of labor and trade have had spectacular ecological success and have radiated into a vast range of habitats. Multicellular organisms arose when groups of single-celled creatures evolved specialization and exchange and, as a result, have occupied a dramatically large number of niches. Their success indicates that the benefits of cooperation among cells were present in niches as different as those occupied by plants and animals, ecologies as different as aquatic, terrestrial and subterranean habitats, and climates ranging from tropical to tundra to the ocean floor. Although the total biomass of single-celled organisms could exceed that of multicellular organisms, there is still extensive cooperation among single-celled organisms, including toxin production, aggregating to produce reproductive bodies, reduced virulence, and labor division (Rainey and Rainey 2003; Velicer and Yu 2003; Crespi 2001; West et al. 2006). Thus, cooperation among cells implies that if free riding can be tamed, cooperation would be beneficial in virtually every niche.

Similarly, eusocial insects are not restricted to a narrow ecological domain, but instead occupy a vast breadth of foraging ecologies ranging from carnivory to gardening, again suggesting that cooperation could provide benefits in a wide range of environments. Furthermore, eusocial insect colonies cooperate in multiple domains. For instance, army ants breed cooperatively, work together to build bridges, defend the colony, manage traffic, and have several castes of workers specialized for different tasks (Couzin and Franks 2003; Franks 1986).

The scope of economic exchange in humans also suggests that gains from specialization and trade are everywhere. Specialization within firms exists because it is more efficient to subdivide labor among individuals that specialize in one or a few specific tasks. This was recognized by the earliest economists (Smith 1776). Trade also offers a way to achieve division of labor and thereby increase efficiency in production. If one individual specializes in building houses, a second in farming, and a third in making music, and they trade their products, all three will typically enjoy better housing, food, and music than if they would produce everything themselves. Whether a firm or collective enterprise is small or large, and whether the web of exchanges is small and simple or large and complex, the fact that there are firms, as well as the fact that there is exchange, is evidence of the presence of gains from specialization.

Finally, people in small-scale hunter-gatherer societies cooperate in a wide range of activities, not just in cooperative hunting and meat sharing. The Aché cooperate in acquiring plant resources as well as in hunting, and both types of foods are shared (Hill 2002). Hill (2002:123) also notes:

> Non-foraging cooperation in Aché forest camps includes services such as clearing a camp spot for others; bringing water for others; collecting firewood for others; lighting or tending another's fire; cooking and food processing for others; building a hut that others share; making, fixing, and lending every imaginable tool; grooming others; keeping insect pests away from others; tending others' infant and juvenile offspring; feeding another's offspring; teaching another's offspring; caring for others when they are ill; collecting medicinal plants for others; listening to others' problems and giving advice; providing company for others who must stay behind in camp or go out to forage alone; and even entertaining others (singing, joking, telling stories) when requested.

Thus, the ecology of small-scale hunter-gatherers also contains the potential for gains from cooperation in myriad contexts. Given the rich variety of gains from cooperation that are captured by multicellular organisms, eusocial insects, and humans, we think it is unlikely that other vertebrates do not cooperate very much because there are no potential benefits to be had from specialization and exchange.

Discrepancy Is Unlikely to Be Due to the Detrimental Effect of Local Competition on Cooperation

We have posited that the differences between cooperation in humans and other vertebrates are not consistent with simple kin selection models. It could be that cooperation among kin is not more prevalent because local competition prevents the evolution of cooperation among relatives. If there is local competition, then even if large gains are possible from cooperation and relatedness between individuals is high, the conditions for kin selection to favor cooperation may not be met. However, local competition is unlikely to explain satisfactorily the *difference* in the levels of cooperation between humans and other vertebrates.

A local interaction model typically implies that neighbors not only face more opportunities for gains from cooperation than distant individuals, but also compete more intensely. With local interaction and local competition, interactants will be related. That is good for the evolution of altruism, but they will also compete more, which is bad for cooperation. In stylized, simple models, the two effects actually cancel out each other (Boyd 1982; Wilson et al. 1992; Taylor 1992a, b) and cooperation will not evolve at all. Thus kin selection will lead to altruism only if there is a discrepancy between assortment in interaction and assortment in competition. Measuring costs and benefits is typically hard, and determining who competes with whom and how intensely is even harder. There is no conspicuous reason why human social organizations and life histories are different from other mammals in a way that would alleviate the effect of population regulation and allow kin selection to favor cooperation in humans but not in other mammalian societies.

The ability to recognize kin and various dispersal mechanisms offers ways to get around the detrimental effect of local competition. If individuals condition their altruistic behavior on cues of relatedness, they confer benefits not just on anyone with whom they interact and whose offspring will compete with theirs relatively intensively, but on a subsample of those with whom relatedness is extra high. Many life cycles can do the same; after an interaction between relatives, for example, all offspring go to a migrant pool, where they compete with each other equally intensely. In addition, a species that cooperates in sporulating when the local food source runs out can be stable because the individuals locally can be related, whereas the spores compete more globally. Again, there is no obvious reason why kin recognition or dispersal is decisively different between humans and other vertebrates.

It is plausible that local population regulation prevents the evolution of cooperation among relatives in humans as well as in other species. Norms and cultural group selection then serve as an alternate mechanism that can allow cooperation to evolve in humans but not in other animals, because of the lack of language and cultural capacities in other species. Another possibility is that the conditions for kin selection are met in both, but norms help implement the behavior for which kin selection would select.

Discrepancy Is Unlikely to Be Due to Cognitive Constraints

One view holds that humans are able to gain the benefits of cooperation because we have more sophisticated cognitive abilities (Pinker 2010). However, complex strategies do not necessarily yield more cooperative outcomes, and thus human cognitive abilities do not explain the greater degree of cooperation in humans—at least not in any obvious way.

In simple evolutionary models of repeated interaction, all positive equilibrium levels of cooperation require reciprocity, and strategies that reciprocate are more complex than ones that always cooperate or always defect. Therefore, every positive level of cooperation requires more complexity than no cooperation at all. Beyond that, there is not much of a link between complexity and the level of cooperation: equilibrium strategies can be very complex and only mildly cooperative, or only a bit more complex than unconditional defection, and already fully cooperative.

When there are multiple equilibria, a sequence of transitions between equilibria can take the population from a mildly complex, but very cooperative state to a very complex and only somewhat cooperative state (van Veelen et al. 2012; van Veelen and García 2010). In models of repeated interaction, which tend to have many equilibria (Abreu 1988; Axelrod and Hamilton 1981; Bendor and Swistak 1995; Fudenberg and Maskin 1986; van Veelen et al. 2012), each transition which increases cooperation increases complexity and vice versa. That seems to contradict the claim that a more complex state is not necessarily more cooperative. However, a possible sequence of transitions is that first a complex cooperative equilibrium is upset and replaced by a simple defecting equilibrium, which in turn is replaced by an extremely complex, but only mildly cooperative equilibrium. Thus the last equilibrium is more complex, yet less cooperative than the first equilibrium. Furthermore, if (genetic) assortment is added to the model, then even the relation between increases in cooperation and complexity for specific transitions between equilibria no longer holds—a transition from one equilibrium to the other can lead to an increase in complexity and a decrease in cooperation (van Veelen et al. 2012).

The predicted levels of cooperation also depend on the set of strategies considered in the model (van Veelen and García 2010), but allowing for more complex strategies does not imply an increase in average levels of cooperation. The likelihoods of transitions between equilibria are sensitive to the assumed distribution of mutation probabilities. Restricting attention to a specific set of strategies is a special and extreme choice for mutation probabilities; it sets all mutation probabilities from strategies within the set that is considered to strategies outside it to 0. It is, of course, not clear what the right strategy set is, or what the right assumptions concerning mutations are. However, if we allow for strategies of any complexity (and mutation is unbiased), the average level of cooperation is not necessarily higher than in models with smaller sets of strategies of limited complexity

It is worth noting, however, that cooperation can require more complexity in behavior than defection for reasons that are abstracted away from in typical models. The cooperative behavior itself will involve performing some task, whereas the defecting behavior of not performing the task is typically simple by default. With human interactions, many cooperative behaviors would not be possible without linguistic communication. Coordinating who does what in collective efforts or negotiating terms of trade is very hard without language and impossible without communication (Smith 2010; Pinker 2010). Thus actual differences in complexity may come from things outside the model, depending on how involved the task is that would benefit the other.

Norms Affect Small-Scale Human Cooperation

Behavior between Kin in Humans Is Subject to Norms and Third-Party Enforcement

Many kinds of small-scale cooperation among kin are regulated by norms in human societies, including core domains like parent-offspring relations and pair bonding. Pair bonding is likely one of the earliest forms of cooperation that characterized our hominid ancestors. Yet this relationship is so regulated that almost every society has the concept of "marriage"—an institutionalized form of pair bonding with normative rights and obligations. Exogamy and endogamy rules are widespread. Such norms proscribe pair bonding with members within a social unit and prescribe pair bonding within another social unit, usually trumping the interests of potential marriage partners. There are additional layers of proscriptions regarding how the particular match is made between two individuals. Societies with arranged marriage prohibit the parties involved from choosing their own spouse and often restrict courtship or other forms of direct interaction between opposite-sex youth. Such norms are upheld not only by the community, but also by family members who may disown their noncompliant children. Norms specify how many persons one can marry. A man and woman cannot choose to marry polygynously in a society that is normatively monogamous, even if they think it is in their mutual, long-term best interest. Norms regulate the direction of wealth transfer at the time of marriage: in some societies, men pay bride price whereas in others the woman's family is expected to pay dowry. Postmarital residence is often regulated by norms. A man who hails from a patrilocal culture cannot decide to live at his wife's family's place for economic reasons without losing face. Sexual relations are regulated through norms that restrict the number of sexual partners people can have and norms that restrict premarital sexual behavior. Norms regarding premarital sex are cross-culturally so variable that in some societies a woman who has had sex before marriage can be stoned and killed whereas in other societies it damages one's reputation to be a virgin at the time of marriage.

Norms also regulate how individuals can raise their own offspring. In state societies, violations like child neglect, infanticide, and corporal punishment of children are within the purview of the law. There is much cultural variation in these norms, ranging from places where it is illegal for parents to beat their children to ones where parents who do not discipline their children are considered negligent. Foot-binding in China was strongly moralized even though it was a child-rearing decision. Community pressure maintained the norm and, correspondingly, condemnation of the norm from other societies led to its demise (Appiah 2010). Sending daughters to school, another child-rearing decision, was not normative in many societies until recently, causing a substantial fraction of women to be excluded from the market labor force. Parental investment is regulated by inheritance norms that specify how wealth should be distributed among one's children—only to sons or only to daughters, to the oldest, the youngest, or equal division among all the children. It would be wrong to practice primogeniture in a society where norms specified an even division or to deny daughters an inheritance in societies where all children should be given a share. The extent of paternal investment in offspring is also moralized. In societies where biparental care is expected, a father who abandons his children with their mother will be stigmatized. In other societies, uncles take up the paternal duties and absent fathers suffer no loss in status.

Pairwise Interactions among Nonkin Are Affected by Norms and Third-Party Interventions

Pairwise cooperation between unrelated people is also regulated by norms, and third parties often intervene to either mediate disagreements or to enforce norms through indirect sanctions. Thus pairwise exchange in humans often involves both direct and indirect reciprocity. In indirect reciprocity, cooperation in a pairwise exchange is maintained because defections damage the violator's reputation, allowing other people to defect when interacting with a violator without damaging their own reputation. In direct reciprocity, cooperation is maintained by the threat of defection by the partner herself.

Third-party monitoring and sanctioning govern many aspects of pairwise relations among the Turkana, a nomadic pastoral society in East Africa. For instance, if a woman refuses to give water to a man who asks for it, she can be criticized. However, if she heard that this man had abandoned his injured friend when the two of them went into enemy territory to steal cattle, then she can refuse him water without facing disapproval from her peers. The relationship between a man and his friend should be ideal for direct reciprocity—they are likely to have known each other for a long time and to have interacted often when herding, attending dances, patrolling, and raiding. Despite this, community approval and disapproval is vital scaffolding in sustaining the cooperation between these two men. In fact, the closer their friendship, the stronger the community's disapproval will likely be. Similarly, a herdsman who loses animals from his flock can expect to be hosted by a Turkana household when he travels in search of his stray livestock. Suppose, however, that this man stole his neighbor's camel. Then, rather than being invited in, he may be censured for cheating his neighbor and told to go his own way. Again, a relationship between neighbors, rather than being in the purview of direct reciprocity, is regulated by community sanctions.

Furthermore, when there is a disagreement within a pairwise relationship, third parties often adjudicate the dispute. For instance, one of the primary functions of the judiciary of state societies and informal courts in politically decentralized societies is to handle disputes such as theft, homicide, battery, and violations of contractual agreements between two individuals. Among the Turkana, if a man's goat is stolen, he reports the matter to the elders or members of the respective families. They summon the alleged thief to determine what happened, and if the accused is at fault they instruct him to compensate the victim. Likewise, men often need to rebuild their herds after loss from raids, droughts, and epidemics, and they call on their friends to loan them animals. If a dispute arises later about repaying the debt, the donor is assured that people knowledgeable of their transaction will testify that he had indeed lent this friend an animal.

Culturally Evolved Norms May Have Enabled the Evolution of Small- and Large-Scale Cooperation in Humans

Thus far we have argued that:

- Levels of small-scale cooperation differ greatly between humans and other vertebrates. It seems hard to explain this difference by disparities in typical ingredients of kin selection models and/or models of direct reciprocity or as the result of ecological or cognitive differences.
- In human societies, culturally transmitted norms regulate many aspects of life, including kin relations and repeated pairwise interactions.

Given that culturally evolved norms are absent in other species, these observations lead us to explore the possibility that norms were essential to the evolutionary transition that led to cooperative breeding in humans. We hypothesize that culturally transmitted norms allowed for extensive small-scale cooperation in early human societies and may have been what helped bands comprising a few nuclear or extended families to seize benefits from social exchange. This led to the evolution of a moral psychology which then allowed the evolution of larger-scale cooperation through cultural group selection.

Cultural group selection models have typically assumed that individuals acquire complex normative behavior by copying successful or prevalent behavior; that is, they can use the same social learning machinery which they use to learn other kinds of locally adaptive behavior to acquire and implement the local moral rules. However, complex moral behavior may be hard to acquire without some kind of innate scaffolding already in place. Many of our normative concerns are somewhat abstract, as we recognize similarities in situations where there are aligned and opposed interests. That gives our moral machinery a grammar-like structure: we recognize common causes and conflicts of interest in novel situations never experienced, and we link them. We do not, for instance, just take turns in doing the dishes, but balance such duties over a larger set of chores. It may be difficult to maintain such rules with only a general cultural learning mechanism. However, if small-scale cooperation maintained by norms and third-party enforcement caused humans to already have an innate moral grammar, then cultural group selection can lead to large-scale cooperation more easily by exapting this moral psychology.

The transition we describe would not immediately yield large-scale cooperation. Our hypothesis is that moral sentiments are essential to supporting costly cooperation at the domestic scale and these then are exapted for larger-scale cooperation. However, simple misfiring of this psychology would be extremely costly. For instance, the Turkana mobilize hundreds of unrelated and unfamiliar warriors to participate in cattle raids against members of neighboring ethnolinguistic groups, and norms and enforcement shape behavior in this common endeavor (Mathew and Boyd 2011). Twenty percent of all male deaths in one of the Turkana territories along a hostile ethnic border are due to warfare (Mathew and Boyd 2011). From archaeological records, ethnographic accounts, and oral histories of various societies, we know that tribal-scale warfare is very old (Willey 1990; Bamforth 1994; Keeley 1996; Lambert 2002; Gat 2006). It is thus clear that if large-scale conflict resulted from a misfiring of mechanisms designed to regulate small-scale cooperation, there would then have been a long history of strong selection acting to correct this mistake. Instead, we think that the availability of a small-scale psychology, which can be extended if the needed adaptive pressure for large-scale cooperation exists, could make the transition to larger-scale cooperation far more likely.

This hypothesis predicts both that species with culturally transmitted norms will be more cooperative and that small- and large-scale cooperation in such species should rely on norms in similar ways. The absence of third-party enforcement of pairwise cooperation and adjudication of disputes should be detrimental to pairwise cooperation. Cooperation within the extended family unit should be sustained through norms and their enforcement through sanctions by family and community members. There should be cross-cultural variation in the norms dictating small-scale as well as large-scale cooperation. The psychology of norm compliance and enforcement for small- and large-scale cooperation will be sensitive to different cues as a consequence of the distinct selection pressures which shaped them. Norm compliance and enforcement in small-scale cooperation should depend on cues of family membership, or cues of repeated interactions. Such cues are rooted in individual identities: Should I help Joe? Is he my relative? Has he helped me when I needed him in the past? Negotiation, deliberation, and consensus among known individuals should be important in achieving norm compliance in small-scale cooperation. In contrast, cultural group selection on institutional variation should be more important in shaping norms governing large-scale cooperation. Norm compliance in large-scale settings should depend on cues of group identity: Should I help Joan? Is she a member of my ethnic group? Is she an American? In such settings, norm compliance can be achieved even without negotiation and consensus.

An alternative view of why we have norms concerning interactions between kin and repeated pairwise interactions is that they are just an after-the-fact reflection of equilibrium behavior. Such a view would be somewhat similar to the role of norms presented by Binmore (1994, 1998), and it would imply that if a norm or its enforcement were to go away, the behavior would not change.

Why Are Norms Necessary to Get the Benefits of Cooperation?

Our argument has thus far been empirical: humans exhibit many forms of small-scale cooperation not seen in other species, like specialization and exchange; norms play a crucial role in small-scale human cooperation; therefore norms potentiate the evolution of small-scale cooperation. However, it is not obvious how the evolution of norms and third-party enforcement should interact with ingredients from typical models assuming kin selection and direct reciprocity. One would expect that norm-enforcing strategies that support cooperation can invade more easily when there is relatedness in the social group, but also that norms are not absolutely necessary for cooperation to evolve. If the latter is true, we are still left with the question of why there is relatively little small-scale cooperation in other vertebrates. Below we sketch a few, admittedly speculative, possibilities of how the evolution of norms and kin selection or repeated interactions might relate.

Norms and Third-Party Enforcement May Resolve Problems of Errors

Errors may help account for why direct reciprocity may not lead to very much pairwise cooperation. Models without errors are typically much too positive about the possibilities for the evolution of cooperation (Hirshleifer and Martinez Coll 1988; Wu and Axelrod 1995). If norms could somehow reduce the detrimental effects of errors, then that could explain the higher levels of reciprocity among people. Still, it is not immediately obvious why this should be the case, because the evolution of indirect reciprocity is typically more sensitive to errors than direct reciprocity. With direct reciprocity actors only need accurate knowledge about the past actions of their partners; with indirect reciprocity they need accurate knowledge about the past behavior of all individuals in the group and whether that behavior was justified.

There are two possible resolutions of this conundrum, neither completely convincing. First, by linking the behaviors of many different individuals, indirect reciprocity increases the expected number of future interactions for each actor and therefore the opportunity costs of defection. However, given the density of pairwise interactions in small primate social groups, it is hard to see how this will have a big effect. Second, the evolution of reciprocity is especially sensitive to "perception errors" which occur when actors have different beliefs about whether a defection occurred. Strategies, like Pavlov's (Nowak and Sigmund 1993), can be stable when perception errors are common, but these strategies cannot easily increase when rare.

Adjudication by third parties can solve this problem, even if the adjudication process is also error prone (Mathew and Boyd, in preparation), because it aligns the beliefs of interactants. Adjudication of pairwise exchange is easiest if rules of behavior are shared within a community, and not restricted to a particular partnership. Otherwise third parties have no basis for evaluating deviations from these norms and arbitrating. However, adjudication does not necessitate enforcement, so why we end up with indirect sanctioning of defectors is unclear. Nonetheless, it can help explain the dearth of pairwise cooperation among frequent associates in other animals.

Norms Can Help Identify How to Get the Benefits of Cooperation in the Local Ecology

It is possible that the conditions for kin selection to favor cooperation are met in humans and other animals, but that different patterns of cooperation are favored in different environments and culturally transmitted norms are needed to adapt to local conditions. Most models do not explain how actors identify when and what to exchange with whom to reap the benefits from the exchange. In principle, this can be solved without norms—as has been done in cells, multicellular organisms, and eusocial insects through genetically evolved specialization, communication, and rules governing exchange. This solution allows extremely complex adaptations like those observed in social insects. However, the ability to adapt to different environments is limited to what can be acquired through individual learning and other individual forms of phenotypic plasticity. Humans occupy a very wide range of environments, and individuals may not have the ability to invent locally appropriate rules for governing cooperation. This problem will be particularly acute for contingent cooperation because sanctions require behaviors of others to be accurately judged.

Norms may solve this problem because cumulative cultural evolution allows rapid evolution of complex adaptations to local conditions. We know that cultural evolution leads to the gradual evolution of complex tools (e.g., kayaks and composite bows) which are superb adaptations to particular local environments, but beyond the inventive capacity of individuals. Norms are social tools that allow complex cooperative behavior in a wide range of environments with a wide range of behaviors, payoffs, and contingencies. Norms of Turkana society dictate, for instance, that one must share certain parts of the animal, lend a goat to a man who has lost his wealth in a raid, feed a traveler, ridicule a man who let his friend down, help your son acquire his first wife with part of your livestock wealth, offer water to a passerby who asks, or give a share of animals to your brothers and uncles after returning from a raid. The nature of cooperation and the conditions under which it would be favorable are quite different among the Aché or the Netsilik. These norms oblige certain behaviors and allow its practitioners to derive some benefits from mutual insurance, division of labor, and exchange. Practitioners may not necessarily recognize what the real benefits to each act are. However, as long as some process can make norms track the socioeconomic environment roughly accurately, norms, if not individuals, can recognize in which kinds of exchange the benefits of cooperation lie.

However, the argument that norms are needed to recognize where the gains from cooperation lie is not entirely convincing either. Vertebrates have repeatedly arrived at solutions to derive the benefits of mutualistic and low-stakes cooperation. For instance, herding, predator mobbing, and territorial chorusing have evolved in several lineages. Sexual reproduction is even a more complex coordination problem involving mate searching, consortship, signaling, and mate guarding. Yet, all mammals are able to orchestrate it carefully. Individuals can also easily identify social situations in which they can gain from engaging in self-serving behavior. Of course, people live in a wider range of environments and use a much wider range of subsistence strategies than any other vertebrate. This may make human cooperation more difficult to achieve without culturally evolved social tools.

Norms and Third-Party Enforcement Can Eliminate Inefficiencies Caused by Asymmetric Interests among Pairs of Individuals

Norms can enforce pairwise exchange that may be beneficial only at a scale larger than that of the dyad. For instance, parents have a greater interest in cooperation between their children than do the children themselves. Examples of sources of gains from cooperation are increased efficiency from labor division, transfers between individuals that are at different stages of their life cycle, and co-insurance. Suppose, for example, that two sisters agree to help each other, if the other sister gets hit by bad luck. Among siblings, Hamilton's rule predicts that they will help as soon as the benefits of a transfer to the one are more than twice the costs to the other. Parents, however, have an interest in their offspring being more helpful than that, and if they could, they would bind their children to help whenever the benefit exceeds the cost. Conflicts of interest like that typically lead to a tug of war between parents and offspring, where the outcome depends on the mutations that both sides have at their disposal. However, a norm that siblings should help each other could work even better if it is shared by a larger community, as everyone ex ante is better off living in a group that has a higher level of between-siblings insurance. Such a norm could then later be exapted to also include nonkin. One could imagine similar scenarios for norms that concern labor division or delayed exchange within the family.

Norms and Third-Party Enforcement May Resolve Information Asymmetries to Inhibit the Evolution of Cooperation

In many situations, needy individuals must communicate their state of need to receive help, and this restricts the conditions under which cooperation can evolve, even among kin. There are situations in which reliable cues will allow potential donors to determine whether the benefits to their kin are sufficiently large. For instance, when the potential recipient is a newborn infant, the mother will have no doubt that the gains to the offspring from receiving care are much larger than the cost to the parent of providing care. Often times, however, the party that would benefit from a transfer is better informed about its state of need than possible donors. For example, it may be difficult for a mother to determine whether an older offspring is really ill or just malingering, because any signal used by a sick offspring to transmit information about its condition or needs can also be sent by a healthy offspring who just wants special treatment.

This problem can be solved if the cost of signaling is high enough to deter the less needy from dishonest signaling (Zahavi 1975; Grafen 1990a, b; Kreps and Sobel 1994; Bergstrom and Lachmann 1997, 1998; Lachmann and Bergstrom 1997). With such costly signals there can be transfers, but the cost of the signal consumes some (or even all) of the potential efficiency gains. When need varies continuously, "partially pooling equilibria" exist, in which ranges of neediness are lumped together under the same signal. These reduce, or even eliminate, the costs of signaling, but the information transmission becomes coarser. Moreover, there are also always equilibria where no transfer gets made (Johnstone 1999). In those equilibria either no signal is sent and no transfer is performed, or, if signaling is free, signals are sent but the reaction to all signals is to not transfer. This means that even if the combination of relatedness and the costs and benefits of the transfer is right, and kin selection suggests that not giving is not an equilibrium under perfect information, it can be that with information asymmetry, not giving is in fact an equilibrium.

Shared social norms and group enforcement may help resolve this problem. Without community monitoring and enforcement, a juvenile who signals to his mother that he is sick needs only to provide the right cues in her presence. He can go out and play when she is not looking. With community monitoring and enforcement, the cost of malingering will be much greater—he can't play when anyone is looking. More generally, the coupling of this "tangled-web-of-lies" phenomenon to shared norms about appropriate behavior can reduce the cost of discriminating honest and dishonest signals, and this in turn can expand the range of conditions under which mutual aid can occur.

Conclusion

Theoretical work on the evolution of cooperation in humans has focused on how people maintain cooperation in large groups comprised of unrelated individuals, a form of cooperation that is rare in other animals. However, human cooperation stands out in yet another aspect: people engage in costly cooperation among kin and close associates to a much greater degree than is seen in other animals. We have argued here that current theories do not adequately explain this difference in levels of small-scale cooperation between humans and other animals. We posit that, like large-scale cooperation, cooperation among kin and friends also depends on norms and third-party judgments. We lay out tentative explanations for why this may be, but more theoretical work is needed to determine precisely how norms and third parties can aid the evolution of cooperation among kin and close associates, and why kin selection and reciprocity does not suffice to attain elaborate small-scale cooperation in many animals.

4

The Puzzle of Human Ultrasociality

How Did Large-Scale Complex Societies Evolve?

Peter Turchin

Abstract

After a long and turbulent history, the study of human cultural evolution is finally becoming comparable to the study of genetic evolution, with human history the counterpart of the biological fossil record. One of the most remarkable products of cultural evolution has been an increase in the scale of human societies by many orders of magnitude. Today, the great majority of humans live in complex societies, which can only exist due to extensive cooperation among large numbers of individuals. *Ultrasociality*, the ability of humans to cooperate in large groups of genetically unrelated individuals, presents a puzzle to both evolutionary and social theory. Although much theoretical effort has been devoted to understanding the evolution of cooperation in small-scale groups (hunter-gatherers living in societies of hundreds to a few thousand individuals), the same cannot be said about the next phase of human evolution, the rise of complex societies encompassing tens and hundreds of millions of people. Evolutionary biologists, political scientists, anthropologists, and others have proposed a multitude of theories to explain how complex societies evolved. However, scientific study has suffered from two limitations. First, with a few exceptions, theories have relied on verbal reasoning; formal models tend to focus on the evolution of cooperation in small groups, whereas the transition from small- to large-scale societies has been mostly neglected. Second, there has been no systematic effort to compare theoretical predictions to data. Human ultrasociality has evolved repeatedly around the world and across time, reflecting both common selection pressures and the unique contingencies affecting each case. An enormous amount of archaeological and historical information exists but has not been studied from an evolutionary perspective. Thus, explicit models that will yield specific and quantitative predictions are needed as well as databases of the cultural evolution of human ultrasociality. Furthermore, a research program combining explicit models with empirical testing of predictions is not only an academic endeavor. Understanding conditions that either promote or inhibit human ultrasociality is highly relevant for addressing the challenges of large-scale cooperation and conflict in the modern world.

Introduction: The Theoretical Background

The great majority of humans today live in large-scale complex societies, which can only exist on the basis of extensive cooperation among large numbers of individuals. Such cooperation can take many forms: volunteering for the army when the country is attacked, willingly paying taxes, voting, helping strangers, refusing to take bribes, etc. In each case, the result of cooperation is production of a *public good* (i.e., no one can be effectively excluded from using the good), while the costs of cooperation are born privately (e.g., one can be killed defending one's country). *Ultrasociality*, the ability of humans to cooperate in huge groups of genetically unrelated individuals (Campbell 1983), is a great puzzle in both evolutionary and social sciences (Richerson and Boyd 1998). We now understand that neither the "selfish gene" perspective (Dawkins 1976) nor rational choice theory (Becker 1978) is capable of resolving this puzzle (Turchin 2006, chapter 5).

Human ultrasociality represents a major evolutionary transition. Other transitions include those from independent replicators to chromosomes, from a prokaryotic to a eukaryotic cell, from unicellular to multicellular organisms, and from solitary individuals to eusocial colonies (Maynard Smith and Szathmáry 1995). A powerful conceptual framework for understanding major transitions is the multilevel selection (MLS) theory (Sober and Wilson 1991; Okasha 2007; Wilson and Wilson 2007). Generally speaking, major transitions involve several interacting processes: evolution of cooperation among lower-level units, selection which acts on higher-level "collectives," policing mechanisms which suppress "free riders" and competition among lower-level units, and increased functional integration of collectives which makes them increasingly organism-like. Eventually, higher-level collectives become so well integrated that they can be treated as "individuals" in their own right (and can serve as lower-level units for the next evolutionary transition).

Evolution of human ultrasociality fits quite well into this scheme, but with one important twist: it occurred in several stages. Thus, it is perhaps best to think of multiple transitions instead of a single one. The first stage was the evolution of cooperation in small-scale groups (i.e., groups of hundreds or, at most, a few thousand of people). Our theories of how small-scale sociality evolved in humans are rapidly maturing (Boyd and Richerson 1985; Sober and Wilson 1991; Richerson and Boyd 1998; Wilson 2002, 2005; Bowles 2006; Turchin 2006; Choi and Bowles 2007; Lehmann and Feldman 2008). Mechanisms involved at this stage were:

- Increasing returns to scale (see Jordan et al. in this volume): examples include big-game hunting and coordinated defense against predators, risk pooling through extended networks, economic returns from trade and division of labor, and ability to generate new and retain existing knowledge. Probably the most important mechanism capable of generating increasing returns to scale is warfare (or parochial altruism, Bowles 2009). Warfare is particularly important because it generates increasing returns at all social scales. Thus, a village which has more warriors will be favored in a conflict against another village, and an empire which collects more taxes and raises more recruits will be favored in a conflict against another empire.

- Inequity aversion and other leveling mechanisms: examples include food sharing, monogamy, and social control of "upstarts" (Boehm 1999, 2011). These mechanisms reduce within-group variation in fitness and, thus, the strength of individual-level selection relative to between-group selection.
- Moralistic punishment and other mechanisms that control free riding (see Jordan et al., in this volume).
- Culture (Richerson and Boyd 2005), which (via conformist transmission) reduces within-group variability and enhances between-group variability.

For a more extended discussion, see Jordan et al. (this volume). Evolution of small-scale sociality operated in both genetic and cultural modes; in fact, the key process was gene–culture coevolution (Richerson and Boyd 2005). Because cooperation in small-scale societies relies on face-to-face interactions, large brains were required to store and process social interactions data (Byrne and Whiten 1988). However, once a human group attains the size of 100–200 individuals (Dunbar 1992; Dunbar and Shultz 2007), even the hypertrophied human brain becomes overwhelmed with the complexity of social computation. Thus, for group size to increase beyond the few hundred individuals typical of small-scale human societies, evolution had to break through the barriers imposed by face-to-face sociality.

The second stage—evolution of large-scale sociality (ultrasociality)—was enabled by several additional key adaptations. First, humans evolved the capacity to demarcate group membership with symbolic markers (Shaw and Wong 1989; Masters 1998; Richerson and Boyd 1998); the first symbolic artifacts appeared around 60,000 years ago (Marean et al. 2007). Markers such as dialect/language, cult/religion, clothing, and ornamentation allowed humans to determine whether someone personally unknown to them was a member of their cooperating group or, vice versa, an alien and an enemy. Second, hierarchical organization allowed unlimited growth in the scale of cooperating groups, simply by adding extra organization levels. Centralized hierarchies are also much more effective in war, which is why all armies have chains of command (Andreski 1971). However, the downside of hierarchical social organization is that it inevitably leads to inequality (Michels 1915; Mosca 1939). As a result, evolution of complex societies reversed the trend to greater egalitarianism that had previously characterized human evolution; this trend reversal is sometimes referred to as "U-shaped curve of despotism" (e.g., Bellah 2011). Other key innovations include literacy and record keeping, formal legal systems, bureaucracies, organized religion, urbanization, and states. The primary mode of evolution during this stage was clearly cultural, although recent analyses indicate that genetic evolution did not cease with the rise of civilization (Hawks et al. 2007).

Cultural evolution has had a turbulent history and it remains a controversial field. Even the nature of cultural variation is contentious, with rival approaches (evolutionary psychology, memetics, and the dual inheritance theory) each offering a variant view (Richerson and Boyd 2005). Nevertheless, the study of human cultural evolution is gradually becoming comparable to the study of genetic evolution, as evidenced, for example, by the successful deployment of the methods of phylogenetic analysis (Mace and Holden 2005; Fortunato and Mace 2009). Thus, evolutionary theory provides a powerful framework for the study of human ultrasociality. It can serve as a unifying conceptual framework for the multitude of theories proposed by political thinkers, anthropologists, and social biologists to explain how complex societies and, in particular, states evolved (for reviews, see Mann 1986; Chase-Dunn and Hall 1997; Sanderson 1999; Grinin and Korotayev 2009). These theories invoke a variety of mechanisms: population pressure, warfare, class struggle, economic exchange, large-scale irrigation works, and information-processing capacity.

However, most of these theories have been formulated only as verbal, rather than mathematical models. Currently, only a handful of modeling studies have focused on the transition from small- to large-scale societies (Dacey 1968; Bremer and Mihalka 1977; Cusack and Stoll 1990; Cederman 1997; Spencer 1998; Cioffi-Revilla 2005; Cederman and Girardin 2010); these studies primarily use simple models borrowed from mathematical ecology or complex agent-based simulations. Together with Sergey Gavrilets, we are in the process of adding to this growing theoretical corpus. One theoretical approach was based on a central mathematical result in MLS theory, the Price equation (Price 1972), which proposed that evolution of prosocial traits is favored not only when group-level selection is strong but also, and most importantly, when between-group variability is maximized (Turchin 2009, 2011). A second approach employed a spatially explicit agent-based model of the emergence of early centralized societies via warfare (Turchin and Gavrilets 2009; Gavrilets et al. 2010). Despite progress, an enormous amount of work remains to be done to develop a cohesive body of theory on the transition from small-scale to large-scale societies.

Thus far, the theories have not been confronted with data in a systematic way, a procedure which would allow us to reject some in favor of others. A major stumbling block has been a lack of good databases codifying information on cultural variation in a broad spectrum of societies. As a result, previous empirical tests have been ad hoc and haphazard, tending to focus on those aspects and regions with which individual authors were familiar. To make further progress we need to start testing theories, and that requires a much better empirical base than is currently available. We are fortunate to have such resources for cross-cultural comparative ethnography as the Ethnographic Atlas (Murdock 1967) and the Standard Cross-Cultural Sample (Murdock and White 1969). However, these static datasets tell us about a culture at a particular point in time. More decisive empirical tests of various theories of social evolution require dynamic datasets that describe cultural trajectories through time.

Experience in other fields (e.g., many natural sciences) suggests that general theories about the functioning of complex adaptive systems (such as human societies) cannot be directly tested with data; we need an intermediate step—specific models. First, general theories formulated verbally need to be translated into explicit mathematical models (in whatever form it takes, systems of equations and agent-based simulation being the most common approaches). As a practical matter, usually we cannot come up with a single "best" model, and thus need to develop a suite of models, each capturing different aspects of the theory. The next step is to extract quantitative predictions from the models which can be then compared to data. Ideally, we want to test not a single theory in isolation, but compare how well rival theories compare to each other, using the data as arbiter. Thus, models play multiple roles in this process: they serve to test the logical coherence of theory (when the expected dynamics indeed emerge from the postulated mechanisms), they generate sharply defined, quantitative predictions, and they can suggest novel ways to test theories.

We need to do two things to answer the question raised in the title of this paper: How did large-scale complex societies evolve? First, we need to translate verbal theories into mathematical models (understood broadly to include agent-based simulations) that will yield specific and quantitative predictions. Second, we need databases of the cultural evolution of human ultrasociality that will allow us to test model predictions empirically. Current developments on both the modeling (Turchin et al. 2012b) and data (Turchin et al. 2012a) fronts indicate that such a research program is not entirely unrealistic.

Conceptual Issues: Cultural Evolution of Ultrasocial Institutions

If cultural evolution provides a powerful framework for the study of human ultrasociality, we must address the following questions: What is it that evolves? How does it evolve? In this section, I discuss conceptual issues and begin by clarifying the relationship between ultrasociality and social complexity.

Ultrasociality and Social Complexity

Ultrasociality is the ability of humans to cooperate with huge numbers (millions and more) of genetically unrelated individuals. As far as we know, it is unique to humans. Ultrasociality is the term used by evolutionary scientists; another closely related term is social complexity. These two concepts may in fact be thought of as simply different approaches to the same general phenomenon by different scientific disciplines: evolutionary science (ultrasociality) and anthropology/archaeology, as well as complexity science (social complexity). A third closely related approach within political science and history focuses on the rise and evolution of the state. In this discussion, I will use ultrasociality because it is the most clearly defined concept (there are many rival definitions of the state; a similar terminological uncertainty surrounds the concept of complexity, but it can be operationalized in the context of social evolution, as will be discussed below).

The defining feature of ultrasociality is the social scale (the size of the cooperating group), which raises the question of what sort of group we wish to address. At the most general level of inquiry, it is best to leave such issues unspecified, because we are interested in cooperation at many different levels involving many different kinds of groups: trading networks, ethnic diasporas, religious cults, alliances of states, as well as, at the highest level, the whole of humanity. For conceptual clarity and empirical relevance, however, I will focus on one kind of cooperating group: the polity.

Polity is defined as an independent political unit. Types of polities range from villages (local communities) to simple and complex chiefdoms to states and empires. A polity can be either centralized or not (e.g., organized as a confederation). What distinguishes a polity from other human groupings and organizations is that it is politically independent of any overarching authority; it possesses sovereignty.

Different types of polities have different characteristic sizes, so they can be arranged along an approximate scale (Table 4.1). At the higher end of the scale, "mega-empires" are instantiated by such polities as the Achaemenid Empire (550–330 BCE), Maurya Empire (322–185 BCE), Han Dynasty China (206 BCE–220 CE), and Roman Empire under the Principate (27 BCE–284 CE) with peak areas of 5–6 million km^2 and peak populations around 50–60 million (or somewhat fewer in the case of the Achaemenid Empire). While populations in tens of millions were characteristic of the largest empires during the ancient and medieval eras, several modern nation-states have populations in hundreds of millions (and in two cases even more than a billion).

The main message of Table 4.1 is that over the last 10,000 years, the scale of human polities has increased by six orders of magnitude, a truly astronomic number. Thus, the second phase of the evolution of human sociality (from small-scale to large-scale societies) actually involved several fairly major evolutionary transitions. The first one was the rise of centralized hierarchical societies, chiefdoms (first chiefdoms appeared around 7,500 years ago in the Near East). The second was the appearance of first urban state societies (ca. 5,000 years ago), followed by the rise of large multiethnic territorial states,

Table 4.1 Social scales of political organization for agrarian polities. Note that population in the 1,000s could mean any number between 1,000 and 9,000. Furthermore, some simple chiefdoms could have populations above or below this range. A similar caveat applies to other levels.

Population	Area (km²)	Polities
10,000,000s	1,000,000s	Mega-empires
1,000,000s	100,000s	Macrostates
100,000s	10,000s	States ("archaic"), supercomplex chiefdoms
10,000s	1,000s	Complex chiefdoms, city states ("microstates")
1,000s	100s	Simple chiefdoms, acephalic tribes
100s	–	Local communities ("villages")

mega-empires (ca. 2,500 years ago). Finally, the last 200 years have seen the evolution of the modern nation-state.

Ultrasocial Norms and Institutions

Although details of the social evolutionary transitions, discussed in the previous section, varied with locality and time period, they also shared certain generic features. This observation leads to the question: What is it that evolves? Cultural evolution can be defined analogously to biological evolution as the change with time in the frequency of cultural traits. Of particular interest to the question of the evolution of complex societies are such cultural traits as social norms and institutions.

Institutions are systems of culturally acquired rules that govern behavior of individuals in specific contexts. Individuals internalize aspects of these rules, termed *norms*. Institutions and norms are the foundation upon which a distinctively human form of society is based. Institutions can be thought of as self-reinforcing, dynamically stable equilibria that arise as individuals' norms converge and complement each other over time (Richerson and Henrich 2012).

It is now well understood that sustained cooperation requires a solution to the collective action problem stemming from the tension between the public nature of benefits yielded by cooperation and private costs borne by cooperating agents. Social norms and institutions are among the most important ways of solving this problem.

Although the usual focus of theorists is on solving cooperative dilemmas within groups of individuals, collective action problems can arise at all levels of organization. For example, a complex chiefdom typically arises when several simple chiefdoms are unified (forcibly or otherwise). For the complex chiefdom to function well and preserve its integrity, its constituent units (formerly, simple chiefdoms) must cooperate with each other and the center. This argument suggests the following definition:

Ultrasocial institutions are institutions that enable cooperation at the level of larger-scale human groups. They are characterized by the tension between benefits they yield at the higher level of social organization and costs borne by lower-level units.

Of particular interest are ultrasocial institutions which play a role in the integration of largest-scale human groups; institutions that enabled the transition from middle-scale societies (simple and complex chiefdoms) to archaic urban states and subsequently to large-scale empires and modern nation-states (and perhaps beyond, to such multinational entities as the European Union). The above definition implies an interesting diagnostic feature of ultrasocial institutions. Since their benefits are only felt at larger scales of social organization and costs are borne by lower-level units, fragmentation into lower-level units should typically lead to a loss of such institutions. For example, when a territorial state fragments into a multitude of province-sized political units (organized as complex chiefdoms), we expect that such ultrasocial institutions as governance by professional bureaucracies, or education systems producing literate elites, would be gradually depleted from the system. Since fragmentation into smaller-scale units is something that has occurred repeatedly throughout human history, this observation provides us with an empirical basis for distinguishing ultrasocial institutions from others. However, given that institutions are locally stable equilibria, we should not expect an immediate effect of fragmentation. Rather, loss of ultrasocial institutions should be a long-term and stochastic process, with different lower-level units "flipping" from one equilibrium to another at random times.

To make this discussion less abstract, as well as to guide future empirical research, the following examples of ultrasocial norms and institutions are provided.

Generalized Trust

Propensity to trust and help individuals outside of one's ethnic group has a clear benefit for multiethnic societies. However, ethnic groups among whom this ultrasocial norm is widespread are vulnerable to free riding by ethnic groups that restrict cooperation to co-ethnics (e.g., ethnic mafias). Other norms that have the same structure (providing a benefit for cooperation at large social scales, but costly for lower-level units) are willingness to pay national taxes, obeying laws even when there is no chance of being caught, refusing bribes and not offering bribes, and volunteering for military service in times of war. A more extended discussion of this point is made by Gintis and van Schaik (this volume).

Government by Professional Bureaucracies

One of the most thoroughly discussed institutions, government by professional bureaucracies provides the basis for a common definition of the state. The benefits for sustaining large-scale societies that result are generally accepted, by social scientists and nomadic conquerors alike. In fact, it is probable that no really large polity (with populations of over a million) has managed to maintain itself for any appreciable time (longer than a human generation) without acquiring bureaucracies. The costs are also significant. Some are direct (e.g., training and paying bureaucrats) but possibly a more important cost is indirect, stemming from the agent-principal problem. Thus, governing a chiefdom can be readily accomplished by the ruler with the aid of relatives, companions, and clients. The last thing such a ruler needs is to share power with a bureaucracy.

Systems of Formal Education

The benefits from this institution include supplying trained scribes, administrators, judges, priests, and other government specialists. In addition, this institution generates a common language as well as a common set of social norms, at least among the elites, if not throughout the population as a whole (Lieberman 2003, 2010). Examples include the Greek and Latin elite education system from Classical to early modern times in Europe, Islamic education from Medieval to modern times, the Mandarin educational system in China, and the modern mass-literacy educational systems. Costs of this institution may not appear to be very significant, because getting education confers individual-level rewards. However, many imperial collapses are followed by "Dark Ages," so called because of the dramatic drop in degree of literacy and the rate of text production. In some cases, literacy has even been lost; for example, as in the loss of Linear A and Linear B during the Greek Dark Age (early first millennium BCE) and the general decrease of literacy following the disintegration of the Roman Empire in the West.

Universalizing Religions and Other Ideological Systems

Also known as "world religions," such integrative ideologies first appeared during the Axial Age (ca. 800 to 200 BCE). They provided the basis for integrating multiethnic populations within first mega-empires, such as Achaemenid Persia (Zoroastrianism), Han China (Confucianism), and Maurya Empire (Buddhism). There is debate whether the presence of moralizing gods belongs here as well (see Norenzayan et al., this volume). My inclination, however, is to treat moralizing gods as a general prosocial institution, rather than a specifically ultrasocial one, because a moralizing god is useful to stabilize cooperation even at the village level, not only at the level of an empire. Costs: an argument can be made that smaller-scale units should abandon a universalizing religion because it is likely to blur an ethnic boundary between it and other similar-sized societies with which they compete. Such abandonment does not need to involve conversion to a different religion, but perhaps development of a distinct sect. Thus, the splitting of Christianity in the post-Roman landscape into Monophysite and Chalcedonian varieties, with Chalcedonians later dividing into Catholics and Orthodox, may be an example of such a process.

Other examples of ultrasocial institutions may be ideological systems for legitimizing power and for restraining rulers to act in a prosocial manner and professional police, judiciary, military, and priesthoods. The latter can be thought of as variations on the theme of professional administrators (bureaucracy).

Cultural Multilevel Selection as the Theoretical Framework

The final conceptual issue is to elucidate the mechanisms of the evolution of ultrasociality and, more specifically, the evolution of ultrasocial norms and institutions. It is not sufficient to point to their benefits for integration of large-scale societies. Such institutions have significant costs, and the historical record indicates that they repeatedly collapsed in past societies. In other words, we need an evolutionary mechanism to explain the spread of such traits despite the costs.

Although other approaches are certainly possible, I believe that the most fruitful avenue for resolving the puzzle of ultrasociality is provided by the theoretical framework of cultural group (or, better, multilevel) selection. MLS is a powerful theoretical framework for understanding how complex hierarchical systems evolve by iteratively adding control levels. It has been very productive as a research program aimed at the understanding of how cooperation in small-scale human societies evolved. Thus, it is a natural next step to apply this framework to evolutionary transitions that lead to large-scale hierarchical societies (Turchin 2011).

Cultural MLS is particularly appropriate for studying human large-scale societies because they have multilevel hierarchical organization (unlike large-scale societies of eusocial insects). This form of internal organization is partly due to the evolutionary history and partly due to the constraints on human cognitive abilities, which make hierarchies an efficient way of organizing human societies. Thus, in the simplest form of a centralized society, a simple chiefdom, the subordinate agent is a village (a local community) and the superior is a chiefly village, where the ruling lineage resides (Carneiro 1998). Adding more levels results in complex chiefdoms, states of various kinds, and empires. A review of such diverse historical states and empires, such as Ancient Rome and Egypt, Medieval France, and imperial confederations of Central Asian nomads, suggests that all of these polities arose in such a multilevel fashion (Turchin and Gavrilets 2009). In other words, lower-level units combined into higher-level units which themselves combined into yet higher-level units, and so on. Internal organization of states and empires often reflected this process of multilevel integration, similarly to biological organisms retaining vestiges of their evolutionary history.

In the MLS framework the central question is: What is the balance of forces favoring cooperation of lower-level units and, therefore, their ability to combine into higher-level collectives? Here "units" and "collectives" are social groups at different levels of hierarchical complexity. For a society to grow in size, it has to make repeated transitions from the *i*-th to $(i + 1)$-th level. The success of each transition depends on the balance of forces favoring integration versus those favoring fission.

A major mathematical result in MLS theory, the Price equation, specifies the conditions concerning the structure of cultural variation and selective pressures that promote evolution of larger-scale societies. Cultural traits promoting cooperation at the $i + 1$ level will spread if

$$\frac{V_{i+1}}{\overline{V_i}} > \frac{-\beta_i}{\beta_{i+1}}, \tag{4.1}$$

where V_{i+1} and V_i are, respectively, cultural variances between higher-level collectives and between lower-level units, respectively; $\overline{V_i}$ indicates a weighted average over all groups. The coefficient β_{i+1} measures the strength of selection on collectives; similarly, β_i measures the strength of selection on lower units. According to the definition of an ultrasocial trait, $\beta_{i+1} > 0$ and $\beta_i < 0$. Thus, evolution of traits promoting integration at the $i + 1$ level is favored by (a) increasing cultural variation among collectives and decreasing variation among lower-level units (the left-hand side), and (b) increasing the effect of the trait on the fitness of collectives and reducing the effect at the lower level (the right-hand side).

The next step is to identify conditions under which the ratio on the left-hand side increases, while the ratio on the right-hand side declines. Ideally, we would like to measure directly the relevant quantities, but the historical record, unfortunately, is not detailed enough to enable us to do so. The alternative approach is to rely on proxies, which requires making assumptions about which observable variables are best correlated with the quantities of theoretical interest (cultural variation and selection coefficients). The general logic in this step is essentially Lakatosian: to test the theory empirically, we first need to construct the "protective belt of auxiliary hypotheses."

Specifically, large states should arise in regions where very different people are culturally in contact, and where interpolity competition (i.e., warfare) is particularly intense. In previous work I have applied this insight to the period of human history from the Axial Age to the Age of Discovery (ca. 500 BCE–1500 CE). I argue that within Afro-Eurasia, conditions particularly favorable for the rise of large empires were obtained on steppe frontiers, contact regions between nomadic pastoralists and settled agriculturalists. An empirical investigation of warfare lethality, focusing on the fates of populations of conquered cities, indicated that genocide was an order of magnitude more frequent in steppe frontier wars than in wars between culturally similar groups. Furthermore, an overall empirical test of the theory's predictions showed that over ninety percent of the largest historical empires arose in world regions classified as steppe frontiers (Turchin 2011). Thus, taking the abstract theory of MLS and constructing the auxiliary belt that relates it to concrete historical processes during the ancient and medieval periods of Afro-Eurasian history allows us to both generate specific predictions and test them with data.

Conclusion: War, Peace, and the Evolution of Social Complexity

One of the principal threats to peace today originates from failed or failing states: since the end of the Cold War, such internal conflicts have claimed far more victims than old-fashioned wars between established states. Over the last two decades there has been a dramatic increase in the frequency of UN peacekeeping missions and U.S.-led multinational military interventions, while the roster of failed states has also increased. Increasingly, the goals of both peacekeepers and development programs have morphed into what is now called nation-building.

The track record of nation-building is, however, not particularly impressive. Why is it that what works in some countries fails in others? I suggest that a major part of the problem is the lack of a theoretical framework that could guide concrete actions. Here is where evolutionary science can be of tremendous use. We need to understand the nature and evolution of war better as well as its converse, large-scale cooperation. The focus on cooperation is important because peace is not simply an absence of war; lasting peace can be achieved only on the basis of humans cooperating with each other.

As a more robust theory of state formation (or, in more general terms, of organizational forms of large-scale social integration) is developed and tested with cross-cultural data, we will be able to answer such questions as: How do we fix failed states? How can we end civil wars and evolve political structures for nonviolent methods of resolving conflicts? How can we promote integration at the global level and stop interstate wars?

This focus on "nation-building" may strike some as misguided. Indeed, as acknowledged above, the track record of nation-building is not impressive. Yet, human suffering on a massive scale caused by state failure also cries for action. Thus, many a U.S. presidential candidate starts off by decrying nation-building during the election campaign, only to become enthusiastic nation-builders once in the Oval Office; George W. Bush exemplifies the most striking example of such a reversal. If we inevitably end up getting involved in nation-building, wouldn't it be a good idea to have a valid theoretical framework to guide practical actions and allow us to learn from previous mistakes?

Furthermore, "nation-building" is often thought of as something that external agents (formerly, imperial or colonial powers, today the "international community") impose on societies who failed as states. An alternative approach (one with which I find myself in agreement) is that the proper way of doing nation-building is "auto-nation-building." In other words, nobody has a better chance of building lasting structures of large-scale governance than the affected societies themselves. In such well-known examples of nation-building as the economic and political reconstruction of post-World War II Germany and Japan, there is little doubt that the primary role was played by the Germans and the Japanese themselves. However, these two nations had already evolved strong and well-functioning states before World War II and the military defeat which "decapitated" them. What about new nations, such as former Soviet Union republics, or failed states, such as Haiti or Somalia? These societies need to make a number of collective decisions. Should they adopt a parliamentary or a presidential form of governance? A confederated or a unitary state structure? These questions are really about what kinds of ultrasocial institutions work best at promoting cooperation at the scale of the whole polity/society. In other words, the research program on the evolution of ultrasociality can be of significant practical use in cases of "auto-nation-building."

Finally, improved understanding of factors that enhance versus impair our ability to cooperate in large-scale societies has practical implication, not only for failed or failing states or for new countries emerging when larger states fragment, but also for established democracies. Just because the latter societies have functioned reasonably well since the end of World War II does not guarantee that they will continue to do so in the future (70 years is not a particularly long period by historical standards). It should be disquieting that according to a broad spectrum of measures, "social capital" (really, the capacity for cooperation) has been declining in the United States and several other old democracies (e.g., Putnam 2000). The reasons for this decline are not understood, and we have no idea how to reverse the trend.

Understanding conditions that either promote or inhibit human ultrasociality is not only a major theoretical puzzle. Such understanding is also highly relevant for addressing the challenges of large-scale cooperation (and its converse, large-scale conflict) in the modern world.

Acknowledgment

I am grateful to several colleagues for discussions and comments on the mansucript: Joseph Bulbulia, Herbert Gintis, Joseph Henrich, Peter Richerson, and Polly Wiessner. This work was supported by an ESRC Large Grant (REF RES-060-25-0085) entitled "Ritual, Community, and Conflict."

5

Like Me

A Homophily-Based Account of Human Culture

Daniel B. M. Haun and Harriet Over

Abstract

This chapter presents a homophily-based account of human social structure and cultural transmission, wherein a tendency to favor similar others (homophily) is a key driving force in creating human-unique forms of culture. Homophily also accounts for observed striking differences between human groups. From early in development, evidence demonstrates that humans show a strong tendency to interact with, and learn from, individuals who are similar to themselves. It is proposed that homophilic preferences of the group, in general, creates a feedback loop to ensure that children engage in high-fidelity copying of the group's behavioral repertoire. This allows children to reap the benefits of others' homophilic preferences and so maintain their position within the group. In consequence, homophilic preferences have transformed a number of mechanisms which humans share with other species (e.g., emulation and majority-biased transmission) into human-unique variants (e.g., social imitation and conformity). Homophilic preferences have, furthermore, spawned a new tendency to interpret the structure of actions as social signals: norm psychology. The homophily account thus connects previously disparate findings in comparative, developmental, and social psychology and provides a unified account of the importance of the preference for similar others in species-specific human social behavior.

Introduction

In many ways, the stability of human cross-cultural variation is surprising since high rates of migration (Hill et al. 2011) and visitation (Chapais 2008) should, over time, reduce differentiation across groups (Yeaman et al. 2011). Assuming a long enough time period, any difference between human groups should inevitably fade by means of these processes (Boyd and Richerson 2005, 2009; Henrich and Boyd 1998).

Independent of intergroup migration, there is another parallel migration into any group at any given point in time: newborns. Every new generation of children confronts the group with a number of individuals that do not act according to the group-specific behavioral repertoire (Harris 2012). Thus, in addition to immigrants entering the community with conflicting behaviors and norms, there is also a constant influx of individuals who enter the community without any established behavioral patterns or sometimes even with predispositions that are counter to the local cultural variant of a particular behavior (e.g., Haun et al. 2006).

How then do children acquire the appropriate group-specific beliefs and behaviors? Previous accounts of cultural transmission have emphasized the role of learning mechanisms, such as high-fidelity imitation (Lyons et al. 2007; Whiten et al. 2009), or cognitive abilities, such as perspective taking (Tomasello 1999) and sensitivity to ostensive cues (Gergely and Csibra 2006). In contrast to these accounts, we emphasize the importance of more social processes, in particular, homophily (a preference for others we perceive as similar to ourselves). The homophily account is based on two closely related claims. First, children preferentially affiliate with and learn from similar others. Second, and more importantly, the homophilic preferences of the group, in general, creates a feedback loop that ensures children engage in high-fidelity copying of the group's behavioral repertoire. This allows children to reap the benefits of others' homophilic preferences and so maintain their position within the group. This homophily-based account thus unites research on the social functions of imitation (e.g., Carpenter and Call 2009; Over and Carpenter 2012; Nadel 2002; Nielsen 2009; Užgiris 1981) with that on group membership (e.g., Dunham et al. 2011; Kinzler et al. 2007; Turner 1991) and normative behavior (Kallgren et al. 2000; Rakoczy et al. 2008).

We do not claim that the homophily account provides an exhaustive description of how social motivations influence cultural transmission. Other social motivations and preferences (e.g., for prestigious others and competent others) and the interactions between them are also important in explaining social learning in humans (Laland 2004). We simply wish to highlight that the preference for similar others is one key factor in explaining cultural transmission and that species differences in this tendency might be one factor in explaining the origins of species-typical features of human cultural transmission.

Below, we outline our homophilic account in more detail, beginning with a discussion on the importance of homophilic assortment from an evolutionary perspective. Thereafter, we review the available evidence that, from early in development, humans have a strong preference for similar others. Finally, we present evidence that this preference for similar others has transformed a number of preexisting cognitive mechanisms (e.g., emulation learning and majority-biased transmission) into a suite of human-unique traits that includes social imitation, conformity, and a norm psychology.

Homophilic Social Preferences from an Evolutionary Perspective

For cooperation to be maintained within a group, it is essential for group members to be able to distinguish cooperators from defectors. In stable, personalized groups, familiarity serves to reduce aggression and to create a tolerant context; the foundation of any cooperative exchange. As groups increase in size, so does the frequency with which individuals have to interact with others less familiar. Eventually, personal interaction history can no longer accurately account for the reliability of a partner.

At some point during human evolution, social networks increased to a size where group members were increasingly more likely to encounter others that were only vaguely familiar. For instance, even the most mobile extant forager groups live in networks that typically exceed several hundred individuals (Hill et al. 2011; Apicella et al. 2012). Under such conditions, familiarity itself remains important, but is no longer as effective. Thus a proxy measure for familiarity is required that reliably correlates with familiarity. Similarity in aspects of the phenotype (morphology and behavior) provides one such measure. Individuals who grow up within the same community are likely to be similar on a number of dimensions, thus making phenotypic similarity an honest signal of group membership.

We argue that a preference for similar others allowed humans to categorize strangers and identify in-group members who were not personally known to them. Choosing to interact and cooperate with more similar strangers maximized the chance of successful cooperative interactions, because similar individuals were more likely to share relevant behavioral tendencies (McElreath et al. 2003; Cohen 2012). As a result, humans were able to function within qualitatively different forms of social organization available to other primates, and tap into the cooperative potential of strangers. Formal models have shown that such a pattern of cultural transmission, in which individuals are disproportionally influenced by those who are similar to themselves, is adaptive since a homophilic preference causes subpopulations to become culturally isolated. This, in turn, allows the mean value of locally adaptive traits to converge to the optimum. A transmission strategy based, for example, on success would only adapt very slowly to a variable habitat (Boyd and Richerson 1987b). In other words, "the preference to interact with people with markers like one's own may be favored by natural selection under plausible conditions" (McElreath et al. 2003:123).

We now shift our focus to empirical evidence supporting this hypothesis. We center our discussion on the developmental and comparative data demonstrating that the human preference for similar others is much stronger than that seen in other primate species.

Like Me? Homophilic Social Preferences from a Comparative Perspective

Homophilic Social Preferences in Nonhuman Primates

Interpersonal relations in chimpanzee groups are characterized by tolerance of in-group members and hostility toward out-group members (Wrangham 1999; Wilson et al. 2012). Members of other groups detected within the home range are typically killed, the one exception being migrating females (Kahlenberg et al. 2008). This preference for in-group members over outgroup members is almost certainly based on familiarity rather than similarity as chimpanzees typically encounter all the members of their own group on a fairly regular basis.

A recent study, however, raises the possibility that some nonhuman primates also use similarity as a means by which to assort between others. Paukner et al. (2009) reported that capuchin monkeys who were presented with two human experimenters—one who imitated them and another who just performed monkey-like movements—sat closer to the imitator and exchanged more tokens with him. Hence a transient increase in behavioral similarity (social mimicry) made capuchins prefer one human to the other.

There are thus some hints that nonhuman primates utilize similarity in their social judgments (at least to some extent) and, in consequence, that the common ancestor of humans and other primates had rudimentary preferences for similar others. This may have provided the evolutionary starting point from which homophilic social preferences in humans could emerge. However, as we will see below, the evidence for homophilic preferences in humans far exceeds that of any other primate.

Homophilic Social Preferences in Children

In contrast to nonhuman primates, the evidence that humans assort unfamiliar others based on similarity is quite substantial (e.g., Gruenfeld and Tiedens 2010; Jones et al. 2004; Tajfel et al. 1971). This preference for similar others appears to structure social interactions from early in development. For example, six-month-old children prefer to look at individuals who speak their own versus a different language, and ten-month-olds prefer to accept toys from speakers of their own language (Kinzler et al. 2007). This preference for native language speakers structures social interactions also later in development: five-year-olds preferentially choose native language speakers over foreign language speakers or foreign-accented speakers as friends (Kinzler et al. 2009). However, in all of the above-mentioned studies with children, it is not possible to separate a preference for similar others from a preference for individuals that children find easier to understand.

Fawcett and Markson (2010) have provided evidence that young children's social preferences are, at least at times, based on self-similarity alone. Fawcett and Markson demonstrated that three-year-old children prefer to play with a puppet who expresses the same food preference as them as opposed to a contrasting preference, and a puppet whose physical appearance matches rather than mismatches their own. Other evidence comes from research on the effects of being imitated. One of the consequences of being imitated is a momentarily increased level of perceived similarity between social partners (Chartrand and Bargh 1999). As of early in development, children appear to prefer individuals who imitate them to individuals who engage in independent behavior. For example, 14-month-old infants look toward and smile more at an experimenter who imitates them than at an experimenter who engages in equally contingent but nonimitative behavior (Agnetta and Rochat 2004; Asendorpf et al. 1996; Meltzoff 1990). Furthermore, infants and toddlers are more likely to help an experimenter who has imitated them than an experimenter who has engaged in contingent but nonimitative behavior (Rekers et al., submitted).

Further evidence for children's preference for similar others comes from the so-called minimal group paradigm (Tajfel et al. 1971), in which individuals are randomly allocated to one of several groups that are only identified by an abstract, seemingly uninformative symbol. In this way, similarity between members of a minimal group is not indicative of any shared behavioral characteristic of the individuals composing the group, but only of shared group identity. Five-year-old children prefer individuals allocated to the same minimal group as them over individuals allocated to a different minimal group. Furthermore, children not only prefer individuals belonging to the same minimal group; they also have more positive expectations about the behavior of in-group members (Dunham et al. 2011).

This preference for similar others seems to occur across cultures (Kinzler et al. 2012; Cohen and Haun 2013). Children's relative reliance on particular cues, however, varies depending on the particular sociocultural context. Recent studies comparing children from different townships along the Brazilian Amazon have demonstrated that children's preferences for certain cues are likely tuned according to locally relevant cue variation. For example, children from accent heterogeneous populations rely more strongly on accent as a similarity cue than children from accent homogeneous populations (Cohen and Haun 2013).

Children Prefer to Learn from Similar Others

Children's preference for similar others not only indirectly channels their own input by creating interaction bubbles of similar others, it also has more immediate implications for children's social learning. Kinzler et al. (2011) demonstrated that five-year-old children are more likely to learn the function of a novel object from an individual who speaks with the child's native accent than from an individual who speaks the same language with a foreign accent. A more recent study claims that even infants preferentially learn from similar others (Buttelmann et al. 2013). In this study, 14-month-old infants listened to a story either told in their native language or in a foreign language. Children subsequently imitated the actions of the speaker of their native language more closely. Caution must be taken when interpreting this result, however, because the design confounds similarity with other factors, such as the relative comprehensibility of the stories. Nevertheless it raises the possibility that children select their models by similarity already in the second year of life.

Other evidence that children preferentially learn from similar others comes from research on the effects of being imitated. In a recent study, Over, Carpenter, Spears, and Gattis (2013) found that five- to six-year-old children were more likely to adopt the preferences and novel object labels of an experimenter who had previously imitated their choices than those of an experimenter who had previously made independent decisions.

Summary

From the evidence presented above, it appears that the human preference for similar others likely far exceeds that of any other primate. This preference is present early in development and structures children's learning as well as their social interactions.

Like Me! The Consequences of Homophilic Preferences

If we prefer similar to dissimilar others, it follows that increasing the similarity between self and other can be a useful strategy for directing others' positive social activities toward the self. We contend that homophilic preferences in humans have interacted with the social-learning mechanisms inherited from our common ancestor with the other great apes and transformed them into species-unique forms of copying behavior which serve to maintain an individual's position within the group.

In contrast to previous accounts (e.g., Carpenter and Call 2009; Užgiris 1981), the homophilic account does not require children to have the goal of making themselves more similar to their social partners. Although children may, at times, actively seek to be like others (Carpenter 2006; Over and Carpenter 2013), the more typical pattern may be for children to learn through experience that imitation is successful in improving social relations without any explicit awareness of this connection. In consequence, their only goal within the social situation may be to get along well with others.

Below we discuss evidence that social-learning mechanisms which we share with other species—emulation and majority-biased transmission—have been transformed by homophilic preferences into a suite of human-unique social-learning processes including social imitation, conformity and a norm-psychology.

Emulation Becomes Imitation

Chimpanzees use a range of social-learning strategies including, most prominently, emulation (Call et al. 2005; Nagell et al. 1993). In emulation learning, an animal focuses on the outcome that is achieved in the physical world rather than on the particular actions used to achieve it (Tomasello et al. 1993). If chimpanzees copy the particular actions of their conspecifics (i.e., imitate), they appear to do so infrequently and with relatively low fidelity (Tennie et al. 2009).

Children, in contrast, show a strong tendency to copy actions faithfully. In fact, imitation by children is sometimes so precise that they even copy actions that are superfluous or disadvantageous to solving the task at hand (Horner and Whiten 2005; Nagell et al. 1993; Nielsen 2006). For example, children from three to five years of age, who have been trained to identify the causally irrelevant parts of novel action sequences, still reproduce causally irrelevant actions, and they continue to do so even when specifically instructed by the experimenter to copy only necessary actions (Lyons 2009; Lyons et al. 2007). This phenomenon has come to be called "overimitation" (Lyons et al. 2007, 2011). It emerges in the second year of life (Nielsen 2006) and becomes increasingly pervasive throughout the preschool period (McGuigan and Whiten 2009; McGuigan et al. 2007).

The homophilic account presumes that these differences in social learning between chimpanzees and humans have been driven, at least in part, by human homophilic preferences. The increased importance of "how something is done" is owed to the significance of behavioral similarity among individuals in a group. Finding a different way to achieve the same ends is no longer functionally equivalent to copying others' actions exactly, since the former decreases similarity with others whereas the latter increases it. In humans, imitation could thus serve new social purposes. This added social dimension effectively turned emulation learning into faithful imitation.

Consequently, it is misleading to refer to high-fidelity imitation as "overimitation," since the term implies that children copy unnecessary parts of action sequences. Under the homophily account, these parts, while being causally irrelevant, still serve an important function for the learner: they produce a high level of similarity between the demonstrator and the learner.

Evidence in favor of the proposal that high-fidelity imitation is used to achieve social goals comes from data which suggest that children increase their tendency to imitate when affiliation is important to them. Over and Carpenter (2009) demonstrated that five-year-old children who have been given the goal to affiliate (through priming with social exclusion) imitate the actions of a model more closely than children who have been given a neutral prime. Further evidence in favor of this hypothesis comes from work demonstrating that children are more likely to copy the specific actions of a model when that model is in the room and so are able to watch their imitation (Nielsen and Blank 2011).

Imitation is also closely associated with social factors in younger children. Nielsen et al. (2008) demonstrated that two-year-old children are more likely to copy the specific actions of a model who engages in a contingent social interaction with them than those of a model whose behavior is not contingent on their own.

The tendency to make the self similar to others can also be used more strategically within social settings. That is, imitation can serve Machiavellian ends (Over and Carpenter 2012). Research with older children has shown that they are able to use imitation to increase their influence over others. For example, Thelen et al. (1980) demonstrated that ten-year-old children are more likely to copy the specific actions of a peer when they will later need to persuade that peer to do something.

Majority-Biased Transmission Becomes Conformity

Homophilic preferences, we argue, have not only influenced how humans interact with individual social partners, but how they respond to the group in general. One way in which humans interact with the group as a whole is through consideration of the majority.

If, due to any combination of underlying mechanisms, an individual is more likely to acquire the behavior displayed by the majority, we refer to it as a majority-biased transmission (Haun et al. 2012). A recent study in chimpanzees showed that naïve individuals copy the behavior of the majority over alternatives, even if those are equally frequent, equally familiar, and equally productive behaviors (Haun et al. 2012).

Thus, chimpanzees follow the majority when they have no prior information available. However, they do not follow a majority if they have to forgo their own behavioral tendencies to do so (Haun et al., submitted). We refer to the tendency to forgo personal preferences in favor of copying the majority as conformity (Haun et al. 2013; van Leeuwen and Haun 2013). In another study, Hopper et al. (2011) argue that chimpanzees conformed against their own preference, based on the finding that individuals retained their socially acquired strategy even though the alternative yielded more preferred rewards. However, because individuals only very rarely experienced the alternative strategy yielding more desirable foods, it remains highly questionable whether individuals were, in fact, fully aware of the alternative.

Similar to chimpanzees, human children follow the majority if they have no relevant information available (majority-biased transmission, Haun et al. 2012). However, in contrast to other primates, human children also adjust their behavior to the majority, even when an equally effective but individually acquired strategy is already available: under one situation, in which a child who has a high level of performance on a certain task is confronted with a majority of peers who unanimously give a false response, children often choose to abandon their own judgment to adjust their behavior to the majority's response

(Berenda 1950; Corriveau and Harris 2010; Haun and Tomasello 2011; Walker and Andrade 1996). Furthermore, children appear to consider the social consequences of conforming versus dissenting. Haun and Tomasello (2011) varied the privacy of the subjects while giving their response and found lower rates of conformity when preschool children were allowed to keep their response private from the majority. Most strikingly, children adjusted their level of conformity from trial to trial depending on the privacy of their response; they conformed more often when they gave their response in public. The authors concluded that the reduction in conformity in the private condition demonstrated a partial contribution of social motivations for children's conformity on the public trials. Hence, children, in contrast to other primates, are additionally guided by social motivations (Haun and Tomasello 2011) when conforming to a majority.

In the absence of a social function, copying the majority when acquiring a new skill is adaptive on an individual level, but there is no reason to follow the majority when the learner already has a different but equally productive strategy available. However, if conformity also serves a social function, then it pays a learner to forgo their own strategy and adopt that of the majority: since sticking to the former will decrease similarity between the self and the group, whereas conforming to the latter will increase similarity between the self and the group. According to the homophily account, this added social dimension increased humans' tendency to conform to the majority, effectively turning majority-biased transmission into conformity.

The Emergence of Norm Psychology

Nonhuman primates, such as chimpanzees, have "rules of conduct" that are reinforced. For example, subordinates tend to display certain gestures when meeting a dominant individual, and violations of this behavioral pattern will result in aggression (Goodall 1986). Although the superficial structure of these patterns of behavior might resemble that of human norms, they differ from norms in important respects (Tomasello 2008). For example, whereas human norms are often variable across groups, gestures negotiating the relationship between dominant and subordinate individuals in chimpanzees are highly similar across different, unrelated populations, thus suggesting they are not culturally learned (Tomasello et al. 1997). Furthermore, chimpanzee "rules" unlike human norms, are not agent neutral. Subordinate chimpanzees failing to submit to the dominant might suffer aggression from the dominant (the affected party), but not from other (unaffected) group members. Chimpanzees do not appear to punish the violations of third parties (Riedl et al. 2012). Humans, on the other hand, punish the transgressions of others even if they do not concern them directly (Henrich et al. 2006). Hence, it does not seem to be the case that chimpanzees collectively intend to do things in a certain way and do not have normative expectations about their conspecifics' behavior, but that their social interactions are better characterized by behavioral regularities as well as individual and idiosyncratic preferences for certain behaviors.

Human norms are rich in their social interpretation: Norms describe the "right" way to do things, the way things "ought" to be done, the way "we" do things (Bruner 1993). Human children appear to detect such norms spontaneously in many behaviors, even in the absence of normative language (Schmidt et al. 2011). After a single confident and intentional demonstration by an adult, children appear to assume that the way in which an action was demonstrated is normative. Following such a demonstration, children will not only follow that norm, but actively enforce it when later observing someone performing the action "incorrectly," often protesting using normative language about what people ought to be doing (Rakoczy et al. 2008). Thus children readily enforce norms on others even if their violation does not impact upon them directly.

In summary, we argued that the social relevance of similarity among individuals gives previously socially neutral behaviors a new social relevance. The "way something can be done" is effectively elevated to the "way we do something," fitting actions with a social signaling function. This normative dimension to actions which have no intrinsic value (e.g., how to hold a fork) is a direct consequence of the relevance of self-other similarity in cooperative groups of increasing size.

Conclusion

Many accounts exist for the species-unique structure of human social behavior. All of them contain lists of human-specific social abilities and motivations for coordination (Tomasello et al. 2005), social learning (Tennie et al. 2009), teaching (Gergely and Csibra 2006), and norm psychology (Chudek and Henrich 2011). We have provided an account that unites some of these previously unconnected sets of abilities and motivations. According to our homophily-based account, a preference for similar over dissimilar others underlies important aspects of human-unique social behavior.

Evidence suggests that, from early in development, children prefer to interact with, and learn from, individuals who are similar to themselves. This preference for similar others and the potential advantages reaped by being similar to others, ensures that children engage in high-fidelity copying of the group's behavioral repertoire. As a result, seemingly irrelevant parts of actions gain social relevance by serving as a similarity marker. This tendency to interpret the physically irrelevant structure of actions as social signals spawned a human-unique form of interpreting the actions of others: norm psychology.

We argue that species difference in homophilic preferences might be one key factor in explaining the origins of species-typical features of human cultural transmission. We predict that humans are unique among living primates in the extent of their preference for similar others. We also predict that this preference is universal across human cultures, albeit relying on different similarity cues in different populations (Logan and Schmittou 1998; Cohen and Haun 2013). Future studies should further test these predictions from cross-cultural and comparative angles.

Acknowledgments

We would like to thank Malinda Carpenter, Emma Cohen, Emma Flynn, Katja Liebal, Nadja Richter, Carel van Schaik, Claudio Tennie, Marco Schmidt, Peter Richerson and an anonymous reviewer for valuable comments on an earlier draft.

Group photos (left to right, top to bottom)
Fiona Jordan, Carel van Schaik, Polly Wiessner, Herb Gintis, Daniel Haun,
Marco Janssen, Daniel Haun, Herb Gintis, Daniel Hruschka, Sarah Mathew,
Pieter François, Polly Wiessner, Laurent Lehmann, Daniel Hruschka,
Pete Richerson, Peter Turchin, James Kitts, Fiona Jordan, Marco Janssen,
Sarah Mathew, Peter Turchin, and Carel van Schaik

6

Cultural Evolution of the Structure of Human Groups

Fiona M. Jordan, Carel van Schaik,
Pieter François, Herbert Gintis, Daniel B. M. Haun,
Daniel J. Hruschka, Marco A. Janssen, James A. Kitts,
Laurent Lehmann, Sarah Mathew, Peter J. Richerson,
Peter Turchin, and Polly Wiessner

Abstract

Small-scale human societies are a leap in size and complexity from those of our primate ancestors. We propose that the behavioral predispositions which allowed the evolution of small-scale societies were also those that allowed the cultural evolution of large-scale sociality, in the form of multiple transitions to large-scale societies. Although sufficient, the cultural evolutionary processes that acted on these predispositions also needed a unique set of niche parameters, including ecological factors, guiding norms, and technologies of social control and coordination. Identifying the regularities and patterns in these factors will be the empirical challenge for the future.

Introduction

What are the behavioral predispositions that cultural evolution has used, and changed, to facilitate the transition of human societies from small to large scale? Much excellent work has been done on the evolution of complex societies (e.g., Johnson and Earle 2000; Keech McIntosh 2005; Flannery 1972; Turchin 2003; Vaughn et al. 2009; Kristiansen and Larsson 2005). Our contribution in this volume seeks to add to the understanding of the evolution of social complexity, from the perspective of the behavioral predispositions that facilitated the evolution of small-scale human societies, and to stimulate proposals for how these were expanded, elaborated, or repressed by cultural evolution to make the formation of complex large-scale societies possible. A complete answer to this question requires that we (a) specify in detail these behavioral predispositions, (b) explore which are necessary for the evolution of small-scale sociality and cooperation, and (c) explore how they can (and have been) exploited by cultural evolutionary processes in the formation of large-scale societies. As Turchin (this volume) points out, what we refer to as small-scale societies in humans are still huge cooperative endeavors, involving many more individuals, compared to the scale of cooperation in other vertebrates. The identification of a minimal set or sets of predispositions necessary for small-scale societies to arise then gives us building blocks necessary for thinking about the cultural evolution of large-scale societies.

In the discussions that led to this chapter, we were informed by the theoretical and definitional perspectives expressed in the four relevant position papers (see Gintis and van Schaik, Turchin, Mathew et al., and Haun and Over, all this volume). Many of the key contributions to our understanding of human sociality and cooperation are discussed therein and need no further review here. We make a distinction between small-scale (groups of hundreds to a few thousands of individuals practicing mostly hunter-gatherer/foraging ways of life) and large-scale (groups of thousands upward to state-level complex societies of millions) sociality on a fuzzy basis. The importance of subsistence type, or complexity of social relations, means that there were and are many border cases in human history; however, our aim in this chapter is not to typologize. Rather, we aim to recognize a broad and (what is possibly the most) salient distinction in the variety of human social structures, and to consider how cultural evolutionary theory can stimulate research toward understanding the puzzle of ultrasociality. We begin with a phylogenetic and developmental perspective.

Mechanisms Enabling Cooperation in Human Small-Scale Societies

From Primate-Scale to Small-Scale Human Groups

Every primate group contains close and more distant relatives as well as nonrelatives, often immigrants. Whereas tolerance and cooperation among relatives is easily explained by kin selection, similar phenomena among nonrelatives require another explanation. In stable, personalized groups, familiarity among nonrelatives serves a basic function: to reduce aggression and create a tolerant context—the foundation of any cooperative exchange (Preuschoft and van Schaik 2000). Familiarity among nonkin could be a very basic extension of the kin recognition mechanism, which reduces aggression and creates tolerance. Likewise, in cooperative groups, individuals preferentially cooperate (i.e., engage in costly acts that will be reciprocated) with others they can trust to engage in mutually beneficial exchanges and interactions. Long-term social bonds among kin as well as nonkin, some possibly recruiting the same psychological mechanisms among human friendships, enable dyadic cooperation in many primate societies (Hruschka 2010; Seyfarth and Cheney 2012).

Mechanisms Enabling Small-Scale Human Societies

Even the smallest-scale human society is far larger than most primate groups, and it is likely that early hominins engaged in fission-fusion social organization, much like both extant human foragers and chimpanzees. For instance, even the most mobile extant forager societies have a network size (a few hundred to a few thousand) that far exceeds the largest chimpanzee community (Johnson and Earle 2000; Apicella et al. 2012). Most mobile hunter-gatherers live in bands of 15–50 people, but their members interact with kin in some 6–10 nearby bands on a regular basis (Heinz 1979; Lee and DeVore 1968; Wobst 1974; Williams 1974; Peterson 1976). These "maximum bands" gather for infrequent ceremonial occasions, if at all. Personal networks built on marriage ties or exchange ties extend outside of the "maximum band" and tap into a broader surrounding population of up to a few thousand people (Gamble 1999; Wiessner 1986; Yengoyan 1968). Thus at some time during hominin evolution, individuals became more likely to encounter strangers who were the kin or partners of *their* partners, but not directly known to *them*; that is, in-group strangers (Hill et al. 2011). At this point the interaction history with ego could no longer be relied on to estimate the reliability of a partner, and the question is how this problem could be overcome. The reputation of unfamiliar people within spheres of interaction became key for tolerance and cooperation, together with indicators of shared customs, norms, and values.

Preexisting mechanisms may have been pressed into service to solve this problem, and we begin by specifying a candidate list of psychological/behavioral predispositions (mechanisms) that, either in isolation or in combination, can produce the sorts of widespread cooperative social outcomes we see in small-scale human societies (Table 6.1). The candidates in this list may be compared with those in Hill et al. (2009) and Rodseth et al. (1991).

For our purposes we take a working definition of "mechanism" to be (partly) biological processes that shape human behavior in a given situation or environment, including, for example, cognitive capacities, cognitive preferences, and emotional reactivity; it also includes, for example, the ability to digest certain foods or the motoric ability to throw projectile weapons. There is general agreement that these are species-typical (i.e., universal) mechanisms and that they are to some extent (though we do not specify) genetically specified. Conglomerate mechanisms in the traditional anthropological sense which are externalized to cognition (e.g., warfare or religion) may themselves be the result of cultural evolution, but here we focus on *species-typical predispositions and capacities*.

Table 6.1 lists candidate mechanisms and indicates whether they are present in other primate species and/or in the last common ancestor (LCA). This list is not prioritized in order of importance. We either do not agree that such a ranking is possible or, if we do, we disagree internally on what that ranking might be. Instead, Table 6.1 groups together those mechanisms that are shared ancestrally

Table 6.1 Mechanisms that enable cooperation in small-scale human societies and their presence in other primate species and/or in the last common ancestor (LCA).

Mechanism	Presence
Kin recognition, kin bias, nepotism	Common in other species but recognition of patrilateral kin probably absent in LCA
Respect of territory, property, mates	Found in other species
Structured social interaction (assortativity)	Presumed in LCA
"Reverse dominance hierarchy" (Boehm 1993)	Leveling coalitions in chimpanzees (van Schaik et al. 2004a)
Direct reciprocity: Who did what to me?	Some evidence in apes, such as sex for food or grooming
Coalition formation, socially organized aggression	Common in other species
Multilocal residence: flexibility of male/female dispersal	Residence flexibility in bonobos
Cooperative breeding	Not in LCA, but in other species (Burkart et al. 2009; Hrdy 2009)
Marriage, pair bonding	Pair bonding not in LCA but other species
Multilocal or multilevel ties outside the group	Presumed absent in LCA
Leadership by persuasion, authority, or prosocial leadership	Minimal in other species, not in LCA
Moralistic punishment, moralistic rewards	Presumed absent in LCA
Reputation and gossip	No third-party reputation in other species beyond dominance; only in humans is reputation used for communicating behaviors that are good or bad for the group
Norm psychology: norm adherence, norm internalization, institutions	Not in LCA
Lethal force at a distance	Not in LCA
Cumulative culture, cultural variation, social-learning biases	Social-learning biases in other species, but cumulative culture limited or absent in LCA
Language	Not in LCA
Symbolic behavior: expressive and as ethnic marker	Not in LCA
Predisposition to impose categorical distinctions onto continuous cultural differences, leading to group boundaries and identities	Not in LCA
Predisposition for collective ritual and synchronicity	Not in LCA

with other primate relatives, and those that are hominin specific (i.e., derived). Some of this may be married up with evidence presented in Shultz et al. (2011), who use comparative primate data and phylogenetic methods to infer some of the appropriate features of social organization for the LCA.

The Critical Importance of Norm Psychology

Provided with such a list, the immediate question becomes: Which of these mechanisms are *essential* for the evolution of cooperation in small-scale human societies? From a primatological perspective, we can identify the preexisting preference for informational conformity (in chimpanzees, see Haun et al. 2012) that became modified into social conformity and norm psychology. When individuals began to live in larger small-scale societies with a high degree of anonymity, yet needed to associate and cooperate on many occasions, they also needed a reliably correlated proxy measure for familiarity. Similarity in all aspects of the phenotype (morphology and behavior) provides one such measure. Thus, while conformity was previously driven by utilitarian reasons, conformity acted to prevent individuals from being classified as dissimilar. This social conformity[1] is truly normative because individuals benefit from being as similar as possible in all respects to other group members (which brings acceptance), and they benefit from detecting deviations from conformity. Those deviations are then used to estimate reduced similarity, possibly on some threshold of perception below which another individual is classified as belonging to an out group. Thus, behaviors that initially had no normative dimension have now acquired one: from the best way to do things to the way *we* do things (see Haun and Over, this volume). For instance, young children actively extract normative information from actions by adults and reinforce them among peers (Rakoczy et al. 2007). This evolutionary development, we posit, is the origin of norm psychology, which subsequently gave rise to institutions (Chudek and Henrich 2011).

From a developmental perspective, we can posit that observational forms of social learning have moved from the more utilitarian emulation (end copying) in apes toward imitation (means copying) in humans. Imitation will produce fine-grained behavioral similarity. Indeed, humans have a tendency to imitate the details of action that are functionally superfluous but are good indicators of similarity ("overimitation," Lyons et al. 2007). Imitation has been documented rather rarely in nonhuman primates, although many would claim it occurs at least occasionally (Whiten et al. 2009), whereas it is ubiquitous among humans from an early age.

1 This statement is not intended to erase the ubiquity or importance of role specialization (formal or informal) within any particular society. In many small-scale societies, role diversity between people is overtly appreciated and tolerated. Differences between people can promote a complementarity that holds groups together—one person might be a musician, another a storyteller, another a dancer, another a hunter.

Cognitively, humans generally have a tendency to categorize continuous variation into usually discrete categories. Thus, continuous variation in similarity can thus become dichotomized into in-group versus out-group, and human in-group–out-group psychology may be based on categorization. Indeed, humans have created dichotomous similarity markers that go beyond morphology and behavior, a truly novel feature that anthropologists call ethnic marking (Efferson et al. 2008a). For ethnic markers to be stable markers of similarity, they must be socially costly (by increasing similarity to one group, one automatically decreases similarity to another), permanent, or both. Indeed, humans show the hallmarks of this process in which even children actively use a variety of similarity markers, as suggested by experiments which removed all possible familiarity-relevant information and showed that even arbitrary markers can serve to guide similarity judgments (see Haun and Over, this volume).

Importantly, humans do not need functional outcomes like chimps do (Haun and Over, this volume) to change their behavior. We are what Gintis and van Schaik refer to as *Homo ludens*; that is, the only species that can make up new games and follow those rules. Because people can be "programmed" with new preferences, the transaction costs of social exchange are reduced. This means that norms can have flexible regularities in their content, and some of these regularities may have become so important as to be independent mechanisms/processes, such as religion or warfare. The task for scholars interested in understanding how norms change is then to draw upon ethnographic and historical data and, using the frameworks of cultural evolution outlined in this volume, to specify the steps in individual cases. Subsequent generalizations can then be addressed at different levels of explanations (Tinbergen 1963; Oyama et al. 2001). For example, experimental and developmental psychologists can add to our understanding of how norms change through mechanism-based approaches.

Niche Parameters

In the course of our discussions, it became apparent that a complete cultural evolutionary explanation could not consist of purely endogenous factors (mechanisms). Thus we identified a set of "niche parameters" for the evolution of human sociality. These contextual features can be said to form the environmental conditions in (and by) which the predispositions identified above are expressed as behaviors or behavioral complexes in human evolution. These niche parameters include a number of elements, and the following is a nonexhaustive list:

1. Fixed locations for sleeping, cooking, and social interaction (e.g., camps and processing sites).
2. Controlled use of fire for defense and/or cooking. (Note that an adaptation to cooked foods may itself be a mechanism or predisposition.)

3. Hunting and/or scavenging.
4. Resource pooling and communal eating: Wrangham (2009) has argued that humans needed to develop normative systems for the distribution of hunted/gathered food.
5. Savannah living, entailing some necessity for defense against predators.
6. Environmental change (Richerson et al. 2009).

Ethologists would argue that the predispositions evolved to maintain a particular social organization and structure in the context of these niche parameters, which, in turn, were molded by, and further co-constructed the species' lifestyle. Modern evolutionary paradigms, such as niche construction (Laland et al. 2000, 2011; Odling-Smee et al. 2003) and developmental systems theory (Oyama et al. 2001; Fuentes 2009; Gray 2001), may be useful in further elaborating the dynamics of construction and feedback between niche parameters and species characteristics. Sterelny (2012) provides one such example by considering how humans have structured the environments of their conspecifics in such a way as to enable cognitive competence in the face of high informational loads and demanding tasks—in both the social and physical domains.

The importance of considering these niche parameters in combination with mechanisms is demonstrated by a portion of Gintis and van Schaik's account of prosociality (this volume). On this view, our primate ancestors evolved a complex sociopolitical order based on a social dominance hierarchy in multi-male/multi-female groups. A niche for hominins in which there was a high return to cooperative hunting or confrontational scavenging (O'Connell et al. 2002) was created by multiple niche parameter factors: the emergence of bipedalism in the hominin line, environmental developments which made a particular diet (of meat from large animals) fitness enhancing, and cultural innovation in the form of fire and cooking (Wrangham 2009; Wrangham and Carmody 2010). The hominin control of fire cannot be accurately dated, but may have been achieved more than 500,000 years ago (Berna et al. 2012) and was probably habitual by 300,000–400,000 years ago (Roebroeks and Villa 2011). This cultural innovation had strong effects on hominin cultural and phylogenetic evolution. Prior to the control of fire, humans almost certainly took to the trees, cliffs, or caves at night like most other primates, as a defense against predators. Because predators have a fear of fire, the control of fire permitted hominins to abandon climbing almost completely. The control of fire may thus have been a prerequisite for the transition to obligate bipedality. Wiessner adds that by controlling fire, hominins could be gathered in one place at night, thus extending social life into the night. The practice of cooking food is a related cultural innovation with broad gene–culture coevolutionary implications. Cooking may involve a central location to which the catch is transported, and the calorie-distribution phenomena typical of food sharing in nonhuman primate species could have given way to food distribution based on agreed-upon fairness norms. Collective hunting in other species does not require a fairness ethic because participants in the kill simply eat what they can secure from the carcass. However, the practice of bringing the kill to a central site for cooking is not compatible with uncoordinated sharing and eating. Meat is only one part of the story. Cooking is important in freeing time from the processing of vegetable foods which make up at least two-thirds of most hunter-gatherer diets. Cooking makes vegetables more digestible and decreases chewing time (Wrangham 2009). Importantly, cooking incentivizes the sharing of vegetables like tubers at central cooking sites because of the costs of building a fire for the small caloric return from each vegetable *in situ.* Seen this way, the control of fire and the practice of cooking are thus (some of the) cultural preconditions (niche parameters) for the emergence of morality and social organization based on normative behavior.

Conditions for the Evolution of Cooperation in Small-Scale Societies

We now turn to a discussion of the conditions necessary for the evolution of cooperation in small-scale human societies (i.e., those of up to a few thousands of individuals). The diversity of approaches to the mechanisms promoting cooperation was highlighted by Bshary and Bergmüller (2007), who identified distinct classes of criteria: from ultimate fitness benefits, to ecological and life history conditions, to specific game theoretical structures. However, the term "conditions" and "mechanisms" promoting cooperation can have different meaning, depending on the disciplinary perspective. Our approach reflects the various fields from which we originate (anthropology to evolutionary biology, primatology to economics) and encompasses different levels of description, from social to genetic. We identified three main requirements or necessary conditions which, in combination with the mechanisms described in the previous section, could produce small-scale society cooperation:

1. increasing returns to scale with group size,
2. control of defectors, and
3. cultural group selection/assortativity.

These may operate in a hierarchical fashion; cultural group selection/assortativity (and the processes therein) can solve the problem of controlling defectors, which in turn allows for increasing returns to scale. Considering a great variety of issues in both general and explicitly evolutionary collective action models (e.g., heterogeneity in resources and/or interests) reveals a range of conditions where issues like the structure of social interaction can be more important than the population size (Marwell and Oliver 1993). However, there was broad agreement with the suggestion that humans can uniquely "change the rules of the games" such that games resulting in more efficient outcomes (returns to scale) may be favored by cultural transmission.

Increasing Returns to Scale

The basic condition for the evolution of group living is that individuals do better in groups than by themselves. Thus, some kind of fitness or benefit function needs to increase with group size. It might not continue to increase for groups of arbitrary size; in fact, there could be a peak, but there needs to be a region of group sizes for which the benefit function increases.

This is not just a general case of, for example, "why do primates live in groups?" There, fitness benefits are largely derived from reduced risk of predation due to grouping. These benefits gradually level off with group size, and generally do so at fairly small group sizes. Much larger groups than about ten individuals require additional benefits. Similarly, the benefits of cooperative hunting, at least among primates, level off at relatively small group sizes. Thus, the various conventional benefits of grouping in primates or carnivores do not explain why even "small-scale" human societies can contain an order of magnitude more members. To account for this, we need to recognize new functions. Examples include: some types of big-game hunting and/or coordinated defense against predators, risk pooling through extended networks and access to their resources, and economic returns from trade and the movement of labor (Wiessner 1986); warfare and the returns of group size on aggression and defense against aggression (Turchin 2009); and the effect of group size on the sophistication of the culture that can arise and be maintained (Powell et al. 2009; Henrich 2004b; Shennan 2001).

Increasing returns to scale is a prerequisite for large-scale cooperation to evolve, but essentially all this means is that there should be some benefits to cooperation for cooperation to evolve. The hard problem in the evolution of cooperation is not whether this precondition is met or not. In this volume, Mathew et al. discuss why it is plausible to suppose that this precondition is almost always met, in most species, in various domains of activities. The hard problem is how cooperation evolves, given that exploiters will appropriate these benefits causing the cooperation to dissolve.

Control of Defectors: Overcoming the Problem of Collective Action

When groups produce public goods that benefit all group members equally, but individuals must bear the cost of producing the goods privately, the rational strategy is to free ride on the efforts of others. For cooperation to evolve, such defectors must be somehow controlled or eliminated. "Defection" can be controlled within the dyadic context and does not always require sanctioning by the group, but control of those who bully, exploit, or disrupt norms facilitating group cooperation requires responses that are sanctioned by the group. This can be accomplished by means of group selection: groups that have more cooperators will do much better than groups with few cooperators so that, despite cooperators losing to defectors within groups, the frequency of cooperators will globally increase. However, such "naked" group selection is very inefficient. Adding mechanisms for the control of defectors, such as moralistic punishment, allows cooperation to evolve under a much broader range of parameters and conditions (Boyd et al. 2003).

Social Norms and Institutions

Researchers have found that long-lasting communities which govern their common resources sustainably are ones that put substantial effort into monitoring and enforcement (e.g., Ellickson 1991; Hechter 1987; Ostrom 1990). To be effective, these norms and rules need to be well understood and accepted. These institutional arrangements start with social norms, learned effectively from infancy (Haun and Over, this volume), and the importance of norms is recurrent throughout this chapter (see also Chudek and Henrich 2011). Norms are essentially statements that apply to the appropriate behaviors for a particular context. Rules are statements with explicit consequences for what happens if the conditions are not met and can therefore be enforced by third parties; for an interdisciplinary perspective, see Hechter and Opp (2001) and Ostrom (2005). Of course, many norms are not oriented toward the control of defectors at all. Norms may be *antisocial* (Kitts 2006), advocating behavior that is harmful to the society in which the group is embedded or even dysfunctional for the very actors who invent and enforce the norm: so-called toxic work cultures provide an informal example. Understanding the content of norms is an important area of research, but here we focus mostly on an important subset of norms that either promote collective action directly or foster social organization of a society that serves as a substrate for collective action.

Human societies are organized by systems of norms and rules that we call institutions. Marriage is an example. In any given society, norms define proper behavior for husbands, wives, children, and other people who interact with the married couple as a married couple. In general, norms differ somewhat for the different roles in the institution. People, of course, do not conform perfectly to the norms attached to roles: spouses may, for example, be unfaithful. People affected by norm violations may directly sanction violators, and typically sanctions are graded (Radcliffe-Brown 1952). A first offense, especially if minor, may provoke only the mildest verbal complaint. If norm violations become habitual or serious, sanctions typically increase in severity in a graded fashion. Third parties frequently become involved at this stage. An extramarital affair may result in the termination of a marriage or even violent retribution by the relatives of the offended spouse. Formal legal institutions may intervene in a complex society. We normally think of norms and rules as making it possible to realize gains from increasing returns at a fairly large scale. Certainly, institutional arrangements (like markets) or organizations (like armies) are used to realize gains at huge scales. Think of the institutional arrangements that make modern international trade possible. However, consistent with sociological research on norms and enforcement in families and small groups (e.g., Hechter 1987), Mathew et al. argue (this volume) that we deploy the normative system to increase cooperation at quite small scales. The institutionalization of mating that we call marriage, all but universal in human societies, is an illustration. Rather than depend upon kin selection and reciprocity to manage mating unaided by culture, we engage rather elaborate institutions even in this intimate and personal sphere.

Postmarital residence norms provide an example. Societies have stated norms that concern where couples will reside after marriage. Although adherence to these norms can vary greatly, they provide the basis for certain preferred types of association and cooperation between different sorts of relatives. A few societies allow married couples to practice natolocal residence (both with their own kin group), but most involve the transfer of one or the other spouse to a new place of residence, thus providing a small increase in the returns to scale on, for example, household or reproductive labor. Further norms indicate the types of cooperation that are expected. In otherwise virilocal systems, for example, where a woman will move to live with her husband and his kin group, initial periods of uxorilocal residence with the woman's kin can require a new son-in-law to provide labor to his wife's family. That there are regularities in the evolutionary transitions of norms of residence strongly suggests that these norms have adaptive value (Fortunato and Jordan 2011).

Norm Regulation: Internalization, Rewards, and Punishments

Some members may comply with and support norms because they have internalized those norms through processes of socialization. Norms, however, are also explicitly enforced by both rewards and punishments. Explicit punishments are leveled with care because costs of losing an otherwise highly productive group member are high, as are risks of later direct retaliation by the punished (or allies of the punished), as well as resistance against the norm itself in reaction to punishment. To avoid some of these dysfunctional consequences of explicit punishment, groups may instead reward those who provide exemplary service to the group by giving them esteem, status, or social approval. If groups prefer exemplary contributors as partners in economic exchange, political alliances, or marriage, this creates models of good behavior for other members. Of course, it also implicitly punishes those who are unproductive, stingy, or noncooperative by leaving them without partners or with less desirable partners or terms of exchange. In applying more explicit punishment, groups often attempt to corral the offender back to good behavior, first by gossip, shaming, and withholding assistance (Boehm 2011; Wiessner 2005). In extreme cases, those who engage in serious norm violations may be repeatedly shunned, ostracized, or subjected to violent punishment at greater cost to the group.

It is an open question as to whether the implementation of norm regulation is qualitatively different between (a) small-scale societies in our hominin past and present-day societies, and (b) contemporary small- and large-scale societies. For example, with respect to the former, did we evolve a "new" mechanism that could be called "respect for authority"? By what means? With respect to both, what coevolutionary feedback processes have been responsible for new forms (both processes and mechanisms) of norm regulation?

Assortativity

For cooperation to evolve, cooperators must assort in some ways with other cooperators (Frank 1998; Hamilton 1971; Eshel and Cavalli-Sforza 1982). In other words, cooperators need to interact with other cooperators more frequently than by chance alone. A number of processes potentially lead to assortivity. For example, accurate recognition of cooperators, using tightly linked phenotypic characteristics—the "green beard" effect (Hamilton 1964)—would allow cooperators to interact preferentially with other cooperators, leading directly to assortativity. Limited dispersal of offspring leads to assortativity in space and the evolution of cooperation by kin selection. Kin recognition allows for the same even if offspring disperse broadly. Cultural transmission biases (Richerson and Christiansen, this volume) can do the same. For example, conformism (i.e., adopting the cultural trait possessed by the largest number of individuals) will result in some groups consisting only of cooperators and others of noncooperators.

Relevant Regularities in the Dynamics of Assortativity

Given the crucial role of assortativity, any pervasive features of the dynamics of sorting and mixing in social interaction networks may prove consequential for the evolution of cooperation. Research across many different kinds of networks has revealed that the following two regularities are extremely pervasive:

1. Social interaction partners tend to be disproportionately similar to one another, a pattern called assortative mixing or homophily (Kandel 1978; McPherson et al. 2001).
2. Partners of partners tend also to be partners, a phenomenon called transitivity or triad closure (Holland and Leinhardt 1970; Rapoport 1957). For example, if A and B are allies, and B and C are allies, then A and C tend also to be allies.

These two regularities jointly produce clusters of culturally similar individuals with high local network closure. By network closure, we mean that actors within a cluster interact with each other more than outsiders do; as a consequence, social interaction between any two cluster members is observable to third-party cluster members that are tied to both of the interaction partners. Clustering of culturally similar individuals with high local network closure thus facilitates cooperation directly, as well as development and maintenance of norms. For example, assortative mixing and network closure lead to greater agreement and clarity for the development of norms as well as greater visibility which leads to more efficient enforcement of norms by third parties (Hechter 1987; Coleman 1990).

As noted earlier, there is evidence for behavioral dispositions leading to patterns of assortativity and closure, and evidence of preferences for homophily (Haun and Over, this volume). Many researchers have inferred a behavioral predisposition toward network closure from structural balance theory (Cartwright and Harary 1956), which posits that unbalanced triads (where A and B are friends, B and C are friends, but A and C dislike each other) are aversive and thus transient, and so tend to resolve into balanced triads (e.g., where A, B, and C are all friends, or A and B are mutual friends but both enemies to C). This pattern can yield homogeneous clusters as well as division into mutually antagonistic factions.

Although assortative mixing and network closure are pervasive and widely believed to follow from behavioral dispositions, recent research has shown that either homophily or triad closure may be largely a byproduct of the other; both may result from features of the environment (e.g., physical space, event timing) or simply from heterogeneity in the baseline tendency toward sociality (Goodreau et al. 2009). Assortative mixing may also result from social influence among network neighbors. Further research (particularly experiments) is needed to elucidate the underlying social dynamics and how these play out in different social and cultural contexts.

Small-Scale Society Cooperation in Human Evolution: Inspiration from Darwin

Darwin (1871) argued that the evolution of human cooperation evolved in two phases. In the "primordial" stage, some stretch of time in the Pleistocene in modern terms, group selection on tribal-scale variation favored the evolution of "social instincts" such as sympathy and patriotism. Tribes which had such prosocial predispositions to a higher degree would prevail in competition with tribes who had them to a lesser degree. By some time deep in the past, all humans came to have more or less the same prosocial "instincts."

After this primordial time, the prosocial dispositions came to act as forces in cultural evolution. As Darwin put it, the "advance of civilization" (in the Holocene in modern terms) depended not only on ongoing natural selection at tribal or larger scales but on advances in laws and customs guided by sympathy and patriotism favoring superior norms and institutional arrangements. Innovations by moral leaders, and the diffusion of these innovations by other moral leaders, aided by the pressure of public opinion, have become the main motors of contemporary institutional evolution. Darwin was quite aware that patriotism could trump sympathy and lead to the evolution of such institutions as slavery. Richerson and Boyd (2005) and Bowles and Gintis (2011) used contemporary gene–culture coevolution models to modernize Darwin's two stage idea, albeit in rather different ways. Selection—either directly on genetic variation (Bowles and Gintis 2011) or indirectly via culturally mediated social selection on genes within groups (Richerson and Boyd 2005)—remodeled ape/hominid social psychology to be much more prosocial during the Pleistocene. In Bowles and Gintis's model, culturally mediated reproductive leveling allows relatively weak group selection for "parochial altruism" to trump within-group selection for selfish behavior. In Richerson and Boyd's "tribal social instincts hypothesis," natural selection acts on cultural rather than genetic variation to favor primitive prosocial norms and institutions (Boyd and Richerson 1985). In both proposals, the initial prosocial norms and institutional arrangements exert social selection which may strengthen genetic predispositions for in-group cooperation and act to guide further institutional innovation and evolution.

From these two models issues arise in considering the necessary conditions for the evolution of cooperation in small-scale societies: the importance of cooperative breeding, and debates about coordination and cooperation.

Cooperation and Coordination

Cooperative breeding has been hypothesized to be foundational for the evolution of small-scale societies and can be seen as one mechanism to increase returns to scale. Human infants are relatively helpless and our juvenile period is long. Our large brains are energy and protein hungry. Burkart, Hrdy and van Schaik (2009) argue that infants cannot be successfully raised by human mothers in the manner of the other apes. Even with less-dependent young, the great apes have very long interbirth intervals and are barely viable demographically (see also Hrdy 2009). In humans, the contributions of pre- and post-reproductive women and adult men to the care and feeding of children can shorten interbirth intervals to an unprecedented extent. Effectively this meant that humans can achieve robust population growth rates, despite having infants that are so costly to nurture that unaided mothers could not raise them alone. Burkart et al. suggest that capturing the increasing returns to scale in infant quality may have been the foundational step in the human cooperative syndrome. Large brains and a long period of juvenile dependence seem to be necessary to support the acquisition of a large, complex cultural repertoire. This repertoire includes both foraging and processing skills and our norms-and-rules social systems and allows us to flexibly exploit myriad activities which exhibit increasing returns to scale. Indeed the creation and maintenance of complex culture itself has increasing returns to scale (Henrich 2004b; Kline and Boyd 2010; Powell et al. 2009; Shennan 2001). Beyond the returns to scale, Hrdy (2009) has argued that through the development of "other-regarding impulses," cooperative breeding set the stage for advanced social learning and cumulative culture, teaching, and language to evolve. Importantly, cooperative child rearing had knock-on effects on the cognitive and emotional development of infants, who looked not only to their own parents but also to alloparents to get the costly care they needed. In effect, babies (and the adults those babies grew up to be) were the products of selection pressures that favored social communication, perspective taking and mutual tolerance—even toward others who might not be close kin.

Hrdy's view is that attention to the *novel* conditions of human development could inform our understanding of human sociality. Others, however, have proposed that our unique levels of cooperation may have roots in simple "coordination," as in the "stag-hunt" game (Tennie et al. 2009). This is in contrast to general public goods games that have free-rider problems: we all gain if we all cooperate, but individuals can benefit from defection if others cooperate; therefore, the outcome for selfish rational actors is that nobody cooperates. However, often the interdependency assumption/stag-hunt payoff assumptions do not match real life. If human warfare were actually like that, there would not be a problem of cowardice and desertions on the battlefield. Each person should have sufficient incentives to contribute if their marginal contribution is what ensures victory. Yet, cowards and deserters are a problem in even prestate raiding, and various forms of sanctions are deployed to motivate warriors to fight (Mathew and Boyd 2011). Moreover, other animals are able to solve various coordination problems like herding, mating, etc., but this has not led to much cooperation. This would be puzzling if being able to engage in games with interdependency-type payoffs was indeed the key factor in making humans cooperative.

At this point we are armed with some idea of the behavioral predispositions that are necessary for the evolution of small-scale sociality and cooperation, as well as some idea of the importance of considering niche parameters. The cultural evolutionary perspective (Richerson and Christiansen, this volume) then allows us to hypothesize how those features can be exploited in the transition from small- to large-scale societies. As emphasized earlier, humans are able to acquire vast amounts of nongenetically encoded behaviors and/or information during their life span. Hence, both genetic and nongenetic change is likely to have affected the emergence of large-scale sociality. Next we discuss the main evolutionary processes or "engines" behind such changes.

Evolutionary Processes Relevant to Understanding Human Sociality

Types of Learning and "Engines of Change"

Individual learning is a generic term for the cognitive processes that allow individuals to acquire novel behaviors and/or select novel actions among alternatives during their life span in the absence of interactions with conspecifics (Boyd and Richerson 1985; Rogers 1988; Dugatkin 2003). It comprises processes such as trial-and-error learning, inference, induction, and deduction, or insight. Individual learning is the generator of novel behaviors. On the other hand, social learning is the generic term for the cognitive processes underlying the acquisition of information when interacting with conspecifics (Boyd and Richerson 1985; Rogers 1988; Dugatkin 2003; Enquist et al. 2007; Cavalli-Sforza and Feldman 1981). Social learning involves processes such as imitation, copying, teaching, and local enhancement. It is the engine of transfer of behavior between individuals in a population.

As individual and social learning tend to occur on a local scale (between individuals within groups), different groups of individuals are likely to innovate and express different combinations of trait values. If different combinations of norms/institutions are associated with differential reproduction and/or payoffs to individuals, beneficial trait combinations may spread in the population. Thus the interaction between individual and social learning causes changes in nongenetically inherited behaviors during an individual's life span, and leads to potential changes in the population-wide distribution of behavior(s). These changes are driven by two factors: cultural group selection and endogenous social change.

Cultural Group Selection

Cultural group selection refers to a competitive advantage for a group as a whole that arises from within-group norms, practices, etc. Cultural group selection can favor group-beneficial outcomes on very large scales, including among thousands of genetically unrelated individuals (Henrich 2004a; Boyd and Richerson 1985). Thus it constitutes a crucial process in understanding how human societies went from relatively egalitarian foraging bands to complex states comprising millions of people. Although many features accompany such a rise, more complex societies generally manage cooperation at a larger scale and/or more efficiently than less complex ones. To account for this, we need an evolutionary process that can favor norms and institutions that increase the scale of cooperation, and which create more efficient outcomes at this new scale. Cultural group selection is such a process.

Selection creates adaptive behavior at any level upon which it operates, and thus group selection can explain group-functional outcomes. Conversely, selection at a lower level does not lead to functional outcomes at a higher level. Genetic group selection cannot explain cooperation observed in large-scale human societies, and most animal and human societies do not have sufficient between-group genetic variation for it to be an important force. However, because humans acquire locally adaptive behavior through social learning, there is a great degree of between-group cultural variation across societies (Bell et al. 2009), thus making cultural group selection a much more plausible mechanism for humans than genetic group selection in humans and other animals.

Empirical studies also support the view that cultural group selection has played a role in shaping human societies. Soltis, Boyd, and Richerson (1995) show that group functional behaviors were able to spread through cultural group selection on a timescale of a few hundred years in New Guinea. Mathew and Boyd (2011) demonstrate that norms governing warfare among Turkana pastoralists in East Africa generate group-beneficial outcomes at the scale of cultural variation. Turchin (2006) shows that empires emerged at the point where there is maximal between-group cultural variation, such as along the boundaries that separated herders and agriculturalists.

Competition between cultural groups will lead to larger and more complex societies with more efficient social institutions to manage production and warfare. Between-group competition can occur through a number of means. One is through *warfare*, as exemplified in the Nuer expansion into Dinka territory (Kelly 1985), and another is through differential *population growth*, as when agriculturalists outcompete hunter-gatherers in reproduction. Additional means for between-group competition include *immigration* into perceived "successful" societies (e.g., migration into the United States), *adopting the social institutions* of successful groups, as exemplified by Enga bachelor cults that were widely borrowed from innovating clans (Wiessner and Tumu 1998), or the spread of democracy in the modern world.

Endogenous Cultural Change

Cultural change can also arise endogenously, from within-group processes that generate variation. Endogenous change can result from prosocial preferences, such as a regard for equitable, or fair, or parochial outcomes that have resulted from a longer history of cultural group selection. Such preferences—combined with abilities for persuasion, leadership, or deliberation—can allow societies to adopt norms that are consistent with these preferences. Democracies, or jury systems, may be the result of preferences shaped by cultural group selection (like fairness and peer sanctioning, respectively). It is important to note that on longer timescales, these institutions will persist only if they also lead to groups that adopt these social arrangements to fare better than other groups. However, on shorter timescales, some of the change that we see in human societies can be the result of people tinkering with their social institutions in accordance with their preferences and their contexts, rather than due to between-group selection itself. Much social/cultural anthropology is concerned with the diversity of these creative processes and their outcomes in a particular cultural milieu, and it is here that cultural evolution scholars can engage with other anthropologists on topics of agency and innovation in creating behavioral and cultural variation. However, as change comes about endogenously, such processes may produce differentially "channeled" or biased types of innovations so that we may see only a subset of all possible types of cultural behaviors and societies (see the "design space" questions in language evolution, Dediu et al. this volume).

Genetic and Cultural Coevolutionary Circuits

Learning rules and/or preferences that support cultural evolution and cultural group selection may themselves evolve and be influenced by genetic evolution. The full coevolutionary feedback between nongenetically inherited phenotypes—including memes, variants, traits, norms, and institutions—and the cognitive machinery which supports them is gene–culture coevolution, or dual inheritance (Cavalli-Sforza and Feldman 1981; Boyd and Richerson 1985). The selection pressure on genes involved in this coevolutionary circuit must be consistent with the principles of natural selection. These can be framed in terms of selection at the individual level by way of inclusive fitness costs and benefits (Frank 1998; Gardner et al. 2011). The equivalence between group- and individual-level selection perspectives is true for any phenotype, regardless of whether the source of variation under study is genetic, cultural, or a combination of both (Frank 1998). As such, any cultural group selection process can also be expressed in terms of selection at the individual level and could be framed in terms of cultural inclusive fitness costs and benefits (André and Morin 2011).

Predictions can arise from considering these different sources and engines of evolutionary change. One implication is that the rates and types of change will differ. For example, we can ask where and when in the historical record we should see large-scale societies arise. With endogenous social change we might expect multiple independent origins of cultural features, each differing somewhat, whereas cultural group selection might be expected to produce spread or diffusion of the same basic phenomena (perhaps with graded differences predictable from, e.g., geography or ecology). With endogenous social change we might see small incremental steps, whereas cultural group selection might produce large changes. To consider how a research program might approach these predictions empirically, we need to have some idea of the "target" state of what can be variously termed social complexity, or (types of) large-scale society. Next we delineate some defining characteristics.

Social Complexity: What Is the "Phenotype" of Large-Scale Societies?

"Social complexity" is a fairly slippery concept with no standard definition and with historically problematic implications for many anthropologists and archaeologists (e.g., Yoffee 1993; Flannery 1999). Demographers, psychologists, historians, and biologists, as well as complexity theorists, may have different phenomena in mind when considering social complexity. The central issue is whether social complexity can be represented by a single principal component (plus "noise"), or whether the notion is better served by multidimensional structures, and, if so, what evidence should be considered in such a description. A well-known multidimensional operationalization of social complexity or cultural complexity is the one provided by Murdock and Provost (1973), which is based on the widely used Standard Cross-Cultural Sample (Murdock and White 1969). The most convincing single measure of social complexity is to use the largest settlement size as proposed by Naroll (1956) and repeated by Chick (1997). A new approach, discussed extensively at this Strüngmann Forum, is one advanced by Turchin, François, and Whitehouse, who are developing a dynamic historical database toward this end (for details, see http://www.cam.ox.ac.uk/ritual/). Instead of trying to define a single metric for measuring social complexity, this practical, empirically based approach uses a number of measures that address different aspect of social complexity. By coding these aspects for a variety of past and present societies, the resulting database can be analyzed with multivariate statistical tools, such as principal component analysis. Many of these variables also act as processes which stabilize social complexity. Here we highlight those measureable features that can index social complexity.

A Multivariate Approach to Social Complexity

We begin with the demographic basics of *scale*. This includes the population size of an independent unit or polity, the territorial extent of the polity, and the population and density of the largest settlement (often, but not necessarily, cities). Populations in large-scale societies have *hierarchy* by which we can identify the jurisdictional levels in administration: the segmentary, modular, or nested structures of organizations. There are within-sector hierarchy structures, such as found in military, bureaucratic, legal, and religious orders, and these involve professional officials, such as military leaders, priests, and judges, whose presence is often used to define a state. *Economic extent* and *specialization* are well developed in large-scale societies; the total number of novel professions extends far beyond the division of labor seen in small-scale societies, which is based on sex, age, and expertise. The degree of specialization and/or exclusivity (i.e., who may practice certain professions) is thus more marked. In addition, there is a greater extent, and often complexity, to the trade networks in large-scale societies. From these three factors emerges *institutional complexity*, composed of both hierarchical (vertical) complexity and the orthogonal feature of horizontal complexity.

Large-scale societies tend to support more and different types of information, especially *cultural information*. Much of this may be "stored" culture in the form of literature, art, and other material information, usually in excess of what can be maintained in a small-scale group. In addition, there is usually *monumental culture* in the form of buildings and architecture, and large public spaces (often dedicated and/or built) for ritual, performance, economics or politics. Some forms of *religion* and *religious practices* (discussed in Bulbulia et al., this volume) are roughly identifiable with large-scale societies; in addition, religion and religious beliefs themselves may have been key causative elements in the evolution of large-scale sociality. Niche parameters that are certain to have had massive feedback effects in the evolution of large-scale sociality are what we term *management technologies*: technologies for coercion, coordination, and production. These include systems of tribute and taxation; environmental modifications such as permanent roads, outposts and observational stations; recording technologies such as writing and accounting; and weapons for large-scale violence.

Some further elements constitute the "dark side" of social complexity. In particular, *inequality* is rife in large-scale societies. Inequality can be economic, and therefore measurable in, for example, the ratios of the largest private fortune to the median. It can also be structural and characterized by features such as human sacrifice, slavery, castes, legal distinctions such as aristocracy, and the deification of rulers. *Urbanization* itself is complex and variable with respect to impacts on human well-being, but there is good cause to see cities, particularly those before the nineteenth century, as "death traps": preindustrial cities sucked in populations, acting as a sink, and went through boom or bust extinctions. Why would we willingly live in a sick, smelly crowd of strangers? Finally, it has been argued that *too much social complexity itself* leads to higher costs of maintaining its structure and can lead to collapses (Tainter 1988). Although such a general statement is debatable, some elements of social complexity can challenge the system attributes that maintain the stability of small-scale societies. Increasing scale affects the ability to monitor behavior and derive information to maintain reputations. Complexity may lead to a loss of local stability of equilibria in dynamical systems (Mayr 1970). Increasing interactions are between strangers and incomplete information that may make the system vulnerable to defectors.

The Transition from Small-Scale Societies to Large-Scale Societies

Increasing Returns to Scale

The major evolutionary transition from small-scale to large-scale societies involved an increase in social scale by five or more orders of magnitude (from hundreds to a few thousands, up to hundreds of millions and more; see Turchin this volume). As discussed earlier in this chapter, a necessary condition for enabling such an evolutionary shift is that the increasing returns to scale (IRTS) function must reach a peak at much higher population numbers, or at least need to increase for a region of group sizes that includes tens and hundreds of millions. What processes can account for such an enormous expansion of increasing returns to scale? Anthropologists, economists, political scientists, and sociologists have contemplated a range of explanations, roughly divided into: (a) warfare, (b) economic efficiency, (c) information-processing capacity, and (d) demographic diversity. The first of these, *warfare*, is easy to understand. Larger societies can mobilize more resources and field larger armies than smaller societies. An interesting feature of this explanation is that it suggests that there is no maximum in the IRTS curve: it continues to increase without limit (a population of a trillion is better than a hundred billion, but ten trillion is even better). This does not mean that we will see societies of ten trillion people any time soon; other processes limit such runaway growth, most obviously the problems with maintaining fighting forces of massive size.

Economic efficiency invokes a variety of mechanisms. For smaller-scale agrarian (or even hunter-gatherer) societies, it has been proposed that they can greatly benefit from extended social networks that allow buffering against variable environments or access to novel resources (Hruschka 2010). For larger-scale societies, including those with modern economies, economists generally agree that there are substantial returns on the scale, resulting from the division of labor between different regions and groups. This idea dates back at least to Adam Smith, more recently developed by Paul Krugman and others (Fujita et al. 1999; Krugman 1991). An *information-processing* hypothesis suggests that the ability of societies to generate new knowledge is not simply a linear, but an accelerating function of its size. Some models (Henrich 2004b; Powell et al. 2009) suggest that there are nonlinearities, because when the numbers or population density of interacting human groups fall below a threshold, such groups start losing technology, rather than cumulating it. Such models should be augmented by accounting for not just the evolution and effects of endogenously produced behaviors, but the niche-constructive effects of material technologies and learning environments as well (Laland et al. 2011; Powell et al. 2009; Sterelny 2012; Mesoudi et al., this volume). When problem solving acts to structure knowledge (or "chunks" it, in psychological terms), not only does new knowledge increase the information-processing capacity of the group, the structuring itself also affords greater capacity for the cognition of new problems. New problems can lead to new knowledge in which more people will have participated in the creation or processing of knowledge or skills, through, for example, phenomena such as formal teaching or semiformal-structured learning environments (Sterelny 2012). Continued cycling of knowledge aggregation can then have positive feedback effects on information-processing group size.

Sociological research on demographic diversity in networks, groups, and organizations reveals that assortative mixing leads social interaction to transpire within culturally similar relationships, a phenomenon called *sociodemographic clustering* (Goodreau et al. 2009). Increasing the size and diversity of the population (subject to these local mixing dynamics) leads to greater cultural homogeneity at the level of social interaction, even as the overall population grows more diverse. In the transition to large-scale societies, for example, this social organization leads exchange and other interaction to occur within dyads that are more culturally similar, while neighborhoods, groups, and formal organizations also become more internally homogeneous in culture. Cultural diversity becomes increasingly compartmentalized as scale increases. If social dilemmas of various kinds (opportunities for individually costly and mutually beneficial cooperation) are faced by people who are more culturally related to each other, this structuring of interaction will enhance cooperation at the level where social interaction typically occurs. Groups are comprised of increasingly compatible members—members who are also *relatively* similar to one another (*vis a vis* neighboring groups). Of course, as increasing scale leads to more culturally homogenous relationships and groups, it also leads to cultural differences between groups. Thus, increased cooperation at a local level may result in tension or conflict at a higher level.

From Small to Large: Which Mechanisms Maintain Large-Scale Sociality?

One way to understand small- to large-scale transitions, of which there have been many in human history, is to ask which of the behavioral mechanisms discussed above, in interaction with the niche parameters and contingent historical facts, were factors in the maintenance of large-scale societies? They may have inhibited (–), were irrelevant (○), facilitated (+) or were crucial (++) in these pathways (see Table 6.2). By asking which are necessary or not, we generate a set of testable hypotheses that can then be compared (in the future!) against the available ethnographic, archaeological, and historical data. One could also consider the transitions from small-scale societies to various types of large-scale societies, such as acephalous tribes, chiefdoms, small states, empires, and modern industrialized states. These pathways will be context specific. For example, in chiefly societies and royalist states, elite marriage alliances may be incredibly important (such as in the case of dynasties), but in modern industrialized societies marriage is less crucial. Among several acephalous pastoral societies of East Africa, age sets crosscut other social groupings of the society and enable large-scale social organization without political centralization (Baxter 1978). It is also revealing to ask what can be removed from large-scale societies today without causing them to collapse; this provides an excellent tool for thinking through case studies. Examples like the Turkana, the Nuer (Kelly 1985), and the Comanche (Kavanagh 1996) illustrate how even quite rudimentary political institutions can allow societies of considerable scale to emerge. These societies were able to coordinate warfare and enforce internal peace among tens to hundreds of thousands of people without hierarchical leadership.

A crucial point in our debate, and for future research, was whether humans need extra (psychological) mechanisms to go from small-scale societies to

Table 6.2 Mechanisms from Table 6.1, identified as to their role in the maintenance of large-scale societies. Key: inhibited (–), were irrelevant (○), facilitated (+), or were crucial (++).

Mechanism	Role in maintenance of large-scale societies
Kin recognition, kin bias, nepotism	+ elites ○ commoners
Respect of territory, property, mates	+
Structured social interaction (assortativity)	++
"Reverse dominance hierarchy"	– or + depending on functional organization of society
Direct reciprocity: Who did what to me?	○
Coalition formation, socially organized aggression	+ for midlevel complexity – can degrade social organization (e.g., revolution, trade unions)
Cooperative breeding	○
Marriage, pair bonding	+ elites ○ commoners
Multilocal residence: flexibility of male/female dispersal	○
Multilocal/multilevel ties outside the group	++
Leadership by persuasion, authority, prosocial leadership, or prestige	++
Moralistic punishment, moralistic rewards	++
Reputation and gossip	+
Norm psychology: norm adherence, norm internalization	++
Lethal force at a distance	++
Cumulative culture, cultural variation, social-learning biases	++
Language	++
Symbolic behavior: expressive, and as ethnic marker	++
Predisposition to impose categorical distinctions onto continuous cultural differences, leading to group boundaries and identities	+
Predisposition for collective ritual and synchronicity	+

large-scale societies. Two proposals on the table are: (a) the religious "bundle," including mechanisms such as agency detection, sacred values, etc. (see Bulbulia et al., this volume), and (b) respect for authority. Although respect for authority might be quite highly heritable in a gene–culture coevolutionary sense, there is no evidence that such an "authoritarian mind" is fixated across our species. For example, while hunter-gatherer groups might have respect for age, and/or respect for knowledge, there does not appear to be universal respect for *command.* These require further conversation and detailed proposals for hypothesis testing. Richerson and Boyd (1999) review data that suggest that even in modern mass armies, where this is a highly organized hierarchical chain of command, combat efficiency is highest in those armies that use prestige as a tool for leadership and least in those that depend more heavily on coercion; see also Turchin's (2006:8–9) discussion of the fluctuations of *asabiya* (roughly a society's spirit of common purpose) in agrarian states. On this view, the same counterdominance impulses that resulted in highly egalitarian small-scale societies remain an important check on elite expropriation, which, when unchecked, can destroy a society's *asabiya.* It is unlikely, however, that any faint population-level biases in genes, such as postulated for the learning of tone languages by Dediu and Ladd (2007), would be important here: they would be utterly swamped by the effects of cumulative cultural evolution of population-level differences on developmental environments.

Drivers of Social Complexity

One useful way to review these potential mechanisms, and to develop a comparative perspective on their relative importance, is to conceive of larger complexes in which they sit as drivers of social complexity. The chapters throughout this volume discuss a number of such complexes, such as religion (Slingerland et al. and Bulbulia et al.), technologies (Mesoudi et al. and Boyd et al.), and warfare (Turchin). Here we consider homogenization and incorporation, and the management technologies of large populations.

Homogenization/Incorporation

A key challenge of administering large-scale societies is coordinating their multiple subunits, whether these are provinces, settlements, cities, or tribes. One factor which can facilitate the emergence and spread of large-scale societies is the prior existence of a set of social units that already share a common language, culture, or administrative structure. For example, the relative homogeneity of Greek city-states may have facilitated the higher-level aggregration of Greek leagues and the early expansion of the Macedonian Empire (Malkin 2011). In other cases, such homogenous administrative units must be reproduced to extend a territory, as was the case with the construction of Roman cities during imperial expansion (Boatwright 2000) or European colonial imposition and formalization of tribal chiefs in Africa to serve as points of control for long-distance administration (Leeson 2005).

If the erosion of strict boundaries between units allows the transition between small-scale societies and large-scale societies, what mechanisms are co-opted to make boundaries porous; that is, how can this homogenization take place? It could be that any dimension of similarity taps into our preexisting psychology for homophily and ostracism aversion (Haun and Over, this volume). However, some are differentially effective, and some candidates seem to warrant special attention, such as warfare, or a common enemy, and religion, which can expand identity through fictive kinship (e.g., my brothers-and sisters-in-arms). In some cases, religious identity becomes more important than ethnic identity, and this "super-effectiveness" of religion is of note because religions explicitly contain norms and rules, characterize the nature of social bonds, and provide social support.

Furthermore, such explicitness can help homogenize when, as in more complex societies, relationships are increasingly defined by position instead of personal relations. Named positions such as a guard, an accountant, or a chairperson can be derived by appointment, election, or other mechanisms. An institutional structure based on positions requires collective choice mechanisms, such as voting procedures, at different levels with clearly defined positions (Ostrom 2005). In more complex societies, formal rules start to define who has access to the public goods of society (i.e., which groups have access and how group membership is defined). For example, following warfare, will subjugated groups be absorbed into the victorious group? Some rights need to be given to those people to make them active members of society.

An open question is then: What are the consequences of the rights given to the "losers"? Denying them access to public goods may be ineffective for the stability of society. How frequent are situations where there are true conferments of rights, as opposed to situations where subjugated people form coalitions to agitate for rights, or rebel? Many characteristics of large-scale societies (discussed above) are what is in essence population substructure (hierarchy, division of labor, specialization, etc.), and this then begs the question of whether substructure can ever be anything but unequal. We lack space to develop these notions here, but there are empirical implications to this question that are relevant to the evolution of the Axial religions and are explored in detail by Turchin (this volume).

Technologies of Coordination/Coercion and Management of Large Populations

As polities comprise larger populations over ever-wider territories, new technologies play an important role in managing people and resources: Engineered roads facilitate communication, trade, and faster deployment of military power. Strategic administrative settlements and ritual centers permit more direct control of far-flung populations. Improved military technology can inflict lethal force on larger groups. External representations, such as clay tablets in southwest Asia or knotted *khipu* strings in the Inca Empire, permit impersonal accounting for finance and trade (Luttwak 1976; Headrick 1981; Basu et al. 2009). Physical infrastructure also plays an important role in storing, protecting, and transferring surplus production. Surplus production has long been suggested by archaeologists to be exceptionally important in the transition from small-scale to large-scale societies (Earle 1991) and is one obvious way in which human niche construction can change the adaptive landscape of cultural evolution. Once a surplus storable energy source arises, payoffs for phenotypes can change, and the forces of cultural evolution are liable to act in different ways. Not all surplus has the effect of increasing the human population directly: some can be used nonnutritionally to "do culture" that can alter the niche in ways that further persist over generations and become selective parts of the environment, for example, to build monuments. A further suggestion is that surplus not only allows large societies to be maintained but allows elites to control them.

At what social and geographical scales do such technologies become necessary for binding polities together? As discussed earlier, human societies on the scale of hundreds of thousands of individuals can organize without much requirement for such physical capital. Niche construction models that incorporate multigenerational investment in roads, fortifications, long-term settlements, storage centers, weaponry stores, and other infrastructures should help us understand the conditions under which long-range feedback between built environments and social organization plays a role in the emergence of large-scale societies.

Case Study: Enga of New Guinea

Using Wiessner's long-term fieldwork with the Enga of New Guinea (see Wiessner and Tumu 1998), we discussed the importance of identifying the "package" of processes/mechanisms that were (and were not) important in the transition from small-scale societies to large-scale societies. Trade, warfare, and ritual were identified and were found to encompass a host of the elements discussed in this chapter. In the Enga, both cultural group selection and endogenous cultural change were engines of change and creative innovation.

The Enga of Papua New Guinea are a highland horticultural population who formerly lived as hunter-gatherers and subsistence horticulturalists with clans of some 500 people. Warfare served to split up groups that had become too large to cooperate, and long-distance trade formed via marriage ties. Some 350 years ago, the South American sweet potato was introduced along local trade routes, releasing constraints on production and allowing the Enga to produce a substantial surplus for the first time in their history in the form of pigs. First contact with Europeans occurred some 70 years ago.

After the arrival of the sweet potato, large-scale wars redistributed the Enga over the landscape as groups sought to take advantage of the new crop.

Postwar population movements greatly disrupted the flow of trade, cooperation, and exchange; the Enga sought to bring order to chaos through the development of large ceremonial exchange systems. By first contact, one of these systems had grown to incorporate some 40,000 people and the exchange of over 100,000 pigs per four-year ritual cycle. To engineer these large systems of ceremonial exchange, Big Men initiated or imported bachelor cults to create uniformity in the norms and values regulating courtship and marriage, so that networks could expand by intermarriage between clans. Similarly, they manipulated ancestral cults to elicit the cooperation of several tribes and provide a forum for planning cycles of ceremonial exchange. Feasting was a key component of all events. Warfare followed by peacemaking served to recreate balance of power in the face of insult or injury so that exchange could flow between clans. Big Men who managed the large cults and ceremonial exchange systems gained great prestige; the public looked to sons of Big Men to replace their fathers so that ceremonial events, which provided benefits to most, would not be disrupted. Big Men drew status from the management of wealth, enjoyed the privilege of polygyny, and controlled the information necessary to arrange ceremonial exchange, but they did not accumulate wealth.

How, then, do norms actually change? For example, when a Big Man coopts a "successful cult" specialist from another group, the norms of the first group are altered by within-group processes, and then acted on by cultural group selection. Other examples are apparent in Enga "dehumanizing" and peacemaking sessions. It appears that homogenizing the preexisting networks in the Enga allowed for subsequent expansion and the development of hierarchy. The Enga case also requires us to consider an historically contingent catalyst of a change to large-scale networks, if we consider the introduction of the sweet potato as an exogenous factor that drove the evolution of the system. Thus there can be multiple and contingent layers of causality for each case where small-scale societies have transformed into larger polities. Careful comparative work based on detailed ethnohistorical description can begin to disentangle these questions.

Cultural Mesoevolution: Bridging Individuals, Populations, and Regions

Empirical Studies Will Drive the Field Forward

In the field of cultural evolution, we are not short on theory, but the anthropological and historical literature is a vastly underutilized resource awaiting our renewed attention. What we need is serious coordinated efforts to connect theory and data that neither do damage to ethnographic detail nor become sterile abstractions for beautiful models. What can we learn from cases like the Enga, from those reported by Turchin and Mathew et al. (both this volume), and from the key works on the evolution of societies worldwide (e.g., Keech McIntosh 2005; Vansina 1990; Kirch 1984)? Case studies allow our investigations to become concrete and stimulate potential focus areas for future research, and a positive outcome of this Forum was the suggested set of elements (see Tables 6.1 and 6.2) to formalize both case-study and comparative research. From there, the challenge will be to generalize patterns to the "broad sweep of history." For example, how general is it to have a large network before a hierarchy? Do we need extra mechanisms, or is it just the "old" small-scale society mechanisms in new contexts and combinations that allow the transition to large-scale societies?

Regularities in Process

How regular are the processes that take us from small-scale societies to large-scale societies? Are the same mechanisms acting or do we need new ones? Do we get emergent properties when old mechanisms interact together, or with new facts such as surplus, increased population size, or warfare? Are there regularities of change? The model of the changing adaptive landscape may be extremely useful here, and there may also be parallels between complexification in social change and the other major transitions in evolution (Maynard Smith and Szathmáry 1995), such as the integration and co-opting of preexisting (functional) entities into larger ones, as in the evolution of multicellularity. These questions have empirical answers and can be addressed in a number of ways (e.g., using the social complexity database mentioned earlier).

Thus far cultural evolution research has spanned two broad areas. Cultural microevolution in the main has adapted theory and modeling approaches from population genetics to uncover the dynamics of cultural transmission between individuals within populations; these dynamics are then increasingly tested empirically using frameworks to study individual behavior from within psychology and cognitive science (for a review, see Mesoudi 2011a). Cultural macroevolution has focused on the population level to explain why norms differ between groups, using the analogy of testing species differences from biology. In this paradigm, predictions are tested using comparative phylogenetic methods that control for the effects of shared ancestry (Galton's Problem) on ethnographic, linguistic, ecological, and archaeological data. A recent relevant example is work by Currie et al. (2010a), which showed regularities in the sequence of political complexity in Austronesian societies. Implemented worldwide or on a region-by-region basis, these approaches can be informative about any regular tendencies in the processes of change, and have been successfully employed to answer questions in the domain of language (see Gray as well as Dediu et al., both this volume), technological change (see papers in Lipo et al. 2006) as well as aspects of social structure such as marriage, residence, and wealth transfers (Fortunato et al. 2006; Holden and Mace 2003; Jordan et al. 2009).

The desirability of bridging these two levels has become apparent in recent years. Not only do both micro and macro approaches suffer from a degree of abstraction that (while necessary) renders them unpalatable to social/cultural anthropologists working in the field, but for a solid and unique theory of cultural evolution to emerge, it is necessary to have mutually reinforcing research programs across the biological, social, historical, and behavioral sciences. A unified approach to social complexity across disciplinary boundaries will be difficult, but "cultural mesoevolution" should consist of work that brings together different fields to carry out in-depth case studies in, for example, language families or cultural regions where the emic status of phenomena permits systematic and quanitative cross-cultural comparisons. The emergence, maintenance, and transmission of norms and their contents would be the target of study, at levels ranging from long-term historical and ecological factors, to the population-level interactions of groups that emerge through cultural group selection, to the behavior of individuals and groups within populations and the endogenous mechanisms of change therein, as well as their development in children (for the latter, see also discussions in Lieven et al., this volume). In addition, there is scope for gene–culture coevolutionary approaches in this mesoevolutionary perspective, for example, where there are subtle population-level genetic biases (Dediu and Ladd 2007). To return to our example of "respect for authority," it is probably true that basic norm psychology mechanisms have produced a human-wide behavioral predisposition toward respect for those in positions of command, with some cultural variation in content but remarkably consistent outcomes. However, enough time may have passed for the Baldwin effect to be acting on any small underlying genetic differences that strengthen any advantage to these behaviors, perhaps at alleles with putative cognitive and behavioral effects such as the D4 dopamine receptors (Chen 1999; Ding et al. 2002). Finally, as Laland et al. (2011) point out, depending on the level and viewpoint at which we conduct our research program, mechanisms at one point may have been outcomes or processes at another point, and debates about the ultimate/proximate dichotomy can become sterile when we speak at cross purposes. Working at a data-rich but still comparative mesoevolutionary level may help us be clearer in this respect. Ideally we would like to extract and generate patterns for ethnographic analysis from the bottom up, rather than impose external categories on ethnographic data. Thus, we need to develop ways to synthesize across individual and demographic data from, for example, psychology, sociology, and behavioral ecology to arrive at norm abstractions needed to model cultural evolution on the macro scale.

A Wish List for Future Research

Throughout this chapter we have discussed a number of specific questions regarding the transitions between small- and large-scale *societies*, and what that can tell us about the transition from small- to large-scale *sociality*. Beyond this central concern, we have identified a wish list of high-level questions about the cultural evolution of human social structure, which we think are ripe for the taking by the scientific community.

- Archaeologists! Why didn't complex societies arise in the last interglacial (ca. 125–85 KYA) among populations of anatomically, and debatably, behaviorally modern humans?
- Ethnologists! What norms are "universal" in *content* at different scales (e.g., small-scale, chiefdom, modern industrial state)? Does scale explain the similarities and differences? Is it the most important context?
- Sociologists! Why hasn't religion or ethnicity disappeared? Why is there increasingly less homogeneity in the age of globalization (the "indigenization of modernity")?
- Anthropologists! Does the ethnographic analogy have legs? Is the notion that modern-day small-scale societies can act as proxies for small-scale societies in prehistory still viable? Pleistocene small-scale societies seem to have been different in their scale (i.e., bigger) and style diversity (i.e., reduced) than ethnographically known populations, but is this just an artifact of the decimated record of durable artifacts? Were there more competition and more pronounced leadership, wider trade networks, and heterarchy in the past? How can we answer these questions?
- Psychologists (especially you developmentalists)! How does norm psychology evolve and develop, as Haun and Over (this volume) have been asking? Does cooperative breeding hold the key to our other-regarding cognition?
- *Everyone!* What should we fund? What data is missing? Do we need more and targeted archaeology? Will massive efforts be required to understand within-population behavioral variation? Is a resurvey of extant ethnography necessary to add longitudinal facts? What is it that is really stopping us from understanding this most basic question of human uniqueness?

These questions are aimed, tongue-in-cheek, at different disciplines, but only a cross-disciplinary effort will properly suffice to further understanding. We look forward to the results.

Technology and Science

7

The Cultural Evolution of Technology

Facts and Theories

Robert Boyd, Peter J. Richerson, and Joseph Henrich

Abstract

The gradual cumulative cultural evolution of locally adaptive technologies has played a crucial role in our species' rapid expansion across the globe. Until recently, human artifacts were not obviously more complex than those made by organisms that lack cultural learning and have limited cognitive capacities. However, cultural evolution creates adaptive tools much more rapidly than genetic evolution creates morphological adaptations. Human tools are finely adapted to local conditions, a fact that seems to preclude explanations of cultural adaptation based on innate cognitive attractors. Theoretical work indicates that culture can lead to cumulative adaptation in a number of different ways. There are many important unsolved problems regarding the cultural evolution of technology. We do not know how accurate cultural learning is in the wild, what maintains cultural continuity through time, or whether cultural adaptation typically requires the cultural transmission of causal understandings.

Introduction

Humans have a larger geographical and ecological range than any other terrestrial vertebrate. About 60,000 years ago, humans emerged from Africa and rapidly spread across the globe. By about 10,000 years ago, human foragers occupied every terrestrial habitat except Antarctica and a number of remote islands, like Hawaii, Iceland, and Madagascar. To accomplish this unparalleled expansion, humans had to adapt rapidly to a vast range of different environments: hot dry deserts, warm but unproductive forests, and frigid arctic tundra.

Technology played a crucial role in this process. Spears, atlatls, and later bow and arrow are used to acquire game; flaked stone tools are necessary to process kills and to shape wood, bone, and process hides; clothing and shelter are crucial for thermoregulation; fire-making paraphernalia are necessary for cooking, heat, and light. Slings, baskets, and pottery facilitate transport and storage; boats expand the ranges of foragers to include lakes and oceans; fish-hooks and cordage make coastal habitats rich sources of protein. In most cases, technological adaptation is specific to local environments because the problems that need to be solved vary from place to place—getting food and regulating body temperature are very different problems in the North American Arctic and the African Kalahari desert.

Humans were able to create this diverse set of tools rapidly because cultural evolution allows human populations to solve problems that are much too hard for individuals to solve by themselves, and it does this much more rapidly than natural selection can assemble genetically transmitted adaptations. In this chapter we attempt to summarize what is known and unknown about this process. We begin with "stylized" *facts*, empirical generalizations relevant to the cultural evolution of technology. We then move to *theory*: there has been a lot of work aimed at understanding the workings of cultural evolution over the last several decades. Here, we summarize some results from those models most relevant to understanding the gradual cultural evolution of complex, adaptive technologies.

We think that these facts and theoretical results indicate that technological change is an evolutionary process. The tools essential for life, in even the simplest foraging societies, are typically beyond the inventive capacities of individuals. They evolve, gradually accumulating complexity through the aggregate efforts of populations of individuals, typically over many generations. People do not invent complex tools, populations do. In this way, the cultural evolution of human technology is similar to the genetic evolution of complex adaptive artifacts in other species, like birds' nests and termite mounds. In both cases, individuals benefit from complex, adaptive technologies that they do not understand. Instead the adaptive design evolves gradually—in the genetic case through natural selection and in the cultural case by individual learning and biased cultural transmission, with natural selection perhaps playing a secondary role. The big difference between these processes is speed. Cultural evolution is much faster than genetic evolution and, as a consequence, human populations can evolve a variety of tools and other artifacts that are adapted to local conditions. In contrast, most animal artifacts are species-typical adaptations to problems which face all members of the species.

Stylized Facts about the Cultural Evolution of Technology

People in Even the Simplest Human Societies Depend on Tools That Are Beyond the Inventive Capacity of Individuals

It is easy to underestimate the scope and sophistication of the technology used in even what seem to be the "simplest" foraging societies. Consider, for example, the Central Inuit of the Canadian Arctic. These foraging peoples occupied a habitat that is harsh and unproductive, even by Arctic standards. Their groups were small, and their lifeways were simple compared to other Arctic foragers. Nonetheless, they depended utterly on a toolkit crammed with complex, highly refined tools. Winter temperatures average about –25°C so survival required warm clothes (Gilligan 2010). In the winter, the Central Inuit wore beautifully designed clothing, made mainly from caribou skins (Issenman 1997). Making such clothing requires a host of complex skills: hides must be cured, thread and needles made, clothing designed, cut and stitched. Even the best clothing is not enough during winter storms; shelter is mandatory. The Central Inuit made snow houses so well designed that interior temperatures were about 10°C. There is no wood in these environments, so houses were lit and heated, food was cooked, and ice melted for water using carved soapstone lamps fueled with seal fat. During the winter, the Central Inuit hunted seals, mainly by ambushing them at their breathing holes using multipiece toggle harpoons; during the summer, they used the leister (a three-pronged spear with a sharp central spike and two hinged, backward facing points) to harvest Arctic char caught in stone weirs. They also hunted seals and walrus in open water from kayaks. Later in summer and the fall, the Central Inuit shifted to caribou hunting using bows that are described in more detail below. We could go on and on. An Inuit "Instruction Manual for Technology" would run to hundreds of pages. And you'd need to master the "Natural History Handbook," "Social Policies and Procedures," "Grammar and Dictionary," and "Beliefs, Stories, and Songs," volumes of comparable length to be a competent Inuit.

So, here is the question: Do you think that you could acquire all the local knowledge necessary to create these books on your own? This is not a ridiculous question. To a first approximation, this is the way that other animals have to learn about their environments. They must rely mainly on innate information and personal experience to figure out how to find food, make shelter, and in some cases to make tools.

We are pretty sure that you would fail, because this experiment has been repeated many times when European explorers were stranded in an unfamiliar habitat. Despite desperate efforts and ample learning time, these hardy men and women suffered or died because they lacked crucial information about how to adapt to the habitat. The Franklin Expedition of 1846 illustrates this point (Lambert 2011). Sir John Franklin, a Fellow of the Royal Society and an experienced Arctic traveler, set out to find the Northwest Passage and spent two icebound winters in the Arctic, the second on King William Island. Everyone eventually perished from starvation and scurvy. The Central Inuit have, however, lived around King William Island for at least 700 years. This area is rich in animal resources. Nonetheless, the British explorers starved because they did not have the necessary local knowledge, and despite being endowed with the same cognitive abilities as the Inuit, and having two years to use these abilities, they failed to learn the skills necessary to subsist in this habitat.

Results from this "lost European explorer experiment" and many others suggest that the technologies of foragers and other relatively simple societies are beyond the inventive capacity of individuals. The reason is not difficult to understand. Kayaks (Dyson 1991), bows (Henrich 2008), and dog sleds (Malaurie 1985) are very complicated artifacts, with multiple interacting parts made of many different materials. The function of these artifacts depends on physical principles known only to engineers during the last two or three centuries. Determining the best design is, in effect, a high-dimensional optimization problem that is usually beyond individual cognitive capacities, sometimes even those of modern engineers (e.g., Dyson 1991). Inevitably, design requires much experimentation, and in most times and most places this is beyond the capacity of individuals (Henrich 2009b).

Tools Usually Evolve Gradually by Small Marginal Changes

Isaac Newton remarked that if he saw farther, it was because he stood on the shoulders of giants. For most innovations in most places at most times in human history, innovators are really midgets standing on the shoulders of a vast pyramid of other midgets. Historians of technology believe that even in the modern world the evolution of artifacts is typically gradual, with many small changes, often in the wrong direction. Nonetheless, highly complex adaptations arise by cultural evolution even though no single innovator contributes more than a small portion of the total (Basalla 1988; Petroski 1994, 1985, 2006).

Two examples (one simple, the other more complex) will illustrate this contention. The simple example is the evolution of the eighteenth-century North American axe. The sharp end of an axe head is called the blade; the other end on the opposite side, with a hole for the handle, is called the poll. The typical "trade axe" introduced from Europe to North America in the seventeenth century had a small rounded poll. This design probably arose from the practice of manufacturing axe heads by bending an iron bar in a U-shape, inserting a piece of steel into the end of the U, welding the two arms and the steel to form the head, and finally sharpening the steel to form the blade (Figure 7.1). The rounded design makes it hard to use the axe as a hammer (e.g., to drive wedges), and the fact that the center of mass of the head is well forward of the handle makes accurate swings difficult (Widule et al. 1978). Over the course of the eighteenth century, a new design, the "American felling axe," was gradually created by North American blacksmiths (Kauffman 2007). This axe had a substantial poll that moved the center of mass backward with a flattened surface, which made it easier to use as a hammer, and is now the standard form of axe heads in Europe and North America. Still, even such a small change took at least a century to emerge and spread.

The evolution of rudders for ships in Europe provides a more complex example of gradual cumulative cultural evolution (Mott 1997). In very small boats, paddles can serve as "rudders." A paddler at the back of the boat tilts

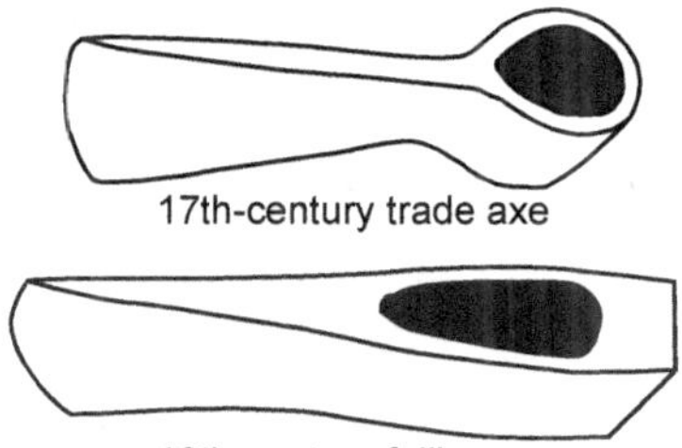

Figure 7.1 (a) Illustration of a European "trade axe" typical of seventeenth-century European axes. This axe has a lightweight, rounded poll. (b) An American "felling axe" of the type which evolved in the eighteenth century in North America and is now used worldwide. The heavier poll makes the axe easier to swing accurately and gives the axe more cutting weight, both tending to increase the "bite" of each swing. The flattened poll allows the axe to be used as a sledge for driving wedges.

the paddle so that it is at an angle to the long axis of the boat creating a torque which causes the boat to turn. However, as boats became larger, the force necessary to accomplish this rapidly became too great. So, paddles became "quarter rudders": a large paddle-like rudder mounted (usually) on both sides of the ship, near the stern, with a long handle at the top end so that the rudder could be rotated around its long axis. Unlike paddles, quarter rudders turn the ship by creating a turning force the same way that a wing creates lift. In classical Greece and Rome, quarter rudders were constructed by fastening a flat piece of wood to a round pole, and were relatively broad compared to their length. Later in the Middle Ages, Mediterranean shipwrights adopted much longer, thinner quarter rudders with a wing-like cross section, a design that greatly reduced drag without reducing turning power. To be efficient, quarter rudders must be about a third as long as the overall length of the ship and mounted so that the long axis of the rudder is at an angle of about 45 degrees to the vertical. As ships became larger, this led to an increasing number of elaborate mounting tackle to handle the very large torques created by the long, heavy rudder. One rudder on a late thirteenth-century Mediterranean trading ship was 18 m long and weighed 11,000 kg. Eventually this led to the invention of the "sternpost rudder," a rudder mounted vertically on the stern using "pintle and gudgeon" hinges (Figure 7.2). This innovation occurred in the Baltic, and it seems likely that sternpost rudders evolved by combining the unusual fixed, quarter rudders used on Norse trading ships and newly developed iron hinges from large castle and cathedral doors. This innovation diffused into the Mediterranean and was applied to the much larger ships common to that region. The first ships that used sternpost rudders in the fourteenth century were otherwise very similar to contemporary ships; they had quarter rudders with a single mast, curved sternposts, and steeply rounded ("bluff") sterns. Because they were mounted in the turbulent wake of the ship rather than the laminar flow along the ship's side,

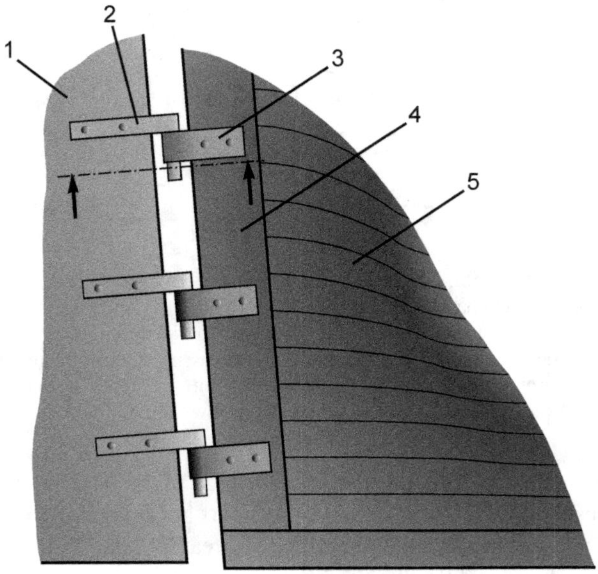

Figure 7.2 A pintle and gudgeon sternpost rudder. The "pintles" are the vertical pins attached to the rudder and the "gudgeons" are the iron loops attached to the sternpost of the hull. The labeled parts are: (1) the rudder, (2) a pintle, (3) a gudgeon, (4) the sternpost, and (5) the hull of the ship. Image created by Eric Gaba for Wikimedia Commons, used with permission.

ships with sternpost rudders were difficult to handle because these rudders created much less turning force than quarter rudders. Gradually over the next several centuries, ship builders added (a) multiple masts which allowed sails to be used to aid steering, (b) a straight, vertical sternpost that allowed more than two pintle and gudeon connectors, and gradually (c) a streamlined stern with more "dead wood" which causes laminar flow around the rudder (Figure 7.3). In this way the modern ship's rudder, and associated design changes, evolved gradually in Europe over a period of more than half a millennium. Interestingly, as Mott (1997) recounts, rudder evolution in China and the Indian Ocean seem to have taken completely independent courses.

Genetic Evolution Leads to Complex, Adaptive Artifacts Often Constructed by Animals with Simple (or No) Nervous Systems

Discussions of animal tool use typically focus on things that animals can carry: stones used by chimpanzees to crush hard-shelled nuts, and leaf tools used by New Caledonian crows to extract insect larvae from holes in branches. The relative rarity of these tools as well as the fact that they are made by animals

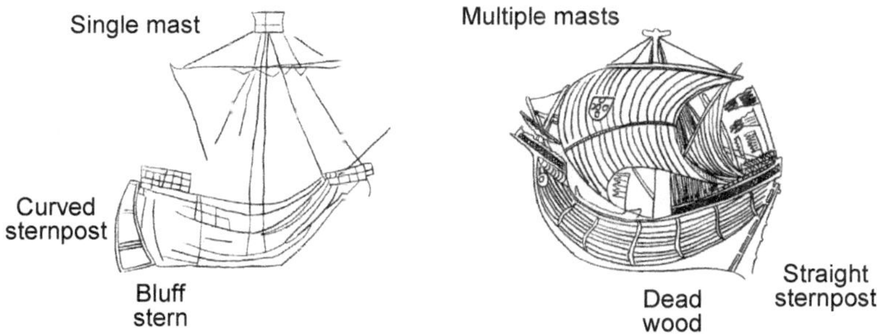

Figure 7.3 Illustration of the development of ship design after the introduction of the sternpost rudder in the Mediterranean region. The left panel shows a tracing of a drawing of a medieval ship from the bell tower of the Cathedral of Palma de Mallorca, which probably dates to the early thirteenth century. The curved sternpost, bluff stern, and single mast were characteristic of contemporary ships with quarter rudders. Note that a very broad rudder was necessary when used with a bluff stern. The right panel shows an early fifteenth-century drawing of a ship with innovations made in response to the introduction of the sternpost rudder, three masts, a straight sternpost carrying a slender rudder and a run of dead wood up to the rudder. Reprinted with permission from Lawrence Mott (1997:131, 139).

like apes and corvids gives the impression that animal artifacts are rare, simple, and limited to clever large-brained creatures, something like ourselves.

Nothing could be further from the truth. Think a bit—you already are aware of many complex animal artifacts. Birds' nests, spider webs, termite mounds, and beaver dams are just a few of the familiar constructions made by nonhuman animals, and a dip into the zoological literature reveals a long list of less familiar artifacts. Many of these artifacts appear highly designed and require very elaborate construction techniques. Take the nests made by the village weaver, one of a number of African weaver birds (Collias and Collias 1964). These hanging nests provide shelter for the brooding young and rival the houses made by many human populations in their complexity. The construction process is highly stereotyped. The bird first weaves a ring, followed by the egg chamber, and finally the entrance. The weaving itself involves elaborate knotting and weaving (Figure 7.4). While practice increases the quality of the construction, social learning plays no role. Birds seem to have some representation of form of the nest, but for the most part it seems that the construction process results from an algorithm which links simple, stereotypical behaviors into a sequence that generates a nest.

The construction of complex artifacts does not require superior cognitive ability. Invertebrates such as termites, funnel wasps, and spiders make complex, highly functional artifacts without any representation of the final form of the artifact (Gould and Gould 2007; Hansell 2005) despite having much

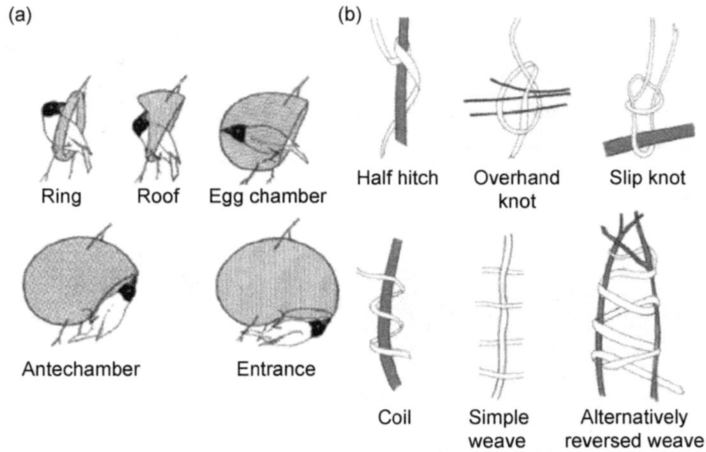

Figure 7.4 (a) Depiction of the construction sequence used by village weavers to construct their nests. The bird first builds a hanging ring by knotting green grass stems onto the fork of a branch and then weaving more stems to make a ring. The ring is extended outward by weaving more stems into the existing structure. (b) A sampling of the knots and weaves found in typical village weaver nests. Reprinted with permission from AOU (Collias and Collias 1964).

simpler cognitive systems than most vertebrates. In fact, complex artifacts can be constructed without a nervous system at all, as demonstrated by Figure 7.5.

The Cultural Evolution of Artifacts Is Usually Faster Than the Genetic Evolution of Morphology

Modern technology evolves with blinding speed. The number of transistors that can be usefully incorporated on an integrated circuit has doubled every eighteen months for almost half a century. The twentieth century saw massive transformations within a few generations. The first author's father grew up in a small town in Upstate New York without telephones, automobiles, or electric lights and now this very same person's grandchildren carry powerful computers in their pockets. These stupendous rates are the end result of an exponentially increasing rate of change that has characterized the technological evolution over most of the last millennium (Enquist et al. 2008).

It is clear that rates of cultural change over the last millennium are much faster than rates of genetic adaptation in a long-lived species like humans. Of course, bacteria can adapt genetically extremely quickly because their generations are measured in minutes. Human genetic adaptation seems to take place on millennial timescales at the fastest. Thus far, the strongest selection

Figure 7.5 The "house" built by the single-celled amoeba *Diffulgia corona*. It is about 0.15 mm in diameter and is made of very small grains of sand. Reprinted with permission from The Natural History Museum, London (Hansell 2005).

signal detected in the human genome by looking for long haplotypes is the gene that allows northern Europeans to digest lactose (Ingram et al. 2009), an allele which has increased to moderately high frequencies in Northern Europe over the last 5,000 years or so.

Until recently it was not so clear that rates of cultural change in less complex human societies were faster than rates of human genetic change, but a recent paper by Perreault (2012) settles the issue: cultural rates are much faster than genetic evolutionary rates. In a famous paper, Gingrich (1983) assembled data from paleontological records which allowed measurement of the rate of change as the percent change in a quantitative morphological character per million years. Gingrich also found that measured rates of change were negatively related to the time period over which the measurement was made. Perreault assembled a sample of 573 cases from the archaeological record (mainly for Holocene North America) and compared the measured rates of change to those in Gingrich's sample of paleontologically measured rates. The effect of the type of transmission on the per generation rate of change estimated in a multivariate analysis is approximately a factor of 50. All other things being equal, the rate of cultural change of the dimensions of pots, points, and houses is fifty times greater than the rate of change in the dimensions of mandibles, molars, and femurs (Figure 7.6).

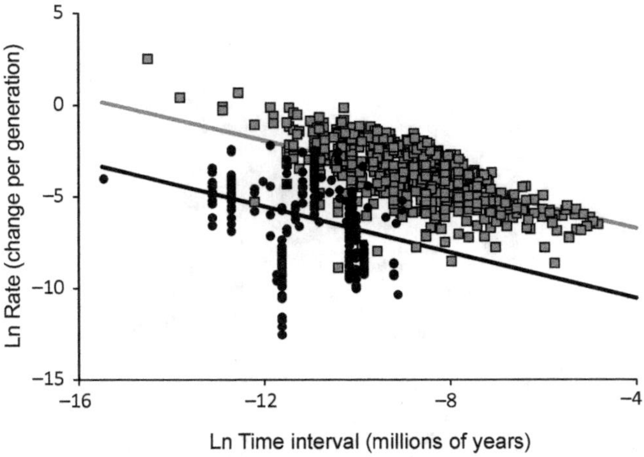

Figure 7.6 The logarithm (Ln) of percent change per generation for genetically heritable morphological traits (black circles) from the fossil record and culturally transmitted traits from the archaeological record (gray squares) plotted against the logarithm of length of time over which the change occurred. The lines represent the best fit in a multivariate analysis of covariance. In both cases, rates decline as the time interval increases, and, interestingly, the per-generation slopes are approximately equal. The distance between the lines gives the difference in cultural and biological traits controlling for other variables. Cultural evolution is a factor $e^{3.91} = 49.8$ times faster than genetic evolution. Reprinted with permission from Charles Perreault (2012).

An Evolutionary Theory of Technology Requires Independent Theories of Function

Understanding the causal relationship between phenotypic variation and reproductive success is a key component of Darwinian theory. Sometimes it is argued that natural selection is a tautology: genes with higher fitness spread (e.g., Bethell 1976):

> Question: How do we know they are higher fitness?
> Answer: Because they spread.

If biologists worked this way, natural selection would indeed be a useless concept. To understand why, consider the following example: A recessive gene causing a severe vision disorder called achromatopsia has spread to roughly 30% of the population on the Micronesian island of Pingelap. Sufferers of achromatopsia cannot see well under any circumstances, but are especially disadvantaged in the bright sunlight of a tropical island (Sacks 1998). Nonetheless, there is no doubt that this gene spread on Pingelap because people who carried it had more descendants than those who did not carry the gene. However, we know that achromatopsia was not favored by natural selection because it did not *cause* their increased reproductive success. Rather the gene was carried by members of a chiefly lineage whose social position allowed them to survive the aftermath of a severe typhoon which struck the island during the 1700s; the spread of the achromatopsia gene was a side effect of other processes, not the result of natural selection.

This kind of functional reasoning is crucial for the inference that complex adaptations were caused by natural selection. For relatively simple characters, it is possible to measure phenotypic variation in nature and connect it to variation in fitness—the study of the evolution of beak morphology in Darwin's finches by the Grants (1986) provides a classic example. However, this tactic is hard to apply to complex characters like the vertebrate eye. Instead, biologists rely on detailed functional analyses which show that many details of the complex adaptation fit with the proposed function of the adaptation. Thus, the lens has to be just the right shape and have just the right index of refraction to form an image on the retina, an exquisitely photosensitive tissue. The iris adjusts the aperture so that the eye works over a wide range of light intensities; three sets of muscles adjust the eye's orientation, up down, right left, and correct for movements of the head. The list of features is long. Moreover, the eyes of different organisms vary in ways that make sense, given the problems they have to solve. Our eyes have "lens-shaped" lenses with an approximately uniform index of refraction, whereas fish have spherical lenses with an index of refraction that gradually increases toward the center of the lens. This difference makes sense, given the optics of living in air and water.

We think that functional analysis should play a similar role in the study of culturally evolved technology. There are good reasons to believe that both payoff-biased transmission and guided variation (Richerson and Boyd 2005) should cause the gradual adaptive cultural evolution of functional artifacts. Thus the careful study of the function of complex culturally evolved artifacts provides evidence that these processes gave rise to the artifacts. The design of bows and arrows provides a good example. Many modern bowyers (bow-and-arrow makers) are interested in recreating designs collected by previous generations of anthropologists. These bowyers include sophisticated engineers, and through their testing and experiments, we have come to know a lot about the design principles of traditional bows and arrows. (For details, see the many papers in the four volumes of *The Traditional Bowyer's Bible*; the paper by Baker [1992] in the first volume provides a good introduction.) Bows used to hunt large game needed to be powerful enough to throw a heavy arrow at high velocity. When a bow is bent, the back (the side away from the archer) is under tension, while the belly (the side closer to the archer) is in compression. This leads to strain within the bow and can result in failure. The simplest way to solve this problem is to make a long bow using some dense elastic wood, like yew or osage orange, a design widely used in South America, Eastern North America, Africa, and Europe. Because a long bow need not be bent very far, this design minimizes the strain on the limbs. In some environments, however, a long bow is not practical. People like the Plains Indians and Central Asian pastoralists, who hunt and fight on horseback, need a short bow. In other environments, like the high Arctic, the right kind of wood is not available. In such environments people make short bows and employ the full range of bowyers' tricks to increase their power. A bow can be made more powerful by removing less wood in shaping the limbs. However, making the bow thicker (front to back) increases the stress within the bow, leading to failure. This problem is exacerbated in short bows because the radius of curvature is greater. To solve this problem, the short bows made by Plains Indians, Inuit, and Central Asian pastoralists are thin front to back, wide near the center, and taper toward the tips. They are also usually recurved, meaning that the bow is constructed so that when it is not braced, it forms a backward "C" shape. Bracing the recurved bow leads to a compound curve (the middle part of the bow curves toward the archer but the tip of each limb curves back away from the archer), a geometry that allows for greater energy storage. Finally, these peoples typically make composite bows. Wood is stronger in compression than tension, so the ability of a bow to sustain strong bending forces can be increased by adding a material that is strong in tension to the back of the bow. Both in Central Asia and Western North America, sinew was glued to the backs of bows to strengthen short bows for use on horseback. The Inuit, however, lashed a woven web of sinew to the back of their bows, probably because available animal glues would not work in the moist, cold conditions of the Arctic. Other components of the bow show similar levels of functional design. Bowstrings need to be strong and should not stretch. In most environments the solution is to make cord by twisting long sinews, often drawn from along the backs of ungulates, and then combining cords into multi-ply bow strings in which the plies twist in opposite directions. In addition, arrows present complicated design problems which have been solved by different peoples in different ways.

The Cultural Evolution of Technology Cannot Be Explained Solely in Terms of Specialized Innate Attractors or Cognitive Biases

A number of authors have argued that the outcomes in cultural evolution are strongly shaped by "inductive biases" created by human cognition (Claidière and Sperber 2007; Boyer 1998; Griffiths and Reali 2011). We agree that such biases probably have important effects, at least in some domains, and have referred to these as "content" or "direct" biases (Boyd and Richerson 1985; Henrich and Henrich 2010). The way that this works is beautifully illustrated by the transmission chain experiments conducted by Tom Griffiths and his collaborators (Griffiths and Reali 2011). For example, in one experiment, subjects are first shown 50 pairs of numbers. Sometimes these are the x, y coordinates of a straight line, sometimes a curve, and other times they are drawn at random (Figure 7.7). Then the subject is given 50 x values and asked to produce the associated y value. These fifty pairs are then used to train a second subject, who is given 50 x values and asked to produce the y values learned during training.

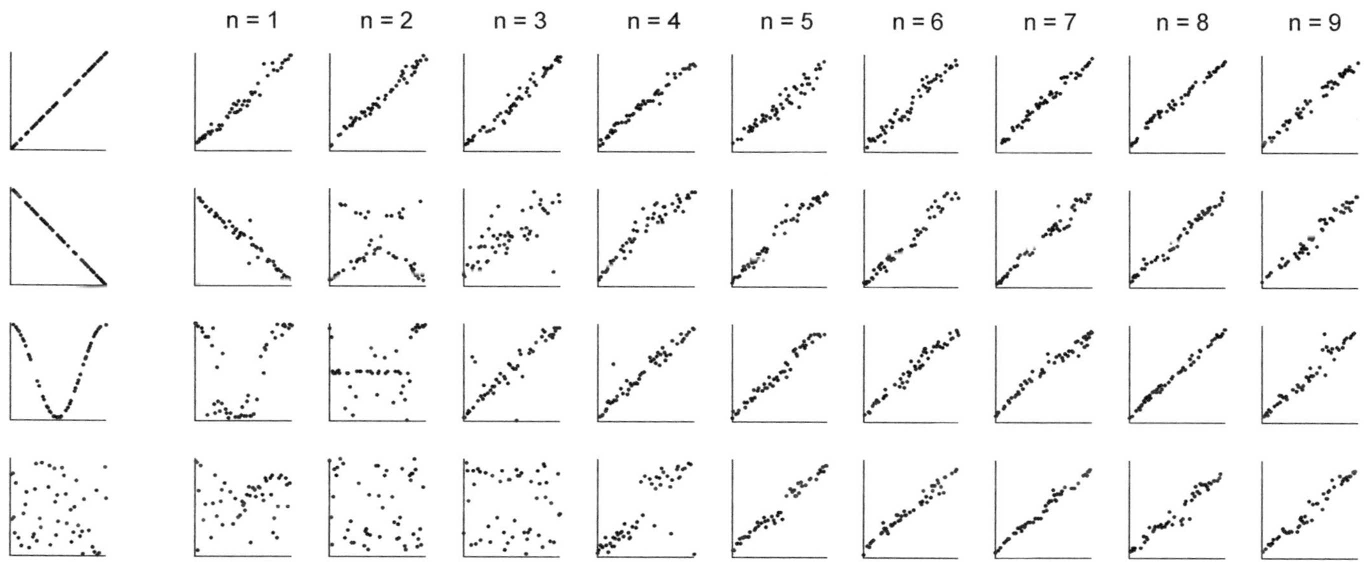

Figure 7.7 Results of four transmission chain experiments which show that an inductive bias in favor of straight line, not the environmental data, determines the final result. The left column shows the data used to train the first subject. The subsequent panel gives that subject's estimates of the y value given an x value. This data was then used to train the next subject in the chain, and the process was repeated for nine subjects. By the fifth subject, each chain had generated a straight line which was stable thereafter. Reprinted with permission from Griffiths and Reali (2011).

This procedure is repeated for eight more subjects. As illustrated in Figure 7.7, transmission is strongly shaped by a bias in favor of straight line relationships with a positive slope. The initial data has no effect on the ultimate outcome. Human learning has an inductive bias that causes people to infer straight lines from data, and when combined with error prone learning, this bias gradually causes people to see straight lines where none existed.

Dan Sperber has argued that such inductive biases, which he calls "attractors," are the main source of cultural stability and thus determine the outcomes of cultural evolution (Claidière and Sperber 2007). Sperber believes that the "frame problem" makes cultural learning extremely difficult. It is difficult, he believes, to copy the behavior of others accurately, where behavior includes things like artifacts. Any real artifact is complex, and both the artifact and the process by which it is made contain many irrelevant details. The learner who is trying to learn how to make an artifact by observation must know what to ignore and what to learn. Inductive biases serve this function. Because these biases shape what is learned and what is ignored, they have a strong effect on cultural outcomes.

Functional thinking suggests that Sperber overemphasizes the importance of such attractors. Perhaps innate attractors would work if humans made only one sort of complex technology, but bows, boats, clothing, and all the other components of technology include a stunning diversity of nonintuitive forms that are often exquisitely designed for a particular environment. The short, flat, recurved composite Plains Indian bow is designed for horse-mounted hunting and warfare. Such complex functional design does not arise by chance. The details matter: its shape, the kind of wood used, the glue used to bind sinew to the back of the bow, the kind of sinew, and the number of plies used in the bowstring, and so on. Moreover, as we have seen, complex cultural design does not usually arise from inventive activities of single individuals. Instead, complex functional human artifacts like bows, dogsleds, and kayaks evolve through a gradual process of cultural accumulation. The cultural evolution of the Plains Indian bow, and its stability through time, however, cannot solely be due to an attractor or inductive bias that causes individuals to make Plains Indian bows. Many inductive biases may, of course, be important. The mind is a complex device with many specialized mechanisms, allowing people to solve problems which they face (Barrett 2013). We have mechanisms that allow us to engage in causal reasoning (Gopnik and Schulz 2004), recognize and categorize objects in the world (Carey 2009; Perfors and Tenenbaum 2009), and learn from observing the behavior of others (Tomasello et al. 2005). We may also have evolved intuitions about the function of artifacts (German and Barrett 2005) and the laws of mechanics (Carey 2009). It seems likely that these mechanisms make it easier to learn how to make some kinds of tools and harder to make others, and this will create cognitive biases that affect the cultural evolution of technology. However, such mechanisms cannot account for the details that are crucial for the function of the Plains Indian bow, because these are specific to the particular adaptive problems faced by mounted bison hunters. There is no "Plains Indian bow attractor" hidden in the recesses of the human mind. The design of these bows must be transmitted sufficiently accurately from person to person so that it remains stable through time, and so improvements can gradually accumulate.

Theory Relevant to the Cultural Evolution of Technology

Gradual Cumulative Adaptation Can Arise from Rare Individual Learning plus Unbiased Transmission

Quite a bit of work has been done on mathematical models that describe how the gradual cultural accumulation of complex cultural adaptations might occur. These models are usefully divided into three types: (a) models in which cumulative adaptation arises from rare individual learning combined with unbiased cultural transmission, (b) models in which adaptations arise from payoff-biased transmission, and (c) models in which cumulative adaptation arises from rare innovations and accurate communication of causal information. We will review in turn work from each category.

Rogers (1988) created an early, and especially simple, model that showed how learning and imitation could be combined to give rise to gradual cultural evolution. In this model, a population lives in an environment that switches between two states with a constant probability. There is a best behavior in each state, and the adaptive problem facing individuals is to determine within which environment they are living. There are two methods for doing this: individuals can, at a cost, learn the best behavior in the environment, or they can copy another individual for free. As long as the net benefit of acquiring the best behavior is greater than the cost of learning, the optimal strategy is a mixture of costly learning and cheap imitation. Gradual cultural evolution occurs when learning is costly and environmental changes are infrequent. Then, at the optimal mixture of learning and imitation, only a few individuals learn and most imitate; thus after an environmental shift, the fraction of the population with the best behavior gradually increases (Figure 7.8).

Barrett et al. (2007a) and Pinker (2010) argue that the main benefit of social learning is that it allows the costs of learning to be spread over a large number of individuals. Information is, in the jargon of economics, a "nonrival" good, meaning that one person's "consumption" does not reduce the value for others. Once produced, valuable information can spread throughout a population at low, or even zero cost, a fact that is at the core of endogenous growth models discussed below (e.g., Romer 1993). However, Rogers's model shows that this argument is wrong when applied to the evolution of social learning. The equilibrium mixture of learning and imitation leads to the same average payoff as a population in which there are no imitators, only learners.

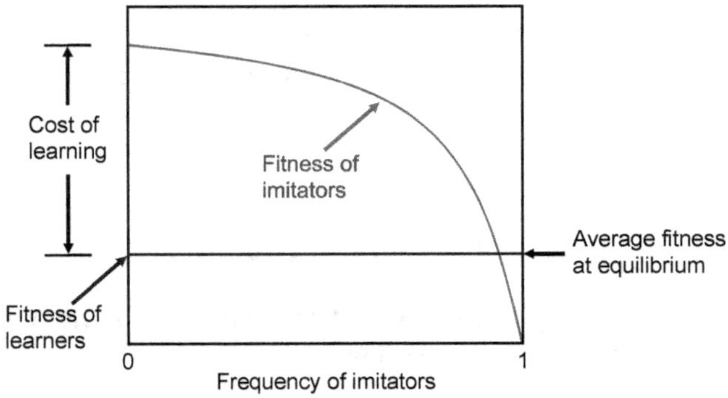

Figure 7.8 A diagrammatic exposition of the model by Rogers (1988). The graph gives the fitness of imitators and learners as a function of the frequency of imitators. Learners monitor the environment and acquire the best behavior at a cost. Imitators copy a random individual for free. When imitators are rare, they have higher fitness than learners because they have the same probability of acquiring the best behavior but do not pay the cost of learning. As imitators become more common, their fitness declines because they increasingly acquire the wrong behavior due to environmental changes. The frequency of imitation increases until both types have the same fitness.

The reason is that imitators do not contribute anything to the population; they just scrounge adaptive information that has been produced by the costly learning efforts of others. This property, often referred to as "Rogers's Paradox," has been the focus of much research (Boyd and Richerson 1995; Kobayashi and Wakano 2012; Lehmann et al. 2010; Rendell et al. 2010; Aoki 2010). So far investigators have discovered three mechanisms that allow culture to increase average fitness.

First, population structure can generate relatedness among interacting individuals, and this in turn alters the evolutionarily stable mix of individual and social learning so that average fitness increases (Rendell et al. 2010; Lehmann et al. 2010). In these models, individual learners are altruists who create benefits for others at a cost to themselves. Thus, simple kin selection arguments predict that when population structure leads to increased relatedness, the evolutionary equilibrium should contain more individual learners than when individuals interact at random. This means that average fitness increases. The work of Lehmann et al. (2010) illustrates how this works in an island model in which local populations exchange genes, but not cultural traits, with the global population. Rendell et al. (2010) simulate gene–culture coevolution on a lattice, and although their results are complex, it seems likely that the increased average fitness which they observe for some parameter combinations is also due to population structure.

Second, cultural learning can allow individuals to learn selectively. The ability to learn selectively is advantageous because opportunities to learn from experience or by observation of the world vary. Sometimes experience provides accurate information at low cost. Think of Goodyear accidentally spilling rubber onto a hot stove, or Fleming observing his mold-contaminated petri dishes. Such rare cues allow accurate low-cost inferences about the environment. However, most individuals will not observe these cues, and thus making the same inference will be much more difficult for them. Organisms which cannot learn from others are stuck with whatever information nature offers. In contrast, an organism capable of cultural learning can afford to be choosy, learning individually when it is cheap and accurate, and relying on cultural learning when environmental information is costly or inaccurate. We have shown (Boyd and Richerson 1987b; Perreault et al. 2012) that selection can lead to a psychology that causes most individuals to rely on cultural learning most of the time, and also simultaneously increase the average fitness of the population over the fitness of a population that does not rely on cultural information. In these models the psychology that controls individual learning has a genetically heritable "information quality threshold" that governs whether an individual relies on inferences from environmental cues or learns from others. Individuals with a low information quality threshold rely on even poor cues, whereas individuals with a high threshold usually imitate. As the mean information quality threshold in the population increases, the fitness of learners increases because they are more likely to make accurate or low-cost inferences. At the same time, the frequency of imitators also increases. As a consequence, the population does not keep up with environmental changes as well as a population of individual learners. Eventually, an equilibrium emerges in which individuals deploy individual and cultural learning in an optimal mix. At this equilibrium, the average fitness of the population is higher than in an ancestral population without cultural learning. When most individuals in the population observe accurate environmental cues, the equilibrium threshold is low, individual learning predominates, and culture plays little role. However, when it is usually difficult for individuals to learn on their own, the equilibrium threshold is high, and most people imitate, even when the environmental cues that they do observe indicate a different behavior than the one they acquire by cultural learning. This analysis assumes selection is weak enough so that only learning affects the frequency of alternative cultural variants. If selection is strong enough to lead to the spread of adaptive cultural variants then, of course, mean fitness will increase for the same reason that it does in genetic models, a fact confirmed by the simulation study of Franz and Nunn (2009b).

Third, the ability to learn culturally can also raise the average fitness of a population by allowing acquired improvements to accumulate from one generation to the next. Many kinds of traits admit successive improvements toward some optimum. Bows vary in many dimensions that affect performance, such as length, width, cross section, taper, and degree of recurve. It is typically more difficult to make large improvements by trial and error than small ones for the same reasons that Fisher (1930) identified in his "geometric model" of genetic adaptation. In a small neighborhood in design space, the performance surface is approximately flat, so that even if small changes are made at random, half of them will increase the payoff (unless the design is already at the optimum). Large changes will improve things only if they are in the small cone that includes the distant optimum. Thus, we expect it to be much harder to design a useful bow from scratch than to tinker with the dimensions of a reasonably good bow. Now, imagine that the environment varies, so that different bows are optimal in different environments, perhaps because the kind of wood available varies. Sometimes a long bow with a round cross section is best, other times a short, flat, wide bow is best. Organisms which cannot imitate would have to start with whatever initial bow design might be provided by their genotype. Over their lifetimes, they can learn and improve their bow. However, when they die, these improvements disappear with them, and their offspring must begin again at the genetically inherited initial design. In contrast, cultural species can learn how to make bows from others after these have been improved by experience. Therefore, cultural learners start their search closer to the best design than pure individual learners and can invest in further improvements. Thereafter, they can transmit *those* improvements to their offspring, and so on down through the generations until quite sophisticated artifacts evolve. Modeling work (Boyd and Richerson 1985; Borenstein et al. 2008; Aoki 2010) shows that this process can increase average fitness.

In an alternative approach, Enquist et al. (2007) argue that "adaptive filtering" can lead to increased average fitness. They, however, incorporate a number of novel features in their model, and this makes it difficult to compare it with other work in this tradition. Most notably, they assume a large number of traits that have two states: present or absent. The present state of some traits increases fitness compared to the absent state, whereas the present state of other traits reduces fitness. Environmental change is modeled by assuming that traits which are currently adaptive when present change to maladaptive at a constant rate. The fitness effects of all traits are independent, so there is no possibility of cumulative evolution in which each step is contingent on the last. Adaptive filtering increases the rate at which individuals switch from the present to the absent state when the trait reduces fitness. Enquist et al. (2007) show that adding adaptive filtering can lead to increased average fitness. They do not provide any model of how it works at the individual level. We think that adaptive filtering is best thought of as a costless, error-free form of individual learning. To determine whether a present trait is maladaptive in the current environment, individuals need to monitor environmental cues and infer whether the present or absent state of the trait has higher fitness. Adaptive filtering must thus entail some kind of inference process. It is error free because it does not lead to any switch from the absent to present state for maladaptive traits. There is no fitness penalty associated with increased adaptive filtering.

Gradual Cumulative Adaptation Can Arise from Payoff-Biased Transmission

If cultural learners can compare the success of individuals modeling different behaviors, then a propensity to imitate the successful can lead to the spread of traits that are correlated with success, even though imitators have no causal understanding of the connection. This is obvious when the scope of traits being compared is narrow. For example, you see that your uncle's bow shoots farther than yours, and notice that it is thicker, but less tapered, and uses a different plait for attaching the sinew. You copy all three traits, even though in reality it was just the plaiting that made the difference. As long as there is a reliable statistical correlation between plaiting and power, the plaiting form trait will change so as to increase power. Causal understanding is useful because it helps exclude irrelevant traits, like the color the bow is painted. However, causal understanding need not be very precise as long as the correlation is reliable. Copying irrelevant traits like thickness or color will only add noise to the process. By recombining different components of technology from different but still successful individuals, copiers can produce both novel and increasingly adaptive tools and techniques over generations without any improvisational insights. An Inuit might copy the bow design from the best bowyer in his community but adopt the sinew plaiting used by the best hunter in a neighboring community. The result could be a better bow than anyone made in the previous generation without anyone inventing anything new.

Consistent with this, laboratory and field evidence suggests that both children and adults are predisposed to copy a wide range of traits from successful or prestigious people (Henrich and Gil-White 2001; McElreath et al. 2008; Mesoudi 2011b; Chudek et al. 2012). Advertisers clearly know this. After all, what does Michael Jordan really know about T-shirts? Recent work in developmental psychology shows that young children readily attend to cues of reliability, success, confidence, and attention to figure out from whom they should learn (Birch et al. 2008, 2010). Even infants selectively attend to knowledgeable adults rather than their own mothers in novel situations (Stenberg 2009). This feature of our cultural learning psychology fits a priori evolutionary predictions, emerges spontaneously in experiments, develops early without instruction, and operates largely outside conscious awareness. Humans have an efficient social learning module, if you like.

Gradual Cumulative Adaptations Can Arise from Rare Innovations Which Spread Rapidly Because Their Benefits Are Understood

Economists have developed quite different models of the gradual evolution of technology in which some rational economic actors innovate at a cost while other actors adopt the innovations because they understand how they work and why they are beneficial. The central problem in these models is to explain why individuals make costly investments in innovation when others will be able to copy these innovations for free: the rational choice version of Rogers's Paradox. There are two families of models that solve this problem in different ways: in "learning by doing models," innovation is a side effect of other economic activities (Arrow 1962). For example, when firms invest in new factories, the design process may yield a better factory as a side effect. This innovation can then be copied by other actors. Endogenous growth models (Romer 1993) assume that actors choose to innovate because they have market power (modeled as monopolistic competition) and because patents prevent others from copying their innovation directly. However, the knowledge that underlies the innovation is not protected and serves as the basis of further innovations. Social learning is usually not modeled explicitly in either tradition; it is simply assumed that new knowledge is available to all decision makers. Moreover, environments are assumed to be constant so that every innovation increases economic welfare. Thus, cumulative economic progress is built into the models by assumption.

The extent to which these models are relevant to the cultural evolution of technology over the long sweep of human history depends on the answers to two questions: First, are most innovations adopted because their effects are understood, or because they are statistically associated with observable, preferred outcomes? Second, are there mechanisms analogous to patent protection and market power that allow innovators to recoup the costs of attempting to innovate? There is evidence that the adoption of new technologies is not always accompanied by the transmission of causal explanation of how they work or why they are beneficial. Fijian food taboos provide an example. Many marine species in the Fijian diet contain toxins, which are particularly dangerous for pregnant women, and perhaps nursing infants. Food taboos targeting these species during pregnancy and lactation prohibit women from eating toxic foods and reduce the incidence of fish poisoning during this period. Although women in these communities all share the same food taboos, they offer quite different causal explanations for them, and little information is exchanged among women save for the taboos themselves (Henrich and Henrich 2010). The taboos are learned and are not related to pregnancy sickness aversions. The transmission pathways for these taboos suggest the adaptive pattern is sustained by selective learning from prestigious women. If this example is typical, rational actor models do not provide a complete account of adaptive cultural traits like the evolution of technology. From classic literature on the diffusion of innovations (Rogers and Shoemaker 1971) we do know that people do use both the properties of practices and the attributes of the people using or promoting practices in adoption decisions, but precise quantitative estimation of the mechanics of these decisions in the field is still in its infancy.

Obviously there were no patents or similar protections during most of human history, but there may be other ways to recoup the costs of innovation. First, innovations may diffuse slowly throughout a population. Thus genes that lead to innovation will have an adaptive advantage during the time period it takes for the innovation to spread widely. It is interesting that something like this seems to have happened in the evolution of blast furnaces in nineteenth-century Pittsburgh (Allen 1983). Innovative firms were copied by other steel firms within the Pittsburgh region, but because the technology did not diffuse rapidly to other cities, Pittsburgh firms as a whole held an advantage, and the share flowing to innovators may have been sufficient to compensate them for their innovative efforts. Second, Henrich and Gil-White (2001) have argued that skillful or prestigious individuals are often compensated by would-be imitators for access. In such cases, the need for access to imitate successfully is analogous to a trade secret and the payments analogous to licensing payments to patent holders (for a detailed discussion and consideration of "innovation-enhancing institutions," see Henrich 2009b).

Rate of Adaptive Accumulation Depends on Population Size and Connectedness

Two models of cumulative cultural adaptation predict that, all other things being equal, large populations will have more diverse and more complex toolkits than small, isolated populations. First, cultural transmission is subject to a process analogous to genetic drift (Neiman 1995; Shennan 2001). This means that cultural variants are lost by chance when their practitioners are not imitated. For instance, the best bowyer may not be copied because he is a poor shot, unsociable, or dies unexpectedly. The rate of loss due to cultural drift will be higher in small populations than in larger ones, where the absolute number of experts is greater. Lost traits can be reintroduced by the flow of people or ideas from other populations, so the equilibrium amount of variation depends on the rate of contact between groups. Second, social learning is subject to errors, and since errors will usually degrade complex adaptive traits, most "pupils" will not attain the level of expertise of their "teachers." In this way, inaccurate learning creates a "treadmill" of cultural loss, against which learners must constantly work to maintain the current level of expertise. This process is counteracted by the ability of individuals to learn selectively from expert practitioners, so that cumulative cultural adaptation happens when rare pupils surpass their teachers (Henrich 2004b; Aoki and Kobayashi 2012; Henrich 2006). Learners in larger populations have access to a larger pool of experts, making such improvements more likely; this means that the equilibrium levels of cultural complexity should increase as population size increases (Mesoudi 2011c). As in the cultural drift models, contact between populations replenishes adaptive variants lost by chance, leading to higher levels of standing variation, and thus more adaptive traits (Powell et al. 2009).

Empirical data provide some support for these models. A number of small, isolated island populations have lost seemingly valuable technology. For instance, the Tasmanian toolkit gradually became simpler after isolation from mainland Australia (Diamond 1978; Henrich 2004b, 2006; but see Read 2006), and other Pacific groups have apparently abandoned useful technologies such as canoes, pottery, and the bow and arrow (Rivers 1926). Elsewhere in the world, the isolated Polar Inuit lost kayaks and the bow and arrow when all knowledgeable people died during a plague, only to have these skills reintroduced by long-distance migrants from Baffin Island (Mary-Rousselière 1996). There have been two systematic tests of this hypothesis: Collard et al. (2005) found no relationship between population size and toolkit diversity or complexity; and neither did a reanalysis of those data by Read (2006). However, neither analysis included any measure of contact between populations, and the sample was drawn mostly from northern continental regions of the Western Hemisphere, where intergroup contact was probably common (Kroeber 1939; Balikci 1989; Jordan 2009), making it impossible to estimate effective population size without much better demographic data than we possess. Kline and Boyd (2010) analyzed data on marine foraging tools from ten societies in Oceania and found a strong relationship of both number of tool types and average tool complexity and population size (Figure 7.9) controlling for a number of other variables. It may have been easier to detect the effect of population size in this analysis because islands were bounded and isolated, thus making population size estimates more reliable, and because it focused on ecologically similar islands with a common cultural history. Higher rates of contact

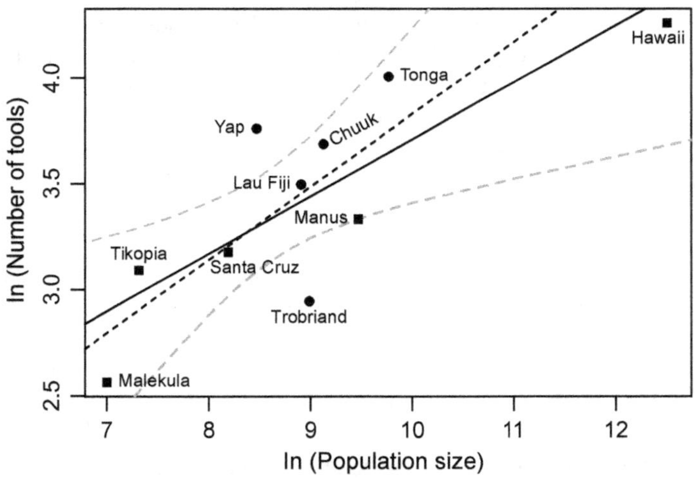

Figure 7.9 Number of tools as a function of population size. Larger populations have significantly more tool types than smaller populations. The trend line is based on a linear regression of the logarithm of the number of tools against the logarithm of population size ($ß = 0.805$, $p = 0.005$, $n = 10$). Four of five low-contact groups (squares) have fewer tools than expected, while four out of five high-contact groups (circles) exceed the expected number of tools. The gray dashed line gives interval estimates. The black dashed line gives the best linear fit when a potential outlier, Hawaii, is removed. Figure courtesy of Richard McElreath.

between groups also increase tool complexity, but the result was only marginally significant.

Conclusion: What We Don't Know

We think that the evidence reviewed makes a convincing case that in most times and places individuals do not invent tools; tools evolve gradually. People everywhere depend on complex tools, many of which are difficult to understand even with the benefit of modern physics, chemistry, and engineering. Consistent with this picture, the history of technology makes it clear that most technological change is gradual, and models of cultural change suggest that gradual accumulation is to be expected when individual innovation is costly or difficult. This leaves two crucial questions unanswered. First, we know that there is heritability of cultural variation at the population level. Technologies and other forms of cultural variation persist in time and in ways that are not related to differences in the external environment (Richerson and Boyd 2005). Without heritability there can be no cumulative cultural evolution. However, we do not know the causes of heritability at the population level. In genetic evolution, heritability at the population level results from heritability at the individual level and restricted gene flow between populations. Genetic transmission is incredibly accurate, and selection is usually weak. This means that in the absence of high levels of gene flow, gene frequencies in populations change slowly. Most models of cultural transmission assume cultural variation is maintained in the same way. However, this need not be the case. Cultural transmission is an inferential process. How demonstrators behave gives evidence about what is going on in their brains, and learners make inferences based on this evidence. However, many inferences are consistent with the same evidence and, as a result, cultural learning may be inherently noisy. To this must be added individual attempts to learn based on environmental cues. It could easily be that cultural transmission is not sufficiently accurate to generate much heritability at the population level (see, however, the developmental evidence reviewed by Haun and Over, this volume). If this is the case, then observed heritability must be due to some kind of frequency-dependent process, like conformist transmission which preserves between-group variation (for a model of how conformist transmission creates group-level heritability, see Henrich and Boyd 2002a), and, as a result, the process of cultural accumulation of adaptive technology might be quite different than that explored in existing models.

In addition, we do not know the extent to which people have causal understandings of the technologies on which they depend. Once again there are two extreme models. On one hand, innovation is the rate-limiting step, but when innovations do occur they are accompanied by causal understandings of how the innovation works, and why it is better than previously used alternatives.

The innovation spreads rapidly because causal understanding spreads with it. Innovation driven by modern science in some domains may approximate this hypothesis. At the other extreme, behavior varies randomly and learners adopt behavior that is associated with prestige or other observable markers of success; as a result, better technologies spread due to a process of selective retention. A variety of intermediate hypotheses are also possible. It may be, as in the models described above, that learning is relatively rare and noisy, and so acts like a high rate of mutation in adaptive directions. In this view, individuals have limited causal understanding which increases the rate of adaptive innovation; thereafter, most spread is due to the correlation of observable behavior with markers of success. There are a rich variety of possible hypotheses that should be explored, both theoretically and empirically.

8

Long-Term Trajectories of Technological Change

Stephen Shennan

Abstract

The study of technology is a field in which generalized evolutionary ideas have been current for many years. However, when we start trying to implement a cultural evolutionary approach more rigorously, it turns out to be more complex than usually supposed. One of the important benefits of taking a cultural evolutionary approach is that it goes beyond relatively simple ideas of competition and technological improvement, and introduces a range of other forces whose impact is not often considered. In the case of technology, the entities that are the subject of variation, inheritance, and selection processes are technological lineages, recipes for techniques, routines, and practices linked by ancestor–descendant relationships. To understand them, we must first address histories of the technologies themselves before we can examine the histories of the human populations through which they are transmitted, which may depend at least partly on the histories of technologies. A number of examples of technological innovation and transmission are examined to illustrate the variety of factors affecting them.

Introduction

The study of technology is a field in which evolutionary ideas have been current for many years. The idea of a kind of Darwinian competition between different means of achieving the same practical ends is one that seems obvious, and indeed has been very fruitful. However, when we start trying to implement a cultural evolutionary approach more rigorously, it turns out to be more complex than expected. It can be argued that one of the important benefits of taking a cultural evolutionary approach is that it goes beyond relatively simple ideas of competition and technological improvement, and introduces a range of other forces whose impact is not often considered. A variety of difficulties also become apparent, not least in terms of the availability of quantitative data to test evolutionary hypotheses when we attempt to go beyond the last 150 years or so—the point in time when information about patents and businesses becomes available. In this chapter, I outline the framework adopted here for viewing technological change as a variation, inheritance, and selection process and discuss several case studies. These are mainly archaeological, not just because I am an archaeologist, but because prior to the last 150 or so years, only archaeology provides even the possibility of obtaining the quantitative diachronic data necessary to test evolutionary models.

What Evolves?

Arthur (2009) defines a technology as "a phenomenon captured and put to use," and distinguishes between "standard technologies" based on physical effects and "nontechnology-like technologies" that impact human behavior or organization. In the case of "standard technologies," which are the focus of this chapter, the entities that are the subject of variation, inheritance, and selection processes are technological lineages, recipes for techniques, routines, and practices linked by ancestor–descendant relationships, which capture and put to use specific phenomena in particular ways. Thus, ceramic vessels, for example, utilize one set of principles to make containers, whereas barrels or plastic bowls make use of very different sets of principles. The replacement of ceramic vessels by plastic ones does not represent a continuous lineage, but a replacement of one lineage by another (though, of course, plastics have their own lineage). Lineages may be regarded as replicators in Dawkins's terms (1976). However, distinctions here are blurred (Lake 1998; see also the extended discussion in Wimsatt and Griesemer 2007, not least their discussion of Sears kit houses). Artifacts are clearly interactors in some respects rather than replicators, but they can also provide a model for people to copy. On the other hand, as Mokyr (2000) points out, techniques themselves can also be interactors, whose effectiveness determines whether they will be reproduced, and the knowledge behind them can be seen as a replicator. But in the past, as Mokyr also points out, using a successful technique did not generally involve knowledge of the principles behind it, and insofar as people formulated such principles, they were often completely mistaken.

The key point is that the object of the evolutionary analysis of technology is technological lineages, not human populations, and because they are reproduced in different ways, there is no reason that they should run in parallel even though they are obviously linked. The variation, selection, and retention processes that underlie cultural evolution were laid out in detail more than 25 years ago (Cavalli-Sforza and Feldman 1981; Boyd and Richerson 1985) and have been extensively explored and elaborated since (e.g., papers in Boyd and Richerson 2005). However, this has mostly been done from an agent-centered perspective, rather than from that of the cultural lineages themselves—the "meme's eye view"—and the two are not the same. This does not mean that we need to subscribe to memetics in the strict sense of assuming a particulate unit of transmission (cf. Henrich and Boyd 2002b; O'Brien et al. 2010; Mesoudi et al. this volume) or to assume that memes are cultural viruses that simply parasitize human brains. It does, however, require us to acknowledge that cultural lineages, including technological lineages, exist and that the transmission of a cultural lineage is by definition vertical with respect to itself, whatever its relation to the human populations through which it is transmitted—a relationship that is very likely to vary over time.

Moreover, the processes that modify what is transmitted also look different, depending on whether one takes the agent- or meme-centered perspective. Thus, in their paper on the evolution of Polynesian canoes, Rogers and Ehrlich (2008) refer to the process which acts on those canoe traits that have a functional significance—technological traits—as natural selection; so it is from the perspective of the traits themselves that traits survive and are copied preferentially as a result of their greater functional effectiveness—something which could, in principle, be tested experimentally. What the authors do not do is to distinguish between *natural selection operating on human agents via cultural traits*, and thus on the *future frequency of those traits*, and *results bias*. In other words, the process could have operated as a result of the makers and users of ineffective canoes drowning more frequently, thus leading to the demise of those designs, whereas groups with better-designed canoes, perhaps different communities, survived and colonized new islands. Alternatively, it could have worked through people observing the performance of different canoe designs and preferentially copying those they perceived as more effective. Making this sort of distinction is actually at the root of some of the most long-standing debates in archaeology; for example, whether the spread of farming into Europe was a process of indigenous adoption (involving results bias) or demographic expansion and extinction (natural selection acting on the bearers of cultural traditions). Despite the numerous attacks on the idea of memes as replicators encouraging their own reproduction, it is emphatically the case in both of the above scenarios that whether or not people reproduce particular traits depends on the specific characteristics of the traits themselves.

In general then, to understand trajectories of technological change we need to do two things:

1. Address histories of technologies.
2. Examine the histories of the human populations through which they are transmitted, which may depend at least partly on the histories of technologies.

With regard to the first, it is necessary to identify histories of technological transmission to show that an ancestor–descendant relationship exists, if indeed it does. Continuities or discontinuities through time may simply reflect contingently optimal responses to stable or unstable local ecological conditions. Moreover, some traits (e.g., the sharpness of a lithic cutting edge) may be so strongly determined by their function that they will contain no signal of their transmission history, even though it is likely that they had one (as opposed to being discovered anew by every novice flint knapper through trial-and-error learning). In any case, assuming that a real technological lineage has been identified, we must then attempt to understand the forces shaping it. With regard to the second, we need to understand independently the histories of the *relevant* populations. The possible linkages or feedback loops between technologies and populations are many and varied. A technology may or may not have an effect on the survival and reproductive success of a human population while replicating successfully itself. Even if it does, its effects may be overwhelmed by other factors, such as the extinction of the population for other reasons. If a technology is vertically transmitted with respect to its human population, it may then be particularly vulnerable in this respect. It perhaps goes without saying that few, if any, studies have taken all of these different elements of the evolutionary analysis of technology into account.

Invention and Connectedness

Some inventions are relatively easy to make and have therefore been made again and again. For example, it seems that wherever people came to depend to a significant extent on collected seeds for food, they invented and used grinding stones, since these are very effective tools for processing seeds into food and the cost of producing them is far outweighed by the subsequent benefits gained. It is also worth using this example to raise another issue: It is now increasingly clear, as I will discuss below, that the probability of invention and innovations occurring is a function of the relevant effective population size. However, finding a correlation between technological complexity and population size should not necessarily lead us toward this particular explanation. In the case of the innovation of grinding stones, there is almost certainly a correlation with increasing population density but this is because in many foraging contexts increasing population density leads to increased exploitation of plant resources, which in turn leads to the use of grinding stones for cost–benefit reasons.

In this connection, it is worth referring to Perkins's (2000) contrast in the space of technological search between "Homing spaces," where there is a clear gradient pointing toward a viable outcome, and "Klondike spaces," where high-yield nuggets are sparsely distributed and there are few signs to indicate when you are near one of them. When technologies involve complex and difficult production processes, they are far less likely to be reinvented if lost, simply because of that difficulty. Moreover, Roux (2010) makes the point that training in any production process that involves the acquisition of high levels of expertise which take a long time to acquire is likely to restrict innovation, because the whole process of learning is designed to fix particular sets of skills and knowledge (cf. Martin 2000); this may be particularly the case with physical expertise and motor skills. Only the most expert, those who have complete control of all aspects of a process and its associated knowledge, are likely to transcend the limits of what they have learned and invent something new. Roux cites (2010:224) the example of inventions in the field of pyrotechnology by traditional craftsmen in India, which were only made by the most skillful individuals who were "exceptional as much for their skills as for their rarity."

Roux's (2010) account of the apparently strange history of the production in the prehistoric Southern Levant of wheel-fashioned pottery (i.e., coil-built, not wheel-thrown, pottery produced on a slowly rotating wheel at around 80 revolutions per minute), using the principle of rotational kinetic energy (RKE), illustrates the issues raised by what she refers to as "discontinuous innovation" that uses novel principles. Using the technique has the obvious advantage that it leads to a halving of manufacturing time, and one might expect this to give it a clear selective advantage over traditional hand-coiling methods. However, in the Southern Levant, the production of wheel-fashioned pottery comes and goes twice before finally becoming permanently established some 3,000 years later. It seems to have been invented in the late fifth millennium BCE and to have been used for over ca. 300 years solely to make a specific type of bowl, which had a ceremonial function, at a time and in a region of significant politico-religious change, involving the emergence of centralized chiefdoms. In the fourth millennium these societies collapsed; 75% of the settlements that had previously been occupied disappeared and wheel-coiling as a technique disappeared, though the RKE principle continued to be used for finishing vessel surfaces in some cases. The technique reappeared in the early third millennium BCE, a time when large fortified towns were built. However, only ca. 3% of vessels were made by this method, indicating that the technique was probably used only by a small number of craftsmen, a situation that lasted for ca. 500 years. When local cities collapsed at the end of the third millennium BCE, wheel-coiling was lost again, only to reappear in the mid-second millennium BCE, after which it came into much more general use.

Roux (2010:228) locates the explanation in the context of the practice and transmission of the craft. For the later fifth millennium, study of the petrology of the pottery and the techniques used to make it, as well as the contexts in which it was found, indicates that only a few individuals made the bowls, that those individuals moved around and were linked to elites. This restricted group of potters within which wheel-coiling was transmitted was distinct from the generality of potters who made the everyday pottery. The same is true of the earlier third millennium, in that wheel-coiling was restricted to a small number of specialists, whose potter's turntables have been found in palace contexts. When the use of RKE took off in the mid-second millennium it was at a time when cities were expanding, probably in the context of a more market-oriented economy, when specialist workshops used the technique to make a wide range of vessels and domestic pottery production declined as a result.

Roux characterizes the wheel-coiling technical system prior to the mid-second millennium as fragile and closed. Fragility relates to the size of the network concerned in terms of the number of interconnected elements. In a precise analogy with genetic drift, any practice that is restricted to a small number of individuals is vulnerable to loss as a result of external circumstances, regardless of its benefits (cf. Rivers 1926). The fact that in the earlier periods few potters were using wheel-coiling meant that the practice disappeared in the face of the socioeconomic collapses which ended the fifth millennium Chalcolithic and third millennium Early Bronze Age periods, because the small number of transmission links that sustained it were broken. In contrast, as the number and spatial extent of transmission links increases, the less vulnerable the technical system becomes to the effects of external historical events, because even if part of the network is destroyed in one place it will survive elsewhere, and this is what happened in the Middle Bronze Age with the expansion of the wheel-coiling transmission network.

The size of the network in itself, however, is only one element in the fragility or robustness of a given technical transmission system. One way in which it can expand is by the transfer of the technique to other areas of production than that in which it originally emerged. Again this did not happen in the Chalcolithic or Early Bronze Age. Wheel-coiling remained restricted to a small circle of specialist potters who only used it to produce specific elite items. This changed in the Middle Bronze Age of the Southern Levant with the application of this innovative practice to a wide range of different items in widespread use.

As noted above, the importance of the effective population size involved in cultural transmission in accounting for the emergence and maintenance or loss of innovations that can build on one another, thus leading to cumulative cultural evolution, is now widely recognized (Shennan 2001; Henrich 2004b; Powell et al. 2009). However, Roux makes a further and more specific point relevant to the evolutionary trajectories of the more complex technologies, not least copper and iron production, that have emerged in the last 7,000–8,000 years. If the transmission of technologies based on novel principles involves a long apprenticeship that excludes the possibility of becoming correspondingly adept at other skills, then it will almost inevitably result in a closed and more or less fragile system, at least initially. First, it will only be possible for relatively small numbers of individuals to undertake the apprenticeship. Second, not all societies will be able to sustain the required division of labor. Third, for various reasons, including people's desire to protect their livelihood, they are likely to keep their knowledge and expertise secret and either transmit it vertically to offspring or charge a significant entry fee.

How such processes work in the case of iron production has recently been examined by Charlton et al. (2010), who have taken an evolutionary approach to understanding preindustrial bloomery iron-smelting technology in a case study from northwest Wales. Despite the technical complications of addressing this subject, it has the advantage that the problem can be clearly framed.

There is no doubt about the goal (at least in general terms), and the conditions required to smelt iron successfully are well understood, arising as they do from universal properties of the materials involved. The bloomery method involves "the solid state reduction of iron oxides into a spongy mass of iron, called a bloom, and the production of a ferrosilicate slag" (Charlton et al. 2010:353), based on the combination of iron ore and charcoal in a furnace supplied with air, where temperatures can reach up to 1400°C, which is still below the melting point of iron. Earlier studies of bloomery iron production in different parts of the world have shown that it is technically extremely diverse.

Once again, as with any study of a technological lineage, the first issue is to identify patterns of cultural descent in the methods used and to distinguish variation arising from transmission from that relating, for example, to the local ore or fuel type; thereafter the forces which affect that variation must be characterized in a situation where, by the very nature of the process, there are only a limited number of successful solutions. The most informative source of information on the processes involved in past episodes of early iron production is chemical variation in the slags produced as a waste product. In this case, the data are quantitative variations in the chemistry of chronologically ordered slag deposits, and the problem is: Can we establish whether or not there is a signal of cultural descent in the chemical variation? If so, what can we infer about the factors that affect the transmission processes which produced it?

Charlton (2009) showed convincingly that a transmission signal could be identified in the slag chemistry, distinct from the effects of ore and fuel composition. In terms of the forces acting on the technical knowledge and practices passed on from one iron producer to another, it is easy to imagine that there might be some more or less random variation in exactly what was done each time. It is also likely that there would be strong selection for those practices that were successful although, given the complexity of the process and its many stages, it would not necessarily be easy to identify precisely what produced a successful smelt on any given occasion. From the point of view of the agents, it is thus likely that transmission would be affected by results bias, albeit in a very noisy form, based on characteristics that the smelters could observe as well as on the connections they were able to make between variation in those characteristics and their techniques of ore preparation, furnace operation, etc. From the point of view of the smelting recipes, this would be a process of natural selection, since recipes would be differentially reproduced depending on their ability to smelt iron successfully. The results of Charlton et al. (2010) suggest that initially all changes related to furnace operation could be accounted for by a drift process, but that at a certain point a second effective procedure was more or less accidentally discovered and a decision was taken to make use of the two distinct procedures, visible in consistently different slag signatures: one of lower yield than the other but producing a higher-quality product in the form of steel. At the same time, there were clear trends in the use of manganese-rich ores with better fluxing capabilities and evidence of decreased variability in reducing conditions related to results bias; that is, iron makers consistently reproduced the airflow conditions that gave the best results for a given recipe.

This, however, was a very small operation, using only a single furnace. At present we do not know if the discovery of these Welsh iron smelters was also independently made by others, or if it was passed on outside this group. Roux's points regarding the closed and fragile nature of transmission networks in the context of preindustrial craft specialization would lead us to hypothesize that innovations made by this group of smelters could easily have been lost.

Juleff (2009) provides another example of reconstructing and explaining a technological lineage in the context of iron production, based on the archaeological discovery of wind-powered furnaces in Sri Lanka. She argues that the recognition by early metal producers of the idea of the "combustion zone" or "effective unit area" within the furnace effectively created a technological meme, an indivisible entity that represented the basic building block of a specific iron-production lineage which, in contrast to the predominant Western tradition (which was based on round furnaces), created linear furnaces in which combustion zones were lined up side-by-side through the regular placement of tuyeres in the front wall to create a draft. This had the advantage of eliminating ineffective zones, such as could potentially arise with getting air to the center of an increasingly large circular furnace. It was further adapted in some favorable hilltop locations in Sri Lanka to the use of monsoon winds, which provided the draft.

Juleff suggests that there is a single evolutionary tree of south, southeast, and east Asian linear furnaces deriving from a single Sri Lankan root (though much more work would need to be done to establish this) and notes that some of them produced high-quality carbon steel for weapons. However, in Sri Lanka itself the lineage seems to have died out in the eleventh century CE. In this case, two potentially opposing forces may have had a bearing on the life of this tradition: (a) its likely vertical transmission in a small population of hereditary iron-smelting groups; (b) its visibility to any nonlocal iron smelters who did happen to visit the areas concerned.

The subtle ways in which selection operated in the context of early complex technologies, leading to optimal solutions, is also illustrated by Jackson and Smedley's work on medieval glass production in northern Europe (Jackson and Smedley 2008). Here bracken was used as a main source of plant ash for glass production. Jackson's analysis of the chemical composition of bracken ashes showed that it changes over the course of the growing season, with the concentration of SiO_2, CaO, Fe_2O_3, and MnO increasing throughout the season and K_2O and P_2O_5 peaking in mid-June and then declining. This makes mid-June the best time to harvest bracken since the alkali concentrations necessary to produce glass are high, and other components that produce less satisfactory outcomes (e.g., coloring) are still low. The knowledge that this was the case was encapsulated in the recommendation to be found in contemporary documents about glass making, which reported that for the best results, bracken should be harvested by the feast of St. John the Baptist (June 24).

Finally, it is worth noting that the distinction between imitation and invention is not necessarily as great as is generally assumed. As Berg (2002) shows, in eighteenth-century Europe the imitation of such materials as Chinese porcelain by local manufacturers was a major source of novel production processes, as evidenced by secured patents (see also Rogers and Ehrlich 2008:175–184).

Innovation Rates

In the case of genetic variation, mutations are basically random, though of course different genes have different mutation rates. In the case of cultural inventions, it seems clear that rates have varied. One tendency has been to take the view that "necessity is the mother of invention": when people are in a situation which they know is not sustainable, they are more likely to take risks and carry out trial-and-error experimentation (Fitzhugh 2001). As Henrich (2009b) points out, however, experimental evidence tends to be against this, and the risks involved are certainly very considerable, from devising alternatives to embarking on them without prior knowledge of the costs and benefits. As we have already seen, the contrasting view—not that they are mutually exclusive—is that they are more likely to arise simply as a result of the existence of larger effective populations, in terms of absolute size, degree of networking or both, since these are likely to increase the innovation rate per unit time as well as the potential for the new combinations of existing elements (what Roux calls "continuous inventions"). As noted earlier, the collapse of populations and social systems leads to the loss of complex technological lineages that have few bearers.[1]

It is also possible that invention rates vary less than innovation rates, in that even if the invention is made, the chances of it becoming a successful innovation will vary. Once again, population size and connectedness are relevant but another relevant factor is the potential, at any given time, for the technological equivalent of an "adaptive radiation," a situation where the diversity of life increases because speciation rates are greater than extinction rates (Lake and Venti 2009). This generally involves the colonization of an empty or newly created ecological space. Of course, the most famous examples are the

1 An alternative theoretical possibility following from the previous discussion of "the meme's eye view" is that the relevant population here is the number of instances of the cultural variant itself; here, the number of wheel-fashioned vessels. This may well be the relevant population in cases where a variant is easy to copy simply as a result of seeing it. However, in the case considered here, it seems unlikely given the difficulty of reverse engineering the process from the product and the amount of time that would have been necessary to acquire the necessary skills.

so-called "Cambrian explosion" and, at a lower taxonomic level, the radiation of Galapagos finches.

O'Brien and Bentley (2011) cite the example of spark arrestors on the smokestacks of U.S. locomotives in the nineteenth century, for which more than 1,000 patents were taken out to solve the problem of trains causing fires while keeping the draft as unobstructed as possible, which they view in terms of fitness landscapes (Kauffman 1993; Kauffman et al. 2000). In this case there was no entirely successful solution, and this new adaptive space remained a rugged fitness landscape with a number of more or less equally fit local optima. Previously empty or newly created niches offer new possibilities for innovation, either because there is no previous history from which to learn or because existing solutions may not work in the new niche. In Kaufmann's adaptive radiation model, whether biological or technological, there is an initial stage in which there are long jumps to new places in the fitness landscape; a second stage when the broad landscape has been explored and local "hill-climbing" occurs, because earlier choices constrain subsequent options in a process of so-called "generative entrenchment" (Wimsatt 1986); and a third stage of relatively little change where adaptations to local peaks cannot be improved but selection between different peaks begins to occur. Lake and Venti (2009) followed through the work of Van Nierop et al. (1997) in exploring Kaufmann's suggestion that the evolution of the bicycle could be seen in precisely these terms. On the basis of a classification of bicycles that takes into account the different levels at which diversity occurs, and thus the process of generative entrenchment, they show through the creation and analysis of a novel kind of taxonomic diversity diagram that the evolution of the bicycle does indeed follow the model of breadth-first search. Lyman et al.'s (2009) analysis of the distribution through time of prehistoric dart and arrow tip diversity in the United States points to a similar conclusion.

Evolutionary Success and Cumulative Culture

As argued at the beginning of this chapter, following the work of Boyd and Richerson as well as many others, the processes at work in the cultural evolution of technology are complex and operate on two main levels: one related to the cultural success of the technologies or technological elements themselves, the other related to the human populations that coevolve with them to varying degrees. Just how "cultural success" is defined is open to discussion; wide-scale prevalence is not necessarily as tautologous an indicator as critics of evolutionary approaches to culture tend to assume. Cultural traits can go to fixation as a result of drift as well as selection, especially in small groups, and small groups have probably characterized most of the past of modern humans and their ancestors. Better evidence comes from cost–benefit comparisons of the type carried out by human behavioral ecologists, based on present-day ethnographic study or experimentation, where the demonstration of favorable cost–benefit balances can make it reasonable to suppose that one technology would have had a selective advantage over another. The assumption is usually made that a selective advantage, from the point of view of the meme lineage, corresponds to a survival/reproductive success advantage, from the point of view of the human population, whatever the mechanism involved in any particular case, in terms of the various biases and forces outlined by Boyd and Richerson and others. However, it is important to insist again that this assumption cannot be taken for granted. In the first place, as they pointed out, there is no reason to expect strong correlation in any particular case given the asymmetry in their lines of transmission. In addition, it is very difficult to actually demonstrate that practicing one version of a behavior rather than another leads to greater reproductive success. If we take the milking of domestic animals as a technology, then the selective advantage of the lactase persistence gene in combination with milking is one of the very few examples where we have direct evidence of a technology contributing to reproductive success (Bersaglieri et al. 2004). Using the development of farming more generally, then recent work demonstrates how complex the processes can be. Bowles (2011) has provided strong evidence that cereal-based agriculture is very unlikely to have been more productive initially than foraging in terms of calorific return per unit labor time; thus, its adoption cannot be explained by a superior cost–benefit relationship on a day-to-day basis. Despite this, it is very clear that it provided greater reproductive success, given archaeological evidence for the dispersal of farming populations at the expense of foragers and evidence from genetics and the age distributions of individuals in prehistoric cemeteries that these farming populations were growing whereas forager ones were relatively static (e.g., Bocquet-Appel 2002; Gignoux et al. 2011).

In this context it is worth raising the issue of the so-called "cultural ratchet" (Tomasello 1994). The evidence that social learning is far more important in humans than in closely related primate species, and that human children show a very strong propensity for high-fidelity imitation (e.g., Dean et al. 2012; see also Haun and Over, this volume), is understandably featured in explanations of the fact that human culture more or less uniquely accumulates modifications over time. However, it is important to be clear that the capacity for high-fidelity transmission was a necessary precondition for this to occur, not that the accumulation of cultural modifications was the selective force behind it. Indeed, there is little evidence for this in terms of technology at least, given its very slow rate of change over most of the 2.5 million years since the first stone tools are recognizable in the archaeological record.

In fact, though the general propensity for high-fidelity social learning is a species property, specific instances of cumulative culture itself are actually a property of specific populations or meta-populations linked by transmission processes. Selection will be operating on them through the cost–benefit dimension but, even in cases where the cost–benefit ratio is favorable, transmission failures that arise from a variety of factors affecting the fidelity of transmission (Lewis and Laland 2012), including the fact that early networks were often fragile, can overwhelm selection and lead to cultural loss. In this sense the "ratchet" metaphor, with its implications of irreversible accumulation, is misleading (cf. Lombard 2012).

The well-known Howieson's Poort archaeological assemblage from southern Africa ca. 65–60 KYA illustrates these issues clearly. It includes a variety of what can be considered complex cultural (including technological) phenomena. Among them is evidence for the heat treatment of tool stone to improve its workability, of multicomponent stone tools, and probably of the bow and arrow (Lombard and Phillipson 2010; Lombard and Haidle 2012), features which seem to disappear from the record with the end of the Howieson's Poort.

The bow and arrow involved the use of arrows with stone tips hafted in a variety of ways and attached using complex adhesives made from plant gums and ochre, which would probably have required heat treatment to dry. It also required a practical understanding of the use of stored energy and the best wood types for making use of it. In this connection, it has been argued that the range, size, age, and behavior of the animals represented in the bone assemblage from the site of Sibudu Cave may point to the use of snares for trapping (Wadley 2010). The use of strong cord or hide strips, with appropriate knots, would also have been required. Lombard and Haidle's (2012) detailed analysis shows that compared with the production of a simple wooden spear, or even a composite stone-tipped spear, the bow-and-arrow combination represented a major increase in the number of material items and different operations that had to be brought together, supporting Lewis and Laland's suggestion, based on their simulation results, that trait combination may be the most important creative process producing cultural accumulation (Lewis and Laland 2012).

Of course, this does not mean that the bow and arrow was bound to continue. Explanations for its subsequent disappearance, and that of the Howieson's Poort generally, vary between those that relate it to loss as a result of decreased population densities or interaction rates and thus a smaller effective population size (e.g., Powell et al. 2009), and those which see it as simply no longer successful in cost–benefit terms in a changed environment. Mackay and Marwick (2011) emphasize the importance of evaluating the costs and benefits of more and less complex technologies in different environments, arguing that we should see the innovations of the Howieson's Poort as adaptations to local circumstances, not as systems that for intrinsic reasons "should have" continued, the perspective, as we have seen, that the "cultural ratchet" idea tends to encourage (again, cf. Lombard 2012). These debates will only be resolved by further data collection. In this context, a recent synthesis of demography and climate in later Pleistocene Africa (Blome et al. 2012) has not found evidence for population decline in southern Africa at the relevant time period, though the temporal resolution remains poor.

Conclusion

A cultural evolutionary approach to technology has a number of important attractions, perhaps the main one being the "population thinking" that it brings. This has had, and will continue to have, a number of important consequences. First, it suggests that population size and connectedness are key factors in affecting rates of cultural innovation and evolution, and that social institutions have an effect on those rates to the extent that they impact transmission fidelity and encourage or discourage connectedness between individuals and groups: who you know is far more important than what you know (cf. Henrich 2010). Early complex or specialist technologies had a number of intrinsic features that would have encouraged small effective population sizes, in terms of both absolute numbers of practitioners and the likelihood of them sharing information. Second, it makes us aware of the importance of characterizing technological lineages and distinguishing them from populations of human agents, thus forcing us to think about the relations between them. More generally, the theoretical work of the last thirty years has demonstrated the wide range of transmission forces potentially operating in any given case, not least the role that can be played by unbiased transmission in finite populations. Far from encouraging the tautologous view (i.e., what is prevalent is by definition that which has been competitively successful), work in cultural evolution has demonstrated the complexity of making inferences about the forces at work while providing the tools to deal with it, including cost–benefit analyses of the type central to human behavioral ecology and mathematical modeling. It also points to the contingency of the supposed "cultural ratchet," despite the undeniable human propensity for high-fidelity social learning.

Acknowledgments

I am grateful to Kevin Laland and an anonymous referee for comments on an earlier version of this paper; to the Ernst Strüngmann Forum for the great privilege of attending the enormously stimulating Forum on Cultural Evolution; and to Julia Lupp and her colleagues for the outstanding arrangements that contributed so much to making it a great success.

9

Neuroscience of Technology

Dietrich Stout

Abstract

Although there is a burgeoning neuroscience of tool use, there is nothing that might be properly called a neuroscience of technology. This review aims to sketch the outlines of such a subject area and its relevance to the study of cultural evolution. Technology is itself an ill-defined term and is often taken to correspond loosely to human action that (a) involves the use or modification of objects, (b) displays a complexly organized multilevel structure, and (c) is socially reproduced. These characteristics may be better understood with reference to neuroscience research on perceptual-motor control, object manipulation, motor resonance, imitation learning, and goal-directed action. Such consideration suggests a number of biases which may affect the cultural evolution of technologies.

Introduction

In the extended analogy developed by Mesoudi et al. (2006), neuroscience is described as the "molecular genetics" of cultural evolution. That is, neuroscience is meant to provide a mechanistic understanding of the way in which cultural traits are instantiated as neural processes that can be replicated across individuals and expressed as (more or less) isomorphic behaviors. Mesoudi et al. recognize that strict analogies with the molecular mechanisms of DNA replication and expression are not likely to be very helpful; rather, the intended analogy appears to be at the level of the explanatory work that needs to be done in each case. Whether or not this genetic analogy proves fruitful with respect to research in social neuroscience, analogies between biological and cultural evolution proposed at higher levels of analysis (e.g., the selection and drift of cultural traits) will stand or fall on their own merit.

As it has matured, neuroscience research has moved away from attempts to identify simple one-to-one structure–function mappings and recognized the need for analyses in terms of dynamic and variable neuronal networks that are soft-assembled in response to context-specific task demands. Neural systems, like genetic and immune systems, are massively *degenerate*, meaning that different structural elements can produce the same functional output (Edelman and Gally 2001; Price and Friston 2002). Conversely, many structural elements also appear to be *pluripotent* (e.g., Anderson 2010), meaning that single structures are capable of supporting multiple, different functions. Such one-to-many and many-to-one structure–function mappings are present across all levels of analysis (Edelman and Gally 2001), invalidating attempts to derive function from structure in a purely bottom-up way. Processes at the neural level cannot be properly understood without reference to higher-order functional and contextual constraints any more than the genetic information "coded" by DNA sequences can be understood apart from the larger processes of cellular metabolism, somatic development (Mayr 1994), and organism reproduction in which it is embedded. In neither case is it possible to assign causal primacy to a "replicator" identified at one fundamental level of analysis. Instead, our objective should be to identify multilevel constraints acting on the reproduction of behavior across individuals including, but not limited to, species-typical learning mechanisms and biases.

Neuroscience can contribute to this enterprise through an iterative research program in which structural and physiological correlates of behavior are used to inform the fractionation of psychological processes, and the fractionation of psychological processes motivates increasingly refined neuroscientific investigation. A neuroscience of technology would seem to be a good place to begin, considering the central place that technology (e.g., Boyd et al. 2011 and this volume) and technological artifacts (e.g., Shennan 2011 and this volume) have had in studies of cultural evolution, as well as the volume of neuroscience research devoted to understanding the perception, execution, and imitation of goal-directed interactions with objects. First, however, it is necessary to step back and consider what exactly we mean by "technology" in this context.

What Is Technology?

Technology is a fuzzy category. There can be little doubt that central examples like laptop computers and atlatls belong, but more peripheral examples raise questions. Is music a technology? What about a martial art or sign language? Is tool use by nonhumans "technology," and how should we categorize complex foraging techniques which do not involve tools (e.g., Byrne and Byrne 1993)? Attempting an exclusive and exhaustive definition of technology is likely to be neither possible nor profitable. It would appear to be more important to identify key dimensions of variation in the "family resemblance" that links exemplars, so that these may become the subject of further study. In other words, we should concentrate less on this question ("What is technology?") and instead openly explore the issue: "What is interesting about technology?" For current purposes, some of the more interesting things about technology are that it (a) often involves the use and modification of objects, (b) is characterized by complexly organized goal structures, and (c) is heavily reliant on social mechanisms for its reproduction.

Objects

From an anthropological perspective, technology's tendency to involve objects is of practical interest because it increases the chance of behaviors leaving physical traces for future study. It is also of theoretical interest because it leads to creation of a durable medium for human action and interaction, broadly referred to as "material culture." From a somewhat narrower cognitive and neuroscientific perspective (e.g., Arbib et al. 2009), the involvement of objects is interesting because it complicates the perceptual-motor and cognitive control of action by introducing a wider array of potential affordances and effectivities to be discovered and coordinated and by requiring mechanisms of perceptual monitoring and internal modeling in the absence of direct somatosensory feedback from the end effector. Furthermore, the potentially greater temporal persistence and causal diversity of object-mediated actions may support the production of more complexly organized and temporally protracted action goals and sequences.

Complex Organization

Complex perceptual-motor and cognitive organization is a basic characteristic of technology, whether or not this complex organization is directly occasioned by the use of tools. For example, weaving a basket is classically "technological" even if no tools are used, whereas sweeping the floor is a more liminal example, even though a tool is being used purposefully to alter the physical environment. This largely reflects the intuition that basket weaving is a more complex and organized activity, but what exactly is meant by "organization" and "complexity" in this context? As recently discussed by Deacon (2012), an information theoretical approach would define organization and complexity as opposite extremes of a scale measuring the redundancy of a system. For example, a pattern of random static on a television screen is maximally *complex* because it is not constrained by any redundant patterning: there is no way to summarize the image on the screen without specifying the state of each individual pixel. It follows that complexity is also increased by increasing the total number of elements (e.g., pixels) or the number of different possible states of each element (e.g., colors). In contrast, many photographs contain *redundant* (i.e., predictable) patterns such as edges, fields, or gradients which allow "lossless" image compression (e.g., the LZW algorithm used to generate GIF, TIFF, and PDF files). These images are more *organized*, but less complex, than an equally sized field of static. A blank white screen is maximally organized and minimally complex because there is only one color option for each pixel. As it turns out, then, maximal complexity and maximal organization are both quite dull.

What is interesting (and characteristic of technology) is *complex organization*. This apparent oxymoron is achieved by opposing complexity to organization at different levels of abstraction; in other words, by exploiting the generative potential of hierarchical[1] systems (Lashley 1951; Mesoudi and O'Brien 2008c; Simon 1962). A classic example from the movement sciences is provided by Bernstein (1996), who described the arm motions of blacksmiths striking a chisel with a hammer. Surprisingly, Bernstein found that the movement trajectories of individual joints in the arms of these expert craftsmen were relatively unpredictable (i.e., "complex") across swings. Nevertheless, these complex movements produced a redundant (i.e., "organized") action outcome across swings in the form of a highly consistent trajectory of the hammer head. Thus, the repeated hammering action is well organized in terms of its consistently reproduced goal but remains complex in terms of its actual kinematic means. Moving upward in scale, one might similarly consider the assembly of redundant action "types" (e.g., hammering, heating, quenching) into complexly contingent action sequences that are again redundant on the still higher level of the standardized artifacts produced (e.g., Japanese swords, Martin 2000). This logic is quite familiar to stone knappers, who must produce standard products "based on raw material which is never standard, and with gestures of percussion that are never perfectly delivered" (Pelegrin 1990:117). In such cases, redundancy in technological outcomes actually *requires* variation in means. Note that it is the increasing abstraction of goals at higher hierarchical levels which supports this generative interplay of complexity and organization by allowing heterogeneous subordinate elements to constitute uniform superordinate goals. This is closely analogous to the way in which standard grammatical units (e.g., noun phrases) can be constituted from an infinite variety of different words. Complexly organized goal hierarchies are not only characteristic of technology, they are also critical in supporting both the adaptive flexibility and the social reproduction of technological behaviors (Byrne and Russon 1998; Wolpert et al. 2003).

Social Reproduction

The heavy reliance of technology on social mechanisms of reproduction is, of course, another one of its interesting features, and the primary focus of the current discussion. Loosely speaking, such reliance differentiates technologies from more biologically determined "instincts," such as the dam building of beavers or the nest building of birds. We must be careful, however, to avoid

---

1 In this chapter, "hierarchy" refers specifically to compositional containment hierarchies in which superordinate behavioral elements are constituted by subordinate elements; causation may be bidirectional (bottom-up and top-down). There is no implication with respect to the organization, hierarchical or otherwise, of neural or cognitive systems implementing these behavioral hierarchies.

simply resurrecting the sterile nature–nurture dichotomy in a new context. Social context plays a role in the development of all human behaviors, and no behavior is completely unconditioned by biologically inherited characteristics. Even a prototypically "innate" human skill like bipedal walking develops through an interaction of physical maturation and socially mediated opportunities for practice (Adolph et al. 2003). The crucial distinction is that the constraints organizing human gait derive almost entirely from the interaction of evolved biomechanical and neural traits with physical substrates rather than from social influences. Social scaffolding serves to motivate the developing system rather than to structure the behavior. Such social motivation may also be critical to technological reproduction (cf. Lave and Wenger 1991), but it is the social transmission of behavioral targets (i.e., multilevel action "goals" as discussed above) that is distinctive.

For example, the remarkable quality and consistency of Japanese swords is achieved by following an elaborate production recipe organized around a vast number of subgoals (Martin 2000), ranging from the kinematics of properly executed hammer strokes, to subtle perceptual cues indicating desired material properties and transformations, to abstract conceptual representations of various production stages or sequences. The consistent outcomes achieved by master sword makers do not simply "fall out" of interactions between some global goal (i.e., "make a sword") and preexisting anatomical and environmental constraints in the way that the regularities of human gait do. Rather they reflect huge amounts of work (cf. Deacon 2012), both individual and social, invested in generating and regenerating the particular array of behavioral constraints that allow practitioners, starting from variable initial conditions and using procedures that are never identical, nevertheless to converge reliably on an astronomically unlikely outcome. As we have seen, this requires a hierarchical structure in which behavioral complexity can be simultaneously preserved and constrained on different levels of abstraction. The interesting question for the study of technology is: How do culturally constructed objects, situations, and social interactions come to constrain individual behavior in precisely this manner?

Low-Level Constraints: Tools and Actions

Tools

Human behavior is constrained not only by inherited somatic and neural structures, but also by an inherited cultural niche filled with predesigned tools. For example, the design of a bicycle affords only a very narrow range of effective actions, and no one needs to show an infant which end of a toy hammer to hold or which end is used to strike an object (Lockman 2006). The constraints imposed by designed objects generate behavioral attractors that can be reproduced across individuals and generations without necessarily involving the sharing of internal action representations or conceptual knowledge. This property of artifacts is exemplified in patients who suffer from ideational or ideomotor apraxias; they have difficulty describing or pantomiming tool actions due to impaired internal representations but often produce appropriate grasps and manipulations when allowed the sensorimotor feedback from actually handling tools (Johnson-Frey 2004).

Seminal studies recording single neurons in the parietal cortex of macaque monkeys (Maravita and Iriki 2004) have shown that the use of simple tools is associated with a modification of the "body schema" to quite literally incorporate the hand-held tool as an extension of the body. Lesion data suggest that a similar mechanism is involved in simple human tool use (Berti and Frassinetti 2000). An extended period (2–6 weeks) of highly structured training (reinforcing successive elements of a behavioral chain) is required to produce such simple tool use in macaques (Iriki et al. 1996; Peeters et al. 2009) and may reflect the experience-dependent formation of new afferent connections from temporoparietal and ventrolateral prefrontal cortex (PFC) to neurons in the intraparietal sulcus (Hihara et al. 2006). Thus, macaque tool use would seem to rely on the adaptive flexibility of bodily representations in an occipitoparietal "dorsal stream" of vision-for-action (Milner and Goodale 2008). However, it is not clear that similar mechanisms can explain the much more diverse, pervasive, complex, and rapidly learned manual tool use of humans. In particular, it seems doubtful that body schema alterations alone are sufficient to explain the use of tools to alter the basic functional properties of the hand (e.g., knives, hammers, potholders) as is commonly seen in humans (Arbib et al. 2009) and perhaps also our closest living relative, the chimpanzee (Mulcahy and Call 2006; Povinelli et al. 2010).

Whereas a causal understanding of tool properties as distinct from the hand may not be necessary to explain macaque use of "simple tools" like rakes or pliers (Maravita and Iriki 2004; Peeters et al. 2009; Umiltà et al. 2008), such understanding is clearly implicated in the human use of "complex tools" which convert hand movements into qualitatively different mechanical actions (Frey 2007). It has been proposed that the human capacity for complex tool use arises from a novel integration in the left inferior parietal lobule of semantic representations of tool function from a ventral, occipitotemporal stream of vision-for-perception with the sensorimotor transformations for action in the dorsal stream, thus allowing for functionally appropriate tool prehension and use (Frey 2007). More recently it has been reported (Peeters et al. 2009) that an anterior region of human parietal cortex (anterior supramarginal gyrus, aSMG) displays a specific response to the observation of simple tool use. This region is located posterior to phAIP, the putative human homolog of the motor part of monkey anterior intraparietal sulcus (AIP) (Frey et al. 2005; Orban et al. 2006), a region involved in visuomotor grip transformations for object manipulation (Fagg and Arbib 1998). Human aSMG has been associated with the planning, pantomiming, and execution of actions with tools (Lewis 2006) and, in contrast to phAIP, is activated by the observation of simple tool use but not of unassisted hand actions (Peeters et al. 2009). It thus appears likely that, in humans, aSMG plays a specific role in coding the casual properties of simple tools as distinct from hands, and it may represent an important convergence point for dorsal and ventral streams. Indeed, training in the use of novel tools results in increased activation of a converging network of ventral and dorsal stream structures, including an anterior portion of left intraparietal sulcus (i.e., in the rough vicinity of phAIP and aSMG) (Weisberg et al. 2007).

Apart from these details of functional neuroanatomy, it is clear that tools constrain action in at least two ways. First, simple tools typically present a small number (perhaps just one) of efficient options for grasping that constrain the way in which they will typically be incorporated into the body schema. These constrained affordances, which themselves reflect the cultural evolution of artifact design, will be reliably and repeatedly discovered across individuals even in the absence of more "active" social transmission. Second, the performance characteristics of "complex" tools, likely represented in ventral stream regions concerned with nonbiological motion (posterior middle temporal gyrus) and object form (fusiform gyrus), will constrain the range of actions for which they are typically used. Given a somewhat longer time of exploration, and perhaps some socially structured motivation, these performance characteristics should also be more or less reliably rediscovered across individuals with minimal other social input, much as the dynamics of bipedal walking are.

The idea that tools constrain human action may, at first blush, seem reminiscent of anthropological arguments that attribute causal agency to artifacts independent of human users (Gosden 2005) or to suggest that artifacts themselves are "active replicators" evolving in the same way that living organisms evolve. However, artifactual constraints on behavior are relational properties which only emerge in the context of goal-oriented action by living agents and, even in this context, typically result in the reproduction of simple behaviors (e.g., particular grips) rather than reproduction of artifacts themselves. This dynamic seems better captured by the concept of ecological inheritance developed in niche construction theory (Odling-Smee et al. 1996) than by strict analogies between artifacts and organisms. Artifacts can indeed embody reproducible information about their own form, use, and construction (e.g., Caldwell and Millen 2009), *but only under the goal-oriented interpretation of a living agent* (cf. Deacon 2012). Although the simplifying assumption that artifact taxa evolve "as if" they were biological taxa has been empirically productive (O'Brien et al. 2001; Shennan 2011), if we wish to understand the actual mechanisms involved or to identify and address cases in which this simplifying assumption might not work, then we must consider the nature and transmission of constraints on interpretation by such agents.

Internal Models

Although an organism's physical environment constitutes a vast array of constraints organizing behavior (or, in positive terms, an array of "affordances" for action), it is also the case that many actions unfold too quickly to be guided by online sensory feedback and error correction. It is thought that this limitation is overcome through the use of *internal models* which predict movements and outcomes in advance of sensory feedback (Wolpert et al. 2003). More specifically, *forward models* predict the sensory consequences of motor acts (i.e., model action outcomes) whereas *inverse models* predict the motor commands necessary to produce a given action (i.e., model bodily states and transformations). It is perhaps more intuitive to refer to these two types of internal model as *predictors* and *controllers*, respectively. Briefly, predictors developed though prior experience can be used to select appropriate controllers in advance of actual sensory feedback, with *post hoc* comparison to actual outcomes allowing for error correction through the elimination of inaccurate predictors. For example, one reaches to pick up a full tea kettle expecting to require a certain amount of muscular force; upon finding the kettle empty there is a rapid reevaluation and correction. The concept of internal models developed out of computational modeling studies of motor control, and the question of their actual neural instantiation is a highly complex and controversial one. Perhaps the most consistently implicated structure is the inferior parietal cortex, which appears to play a key role in the integration of sensory and motor information, for example, during object manipulation (Arbib et al. 2009), the use of subvocal articulation to support speech perception (Price 2010), and the central cancellation of the sensory consequences of self-tickling (Blakemore et al. 1998). Such integration is obviously critical to imitation, in which the sensory consequences of others' actions must be matched to appropriate motor commands for self-execution (Wolpert et al. 2003), and numerous studies have confirmed inferior parietal cortex involvement in imitation (Buxbaum et al. 2005; Chaminade et al. 2005).

Motor Resonance

Internal models thus provide a useful framework for understanding the imitation of simple actions. The neuroscience of imitation and social cognition has been massively impacted by the description and study of *mirror neurons* in the inferior frontal and parietal cortex of macaque monkeys (Rizzolatti and Craighero 2004). These are neurons that respond both to observed actions and the self-performance of a similar action. Neurons with similar properties are thought to exist in humans (although the invasive recording techniques used in monkeys cannot be applied to humans to confirm this directly) and to reflect the direct mapping of motor representations of one's own actions to sensory representations of the actions of others. In the language of internal models, predictors are compared with sensory perception of others' actions rather than feedback from one's own actions and then used to select the appropriate controllers needed to generate isomorphic movements (Wolpert et al. 2003). These activated controllers can then be executed to produce the actual movement (imitation) or suppressed to model the movement without overt action. It is thought that this automatic activation of motor controllers, or *motor resonance*, provides a mechanism of action understanding through internal simulation, and that this basic mechanism of action understanding is the foundation for more sophisticated forms of social cognition, including the understanding of intentions (Wolpert et al. 2003) and theory of mind (Gallese et al. 2009).

Imitation

There are several important issues for this account of action understanding and imitation, particularly with respect to the social reproduction of technology. Most fundamentally, this account requires a mechanism whereby predictors incorporating rich somatosensory or kinesthetic feedback from one's own body can be matched with the purely visual and auditory input generated by the actions of others: input which is not even presented in the same spatial perspective. This is the so-called *correspondence problem* and, although special purpose mechanisms have been proposed, a prevailing view is that it is solved by general purpose mechanisms of associative learning (Brass and Heyes 2005). Thus, internal models are linked to the observed behavior of others through stimulus generalization from one's own visible movements (i.e., recognizing that one's own hand posture is "the same" as that of another individual) and/or associated contextual cues, supported by simple mechanisms such as Hebbian learning (i.e., synaptic plasticity: "cells that fire together, wire together"). An elegant neural network model of this process is presented by Laland and Bateson (2001). However, this simple solution to the correspondence problem raises another issue: if imitation is enabled by the activation of associations with already existing internal models, how is it possible to imitate novel actions? In other words, how is imitation learning achieved?

This is a critical problem for the social reproduction of behavior generally and of technology specifically. The solution likely lies in the hierarchical structure of behavior. Because action goals are abstractions over constituent means (note that the means/goal distinction is a relative one: "goals" become "means" at a different level of analysis), it is possible to assemble new behaviors from familiar constituents (Buccino et al. 2004). Furthermore, because the relative abstraction of goals renders them robust to variation in lower-level means, it is possible to "copy" behaviors without matching these details perfectly. Indeed, it would never be possible to imitate precisely the kinematics of other individuals with differently sized and shaped bodies (de Vignemont and Haggard 2008). Conversely, it is quite possible to imitate a gesture with an entirely different effector (e.g., right vs. left hand). In cases where such variation in lower-level action details does not make a difference to intended outcomes (i.e., is unconstrained), the purely observational learning of novel actions may be possible through statistical parsing of multilevel behavioral regularities (Buchsbaum et al. 2011; Byrne 1999), thus supporting fast imitation learning. However, when action details do matter and these action elements are not already in the behavioral repertoire, it will be necessary to engage in active behavioral exploration (i.e., practice or play) to (re)produce effective internal models through an iterative process matching self-actions to ever closer approximations of observed kinematic and/or environmental outcomes. This hierarchical model addresses the issue of novelty but raises a final issue for resonance accounts of imitation: Exactly what kind or level of information is being shared during imitation learning?

Many of the possible kinds of representations that might be shared during action observation and imitation have been reviewed by de Vignemont and Haggard (2008), who distinguish various types (sensory vs. motor, semantic vs. pragmatic) and degrees of abstraction (from specific motor commands and sensory predictions to abstract *prior intentions* specifying the global goal of an action, as in "to drink from a glass of water"). They conclude that resonance or mirror mechanisms most likely involve pragmatic, motor *intentions in action*, defined as "dynamic sequences of specific movements." These intentions in action are more specific and less interpretive than prior intentions (e.g., "grasp object and bring to mouth" as opposed to "eat"), but are not specified to the level of particular motor commands. This is consistent with work in monkeys which shows that mirror neurons are selectively responsive to action types (e.g., grasp) despite substantial variation in the motor details (e.g., precision vs. power grip) (Rizzolatti and Craighero 2004), and that the response of inferior parietal mirror neurons coding for particular action types is modulated by the final goal of the motor sequence in which the actions are embedded (Fogassi et al. 2005). In humans, putatively homologous regions of inferior frontal and inferior parietal cortex display a similarly selective response to simple action goals (i.e., gasp a particular object) across variation in specific kinematic means (Grafton 2009). Thus it appears that classic motor resonance mechanisms operating in anterior inferior parietal cortex and posterior inferior frontal cortex achieve a best-fit matching of observed actions to "mid-level" internal models (representing goal-directed sequences of elementary actions, Wolpert et al. 2003) that are already in the motor repertoire of the observer.

This mechanism for observational action understanding (Rizzolatti and Craighero 2004), working together with cortical regions that support motor planning (dorsal premotor), spatial awareness (superior parietal), biological motion perception (superior temporal sulcus), object representation (inferior temporal), and working memory (middle frontal gyrus), may support imitation of simple goal-oriented actions (Buccino et al. 2004; Menz et al. 2009; Molenberghs et al. 2009), such as reaching to grasp or strike an object. Interestingly, Hecht et al. (2012) recently reported a pattern of increasing connectivity of "core" mirror-system and related temporal-parietal regions across (nonimitating) macaques, (infrequently imitating) chimpanzees, and humans.

Imitative matching of shared internal models may constitute a basic unit for the social transmission of behavior. As a matter of speculation, it seems possible that particular internal models widely shared in a population for one reason (e.g., manipulating chopsticks) could bias the perception and reproduction of unrelated technical gestures. However, this mechanism by itself is insufficient to explain the transmission of complexly organized technologies, which typically require fidelity both at lower (e.g., embodied skills) and higher (e.g., sequences of goals) levels of action organization. This is likely to implicate additional mechanisms of more abstract goal representation in the PFC, as well as more concrete sensory processing of observed movements. For example, it is increasingly well documented that the "elementary" percussive gesture of stone knapping requires a highly coordinated and precise strike (Bril et al. 2000, 2010). In other words, kinematic variation is highly constrained by desired outcomes. However, the important parameters (e.g., kinetic energy) are not perceptually available to naïve observers nor captured in existing internal models for more generic percussive acts. Thus, the observer must begin by (incorrectly) imitating the observed gesture, checking the outcome against the predicted (desired) outcome, and then embarking on a lengthy process of goal-oriented behavioral exploration or deliberate practice (Ericsson et al. 1993) to (re)discover the relevant task constraints and develop corresponding internal models. Because there are a huge number of variables to be explored, this skill acquisition process may be quite lengthy. This process may be accelerated somewhat by continued observation of expert performance, which provides a sensory model to be matched through processes of stimulus generalization and associative learning discussed above, or through intentional teaching by a mentor, which might involve ostensive cues and/or modified performance (demonstration) to highlight relevant variables, structured coaching (Vygotsky 1978), or explicit semantic information about the task. In this context, social motivation for practice (implicating additional social cognitive and affective mechanisms) may also be critical for technological reproduction (Lave and Wenger 1991; Roux 1990; Stout 2002, 2010) although this is beyond the scope of the current review.

This account of best-fit action matching followed by individual kinematic rediscovery also requires the presence of more abstract goal representations. The initial (mis)matched internal model provides a starting point for generating a range of behaviors; however, subsequent selection on this behavioral variation can only occur with reference to a desired outcome. In line with a widely held distinction between imitation (of means) and emulation (of goals), neuroscience studies of imitation have used tasks in which the goal is simply to produce a particular movement or body posture (Buccino et al. 2004; Chaminade et al. 2005; see also, the do-as-I-do imitation task in nonhuman primates in

Custance et al. 1995). This avoids theoretical complications surrounding the imitation–emulation dichotomy by equating goals and means. However, technological action is characteristically organized (i.e., constrained) with respect to higher-level goals, most typically involving transformations of objects or substrates. How are these higher-order constraints represented and reproduced across individuals?

High-Level Constraints: Objects and Goals

The word "object" can refer to either a tangible physical entity or the goal of an action. This etymological association highlights the simple fact that action goals typically have to do with modifying, manipulating, or using physical objects. This is certainly true of "technological" actions. As noted in the introduction, the systematic transformation of durable objects provides an important medium for the elaboration of temporally extended, complexly organized action sequences. Unfortunately, and in sharp contrast to the growing body of research on (simple) tool use (e.g., Lewis 2006), there has been almost no neuroscientific investigation of such object transformation.

Action Outcomes

Our understanding of object-related goal representations thus comes primarily from studies of simple observation and manipulation. Such studies consistently implicate prefrontal and parietal cortex, with more abstract goal representations generally being associated with greater distance from the primary sensorimotor cortex surrounding the central sulcus (i.e., more anterior in frontal cortex, more posterior in parietal cortex). In monkeys, Nelissen et al. (2005) found that more posterior inferior frontal regions responded to variation in the specific context of observed actions (agents, effectors and actions) whereas the more anterior area 45B (a putative homolog of human area 45, i.e., anterior Broca's area) was responsive to objects. More directly relating to object transformation, Hamilton and Grafton (2008) studied the representation of observed action outcomes in humans using a simple sliding-top box which could be opened in different ways (variation in outcome) and with different motions (variation in kinematic parameters). They found a selective response to outcomes in inferior frontal cortex (area 44, posterior to area 45) and inferior parietal cortex, both in the right hemisphere. This contrasts with a previous study by Grafton and Hamilton (2007) in which variation in the target object for a simple reach-to-grasp was associated with selective responses in inferior frontal cortex and a relatively anterior portion of left inferior parietal cortex (i.e., phAIP, discussed above), both in the left hemisphere. Across experiments, variation in low-level kinematics was associated with response in visual association cortex (implicating sensory matching rather than motor resonance).

These findings suggest the presence of a diversity of goal representations in human parietofrontal cortex (Grafton 2009), with higher levels (e.g., material outcomes) in particular being represented in the right hemisphere.

Involvement of the right hemisphere may seem surprising, considering the extensive evidence that simple tool use is left lateralized (Lewis 2006). However, action outcomes, particularly those which involve object transformation, may unfold on a longer temporal scale and involve larger-scale visuospatial processing, both of which may be preferentially associated with the right hemisphere (Stout et al. 2008). In fact, right hemisphere involvement has been consistently reported in the small number of imaging studies that have actually studied object transformations. Chaminade et al. (2002), who conducted a PET study of subjects imitating simple construction actions using Lego blocks, reported activation of the right dorsolateral PFC in cases where only partial information (goal only or means only) was available to guide action planning. Frey and Gerry (2006) had subjects learn by observation how to assemble different objects from Tinkertoys and found that the right anterior intraparietal sulcus was the only region in which activation correlated with successful imitation of the demonstrated (arbitrary) sequences of assembly. They concluded that this region is important in forming representations of the temporal ordering of component actions. This is consistent with patterns of impairment to complex action sequencing observed following right hemisphere lesions (Hartmann et al. 2005).

Finally, in the only imaging studies to date of actual technological production, a series of PET and fMRI investigations of stone tool making (Stout and Chaminade 2007; Stout et al. 2008, 2011) have consistently reported right hemisphere activation. Stout et al. (2008) found increased activation of right inferior parietal and inferior frontal cortex during skilled handaxe production as compared to simple Oldowan flake production, as result mirrored by fMRI data from the observation of tool production (Stout et al. 2011). This does not appear to reflect the presence of low-level differences in manipulative complexity across the two tasks (Faisal et al. 2010) and is thus attributable to the more complexly organized goal structure of handaxe production.

Goal-Organized Imitation

Though more study of object transformation is clearly needed, the emerging picture suggests that a right lateralized parietofrontal network is involved in representing goals at the level of discrete action outcomes and sequential object transformations. This would be in addition to a better-known left hemisphere motor resonance system involved in representing object-directed intentions in action, as discussed in the previous section. This distinction parallels the differential roles of left versus right hemispheres in rapid (e.g., phonology, syntax) versus slow (e.g., prosody, context) linguistic processing and may reflect a more general hemispheric division of labor between rapid, small-scale action control on the left and large-scale, longer duration integrative functions on the right (Stout and Chaminade 2012). In any case, attention to action outcomes and object transformations is likely to be at least as important as motor resonance in the social reproduction of behavior. As argued above, selection among potential internal models for simple actions can only occur with reference to higher-order goals. It is likely that simple object-transformation targets often provide this constraint on action variation.

For example, Richard Byrne has proposed an influential "string parsing" model of imitation, in which "recurring patterns in the visible stream of behavior are detected and used to build a statistical sketch of the underlying hierarchical structure" (Byrne 1999:63). In this cognitively simple way, observers pick out essential actions or stages (i.e., constraints indicated by redundant patterning) from complex observed sequences, allowing for "program-level" copying (Byrne and Russon 1998) of large-scale behavioral organization. Byrne suggests that these redundancies might be bodily movements or effects on objects, but that the latter are likely to be much more easily observable. As we have seen, more or less separable neural systems exist for representing each of these levels of action organization. Importantly, Byrne's model specifically does not require causal or intentional interpretation of the observed actions; such understandings may be developed later if at all. All that is required are opportunities and motivation for repeated observation. In addition to broader issues of social context alluded to above, this implies the need for at least one additional level of abstraction in goal representation: that of the overall goal of the sequence, the desirability of which motivates attention to and copying of subgoals. In fact, developmental (Bekkering and Prinz 2002; Flynn and Whiten 2008b) and experimental transmission studies (Mesoudi and Whiten 2004) indicate that there is a bias toward higher-fidelity copying at higher levels of hierarchical organization.

Such higher-order goal representations are most likely supported by PFC. Neurophysiologically, PFC is well suited to maintain stable superordinate goal representations over extended subordinate action sequences because prefrontal neurons are able to sustain firing over extended periods of time and across events (Barbey et al. 2009). Connectionally, PFC represents a high-level convergence zone for the brain's sensorimotor systems and is thought itself to be organized in a multilevel fashion, with increasingly abstract representations being instantiated in increasingly anterior regions. The precise meaning of "abstraction" in this context remains controversial, with some nonexclusive alternatives being domain generality (integrating across cognitive domains), relational integration (relations between stimuli, relations between relations, etc.), temporal abstraction (maintaining goals over time), and policy abstraction (representing goals as abstractions over subgoals) (Badre and D'Esposito 2009). This raises the question of exactly what constitutes higher-order goals of technological action, and what needs to be represented.

Classically, technology is associated with achieving economic goals and, ultimately, with increasing the quantity and efficiency of energy capture from the environment (White 1943). In everyday life, however, such ultimate aims may be very far from the mind of individuals. More proximate goals are likely quite diverse, from the pursuit of a valued item or commodity (money, a useful tool) to an assertion of personal identity and pursuit of social status. In the case of technological skill acquisition, the latter is perhaps most often the motivating goal (Lave and Wenger 1991); it is even likely that the "value" attached to many items is itself socially motivated. These are questions largely beyond the scope of this chapter, but it is important to note that they do matter and are quite relevant to understanding the neural systems and mechanisms which may be involved. This is particularly true if one accepts the argument that there is a very important "top-down" or goal-directed element in imitation and technological transmission. For example, a fascinating intergenerational study of Zinacantec weavers (Greenfield 2003) found that the introduction of a market economy in the region changed the goal of weaving from the fulfillment of traditional social relations to commercial profit. This introduced a new subgoal of producing innovative designs, which was in turn realized by an increase in the complexity of conceptualization and manipulation of designs (from unitary blocks of color to a thread-by-thread basis). Thus, changes in high-level goals can have important "trickle-down" effects on all levels of action organization and technological transmission.

Structured Event Complexes

Structured event complex (SEC) theory (Barbey et al. 2009) provides a model of PFC function that might be useful in addressing the fundamental diversity of potential technological goals. An SEC is defined as "a goal-oriented set of events that is structured in sequence and represents event features (including agents, objects, actions, mental states and background settings), social norms of behavior, ethical and moral rules, and temporal event boundaries" (Barbey et al. 2009:1292). This information is "stored" throughout the brain in the form of embodied sensory, motor, and visceral associations or *feature maps* that are integrated into more abstract cross-modal representations in associative *convergence zones*, such as PFC. Barbey et al. (2009) review neuroscience evidence linking major dimensions of variation in SECs to functional gradients and regions in PFC. Thus, it is expected that more complex SECs (e.g., more relations, greater policy abstraction) will be associated with more anterior PFC, multiple event integration with right PFC, mechanistic plans and actions with dorsolateral PFC, and social norms and scripts with ventromedial PFC. This range, coupled with widely distributed representations in posterior cortex, suggests that virtually the entire brain may become involved in different SECs and in different activities we consider "technological."

According to SEC theory, once the complex web of associations that constitutes an SEC is formed, the entire SEC can be activated in a bottom-up fashion by any of its elements. For example, upon walking into a restaurant, a wide array of expectations and intentions become active. Similarly, the need to drive a nail into a board can stimulate an entire SEC having to do with finding and using an appropriate tool. Thus, SECs are substantially similar to the scripts and schemas of cognitive psychology (e.g., Abelson 1981), the cultural models of cognitive anthropologists (Holland and Quinn 1987), and the concepts of *constellations of knowledge* and *umbrella plans* specifically used to describe technological proficiency (Keller and Keller 1996). Their acquisition and refinement may thus be related to cognitive development in children (e.g., Bruner 1990) and the acquisition of technical expertise in adults (Keller and Keller 1996).

The activation of related representations in SECs also resembles the priming of semantic concepts (e.g., "dachshund" primes "poodle" and "leash," Patterson et al. 2007), but involves a wider diversity of representation types (e.g., internal models). Like webs of semantic association, different SECs would be overlapping, with fuzzy boundaries and ambiguous membership (e.g., should instant messaging etiquette be similar to a phone call or email) leading to substantial possibilities for creativity (i.e., a generative system). Indeed, a core function of SECs is thought to be the support of counterfactual thinking (e.g., predicting future outcomes, planning contingencies, and imagining alternatives). SEC theory is thus a promising direction for exploring the little known neuroscience of creativity, and for better understanding the source(s) of technological innovation.

Conclusion

The neuroscience of technology should be of interest to many different people for many different reasons. In this chapter, implications for our understanding of the cultural evolution of technologies have been highlighted. Of specific interest are mechanisms that might (a) bias or constrain the reproduction of technological practices and/or (b) generate new technological variants (complexity). With respect to the former, analysis suggests that there should be biases operating at multiple levels of organization. Starting at the bottom, there are the constraints imposed by designed tools themselves that strongly tend to encourage particular patterns of prehension which, in turn, affects the way simple tools are incorporated into the body schema, and thus their most likely patterns of usage. The existence of such constraints, even in archaeological tools of unknown function, could be directly discovered through experiments with modern subjects. At a somewhat higher level, the performance characteristics of tools also constrain usage. However, these constraints will tend to be less determining relative to the vast range of potential activities that could be imagined: top-down goals and context become more important in narrowing the search space, presenting greater problems for experimental archaeologists.

Research on motor resonance suggests that simple goal-oriented gestures, such as reaching to grasp, place, or strike an object, may be the basic units of imitation and social transmission. The best-fit matching of observed gestures to functionally similar gestures already in the observer's motor repertoire likely presents a common occasion for biased transmission, where a more generic gesture is substituted for a specialized one. Stabilizing selection against such substitutions is likely an important dimension of technological apprenticeship and skill acquisition (cf. results bias; see also Shennan, this volume). This implies a top-down influence of desired goals constraining low-level action. It is likely that the most easily observable and transmissible goals providing such detailed constraints on action sequences are often visible transformations of objects. This creates a possibility for transmission biases against transformations that are subtle or otherwise difficult to observe. This will be the case, in particular, where the observer's causal/intentional understanding of the process is not well developed. For example, novice stone knappers often make mistakes with respect to subtle techniques, like platform preparation, and may not be able to tell that the nice large flake they just removed actually does not get them any closer to their goal One way to counter such transmission bias in poorly understood systems is to slavishly "overimitate" all aspects of production, which can promote technological stasis (Martin 2000).

Indeed, the biases/constraints just discussed can be viewed in two ways. First, all other things being equal, they will tend to bias the kinds of technology that are successfully reproduced (i.e., against those with highly specialized gestures and cryptic elements). Second, however, they provide a key to understanding the types of errors that are most likely. Assuming that some such errors are actually beneficial, this might be seen as somewhat akin to understanding the mechanisms of mutation in biological evolution. In this way, these constraints can also act as an accidental mechanism generating novelty.

A major disanalogy with conventional conceptions of biological evolution is that technological innovation can also be intentional and goal oriented. The neuroscience of such creative action is not well understood, perhaps because creativity is itself hard to define and operationalize experimentally. It also seems likely that the generation and preservation of technological innovations will be more profitably studied as a social phenomenon than a psychological one. Nevertheless, a common conception of innovation by individuals is that it represents the recombination of existing ideas into a new framework. Insofar as this is the case, SEC theory, which addresses precisely the mechanisms of association between elements of goal-directed action, may provide a good framework for investigating the issue.

10

Scientific Method as Cultural Innovation

Robert N. McCauley

Abstract

Consideration of scientific method as a cultural innovation requires examining the philosophy and sociology of science, anthropology, developmental, cognitive, and social psychology as well as the histories of science and technology. Anarchistic philosophical proposals about science set the stage for subsequent endorsements of quite liberal conceptions of science and scientific thinking that root these pursuits in basic features of human—even animal—cognition or in the intimate connection between science and technology. That every methodological prescription has its limits or that science is not uniform does not entail methodological anarchism. Like any other radial category, science includes more and less central instances and practices. Justifications for such liberality regarding science that are grounded in the acquisition of empirical knowledge by infants and other species downplay the sciences' systematic approach to criticizing hypotheses and scientists' mastery of a vast collection of intellectual tools, facts, and theories. Justifications that look to the close ties between science and technology neglect reasons for distinguishing them. Intimate ties are not inextricable ties. Research on scientific cognition suggests that, in some respects, human minds are not well suited to do science and that measures progressively sustaining science's systematic program of criticism and its ever more counterintuitive representations both depend on cultural achievements and are themselves cultural achievements involving what have proven to be comparatively extraordinary social conditions. This richer, epistemologically unsurpassed form of science is both rare and fragile, having arisen no more than a few times in human history.

Introduction

The increasing scope, precision, and sophistication of modern science and its explanatory and predictive successes encompass considerably more than science's barest cognitive essentials. To focus on those at the expense of characterizing progressive scientific traditions downplays the crucial role cultural innovations have played in science's achievements.

Making this case requires clarifying how much about science comes naturally to human minds. I thus begin by outlining arguments for skepticism about the *scientific method* that have set the stage for recent discussions. It also demands situating positions that (a) construe science as the outcome of natural predilections of mind, emphasizing its continuity with commonsense and (b) fixate on the inevitable entanglement of science with technology. Those accounts are incomplete. The first takes insufficient notice of the elaborate measures necessary to insure critical scrutiny in science and the extensive education required for participating in it, and the second minimizes the vital position that cognitive ideals occupy. These matters are discussed in the first section.

Thereafter, cognitive and historical considerations are presented that favor an accounting of scientific method as cultural innovation. The cognitive science of science urges caution about the Cartesian picture of rationality as residing between matched pairs of human ears. Any constructive account of scientific method and rationality, in the face of myriad shortcomings of individual reasoners, dwells, instead, in the special cultural, social, economic, and political arrangements that undergird modern science. Although scientific sparks and brushfires have erupted sporadically in human history, sustained traditions of disciplined inquiry with institutions fostering methodical criticism are recent, refined, and rare.

Integrating Cognitively Liberal Conceptions of Science

Some philosophers and sociologists of science have disputed claims for scientific rationality and posed problems for a uniform scientific method. Some anthropologists, developmental psychologists, and literary theorists have endorsed liberal accounts of scientific cognition, which can also challenge a view of scientific method as cultural innovation.

Against Method

Although Thomas Kuhn (1970) famously assailed the methodological unity of science, no one criticized it more provocatively than Paul Feyerabend (1975). Both were reacting to decades of armchair philosophizing aimed at rationally reconstructing science in terms of observations and mathematical logic. Both stressed how prevailing programs of research influence the acceptability of methods. Feyerabend, for example, maintained that Galileo's arguments on behalf of the telescope's veracity on Earth—when viewing ships too distant to be seen by the naked eye—were for Aristotelians, who distinguished terrestrial and celestial principles metaphysically, reasons for *doubting* the reliability of telescopic images of heavenly bodies.

Both also defied the methodological proposals of prominent philosophers. Feyerabend assaulted Karl Popper's suggestion that aspiring to test persistently and falsify hypotheses empirically is what distinguishes science. Feyerabend insisted that this view was unworkable, since *from the outset* scientists know about evidence that is incompatible with new hypotheses. The neutrino hypothesis would have never gotten off the ground, since its first empirical corroboration came more than two decades after Wolfgang Pauli initially proposed it (Dunbar 1995). Shoving leading formulations off their pedestals, Feyerabend suggested the only plausible account of scientific method was "*anything goes*," though, he noted straightaway that not even that slogan was a methodological recommendation.

That contemporary sciences embrace diverse methods and entertain abstruse theories, which often resist ready interpretation, only increases wariness concerning pronouncements about scientific method. The rise of the "Strong Program" in the sociology of science (Bloor 1991) and nonmodernist variants (Latour 1993), which hold that social arrangements fundamentally shape scientific interests and procedures, combined with philosophers' failure to provide compelling accounts of the influence of the superempirical virtues (e.g., simplicity, consilience, elegance) on theory choice have only exacerbated such reservations about an identifiable scientific method.

Managing Methodological Skepticism: Cognitively Liberal Conceptions of Science

Feyerabend's methodological anarchism and sociologists' challenges to scientific rationality created a milieu that pushed the defenders of science toward more modest accounts of its essential intellectual activities. These comparatively liberal accounts construe scientific cognition so broadly as to include not only everyday thinking but also learning in infants and animals.

The Roots of Science

Some Anthropologists' Views. Noting that the sciences employ no "single universally applicable methodology," Robin Dunbar explores more rudimentary cognitive underpinnings, born of "the natural mechanisms of everyday survival" (Dunbar 1995:94, 96). Science involves learning about the world and its causal structure. Dunbar holds that "all higher organisms" carry out "plain simple learning," equipping them with expectations for predicting things well enough to survive and reproduce (Dunbar 1995:77, 75). Thus, he suggests that science's cognitive essentials (i.e., learning inductively, including hypothesis testing) come as naturally to many animals as they do to humans.

One consequence of such liberality is the reluctance of many anthropologists to differentiate science and religion (e.g., Horton 1993). In small-scale societies, religions provide the frameworks with which people explain events, whereas in most modern, large-scale societies, science has largely usurped that prerogative, increasingly confining religion to matters of morality as well as social and psychological well-being, at least in public discussion. Such liberalism, however, provides little insight about why, with regard to explaining events, religious worldviews are *not* typically overthrown in the first case and why modern science does just that in the second.

Some Developmental Psychologists' Views. Alison Gopnik, Andrew Meltzoff, and Patricia Kuhl (1999) have advanced the stronger and somewhat less liberal view that scientific progress and human cognitive development, in particular, proceed similarly—that babies are "scientists in the crib." They emphasize that, like scientists, infants are active learners who are sensitive to evidence.

The various looking-tasks that developmental psychologists have devised for ascertaining what babies know assume that they recognize violations of their expectations. At six months of age, infants can detect statistical patterns and draw probabilistic inferences from populations to samples (Denison et al. 2013); fourteen-month-olds can predict single-event probability from large set sizes (Denison and Xu 2009). Three- and four-year-old children make causal inferences based on probabilistic evidence, even when it conflicts with information about spatial contiguity (Kushnir and Gopnik 2007). Facing upended expectations, toddlers and preschool children seek evidence in exploratory play and carry out explanatory reasoning (Legare 2012; Legare et al. 2010).

That infants produce new theories, however, is less plausible, certainly if "theories" refer to scientists' linguistic constructions. Still, findings about prelinguistic infants' growing knowledge surely imply that they do develop new expectations. Gopnik holds that "children's brains...must be *unconsciously* processing information in a way that parallels the methods of scientific discovery" (Gopnik 2010:80, emphasis added). Even if babies qualify as theorizers, though, theorizing is not unique to science, as Horton's observations about religion suggest. Theorizing by young children may be necessary, but it is not sufficient for their activities to count as scientific.

The Critical Side of Science

Scientific Pluralism. Methodological anarchists and the Strong Program sociologists of science have overplayed their hands. Given the range of phenomena that human ingenuity has enabled us to study scientifically as well as the serendipity and hubbub of human affairs in general, it is not shocking that, finally, only vague methodological prescriptions ("attend to evidence;" "pursue overall coherence") will plausibly characterize all productive forms of scientific inquiry. "Science" is a radial category that encompasses numerous endeavors that are spread across a vast conceptual space with more and less salient cases along a host of relevant dimensions. Exhibiting scientific rationality in some inquiry may involve conforming to any of a hundred viable principles that collectively cover the central regions of that space well enough to count as proceeding reasonably in empirical investigation. Methodological anarchism hardly exhausts the options for responding to Feyerabend's arguments that no particular, exception-less, methodological recommendation will capture the entire array of activities that we regard as scientific.

Nor do the effects of cultural circumstances on scientific topics, theories, methods, and assessments, let alone training, organization, funding, and institutions, constitute an insurmountable barrier to constructing a case for the reasonableness and epistemic prominence of science. Does anyone contest the suggestion that culture shapes human thought and conduct? *That*, however, hardly establishes that science's progress, empirical findings, or ascendant theories are rationally suspect or that scientists cannot reassess them through further criticism and research. Scientific objectivity resides neither in unimpeachable methods nor in investigators' neutrality.

Situating Cognitively Liberal Conceptions of Science. Dunbar's conjecture that some animals (e.g., rats) carry out hypothetical causal inferences is controversial (Dunbar 1995). Michael Tomasello has argued, for example, that not even chimpanzees recognize underlying causes (Tomasello 1999:22). Dunbar also acknowledges problems about the representational format of hypotheses that animals allegedly adopt (Dunbar 1995).

Introducing a distinction between "cookbook" science and explanatory science, Dunbar signals that, ultimately, the contention that thinking scientifically comes naturally to animals will not bear too much weight (Dunbar 1995:17). The hypotheses Dunbar attributes to animals are about patterns of perceptible events closely associated in time and space. This is cookbook science, which resembles patterns characteristic of human folk physics and folk biology. Following Lewis Wolpert (1992), Dunbar ultimately insists that the factors which have launched the "superpowerful process" of "explanatory science" consist of "features of formal science that do not really exist in the everyday version" (Dunbar 1995:88).

Cognitive liberalism, then, will not account for much that is vital to science after all. Neither inductive capacities nor even the more sophisticated cognition of crib-based scientists explains modern science's wealth of explanatory and predictive accomplishments or the contributions of other eras to the history of scientific knowledge.

Criticism as a Scientific Obligation. What distinguishes science from other explanatory and predictive enterprises is a fixation on *criticism*. Scientists constantly push theories for new empirically testable consequences and for coherence internally and externally with the best theories about related matters (Tweney 2011).

Infants, young children, and people in cultures in which science never flowered understand that evidence matters. That, however, is only the beginning. First, that does not establish that they will discern *relevant* evidence. Researchers must know the ascendant theories, their implications, and their competitors to understand what counts as relevant evidence. Evidence is always evidence-relative-to-a-theory.

Without knowing the theories, people will fail to recognize evidence right before their eyes. Correlations between the proximity of islands, their volcanic activity, size, elevation, and more are not difficult to detect in an island chain, but it requires some understanding of the theory of plate tectonics to grasp their evidential status. Without that theory the role those patterns might play as evidence will be obscure, at best.

Second, scientists must systematically collect and record evidence. Getting more and diverse evidence demands assembling and documenting it conscientiously. For some theories and models (e.g., concerning climate), scientists must examine long-term trends in disparate places with considerable precision. Aiming to build definitive star maps, John Flamsteed made hourly measurements of planets and the positions of various stars for *forty years* (Jardine 2000).

Third, scientists are also experts at generating new evidence. Science's idealized theories identify relevant variables that affect a system's behavior over which scientists seek experimental control, when the systems under scrutiny are not so large (or so small) or so complex or so remote that they preclude such interventions. Complicated experimental arrangements and instruments (whether supercolliders, eye trackers, or electron microscopes) play a vital role in science. These devices furnish opportunities to examine phenomena in unfamiliar environments or in what would typically be the inaccessible provinces of ordinary environments where diverging empirical implications of competing theories can be tested. Scientists become skilled experimentalists, producing conditions that differ from typical circumstances *in theoretically significant ways* and for which human natural cognitive inclinations are uninformative and unhelpful.

Fourth, scientists must also analyze and assess the evidence they amass. Obtaining evidence is one thing; knowing what to make of it is quite another. Scientists need facility with several forms of mathematical representation to comprehend theories and to evaluate evidence. The demands of science for treating data systematically to ascertain their evidential import have led to a variety of mathematical tools for their analysis. Mathematical clarity and precision are crucial for exploring, measuring, and dissecting the dynamics of complex systems.

Liberalism Inspired by Science's Connections with Technology

Science and technology have always been connected, but since the mid-nineteenth century, they have become practically inextricable. Scientific advances routinely depend upon devising machinery for creating special environments for testing hypotheses. More familiar are the increasingly widespread technologies that modern science has created, including everyday gadgets. Teasing theoretical science and its methods apart from technology *conceptually* runs the risk of appearing to underplay this intimate connection.

Is Technology Inherently Scientific?

Technological Grounds for Cognitive Liberalism. Barbara Herrnstein Smith correctly holds that theoretical understanding routinely depends on technologies implementing theories and that new technologies just as routinely provoke new explanatory conjectures. Consequently, she asserts that to separate science and technology so straightforwardly involves a "narrow, historically and culturally quite specific, understanding of 'science'" that results in a distinction that "can only be arbitrary and artificial" (Smith 2009:132, 135). Envisioning technology as inherently scientific also motivates cognitive liberalism about science. Smith's cognitive liberalism includes *as scientific* all production and use of technology by human groups.

Perhaps the distinction is artificial, but that does not mean that it is not useful. A variety of independent considerations demonstrate that it is not arbitrary (see discussion in the next section). Examining science's *cognitive* foundations provides grounds for distinguishing it from technology and for curtailing this version of liberalism too.

An Alternative View of the Intimate Relation between Modern Science and Technology. Ironically, Smith's charge that a sharp distinction between science and technology is "narrow" and "historically and culturally...specific" seems to concede its applicability to modern science, *in which their connections seem more profound than ever*. John Gribbin, who opens his history of modern science with the observation that technological developments are more important than scientific genius in the genesis of science, offers a more nuanced account of their relationship that not only does not preclude a clear distinction between science and technology but, in fact, assumes it (Gribbin 2003:xix). Gribbin (2003:xx) states: "Technology came first, because it is possible to make machines by trial and error without fully understanding the principles on which they operate. But once science and technology got together, progress really took off." He then highlights their autocatalytic relationship, which the industrial, electronic, and digital revolutions have only accelerated. Technology may be a necessary condition for the pursuit of science, but it does not follow that the most noteworthy cognitive features of science depend upon technology.

Science as Cultural Achievement

The constructive case for cleaving science and technology segues into a larger examination of science as cultural innovation. Considerations from across the disciplines suggest that cognitive liberalism regarding science is incomplete at best. In light of liberal proposals, it is ironic that more than three decades of research in the cognitive science of science suggests that not even scientists, when operating in isolation, are wonderfully impressive scientific thinkers! Diverse factors point to the paramount position culture has occupied in the development of science.

Science is one of many knowledge-seeking activities that humans undertake, but as a continuing, systematic endeavor to explain the world, it is *unsurpassed.* It is "science" in this sense that is pivotal from *both an epistemological and an historical point of view*. Consequently, it will prove equally decisive in reflection about its status as a *cultural* innovation.

Teasing Science and Technology Apart

Science is that unsurpassed knowledge-seeking activity not because of what it has in common with material technology but because of what sets it apart.

History Matters

Ancient History. The ties that bind contemporary science and technology make it difficult to envision circumstances without such ties (because, for example, science did not exist). Two historical observations spotlight technology's cognitive independence from science. The first is science's historical scarcity. Even on inclusive conceptions, science has bloomed infrequently and flourished even less. If the list of continuing scientific activity were to include (a) ancient cultures—Chinese, Babylonian, Egyptian, and Mayan—by virtue of their astronomical record keeping and cosmological speculations, (b) ancient Greeks, (c) Arabs and Chinese during the last centuries of the first millennium through the Middle Ages, and (d) Europeans in the sixteenth and seventeenth centuries and the emergence of modern science that their work inspired, that list would include but a fraction of human history in a much smaller fraction of human societies.

Prehistory. This second consideration is the obverse of the first. Science's rarity contrasts starkly with the ubiquity of technology. *Every* culture possesses technology. The birth of science in human history contrasts with technology's prehistoric origins. Prehistoric technologies surfaced independently of science and predate ancient civilizations by a couple of million years among our earlier ancestors. This prehistoric pattern of technology thriving without science has persisted in most places at most times since. That science is required to guide technological progress is a very recent notion.

Natural History. Consider two further facts about *natural history*. First, archaeology has disclosed at least a half dozen other species that produced and used technology. Second, not even the members of our genus, indeed not even primates, have a monopoly on the production and use of tools. Animals—from chimpanzees to New Caledonian crows—both fabricate and use tools (Weir et al. 2002; Kenward et al. 2005). Unlike the pursuit of science, the construction of artifacts is not uniquely human, though, admittedly, the ongoing improvement of tools over generations does seem to be an accomplishment peculiar to species among our genus and a particularly well-established dynamic of human cultural change.

Science as an Abstract Technology

Broad conceptions of technology that include *abstract* intellectual tools as well as implements and structured environments cast science and technology's relationship differently, but justify distinguishing them nonetheless. If written representations count as a technological genus, then science is one of its species. It stands apart from material technology, however, in two notable ways: (a) science, unlike material technology, depends upon literacy and (b) it always includes abstract theoretical interests in understanding nature for its own sake. The latter raises two issues.

Seeking Understanding. Science pursues and explores accounts of the world for their intrinsic interest. If science began with ancient societies' systematic collections of astronomical observations, then it probably arose from practical concerns about calendars. Still, the ancient Greeks differed crucially from earlier astronomers, because they valued reflection about the world for its own sake, regardless of practicalities. The Greeks were the first to discuss theories critically, to marshal empirical evidence, and to advance competing theories. Whatever practical advances it may spawn, science is also always about gaining a deeper understanding of the world.

Toby Huff cites such considerations, when arguing that the Chinese did not develop a scientific tradition, despite their consummate technological innovation and sophistication. Huff holds that their focus remained overwhelmingly practical and that institutions supporting empirical criticism of theories never emerged. Aside from a brief period in ancient China among the Mohists, the Chinese never established a sustained tradition of scientific investigation (Boltz et al. 2003). Although the Chinese had the printing press many centuries before Europe, education focused on memorization of Confucian classics (Huff 1993:279).

Impracticality. Scientific pursuits always involve *speculations* that aim to elucidate the world's workings and no other human endeavor recognizes that fact so self-consciously. Scientific speculations depict idealized worlds (of frictionless planes, classical genes, and rational consumers) that go *beyond* what is known, supplying insights about real patterns behind the appearances that enable us to make sense of the world. Those idealized models also have implications for how unexplored parts of the world should prove to be. In these respects, they take seemingly impractical, intellectual risks. They discuss entities, processes, and relations that are removed from practical problems and all previous experience.

Clarification: Cognitively Unnatural Technologies of Modern Science

Most technologies that modern science engenders are as cognitively inaccessible as its theories. Laypersons are unaware of the theoretical underpinnings of the structures and operations of these technologies. This encompasses both the experimental apparatus of science and familiar machines (e.g., cell phones).

The practical benefits of these technologies play an undeniable role in the cultural prestige of science. Science's epistemic standing rests largely on the fact that the sciences regularly enable us to do things that once seemed impossible: from finding oil miles below Earth's surface to transplanting organs, to sending spacecraft to distant planets. Only with science were these envisioned, let alone realized. All of this is quite removed from what most people do with eggbeaters, elevators, and exit ramps. On these fronts, the technologies that contemporary science spawns also stand apart.

Cognitive Reflections

A tradition of criticizing theories systematically requires that scientists become proficient with the requisite intellectual skills. A decade of scientific training is necessary for novices to gain control of these tools and to begin to appreciate the subtleties of their employment. That is because their acquisition and application call for thought and practices which do *not* come naturally to human minds.

Deductive and Probabilistic Inference

Wason. The Wason selection task famously demonstrated how dismally people perform when carrying out conditional inference (Wason 1966). Around eighty percent of participants go wrong. This, alone, should substantially dampen optimism about the naturalness of scientific reasoning, for scientists are always reasoning hypothetically: exploring a theory's implications, contemplating some mechanism's operation, or pondering some nexus of causal variables. Subsequent research on the Wason selection task seems to corroborate that in nearly all settings, conditional inference is reasoning that most humans do not do well (Cosmides and Tooby 2005).

Tversky and Kahneman. Estimating the likelihood of events about which scientists have incomplete information is pivotal in explanatory theorizing, argumentation, and decision making. Amos Tversky and Daniel Kahneman have shown that humans' intuitive judgments under conditions of uncertainty routinely transgress normative principles of probability. Scores of studies have disclosed that people neglect such considerations as regression to the mean and base rate information, fail to attend to sample sizes when weighing the significance of evidence, and disregard basic principles of probability theory (Kahneman 2011).

A collection of cognitive shortcuts, which humans consistently take, explain these and other failures. Such biased heuristics serve for most purposes, but their inexact solutions are inappropriate for most scientific jobs. Most of the exotic circumstances in which scientific experiments take place contravene the presuppositions of such heuristics; consequently, these heuristics render us susceptible to *perceptual and cognitive illusions* in many circumstances. These heuristics feel so right that not even monetary incentives for correct answers boost participants' performance (Camerer and Hogarth 1999). Similarly, neither substantive expertise nor advanced training in probability and statistics overcome these natural tendencies. For example, there was "no effect of statistical sophistication" in how participants performed on ranking the probabilities of conjunctions and their conjuncts. In Tversky and Kahneman's experiments with such problems, more than eighty percent of "highly sophisticated respondents" provided rankings that violated the dictates of probability theory (Tversky and Kahneman 2002:26).

Other Cognitive and Psychological Obstacles

The cognitive science of science has uncovered an assortment of additional intellectual pitfalls which can trip up those with scientific training.

Intrusive Intuitions often Swamp Science's Radically Counterintuitive Representations. Usually sooner rather than later, the sciences inevitably generate radically counterintuitive representations that do not square with our folk conceptions of the world. Learning scientific models and principles that contradict heuristics' deliverances, however, does not undo those deliverances. We are Copernicans, yet few ever see the sky that way (McCauley 2011). Experimental research with people who have passed physics courses reveals that many retain numerous false assumptions about basic motions (McCloskey

1983); thus, ordinary phenomena pose perceptual, explanatory, and predictive problems *that usually go completely unrecognized* (Liu and MacIsaac 2005). Practice with hundreds of textbook problems does not assure that students overcome the conceptual difficulties associated with basic mechanics (Kim and Pak 2002). Elementary problems do not trick experts, but without opportunities to apply their knowledge of relevant formulae, experts' intuitions for motions like collisions are often incorrect (Proffitt and Gilden 1989). Formal education helps, but the knowledge is remarkably fragile.

Confirmation Bias. Psychological and historical research discloses inquirers' penchant to exhibit confirmation bias, which can take a variety of forms. Besides attending only to confirming evidence, scientists can be disinclined to search for contrary evidence and sometimes disregard it when it appears. History is replete with otherwise distinguished scientists who defended problematic theories, insisting that failures to replicate their positive findings resulted from others' carelessness (Gratzer 2000). Theorists cling to their theories and ignore alternatives, particularly when considering the import of unfavorable evidence. When given the choice, instead of seeking information that would bear on the comparisons of theories, experimental participants would pursue "pseudodiagnostic" information, which would neither support their favored theory as they thought nor support such comparisons (Mynatt et al. 1981).

Motivated Perception. Motivated perception concerns the impact that commitments to theories can have on perception. Adherence to a scientific theory means seeing the world in a particular way. Armed with theories, we find them hard to shake. After the ascendance of Copernicanism, European astronomers observed changes in the firmament that the Aristotelian conception had ruled out as impossible. Chinese astronomers, without telescopes but also without the burden of Aristotelian cosmology, had recognized such changes centuries earlier.

Cultural and Historical Reflections

Unfortunately, nature has not groomed human minds for carrying out science's obligatory criticism of theories. Learning and doing science demand grasping intellectual constructs and procuring cognitive skills that humans find difficult to acquire, onerous to retain, challenging to exercise, and unnatural all around. (Experimental science involves a host of practical skills that are no less challenging.) These psychological findings do not support the Cartesian picture locating Reason within individuals' minds. Science's epistemic prominence does not arise from guarantees about individuals' exemplary thought and conduct but from a host of sociocultural arrangements.

How Have Humans Managed to Do Science?

Science proceeds because of the insistence on the public availability of scientific work and on opportunities to criticize it. To figure in the history of scientific inquiry, sooner or later (sometimes after their death), scientists must offer statements of their positions and the evidence for them for public scrutiny by the scientific community. Copernicus permitted the publication of his *De Revolutionibus* only after his death. That inevitable publicity assures that the criticism of scientific work never need turn on the reliability of any individual's cognitive processing. Individual scientists may be blind to the weaknesses of their theories, the gaps in their evidence, the mistakes in their reasoning, and the errors in their calculations. They may also manifest a decided preference for evidence that supports their hypotheses. Fortunately, the history of science provides ample testimony to the fact that scientists suffer far fewer failings when it comes to assessing positions that compete with their own. It is that public competition in which the partisans and other scientists uncover a theory's failures and problems.

Literacy. That astronomical protoscience ("protoscience" because, among other things, it was subservient to state religions) arose in the first literate cultures is no coincidence. Beyond record keeping, the expectation that scientific work must become publically available links science to literacy. Written symbols last. Literacy permits the storage of ideas, relieving demands on memory. Literate people can return to documents after long delays and retrieve knowledge. The resuscitation of the texts, topics, and theories of ancient Greek science ignited new projects of research that resulted in new scientific developments in substantially different cultural settings, namely in both the Arab world of the tenth century and, again, in Renaissance Europe with the eventual birth of modern science. Documents are critical aids to thought, permitting clarity and precision almost nonexistent in speech but imperative for presenting and testing scientific theories. They are a prerequisite for the careful, systematic, extended criticism that characterizes science. Copied, published, and transported texts introduce the possibility of widespread access to ideas that is beyond their authors' control, which is decisive for the objectivity of science. Scientists discuss the contents of externalized texts, rather than the contents of their creators' mental states. All of these considerations counsel greater caution about what we write than about what we say, and although science is not only about what gets written and published, it is always finally about that.

The opportunity to criticize written, publically available theories occasions the development of intellectual skills that exceed doing arithmetic or the mere decoding of text. Publically accessible exchanges tend toward standardized forms to make positions and reasoning clear. This was as true about the exchanges of the medieval schoolmen as it is about those of contemporary scientists. What the emergence of the empirical sciences adds to these procedures of rational, literate inquiry is a particularly disciplined approach to the collection, generation, analysis, and assessment of empirical evidence and of experimental evidence, in particular. To do that effectively requires years of education and training.

Education. Science depends upon the invention of external linguistic and mathematical symbols and an educational system that engenders facility with such symbols in numbers sufficient to generate a *community* of inquirers. Preserving and transmitting such proficiencies require ample investments in an educational infrastructure. Like literate humans, scientists are made not born. Both call for appropriate materials and years of tutelage. Participating in science at its highest levels routinely requires more than twenty years of formal education. This type of education is a uniquely modern phenomenon, which remains confined primarily to the wealthiest half of the world's nations.

Science has been rare in part because literacy (and numeracy) has been rare. The reinvention of the printing press in Europe predated the rise of modern science by less than two centuries. It introduced the possibility of widespread literacy, the proliferation of schools, and the dissemination of scientific works. Most cultures in history did not possess a system of writing and only a fraction of those that did produced a substantial corpus. An even smaller fraction of those produced science.

How Has Science Achieved Its Celebrated Epistemic Status?

This is not a substantive question about settled scientific views but a procedural one about how science works. *Scientific communities* have erected safeguards to catch and correct errors. In addition to the public availability of scientific controversies, two principles deserve special mention.

Peer Review. Scientific journals make extensive use of peer reviewing. Expert, independent referees provide editors with written reports laying out their reservations about scientific papers. Even published authors must nearly always incorporate additional arguments and analyses to meet their referees' objections.

Ideally, that is how the system works. Research indicates, however, that referees treat papers with which they agree more gently than those with which they disagree, which sometimes leads to inappropriate decisions. The process is by no means perfect (Armstrong 1997).[1] Still, scientific communities retain an unending interest in self-improvement, which has led to innovations such as the Public Library of Science. Science must deal with fraud and deceit, but no human pursuit does remotely as good a job of uncovering deceptions. Science has developed good procedures for smoking such ruses out, at least eventually.

[1] For analyses and extended discussion, see http://www.nature.com/nature/peerreview/debate/

Replication. Science requires the replicability of results. It does not tolerate secret formulas, special sensitivities, or "singularities." Scientists must report on intersubjectively available phenomena. They must describe their experiments at a level of detail that permits other scientists to reproduce them. Failure to replicate findings instantly clouds their credibility. Although its critical examination may wax and wane, until some finding is replicated (ideally, by its critics), its position remains thoroughly provisional. Even often-replicated findings remain susceptible to questioning, which is to say that under at least some circumstances their status is only somewhat less provisional.

Public availability of scientific claims, peer review, and demand for replicability are three important pillars that support the epistemic credibility of scientific methods. The sciences' pattern of explanatory, predictive, and technological triumphs and the accelerated pace of those triumphs over the past century only burnish that standing.

How Has Science Progressed?

The public availability of scientific works insures that science remains a *social* endeavor, which is the key to its long-standing pattern of theoretical and practical triumphs. Although science provides no guarantees, its continuing success depends on its inherently social character. Knowledge, criticism, and decision making are collective accomplishments, distributed across the community (Solomon 2001). Science is inherently social and therefore inherently institutional.

Universities. The gradual development of independent universities proved a critical variable buttressing science's long-term success in Europe (Grant 1996). Late Medieval universities deemed natural philosophy a legitimate component of advanced education, positioning it so that it would be open to upheavals when new theories and methods began to change the terrain three centuries later. They developed standardized curricula, which would eventually serve for credentialing, and supported scientific research.

State-supported, institutionalized experimental science arose in the ninth and tenth centuries in Baghdad and persisted for two centuries in a few locales in the Islamic world (Al-Khalili 2011). Medieval Arabic science, however, never enjoyed a lasting alliance with educational institutions independent of Islam, which has generally proven less congenial than Christianity to scientific education. Without political cover from local rulers, scientific institutions had short lives. For example, Nasr al-Dīn al-Tūsi's observatory and school at Marāgha only thrived for sixty years before falling into disrepair.

Scientific Societies and Disciplines. Institutional arrangements that secure the openness, publicity, and integrity of scientific research were critical to the rise of modern science (Jardine 2000:316). Scientific institutions, such as national academies, articulate and enforce standards.

Experimentation and systematic observation carry crucial implications for social and economic arrangements. Modern science requires vast sums to support exotic infrastructure and to probe unusual environments. By the early eighteenth century, some European governments and companies were investing in expeditions to the far reaches of Earth for strategic advantages and profitable ventures, certainly, but for gathering data and specimens and testing scientific hypotheses as well.

Since the middle of the nineteenth century, science has become a fount of knowledge and technical innovation. New social arrangements and infrastructure have enhanced scientific productivity. In addition to schools providing general science education, diverse organizations (professional societies, university departments, journals, laboratories, research institutes, foundations, government and corporate funding) have enabled large numbers to learn and do science. These arrangements facilitate communication, disseminate scientific work, and institutionalize compensatory strategies for handling individual scientists' fallibility. Not even the resulting bureaucracies have been able to undo the fact that most of the time these measures have insured that the collective outcome in the long run is superior to the efforts of individuals in the short run.

Science's Fragility

Science as an unsurpassed method for acquiring empirical knowledge depends on a combination of cultural elements, including literacy, long-term education, freedom from religious and political repression, many peculiar institutions, and substantial resources for theoretical research. For many reasons, including both its cognitive unnaturalness and the obvious difficulties with sustaining such arrangements, this combination is both historically rare and inherently fragile.

Acknowledgments

I am grateful to Bart Reeder, Tamara Beck, Dorinda McCauley, and Patrick Emery for valuable comments on an earlier version of this paper and to Jürgen Renn for helpful conversations about these topics.

First column (top to bottom): Alex Mesoudi, Victoria Reyes-García, Stephen Shennan, Emma Flynn, Briggs Buchanan, Dietrich Stout, Victoria Reyes-García
Second column: Kevin Laland, Rob Boyd, Alex Mesoudi, Briggs Buchanan, Robert McCauley, a special occasion, Rob Boyd
Third column: Emma Flynn, Claudio Tennie, Jürgen Renn, Dietrich Stout, Kevin Laland, Jürgen Renn, Stephen Shennan

Herzlichen Glückwunsch

11

The Cultural Evolution of Technology and Science

Alex Mesoudi, Kevin N. Laland, Robert Boyd,
Briggs Buchanan, Emma Flynn, Robert N. McCauley,
Jürgen Renn, Victoria Reyes-García,
Stephen Shennan, Dietrich Stout, and Claudio Tennie

Abstract

This chapter explores how the principles and methods of cultural evolution can inform our understanding of technology and science. Both technology and science are prime examples of cumulative cultural evolution, with each generation preserving and building upon the achievements of prior generations. A key benefit of an evolutionary approach to technological or scientific change is "population thinking," where broad trends and patterns are explained in terms of individual-level mechanisms of variation, selection, and transmission. This chapter outlines some of these mechanisms and their implications for technological change, including sources of innovation, types of social learning, facilitatory developmental factors, and cultural transmission mechanisms. The role of external representations and human-constructed environments in technological evolution are explored, and factors are examined which determine the varying rates of technological change over time: from intrinsic characteristics of single technological traits, such as efficacy or manufacturing cost, to larger social and population-level factors, such as population size or social institutions. Science can be viewed as both a product of cultural evolution as well as a form of cultural evolution in its own right. Science and technology constitute separate yet interacting evolutionary processes. Outstanding issues and promising avenues for future investigation are highlighted and potential applications of this work are noted.

Introduction

Aims and Overview

Our aim in this chapter is to explore how the methods and concepts developed in the field of cultural evolution can be applied to the domains of technology and science. Both technology and science are prime examples of cumulative cultural evolution. Technological and scientific knowledge is accumulated over successive generations, with each generation building upon achievements of prior generations. Both have had an inestimable impact on our species' way of life. As Boyd et al. (this volume) argue, the cumulative cultural evolution of locally adaptive technology has allowed humans to colonize and inhabit virtually every terrestrial environment on the planet. Yet there are also numerous examples of the negative consequences of technology, such as the overexploitation of resources, facilitation of large-scale warfare, and increase in wealth inequality. Technology can significantly transform the way we think and act at a quite fundamental level (Stout, this volume); it can also generate novel coevolutionary dynamics between human lineages and the technology that they use (Shennan, this volume). Science, a more recent cultural innovation, has dramatically accelerated technological evolution and represents a unique system of knowledge not seen in any other species (McCauley, this volume). Advances in our understanding of these two phenomena have been achieved across the social sciences and humanities. Here we explore how the burgeoning interdisciplinary science of cultural evolution (Boyd and Richerson 2005; Mesoudi 2011a) might further this understanding.

In the following sections we outline the individual-level mechanisms that are thought to generate population-level patterns of technological change. We then explore the role of external representations and human-constructed environments in technological evolution and examine the factors which determine the varying rates of technological change over time. Discussion follows on how science and technology interrelate and how scientific and technological evolution differ as processes. We conclude by highlighting outstanding issues and promising avenues for future investigation and note some potential applications of this work.

Definitions and Scope

Both technology and science are challenging concepts to characterize, and numerous definitions of each exist. It is, nonetheless, helpful to delineate the scope of the domains of interest here to focus our chapter and distinguish our topic from the other three topics that formed this Strüngmann Forum: the cultural evolution of sociality (Jordan et al., this volume), language (Dediu et al., this volume), and religion (Bulbulia et al., this volume).

Science and technology are both forms of *knowledge*. Knowledge is the potential of an individual (individual knowledge) or a group (shared knowledge) to solve problems by individual or collective action. Knowledge is typically stored in individual brains (internal representation), as well as in social structures, material artifacts, external representations, and environmental structures (external means). Knowledge is socially transmitted from individual to individual via various processes, typically involving many of these external means.

We can therefore distinguish between knowledge itself (the product) and the means by which that knowledge is acquired and transmitted (the process).

Technology refers to goal-oriented shared knowledge together with its external means. It is thus knowledge geared to, and organized around, solving specific problems faced by a group or society (see also Shennan, this volume). For example, the bow and arrow solves the problem of killing animals from a distance; sextants and GPS solve the problem of navigation; and telephones and smoke signals solve the problem of remote communication. These are all examples of external, physical manifestations of technological knowledge External *representations* are a special class of intentional external manifestations of technological knowledge which contain information that is interpretable by human minds, and which can become important for the transmission of that technology. These are discussed in the section, External Representations and Human-Constructed Environments.

Our focus in this chapter is on technological *evolution*; that is, technological change as an evolutionary process. As Darwin noted, evolutionary processes require "descent with modification," by which he meant the gradual accumulation of modifications over time. This aspect of our account, we argue, restricts technological evolution to humans, among extant species, as well as some extinct hominin species. Many nonhuman species use tools, but only humans appear to possess cumulative technological evolution (Boyd and Richerson 1996; Tomasello et al. 1993). Chimpanzees, for example, exhibit regional traditions of tool use behaviors, such as nutcracking or termite fishing, that have potentially spread via cultural transmission (Whiten et al. 1999). Yet none of these behaviors show clear evidence of having been gradually accumulated and improved upon over time. One test for the presence of such "descent with modification" in tool use is the presence of behaviors that are outside the individual learning ability of an organism, or what Tennie et al. (2009) refer to as the "zone of latent solutions." No such behaviors have been unambiguously reported in chimpanzees, and it seems vanishingly unlikely that any widely used human technology—from the bow and arrow to the iPad—could have been invented by a single person alone. This unique aspect of human technology likely arises from the unusually high-fidelity social learning exhibited by humans compared to other species, as we discuss in the section, How Do Individual-Level Mechanisms Generate Population-Level Patterns of Technological Change?

We restrict our attention here to technologies that have clear external, physical means. These range from artifacts and texts to structured environments and collective practices. Whether individual behavior that does not involve objects (e.g., bodily techniques) qualifies as "technology" is controversial. Some researchers within both cognitive science and cultural anthropology have distinguished at least some human behavior as involving "enactive representations" (Bruner 1964), which may seem to constitute sufficient grounds for inclusion. It has also been argued that language can be viewed as the "technology of the intellect" (Goody 1973; see also Everett 2012). We acknowledge these alternative perspectives but restrict our focus here to technology that involves clear external means, leaving language to be discussed elsewhere in this volume (Dediu et al., this volume), and consider non-object-based behavior to be too broad (see Stout, this volume).

Science is defined as means-oriented shared knowledge together with the process responsible for its generation. Unlike technology, scientific knowledge is primarily pursued with no instrumental goals in mind. Science is facilitated by technology, most obviously in the form of scientific instruments (e.g., telescopes or microscopes), as well as by symbolic representation systems such as writing. In addition, science, in turn, facilitates and accelerates technological evolution in a coevolutionary feedback process. As McCauley (this volume) points out, we must be careful to make a clear distinction between knowledge that appears to be scientific, such as folk knowledge possessed by nonliterate societies, and knowledge that results from scientific institutions and practices. Nonliterate societies may have extremely sophisticated folk understanding of the world, such as the astronomical knowledge used by Polynesian sailors to navigate the Pacific islands. However, such knowledge is typically characterized by location-specific features (e.g., valid only for observers close to the equator), is not subject to procedures characteristic of science (e.g., open criticism afforded by publications and their discussion in a scientific community), and is prone to loss without the institutional elements of science that only emerged over the last couple of centuries in literate, large-scale societies. The interrelation between science and technology, and how scientific change can be understood as an evolutionary process, is discussed in the section, Science.

How Do Individual-Level Mechanisms Generate Population-Level Patterns of Technological Change?

One of the key benefits of adopting an evolutionary approach to culture is Darwinian "population thinking" (Richerson and Boyd 2005), in which patterns and trends at the population level are explained in terms of the underlying, individual-level mechanisms of variation, selection, and transmission. For biological (genetic) evolution, these individual-level processes are natural selection, mutation, recombination, etc. Cultural evolution may be determined by similar individual-level processes, but several processes unique to cultural change have also been modeled and explored, thus necessitating a departure from strictly neo-Darwinian assumptions. For example, where genetic mutation is blind with respect to selection, cultural innovation may, to some extent, be directed by purposeful agents (Mesoudi 2008). In this section we attempt to catalog these processes and, where possible, apply them to technological change. Table 11.1 provides an overview of these individual-level processes, along with their population-level effects and presence in nonhuman species.

Table 11.1 Individual-level mechanisms responsible for population-level patterns of cultural evolution.

Individual-Level Processes	Population-Level Effects	Presence in Nonhuman Species
1. Sources of innovation		
• Chance factors (accidents, copy error) • Novel invention (trial and error, insight, or exploration, through personal or group endeavor) • Refinement (modification or improvement of existing variant, through personal or group endeavor) • Recombination (combining existing elements to form a new variant, through personal or group endeavor) • Exaptation (applying existing technology to new function)	All sources generate cultural variation	All sources are observed in humans but little evidence exists for refinement and recombination in nonhuman species
2. Type of social learning		
• Imitation (including "overimitation") • Teaching (including scaffolding, pedagogical cueing supported by language)	Capable of supporting technological evolution through facilitating high-fidelity transmission	Rare or absent in nonhuman species
• Emulation • Enhancement effects (local, stimulus) • Facilitatory effects (social, response) • Observational conditioning	Thought incapable of supporting technological evolution because fidelity is typically too low	Common in humans and other species
3. Facilitatory developmental factors		
• Zone of proximal development • Structuring the learning environment • Apprenticeship, collaboration, and cooperation	Further enhances the fidelity of information transmission by directing and motivating learning	Little compelling evidence for these mechanisms outside of humans
4. Cultural transmission processes		
• Evolved biases ∘ Content bias ∘ Direct/results bias ∘ Context biases (model-based, frequency-dependent, state-based)	Capable of biasing the direction and rate of cultural evolution; differentially affects the distribution of cultural variants and pattern of diffusion	Observed in humans and nonhumans
• Unbiased transmission/random copying	Incapable of biasing the direction and rate of cultural evolution; differentially affects the distribution of cultural variants and pattern of diffusion	Observed in humans and nonhumans
• Guided variation	Causes cultural evolution to shift toward inferential prior knowledge	Observed in humans but presence in nonhumans is contentious

Sources of Innovation

Innovation is a tricky term to delineate. In different disciplines, it has been variously deployed (see O'Brien and Shennan 2010) to refer to (a) a successful novel variant (i.e., inventions that succeed, as used in sociology), a novel variant (characterized independently of whether they propagate, as used in biology), or any kind of variant; (b) as the ideas underlying an invention or its first implementation; and (c) both the process by which variants are generated as well as the product. Within the field of cultural evolution, innovation has generally been thought of as the functional equivalent to mutation in biological evolution (Cavalli-Sforza and Feldman 1981); that is, innovation introduces new cultural variation into the population through copying error, novel invention, refinement, recombination, and exaptation. Hence, innovation is not a synonym of variation, the latter being a broader category that encompasses diverse forms only some of which are novel.

The technological record contains numerous examples of innovation (Basalla 1988; O'Brien and Shennan 2010; Ziman 2000; Petroski 1994). One example of innovation through recombination is given in Boyd et al. (this volume), where door hinges were likely co-opted for use on rudders in medieval ships (this can also be seen as an example of exaptation, given that the function of the hinge has changed). A fruitful area of study in archaeology has been the modeling of copying error due to limitations of the human perceptual system, loosely analogous to random mutation in genetic evolution. For example, Hamilton and Buchanan (2009), building on previous work by Eerkens and Lipo (2005), modeled the population-level effects of small, imperceptible errors in the repeated cultural transmission of artifact shapes or sizes. Imagine an artifact manufacturer intends to make an exact replica of an existing artifact. If the manufacturer's artifact differs from the original artifact by less than a certain amount (e.g., by less than 3%, which is a typical threshold for shapes), then even though the artifacts may appear identical to the manufacturer, the new artifact may in fact be imperceptibly larger or imperceptibly smaller than the original. If these tiny, random errors are compounded over successive generations of artifact makers, then different artifact lineages can diverge randomly within known limits (ultimately set by the magnitude of the copying error). Hamilton and Buchanan (2009) showed that Clovis projectile point size across late Pleistocene North America fit the predictions of this process of accumulated copying error, suggesting that this technology changed solely due to this random, unbiased process. However, other cases do not fit the predictions of this accumulated copying error (Eerkens and Lipo 2005; Kempe et al. 2012), thus showing how unbiased copying error can provide a useful null model for detecting nonrandom, biased cultural transmission (see section below on Cultural Transmission Processes).

Types of Social Learning

Social learning refers to the transfer of knowledge or behavior from one individual to another. Social learning is a necessary prerequisite for technological evolution, and much recent research has focused on the fidelity of different social learning processes. This is because social learning must be of sufficient high fidelity such that technological knowledge, which is often cognitively opaque and difficult to acquire, is preserved and accumulated over successive generations (Lewis and Laland 2012). If fidelity is too low, then technological knowledge is easily lost due to the copying error discussed in the previous section.

Comparative, social, and developmental psychologists have explored in detail various types of social learning and their potential to support the high-fidelity transmission of knowledge, including stimulus/local enhancement, emulation, imitation, and teaching (Whiten and Ham 1992; Hoppitt and Laland 2008). Forms of social facilitation, such as stimulus or local enhancement, do little more than draw attention to aspects of the environment. Emulation provides information about how the environment can be manipulated or about the affordances of different objects. Social facilitation and emulation are typically considered to be unlikely to provide the necessary high fidelity required for successful cultural accumulation (although for experimental evidence that emulation can result in cumulative improvement in lineages of simple artifacts, such as paper airplanes; see Caldwell and Millen 2009).

Imitation (including "overimitation") and teaching (often through verbal instruction) appear to be better candidates for facilitating high-fidelity transmission of knowledge; indeed, experimental evidence links these processes to cumulative cultural learning (Dean et al. 2012). Imitation refers to the copying of motor actions performed by other individuals (as opposed to emulation, in which the result of behavior is copied, but not the behavior itself; for further discussion, see Stout, this volume). With respect to technology, it is likely that complex artifactual knowledge (e.g., how to make a projectile point) can only be transmitted faithfully through imitating the precise actions required to make the artifact or through verbal instruction and other forms of teaching. It is rarely the case that complex artifacts can be reverse engineered from the finished product (i.e., through emulation), at least not without introducing substantial variation into the technique (Dean et al. 2012; Flynn and Whiten 2008a; Tennie et al. 2009).

Overimitation (Lyons et al. 2007; McGuigan et al. 2007) describes the tendency of human infants (and also adults, Flynn and Smith 2012) to copy the actions of others with such high fidelity that they reproduce aspects of what they have seen which are not necessarily causally relevant to the goal of the task. For example, children who observe an adult tapping a tool into a hole on the top of a transparent box before using the tool to unlock the box and retrieve a reward will copy both the causally relevant (unlocking) and irrelevant (tapping) actions. Interestingly, chimpanzees do not appear to overimitate; they only copy causally relevant modeled actions (Horner and Whiten 2005). This may indicate a general lack of high-fidelity social learning in chimpanzees and may, in turn, be related to their lack of cumulative culture (Tennie et al. 2009).

One may wonder, however, how overimitation can result in the adaptive accumulation of effective technological knowledge (see Boyd et al., this volume) when it allows the preservation of irrelevant actions. We think that this is probably an artifact of the experimental tasks typically used to test for overimitation, in which relevant and irrelevant actions are clearly defined. Such a contrast would be difficult to discern for much of the technological knowledge acquired by humans (both children and adults) in nonexperimental settings. This excessive imitation of others' actions may therefore be a manifestation of a hypertrophied human tendency for imitation that is highly adaptive in natural settings where the functional aspects of a task may be ambiguous. Alternatively, overimitation may serve a social function, such as indicating and enhancing affiliation with in-group members (Haun and Over, this volume) or the adoption of normative behavior (Kenward 2012). Teaching, or the "pedagogical stance" (Csibra and Gergely 2009), may further enhance the fidelity of imitation, with experts tailoring their behavior to maximize the likelihood of successful acquisition by the learner and using cues like eye contact to indicate the pedagogic importance of a particular expert act. Language, too, allows the high-fidelity transmission of knowledge necessary for much technological learning, in the form of verbal and written instructions.

Facilitatory Developmental Factors

Various developmental factors may further enhance the fidelity of information transmission. The *zone of proximal development* (Vygotsky 1978) is defined as the difference between what an individual can achieve alone and what that individual can achieve with the support of an expert other. Thus, the acquisition of specific cultural behaviors will be achieved at different times during childhood, depending on the complexity of the behavior and the available social support. Equally, different social learning processes will be more appropriate at different ages. Negotiation or collaboration, for example, may not be an appropriate form of transmission during early childhood (Flynn and Whiten 2012). With age comes cognitive development, and changes in abilities such as theory of mind or inhibitory control may facilitate learning and the ability to use different social learning mechanisms.

Thus, as children get older they acquire more experience with the world—buttons can be pressed, levers pushed and handles pulled or turned. When faced with novel technologies, children can draw on their previous experience and the internal representations associated with that experience (Wood et al. 2013). Experience can help tremendously in dealing with a world that contains numerous artifacts. On the other hand, previous experience can also hinder solutions via "functional fixedness," where the intended use of an object overrides alternative, potentially superior uses (Adamson 1952; see also Barrett et al. 2007b).

Through active (e.g., imitation) as well as passive (e.g., teaching) social learning, the zone of proximal development will also improve cognitive capacities of children over time, leading to what has been called cultural intelligence (Herrmann et al. 2007; Tomasello 1999). This ontogenetic cultural intelligence complements phylogenetic cultural intelligence, such as the species-specific social learning abilities discussed above, which evolve biologically over much larger time frames via natural selection and lead to a species' zone of latent solutions (Tennie and Over 2012). Because of this, we likely share with our great ape relatives some degree of basic phylogenetic cultural intelligence, which could have acted as an important initial impetus for innovativeness in our species, and thus as a potential—and necessary—starting point for cumulative culture (Enquist et al. 2008).

Cultural Transmission Processes

Where sufficiently high-fidelity social learning is present, a further set of cultural transmission biases potentially come into play that describe who, when, and what people copy. Formal models of cultural evolution have identified several such biases that are supported by empirical evidence from psychology, sociology, and other disciplines (Boyd and Richerson 1985; Mesoudi 2009). Content biases relate to the content of the information being transmitted, with some forms of knowledge intrinsically more cognitively attractive or more easily copied (e.g., less complex and/or skill intensive technologies) than others. Direct or results bias occurs when traits are copied based on their observed effects on the world. Context biases refer to factors external to the content or consequence of knowledge, such as the preferential copying of prestigious or powerful individuals (model-based bias, Reyes-García et al. 2008), the copying of traits that are common or rare (frequency-dependent bias), or the preferential copying of traits under certain circumstances, such as copying when uncertain or when the environment has changed (state-based bias; for an overview, see Rendell et al. 2011). Studies of the diffusion of technological innovations suggest that conformist (positive frequency-dependent) bias may be responsible for much technological cultural transmission (Henrich 2001).

Unbiased transmission, or random copying, occurs when learners select models to copy entirely at random (although where different individuals within a population deploy inconsistent transmission biases, the summed effect may resemble unbiased transmission). Archaeologists have borrowed drift models from population genetics to model random copying, showing that certain artifacts change as if they were being copied at random (Neiman 1995), with a lack of fit to such models indicating nonrandom, biased transmission such as anti-conformist frequency dependence (Shennan and Wilkinson 2001; Mesoudi and Lycett 2009).

Finally, guided variation occurs when individuals modify their behavior as a result of individual learning and this modified behavior is then copied by others. If individuals in the population tend to modify their behavior in the same way, this leads to directional change, a process that Boyd and Richerson (1985) labeled guided variation. Guided variation, unlike the various transmission biases, does not depend on the amount of variation in the population and, as a result, it works quite differently to what we normally think of as "selection." To see why this is important, imagine a population in which all individuals are identical, and an environmental change favors a different behavior. Selection will not lead to the spread of the new behavior, because there is no variation to select. In contrast, guided variation can lead to change because it is the result of individual learning, not the culling of existing variation.

Modeling Technological Change

This distinction between selection and guided variation is captured by one of the canonical mathematical representations of evolutionary change, the Price equation (Price 1970), and provides one potential formal framework for modeling cultural, including technological, change (although other modeling frameworks are both possible and useful, see Cavalli-Sforza and Feldman 1981; Boyd and Richerson 1985; Gintis 2000). As we illustrate here, the Price equation allows the formal delineation of these different kinds of processes that drive cultural change, as well as incorporating evolution at multiple levels, and the coevolution of multiple traits, all of which are particularly relevant for technological evolution. It also suggests that a more nuanced definition of "fitness" is required with respect to technological change, compared to biological definitions of fitness (see also Shennan, this volume).

Suppose there is a population of variable cultural entities, for example, different design variants of a tool or weapon or other technological trait, such as two variant bow designs, simple and recurved. Labeling the frequency of one of the variants q, the change in this frequency, Δq, is proportional to:

$$\Delta q \propto \beta \operatorname{var}(q) + E(w_i \Delta q). \tag{11.1}$$

The first term, β var(q), gives the change due to what biologists typically call selection. The β parameter measures how much changing the cultural variant affects the fitness of that variant (i.e., the regression of fitness on trait frequency). For example, if people tended to copy more powerful bow designs, then β would be the effect on power of switching from simple to recurved variants. This is multiplied by var(q), the variance of the trait in the population. If most people use the same design, then the variance will be small and selective processes will have little effect, whereas if both designs are in regular use, comparison by learners can lead to rapid change. Just as the strength of natural selection depends on the amount of genetic variation in a population, the strength of cultural selection depends here on the amount of cultural variation.

The second term, $E(w_i \Delta q)$, is the expected amount of change due to individual-level processes. For example, it could be that individuals experiment with their own bows and sometimes switch from one design to the other as a result of their individual experimentation. This term would give the net effect of this individual experimentation. This individually driven change operates differently to selection because it does not depend on the amount of variation in the population. Thus, if we define cultural fitness in exactly the same way as biological fitness, it will determine the rate and direction of cultural change due to selective processes but it will not capture this latter nonselective change. It seems likely that nonselective change due to individual learning as well as shifts induced by the inferential nature of cultural transmission are more significant in cultural evolution than are nonselective processes in genetic evolution (such as meiotic drive). If so, knowing the cultural fitness of alternative variants alone will not allow prediction of the overall direction of cultural change.

Alternatively, it might be possible to define cultural fitness in terms of the "goals" of the learning processes that govern both individual learning and various forms of biased transmission. This would have to be averaged with the direct effects of selection on cultural variants, for example, due to the fact that these variants affect fecundity of a trait with significant vertical transmission to create a metric that predicts overall cultural change.

This framework can be extended to address evolution at multiple levels. Cultural variation affects the success, prestige, and survival of different levels of social organization. For example, some new fishing technique or invention such as fine nets might allow an individual to obtain more fish relative to other individuals within their group, but groups in which this technique/invention is common may do worse in competition with other groups with less effective fishing techniques due to the former's overexploitation of fishing stocks. The change in this cultural variant can then be partitioned into the average effect of individual variation on the rate of cultural transmission within groups, and the effect of the variation among groups on the rate of, say, group survival. This can be expressed using the Price equation as follows:

$$\Delta q \propto \beta_g \operatorname{var}\left(\bar{q}\right) + E\left(\beta_w \operatorname{var}\left(q_i\right)\right) + E\left(w_i \Delta q\right), \tag{11.2}$$

where β_g gives the effect of differences in frequency of the trait on group survival and $\operatorname{var}(\bar{q})$ is the variance in trait frequency across groups, such that the first term gives the change in frequency due to group-level processes. Analogously, the second term is the average of the changes within groups and the third term is, as before, the effect of individual transformations. Notice that within this framework there are two fitnesses: the average effect on group replication and the effect on individual replication within groups.

It is also possible to accommodate multiple cultural traits. For example, suppose that the usefulness of projectile points depends on both the length and the width of the point. This leads to two new dynamic processes. First, the fitness value of one trait may depend on the value of the other trait. For example, it might be that as points get longer, they must also get wider. This will mean that changes in the frequency of one trait will depend on the other trait, and thus evolution will lead to coherent change linking functional suites of cultural traits. Second, cultural "hitchhiking" may result from accidental correlations between traits. For example, there is evidence that languages have often spread because they are spoken by groups that possess advantageous agricultural technologies (Diamond and Bellwood 2003). For more detailed analysis of cultural change using the Price equation, see Beheim and Baldini (2012).

External Representations and Human-Constructed Environments

External Representations

All technological knowledge has some external manifestation in the form of material artifacts, according to our characterization above. More interesting, perhaps, is the way in which artifacts can both constrain human behavior patterns and embody information about their own production and reproduction (see Stout, this volume). The functional properties of existing artifacts can clearly channel behavior. In this sense, artifacts may be said to embody information about their use that both afford and constrain possibilities for innovation. As the saying goes, "if all you have is a hammer, everything looks like a nail." Note, however, that the respects in which information is "embodied" in features of the external world or in which it is "contained" in artifacts, which is to say the respects in which these items *represent* information, is always relative to an apprehending mind. All such information is inherently relational and dependent on the goal-oriented interpretations of agents. Taking this a step further, it is also possible for human agents to interpret the physical form of artifacts as evidence of the processes by which they were produced. Examples include many prehistoric technologies, like Acheulean handaxes, that became "extinct" but have been reverse engineered by archaeologists from material remains. Unlike the functional properties of tools, which (at least arguably) were intended by the manufacturers, reverse engineering is a process of interpretation purely on the part of the observer. Finally, in considering the cumulative cultural evolution of technologies, we should also remember that it is not just ideas that have accumulated but actual physical artifacts. Many modern technologies (e.g., automobiles) require a vast industrial apparatus that has built up over generations; as a thought experiment, it seems implausible that this apparatus could be reassembled "from scratch" in a single generation, even if all relevant knowledge was somehow preserved.

Thus, artifacts themselves are a medium of cultural transmission. Once this is recognized, it then becomes apparent that researchers may sometimes need to incorporate this factor into their models, as the frequency, longevity, or rate of change of artifacts may not resemble that of their users (see section below on Human-Constructed Environments and Niche Construction). Can we simply treat artifact lineages as if they were equivalent to biological lineages? Some cultural evolution researchers have successfully taken this approach (e.g., O'Brien and Lyman 2003b; Tehrani and Collard 2002), and in the interests of methodological tractability, researchers can in practice often ignore many of the cognitive, behavioral, and social processes through which humans reproduce artifact types. The same approach is taken by evolutionary biologists, who have achieved great success with phylogenetic analyses of phenotypic variation without necessarily understanding the developmental processes that produce the phenotypes. Nevertheless, in other instances it is unlikely that researchers can get away with ignoring the human vehicles (see Stout, this volume); more recent phylogenetic analyses of cultural diversity are indeed more mechanistically and demographically explicit (e.g., Bouckaert et al. 2012).

External *representations* are special cases of material culture: artifacts or design elements whose primary intended function is to convey information. This is accomplished by making use of shared mappings between particular physical signs and specific concepts or items. Classic archaeological examples include art such as cave paintings or figurines, personal adornments like beads, containers with decorative or "symbolic" markings, tally sticks, tokens and so forth. The information conveyed by external representations may be (a) symbolic, in the sense that the shared mappings are more or less arbitrary and involve complex associations between signs (cf. Deacon 1997), (b) indexical (i.e., based on reliable correlations such as that between shell beads and the many hours of labor required to produce them), and/or (c) iconic (i.e., based on physical resemblance as in much artistic expression). Of these, indexical reference is often thought to be important for social signaling whereas symbolic representation may also be relevant for the creation, manipulation, and transmission of technological and scientific knowledge. For example, symbolic systems make it possible to just focus on a particular aspect of the material world, such as the countability of objects, thus giving rise to new mental constructions, such as numbers and arithmetic.

The Evolution of Writing Systems

Perhaps the most prominent system of external representation is writing. As well as greatly facilitating the transmission of technological and scientific knowledge, writing itself is the result of a lengthy process of cultural evolution (Hyman and Renn 2012). The first writing systems emerged around 3300 BCE in Mesopotamia, initially in the form of clay tablets with numerical notations and seals which were likely used in the state administration of taxes and expenditure. From this, a system known as archaic cuneiform or proto-cuneiform evolved as a technology for the administration of centralized city-states. This proto-writing system did not represent the meaning of words or sentences, nor did it reflect grammatical structures of language, but rather disclosed meanings related to specific societal practices such as accounting. Since it was not used as a universal means of communication, it could only represent specific meanings in limited contexts. Nevertheless, it was on this basis that a long-term and stable Babylonian administrative economy developed, which in turn served as a precondition for the second stage in the evolution of writing: a universal means of codifying language. This second stage would have been impossible without the spread and manifold use of the earlier proto-writing.

Early Egyptian writing was more closely associated with representational and aesthetic functions (e.g., in monumental inscriptions legitimizing the authority of priests and rulers). Here, as well, writing gradually assumed an ever greater range of functions, such as for correspondence, historiography, and literature. Writing thus filled an increasing number of niches in the growing knowledge economy of a complex society as well as in new societies with varying socioeconomic structures. Accordingly, it underwent, as in Babylonia, an evolution characterized by an adaptation to these new niches and functions. Thus, writing took on new forms: it transformed from hieroglyphic into hieratic and demotic forms, evolved from a predominantly logographic Sumerian cuneiform into a predominantly syllabic Akkadian cuneiform, and developed into the West Semitic writing systems.

Further development of writing systems is characterized by processes of spread, variation, and selective adaptation to local needs and speciation. Speciation occurs when the adaptation of a writing system to a new niche (e.g., a new domain of knowledge, a new language or a previously illiterate society) leads to changes in the writing system that fundamentally affect the way in which the system functions as an external representation of knowledge. Thus, Minoan writing probably emerged as a result of diffusion from Mesopotamia in the context of the palace economy on Crete around the turn of the third to the second millennium BCE in the form of two different systems: Cretan hieroglyphs and the syllabic Linear A script. The latter was apparently the source of the Cypro-Minoan script, employed on the island of Cyprus in the second half of the second millennium, which in turn was the source of the Cypriot syllabary, which came into use toward the end of the first millennium. Given the significant Phoenician presence in Cyprus and the extensive contact between Phoenicians and Greeks, this syllabary may have influenced the emergence of alphabetic writing in the ninth century. Whereas Phoenician alphabetic writing possessed characters only for consonants, the Greek script adapted certain Phoenician semivowel characters as vowels. A West Greek alphabet became the source for the creation of the Latin and Cyrillic alphabets, two of the most frequently used scripts in the world.

Earlier writing systems suffered from ambiguities, with the same written symbols mapping onto many different spoken forms, but they led to a more concise representation of language and thus became a universal means for representing knowledge. An interesting feature of the evolution of writing systems is that the use of a particular system may expose genetic differences among speakers not seen when using an alternative system (see Dediu et al., this volume). For example, dyslexia, which has a strong genetic basis, is expressed less frequently and less strongly in speakers of languages with a simple mapping between orthography and phonology, such as Italian, compared to languages with more complex orthography–phonology mapping, such as English or French (Paulesu et al. 2001) This may point to gene–culture coevolutionary interactions triggered by the cultural innovation of writing technology.

Human-Constructed Environments and Niche Construction

An important ramification of external representation is that artifacts and features of the environment constructed or modified by human activities can feed back to shape other aspects of technology. This can be regarded as a form of niche construction, the process of environmental modification through which organisms modify patterns of selection acting on themselves and other organisms (Odling-Smee et al. 2003). Insights from niche construction theory (NCT) suggest that where organisms manufacture or modify features of the external environment experienced by their descendants (including artifacts), they can affect the evolutionary process in a number of ways, affecting the rates and direction of change, the equilibria reached, the amount of variation maintained, the carrying capacity of populations, the evolutionary dynamics (e.g., momentum, inertia, autocatalytic effects), and the likelihood that costly traits evolve (Odling-Smee et al. 2003; Laland et al. 1996, 1999; Silver and Paolo 2006; Lehmann 2007, 2008; Kylafis and Loreau 2008). Nonhuman examples of niche construction include beavers creating and inheriting lakes through their dam-building activity and earthworms changing the structure and nutrient content of soil by mixing decomposing organic material with inorganic material, thus making it easier for the worms to absorb water and allowing them to retain their ancestral freshwater kidneys, rather than evolve novel adaptations to a terrestrial environment. It is likely that human-manufactured external representations will generate similar kinds of feedback effects on cultural evolution, at which point explicit models will be required to track environmentally based resources.

Archaeologists have recently begun to use the framework of NCT to investigate the long-term effects of niche construction on technological evolution (Riede 2011; Riel-Salvatore 2010; Rowley-Conwy and Layton 2011; Wollstonecroft 2011; Smith 2007). For example Riel-Salvatore (2010) has used NCT to examine the technological changes in stone tool assemblages from the

Middle Paleolithic to the Upper Paleolithic in Italy. In this study Riel-Salvatore argues that the transition occurred sporadically in time and space over the Italian peninsula depending on the specific traits of the ecological and cultural inheritance system in each region. NCT also has been used in more recent time periods to examine technological changes associated with the domestication of plants and animals. In particular, Smith's (2007) work has emphasized the potency with which sedentary hunter-gatherers and agriculturalists engage in niche construction and the subsequent impact this niche construction has had on technology. Smith (2007) describes a number of technologies associated with the manipulation of plants and animals including the controlled use of fire, the construction of fish weirs, and the use of rainfall collection features and irrigation ditches (see also Wollstonecroft 2011). Smith suggests that the independent centers of plant domestication around the world had the common feature of being located in resource-rich zones. It also follows from NCT that it is within these resource-rich zones that we should expect to find evidence for high rates of technological evolution. Interestingly, this prediction runs counter to evolutionary ecological models of risk which suggest that invention should occur in more marginal environments (Fitzhugh 2001).

Rates of Technological Evolution

Determining Factors

Rates of technological evolution vary widely across time periods and regions. Several factors have been identified that explain different rates of change (summarized in Table 11.2). These range from the intrinsic characteristics of single technological traits to larger social and population-level properties. Rather than seeking a single factor that determines the rate of all technological change, we see these as a list of factors that may apply, singly or jointly, to specific case studies.

Units of Technological Evolution

The range of phenomena listed in Table 11.2 raises the issue of what the appropriate scale is when measuring rates of technological change. This is another way of framing the question—What are the units of cultural evolution?—which has been a source of confusion and contention within the field (e.g., Aunger 2000). We suggest that it is not inherently problematic, with the understanding that the pragmatics of quantification may be problematic in particular cases, and truly universal measures that can be applied across different technologies remain elusive.

Cultural evolution researchers can focus on, count, or model (a) knowledge, (b) behavior, and/or (c) artifacts, which are loosely equivalent to gene,

Table 11.2 List of factors that differentially affect rates of technological change.

Factor	Example
Intrinsic characteristics of the technology	Functional features of Polynesian canoes change less rapidly than stylistic nonfunctional traits (Rogers and Ehrlich 2008)
Manufacturing cost	Japanese Katana swords remained unchanged for centuries because any slight modification disrupted the manufacturing process (Martin 2000)
Fit with prior knowledge	Boiling of water fails to spread as a health practice due to incompatibility with preexisting beliefs of "heat illness" (Rogers 1995)
Generative entrenchment (lock-in of technology due to frequency-dependent adaptive landscapes)	QWERTY keyboard, originally designed to slow down typing in early typewriters to avoid jamming, but still used in computer keyboards where jamming is not a problem (Rogers 1995)
Key innovations (which transform adaptive landscapes and open up new innovation opportunities)	The vacuum-tube radio, which led to a cascade of innovations related to radio design and technology (O'Brien and Bentley 2011)
External representation of knowledge	Written records of medicinal plant use in medieval Italy reduce variation and change compared to regions without written records (Leonti 2011)
Social network structure	Centralized expert hubs facilitate transmission of adaptive food taboos on Fiji (Henrich and Henrich 2010)
Population size	Loss of technology on Tasmania due to reduction in population size (Henrich 2004b); increase in complexity in Upper Paleolithic Europe due to increased population densities (Powell et al. 2009)
Social institutions (e.g., trade networks, guilds, market economies, elite classes, universities)	Market integration in contemporary hunter-gatherer and horticultural populations results in the loss of ethnobotanical knowledge (Reyes-García et al. 2005)
Intergroup conflict	Technology conferring an advantage in intergroup conflict spreads rapidly, e.g., horses in ancient China (Di Cosmo 2002)
Intergroup boundaries and ethnic identity	Weaving techniques in Iranian tribal populations fail to spread due to norms against sharing knowledge with women from other tribes (Tehrani and Collard 2009)
External environmental change	The origin and spread of agriculture in the Holocene due to warmer, wetter, and less variable climate (Richerson et al. 2001)

phenotype, or extended phenotype frequencies within biological evolution. The appropriate level of analysis depends on the nature of the research. Historical reconstruction may benefit from a systemic approach that encompasses all three levels. Empirical researchers tend to focus on the level that is most practical and measurable (e.g., archaeologists are typically constrained to artifacts, whereas anthropologists are able to measure knowledge and/or behavior), which also allows them to consider the factors that affect the dynamics of each of the system's component parts. Mathematical modelers tend to track the smallest indivisible unit in the system. Thus, while debates over units have often received attention in the more philosophical literatures, it has not greatly hindered research in any of these more empirically focused domains.

As discussed above, several modeling frameworks now exist within which to address rates of cultural evolutionary change and levels of selection. As well as the Price equation, there are also other population genetic and game theoretical models of cultural evolution and gene–culture coevolution (e.g., Boyd and Richerson 1985, 2005; Cavalli-Sforza and Feldman 1981). Such methods include models of macroevolutionary change and cumulative cultural evolution (Mesoudi 2011c; Enquist et al. 2011; Strimling et al. 2009; Pradhan et al. 2012; Aoki et al. 2011; Ehn and Laland 2012; Perreault 2012), many of which have addressed issues of rates of cumulative technological change. Phylogenetic methods have also been increasingly applied to technological evolution (O'Brien and Lyman 2003a; Lipo et al. 2006), which has allowed the reconstruction of macroevolutionary patterns and, as a consequence, the rate of change (frequency of branching) over time.

Fitness Landscapes and Technological Evolution

Several of the factors listed in Table 11.2 draw on concepts of fitness landscapes from evolutionary biology (Wright 1932), with the shape of the fitness landscape either speeding up technological evolution (e.g., where a key innovation changes the shape of the landscape resulting in a burst of diversification) or slowing it down (e.g., when a technology becomes locked in due to the frequency-dependent nature of the fitness landscape, preventing further change). Shennan (this volume) provides several examples of how the concept of fitness landscapes has illuminated specific case studies, such as the evolution of the bicycle (Lake and Venti 2009). Although the shape (e.g., ruggedness) of the underlying fitness landscape is likely to be a major determinant of technological evolution (Boyd and Richerson 1992; Arthur 2009; Kauffman 1993), it has tended to be overlooked in formal models and experimental simulations of cultural evolution, which typically make simplifying assumptions about trait fitness (although see Mesoudi and O'Brien 2008a, b). This is likely because modeling cultural evolution on changing (e.g., frequency-dependent) fitness landscapes reduces the tractability of such models. Nevertheless, we see fitness landscapes as a fruitful line of investigation for the study of technological evolution. The important point here is that the fitness landscapes associated with technological innovations should be regarded as dynamic and frequency dependent, rather than fixed, such that the spread of an innovation can both channel the direction of new innovations and open up a suite of new possibilities.

Science

Conceptual and Cultural Foundations

Science is a recent cultural innovation that emerged primarily in large-scale literate human societies (see McCauley, this volume). Theoretically important characteristics of science are as follows (Renn 2012):

1. It is primarily noninstrumental in character and is not just concerned with technology but with explanation for its own sake.
2. It has a sustained tradition of criticism and public scrutiny.
3. It is dependent on literacy as well as, in particular, lasting external linguistic representations that precisely preserve scientific knowledge and allow abstraction of thought.
4. It is, at least partly, intellectually independent from political and religious authority.

Historically, the introduction of science had a major impact on the level of discourse and standards of truth within societies, through the formalization of argument and use of evidence. With the invention of printed publications (scientific pamphlets, journals, and books), a new level of public scrutiny was reached. Increasing costs of acquiring scientific knowledge and the mastering of scientific methods also led to the invention and spread of universities and other scientific institutions.

Historical Evolution

Like writing, science as a process is a product of cultural evolution, and we can again examine the historical precedents and selective pressures that gave rise to it (Renn 2012). Science, in the sense of a pursuit of knowledge for its own sake, emerged in large-scale literate societies when administrative elites required schooling and resulted in a division of physical and intellectual labor. This happened in Babylonia and Egypt by the second millennium BCE, somewhat later in China, and much later in Mayan culture, resulting in the emergence of mathematics, astronomy, and medicine. Throughout this period, science was merely a contingent by-product of other social activities (e.g., administration) and was not a necessary or valued function in these societies. Indeed, most societies prior to the early modern period did not systematically support science. Instead, science was pursued because of individual interests or prestige.

This situation changed fundamentally after the early modern period due to the increasing and sustained economic and ideological significance of science in European societies. Early modern science was characterized by the take-up of technological challenges such as ballistics, ship building, large-scale construction (e.g., of cathedrals), urban infrastructure (e.g., hydraulics), and machines (e.g., mills and other labor-saving technology). The printing press enabled the dissemination of technical knowledge that was previously transmitted orally and by participation. For the first time in history, an extensive technical literature emerged. The rapid development and scaling-up of science led to new forms of institutionalization of good scientific practice which the new academies (e.g., Accademia del Cimento, the Royal Society) articulated and imposed, thus creating a scientific community with its own norms. Early modern science was practiced by a broad network of participants that extended throughout Europe and then, with colonization, globally. Science as practiced within this network and its institutional support by political and ecclesiastical authorities was accompanied by a broadly shared conviction about its practical utility. Its actual technological benefit was initially limited but that hardly affected this societal perception.

In this historical situation, a self-reinforcing mechanism emerged that connected the production of scientific knowledge with socioeconomic growth. The mechanization of labor processes in the late seventeenth and eighteenth centuries provoked new scientific questions and opportunities for science to improve technology. The combination of this mechanization of labor with new ways of exploiting energy, in particular the use of coal for the steam engine, led in the nineteenth century to the Industrial Revolution in England. Together with the driving forces of market economies, this created further challenges and opportunities for science to improve technology. In this way, science and technology became inextricably intertwined with each other and with economic development. From that point on, science ceased to be a contingent by-product of cultural evolution and became one of its driving forces.

Evolutionary Character

The evolution of scientific knowledge itself exhibits all the dynamics characteristic of an evolutionary process, here referred to as "epistemic evolution" (following Renn 2012; see also Hull 1988; Thagard 1992; Renn 1995; Damerow et al. 2004). The exploration of the inherent potential of the means for gaining knowledge in a society gives rise to a variety of conceptual alternatives within a knowledge system, corresponding to mutation in biological evolution. As these alternatives are elaborated and pursued, they lead to internal tensions and contradictions, resulting in the transformation or the branching of a new knowledge system; this can be seen as analogous to speciation. For example, in the early modern period a broad variety of proposals for a new theory of motion was advanced by Galileo, Descartes, Harriot, and others which eventually led to convergence on a new understanding of motion (Damerow et al. 2004; Schemmel 2008). Various selective pressures may act on scientific knowledge systems and theories, such as compatibility with existing knowledge, internal coherence, compliance with methodological and institutional constraints, as well as societal expectations, prestige, fashions, and ideologies. Existing proposals for epistemic evolution are verbal rather than mathematical, and it remains a challenge for the future to construct formal models that successfully incorporate these processes.

The Future

We can also ask about the future state of science from an evolutionary perspective. Despite much talk about the importance of interdisciplinarity, the course of science seems to be one of increasing disciplinary specialization, as fields of study become so complex and knowledge intensive that a single scientist struggles to specialize in more than one domain. In a sense, this is an inevitable by-product of the inherently cumulative nature of science. As more and more scientific knowledge is accumulated in a particular field, it becomes more and more costly for a single scientist to acquire that accumulated knowledge. There is evidence of this from the quantitative analysis of science, scientometrics. First, analysis of scientific accumulation (measured using number of publications or number of patents) shows an exponential increase since records began (e.g., May 1966; Price 1963), thus supporting the assumption that science is cumulative. Second, the length of time it takes for a scientist to become expert in their field has increased; for example, the average age at which Nobel prize winners made their prize-winning discovery has increased from 32 to 38 in the hundred years since Nobel prizes were first awarded (Jones 2010). This is not due to increased life expectancy, but due to increasing training periods in fields such as mathematics and physics (Jones 2010). Mesoudi (2011c) modeled this process, showing that the increasing costs of acquiring ever-increasing knowledge can eventually constrain further innovation, at a point where individuals spend so much time learning what has gone before that they have no time left to discover anything new.

One potential solution to this increasing burden of acquiring prior knowledge is disciplinary specialization, with scientists becoming more specialized as their fields become more complex. However, specialization comes with its own potential costs. As science gets bigger and more specialized, inevitably divisions arise between scientific disciplines, which leads scientists to grasp at conceptual tools to render their activities more manageable. This includes screening off and dismissing domains as the business of other disciplines, treating complex phenomena as black boxes, and regarding certain processes, or sources of variation, as relatively unimportant. As a result, scientific disciplines can effectively become "clubs" in which like-minded researchers share consensus over what is, and what is not, reasonably treated as "cause" and "context."

Although this black boxing or screening off is often initially useful, it becomes a problem when core assumptions become dogma or entrenched. A good example is Ernst Mayr's distinction between proximate and ultimate causation, where an initially useful heuristic has sometimes become an unthinkingly applied convention in which developmental processes are seen as irrelevant to evolution, leading to divisions between academic fields of enquiry, and prompting several major debates within biology (Laland et al. 2011). The danger here is that discipline-based scientific fields emerge which through their core assumptions exclude, or hinder, certain phenomena from being considered as causes, leading to the neglect of relevant processes that contribute to evolutionary change or stasis, and hindering interdisciplinary exchange. One solution proposed by Levins (1966) is to encourage pluralism with regard to model building, with different classes of models screening off different processes, but collectively covering all processes. Levins's (1966) idea focused on formal models but the same point holds for conceptual frameworks. Pluralism is vital to scientific progress (McCauley 2001).

Conclusions

Progress So Far

In the course of our discussions, we were encouraged by the progress being made in the study of technological and scientific change within a number of different disciplines, which use a number of different methods, all inspired by a cultural evolutionary framework. We discussed research from archaeology, anthropology, psychology, the history and philosophy of science and technology, neuroscience, and economics. The methods used to pursue this research have included mathematical models, agent-based simulations, laboratory experiments, ethnographic surveys and experiments, archaeological/historical analysis, phylogenetic methods, comparative studies of nonhuman species, and brain-imaging techniques. We see this interdisciplinary, multi-method approach as one of the key benefits of a cultural evolutionary approach (thus combating the disciplinary fragmentation noted earlier), as empirical findings inform the assumptions of models, which in turn guide empirical work by highlighting key variables upon which to focus (Mesoudi 2011a). The methodological toolkit and theoretical framework are now in place, and thus the hard work of applying these methods to specific empirical case studies can now begin.

Outstanding Questions

Despite our optimism, outstanding questions remain and thus we wish to highlight promising avenues for further study. First, we see potential for greater links to the economic models of the evolution of institutions, which may prove useful for modeling such phenomena as trade networks, guilds, and universities—institutions that affect the course of technological and scientific evolution. Economists conceptualize institutions as self-sustaining normative systems that structure individual behavior within groups. For example, Greif (1993) has studied the evolution of institutions governing international trade in the Renaissance Mediterranean. Merchants from the Maghreb organized their enterprises based on kinship, while Genovese merchants based them on contractual arrangements. Both institutions regulated behavior and thus solved the principle-agent problem inherent in long-distance trade and, as a consequence, competed with each other on an equal basis during the early Renaissance. However, as the volume of trade increased, the Genovese system scaled up more easily and ultimately replaced the kinship-based system. In economics modeling, there is a growing literature on institutional change using this kind of framework.

Second, an empirical question raised by Boyd et al. (this volume) remains unanswered concerning the extent to which technological change depends on a rational causal understanding of problems, which is then transmitted to others along with technological artifacts. This is a typical assumption of rational actor economic models, and is also assumed by some evolutionary psychologists (e.g., Pinker 2010). However, ethnographic and archaeological evidence suggest that people are rarely fully aware of the causal reasons behind why a particular technology works (Henrich and Henrich 2010; Shennan, this volume). If this is the case, then relatively content-free transmission biases (e.g., prestige or results bias) will play a more important causal role in technological evolution than sophisticated and explicit cognitive representations. An additional question concerns how science affects this issue: science provides explicit tools for determining causal explanations for phenomena in the world, at least in theory, enhancing people's ability to adopt efficacious technology.

Third, a repeated theme in our discussion of both technology and science was the notion of "fragility." It is apparent that science, as a cultural system, is highly fragile (McCauley, this volume), originating and persisting only in the presence of a precarious set of social, political, and economic conditions. Technology, too, is often surprisingly fragile, easily susceptible to loss in the face of population reduction or disruption to social networks (see Table 11.2). Certain factors, such as particular forms of long-lasting external storage like writing, can reduce the fragility of technological and scientific knowledge. However, the stability of scientific and technological knowledge in our own industrialized societies should not be taken for granted, as a historical perspective demonstrates the ease with which knowledge can be lost. One potentially fruitful line of study might be to more explicitly conceptualize and model the notion of fragility, incorporating factors that may increase or decrease the fragility of a technological system.

Practical Applications

A final line of discussion centered around the practical applications of the research outlined above. One potential application relates to the predictability of technological evolution: Can we predict, using the mechanisms listed in Table 11.1 and the factors in Table 11.2, whether a particular technology will spread through a population or not, and how rapidly? Although technological evolution, like genetic evolution, is likely to be inherently stochastic and future evolutionary trends may never be predicted with absolute certainty, the criteria listed above may provide some guides to likely general trends. If so, this raises the possibility of intentionally designing technology or creating social conditions to favor the spread of technologies deemed to be beneficial to society and, alternatively, to prevent the spread of technologies deemed to be harmful. This is being attempted in some fields such as marketing (Heath and Heath 2007) and the diffusion of innovations within sociology (Rogers 1995), both of which share substantial overlap with the cultural evolution literature discussed here. For example, Heath and Heath (2007) discuss content biases such as the emotional salience of a particular cultural variant (Heath et al. 2001), whereas Rogers (1995) discusses factors such as the prestige or centrality of actors within social networks equivalent to model-based prestige biases. Ethnographic studies may prove particularly useful here, presenting opportunities to track the spread (or loss) of knowledge within small-scale societies (Reyes-García et al. 2008, 2009). Network-based diffusion analysis (Franz and Nunn 2009a; Hoppitt et al. 2010), originally developed to detect social transmission in nonhuman species, may also prove useful.

However, the study of science and technology itself gives reason to be cautious about imposing practical objectives onto science, given the risk that this may in fact *inhibit* scientific innovation. The history of science amply illustrates that major innovations have rarely been the result of imposing specific societal expectations onto science but rather of serendipity and accident, such as the discovery of X-rays or antibiotics. The very autonomy of science is in potential conflict with its functional role in society as a promoter of technological innovations and economic growth. This intrinsic tension between science and society is becoming more acute because science has become relevant not only to societal welfare, but also to the very survival of the human species. Thus, challenges such as climate change, global energy, food and water provision, global health, and living with nuclear technology require persistent scientific innovation at a global scale, yet remain unbiased by immediate economic and political constraints. Such global basic science has yet to find the societal niche and support that it requires. The concepts and tools of cultural evolution may prove helpful in defining this niche.

Language

12

The Interplay of Genetic and Cultural Factors in Ongoing Language Evolution

Stephen C. Levinson and Dan Dediu

Abstract

This chapter discusses recent advances in our understanding of the complex interplay between cultural and biological factors in language change and evolution. Three "myths" (the independence of biological and cultural evolution, a fixed biological foundation for culture, and the cognitive uniformity of humans) are identified and falsified. Strong genetic biases are shown to affect language profoundly, using the example of village sign languages that emerge and complexify due to persistent high frequencies of genetic deafness in certain communities. Evidence is presented for the genetic bases of language and speech, and the extensive genetic variation within populations affecting them. Finally, it is proposed that in addition to intrapopulation variation, interpopulation differences in genetic biases that affect language and speech contribute to the emergence of linguistic diversity, through iterated cultural transmission across generations as well as communication and alignment within them. Thus, biological and cultural processes cannot be meaningfully separated when studying the cultural evolution of language.

Cultural Evolution and Biological Evolution Are Two Sides of One Process

This chapter is about the relationship between cultural and biological evolution, as evidenced in the domain of language. Many scholars with an interest in cultural evolution operate with a set of myths or fictions, tacitly holding something like the following:

1. *Fiction of independence of biological and cultural evolution.* Biological and cultural evolution are for practical purposes now independent processes, and despite the "curious parallels" between the diversification

of species and of languages (as noted by Darwin 1871), the underlying principles are fundamentally different.

2. *Fiction of a fixed biological platform for culture.* Nevertheless, on a deep and ancient timescale, the two evolutionary tracks were in fruitful interplay and coevolved a now fixed platform for cultural liftoff: the big human brain, the language capacity, and the manual dexterity that made technology possible.
3. *Fiction of the cognitive uniformity of the species.* Since, generically, all humans have complex cognition, use language and make things with their hands, these capacities can be taken to be near uniform across the species: they form a constant background to cultural evolution.

In contrast, this chapter advances the following propositions:

1. Cultural and biological (genetic) evolution constitute twin tracks of an evolutionary process.
2. There are two-way feedback relations between the tracks.
3. These relations are ongoing.
4. There is significant variation within populations both with regard to genes and cultural variants, which supplies the "fuel" for evolutionary processes.
5. Even slight differences in the distribution of gene frequencies within populations can bias vertical and horizontal cultural processes and thus seed cultural evolution.

Propositions (1) and (2) have been much discussed. The general consensus is that, with a certain latitude over models, there is essentially no other way to explain the evolution of the capacities for culture (see, e.g., Boyd and Richerson 1985). Proposition (3) merely states Lyells's principle of Uniformitarianism: processes which used to operate on geologic timescales are still in operation now. It runs, though, directly into conflict with Fiction (2), the presumption that the biological platform for culture was achieved in prehistory and now remains essentially static. The fiction has currency because it is built into our cultural baggage, with humankind at the apex of the tree of life. In fact, of course, our rapid biological adaptation to cultural innovation is well attested (as in the development of adult lactose absorption in response to the culture of dairying, or the adaptation of the immune system to the diseases we have brought upon ourselves by migration, farming, behavior, sexual mores or misuse of antibiotics; Laland et al. 2010). Arguments that the different timescales of cultural and biological evolution sever the connection are unfounded; if one likens the twin tracks of culture and biology to obligate symbiosis (as between pollinators and flowering plants), it is obvious that differential rates of evolution are not an impediment to coevolution (Levinson 2006). Indeed, it is sobering to realize that the speed of biological change in our species is ever increasing (Hawks et al. 2007).

Proposition (4) should be self-evident, but it runs against Fiction (3), which is the presupposition (or idealization) adopted at the birth of the cognitive sciences; namely that humans are, from a cognitive point of view, more or less clones, so that studying the minds or brains of, for example, highly educated Western undergraduates is to have sampled the entire variation (Henrich et al. 2010b; Levinson 2012a). However, it is Proposition (5) that has perhaps the most resonance for this volume, for it contains the following little time bomb: Cultural variation may not be wholly *sui generis*, for it may be seeded by small genetic differences across populations.

What this chapter explores, then, is the positive feedback relations between the cultural and biological tracks of evolution in the language domain, on the presumption that they continue to influence the direction of change.

Village Sign Languages

A compelling example of this mutual influence between biology and language is represented by village sign languages (Zeshan and de Vos 2012). Around the world, in developing nations without elaborate state institutions for handling special educational needs, population pockets can be found with a high incidence of congenital deafness. These are likely to be small village populations with considerable inbreeding, where the incidence of deafness may be tenfold higher than in the surrounding population (Winata et al. 1995; Scott et al. 1995). In these circumstances, sign language spontaneously arises, which over several generations begins to acquire the full expressivity we associate with spoken languages (De Vos 2012; Sandler et al. 2005). The hearing population also acquires the sign language, since relatives and friends are likely to be deaf. Deaf villagers thus become fully competent members of society, and marriage and reproduction takes place with little prejudice, so maintaining the high incidence of deafness in the village. Here in microcosm we see the interdependence of a cultural form of language, a specific sign language, and the genetic basis that perpetuates it. There is a positive feedback loop between cultural and genetic evolution.

In contrast, the major sign languages have arisen in the context of institutions for special education of the deaf. The exact history of their origins is often complex and little understood. French sign language is three hundred years old, but some are of recent origin. For example, Israeli sign language, ISL, arose during the formation of the state of Israel, when immigrants from different national institutional backgrounds and small-scale sign communities were brought together (Meir et al. 2010). The celebrated case of Nicaraguan sign language emergence is rather similar, where urban institutionalization of deaf children from different villages with some input from other national sign languages led to the rapid development of a new sign language (Senghas et al. 2005). The institutionalization of the deaf and the associated use of sign language and inter-deaf marriages represent another case of biology shaping culture, and, at least for some prevalent forms of genetic deafness, this cultural trait might have fed back on biology by increasing the frequency of the genes involved (Nance and Kearsey 2004; Arnos et al. 2008).

Village sign languages, however, are distinct (Meir et al. 2010): they have spontaneously arisen without input from another sign language wherever there is a high local incidence of deafness and no institutionalization of the deaf. The communities in which they are used may number a few thousand of whom up to 4% may be congenitally deaf (in the United States, the percentage in the general population is ca. 0.07%). Nevertheless, up to two-thirds of the entire community may be sign language users, providing the critical mass of individual users necessary for language evolution to work (Senghas 2005). The village sign languages that have been researched seem to be of no great antiquity, with a depth of between 4–10 generations, although there is also circumstantial evidence for recurrence in the same populations over greater time spans. There is enormous interest in what levels of linguistic complexity can arise on that kind of timescale (75–200 years) in such isolated settings without input from other signed languages (although, *nota bene*, with one-sided participation of hearing speakers of a spoken language).[1]

Here we focus on two of the better studied such languages. Al-Sayyid Bedouin Sign Language (ABSL) has arisen in a community of 3,500 with ca. 130 deaf adults over a period of ca. 75 years (Sandler et al. 2005). ABSL does not show any clear evidence of Hockett's "double articulation" (i.e., the recurrent use of meaningless elements like phonemes to construct higher-level units like words) and in that sense lacks a "phonology." Words themselves tend to have multiple conventional variants, with over a quarter having more than three acceptable alternate forms.

None of the village sign languages investigated so far shows any inflectional morphology, unlike the established national sign languages. However ABSL does employ noun compounding to modify nouns. Syntactic structures are simple, and despite having established an SOV (subject–object–verb) word order, clauses with two or more nominals are avoided in favor of a string of simple subject-plus-verb structures (so "GIRL STAND; MAN BALL THROW; GIRL CATCH" for "The man throws the ball to the girl"). As in other village sign language systems, there is *no evidence for syntactic subordination*, thus no evidence for the recursion or Merge expected by nativist generative grammarians.

Less grammatical work has been done on Kata Kalok (KK), a village sign language used for at least seven generations in Bengkala, northern Bali (Winata

1 There is, however, no significant structural borrowing from the spoken to the signed languages (e.g., word orders are distinct, vocabularies differently structured, syntactic and semantic calquing absent). Where there is "contamination," as it were, it is from the surrounding spoken gesture systems, which are likely to be incorporated and partially grammaticalized in the sign languages.

et al. 1995; De Vos 2012). Again there is little evidence for double articulation or "phonology." Inflectional morphology is largely absent. Contrary to earlier reports, there appears to be no canonical word order, with avoidance of explicit multiple noun phrases, but when present all three orders (SVO, OVS, SOV) are equally likely. Nominals may function as modifiers and predicates, and order within phrases is very variable. Nominal semantics is more general than in the surrounding spoken language Balinese; for example KK has four color words, Balinese eleven, and KK has much less detailed kinship terminology.

Thus, village sign languages display simple structure, show much internal variation in allowable forms, do not show recursion, may not exhibit double articulation, and avoid multiple noun phrases that might cause confusion about "who did what to whom." Despite this, they have been put through the processes of recurrent learning by new generations and communication within generations. The evidence available so far from village sign languages is that it takes a great deal more than a half dozen generations to evolve a highly structured language. Experiments with iterated learning suggest that 6–10 generations might be required for highly structured artificial languages to evolve (Kirby et al. 2008), but these must be taken as very rough guides. Village sign systems are the only plausible cases of languages evolving *de novo*, since all *spoken* languages, as far as we know, descend from one or more other languages by direct descent or hybridization. They are also one of the few cases where we can readily detect a feedback from culture to biology (on top of the more general influence of biology on culture), although the general role of language skills in sexual selection and biological fitness is highly plausible.

From a genetic point of view, in both the KK and ABSL cases, a mutation was introduced in the population relatively recently (probably 7–8 generations ago for ABSL and 7–20 for KK) either from outside or through a spontaneous event in the population. Interestingly, these two mutations are different: for ABSL, the mutation affects the locus *DFNB1* on chromosome 13q12, most probably the genes *GJB2* and *GJB6* (OMIM[2] 220290), while for KK the mutation disrupts the *DFNB3* locus and more precisely the *MYO15A* gene on chromosome 17p11.2 affecting the hair cells in the cochlea essential for hearing (OMIM 602666) (Liang et al. 1998). However, both mutations, even if they affect different genes, result in a *nonsyndromic hearing loss* (i.e., deafness not accompanied by other problems or defects) starting before the onset of language acquisition (*prelingual*) in individuals carrying two copies of the mutation (*homozygous*). Individuals carrying a single copy of the mutation and a copy of the normal gene (*heterozygous*) have normal hearing, making the mutation *recessive*. Because both affected loci are on the nonsex chromosomes (*autosomes*) 13 and 17, respectively, a child will inherit a copy from its mother and one copy from its father. Thus, a child will be affected by hearing loss only

[2] Online Mendelian Inheritance in Man (OMIM), http://www.ncbi.nlm.nih.gov/omim, uses unique numeric identifiers for genetic disorders and genes.

if the two copies inherited from both parents are mutated, meaning that both parents must be heterozygous carriers (with normal hearing) or homozygous deaf. In both communities, therefore, the probabilities of having a deaf child given the genetic structure of the parents are given in Table 12.1.

In both communities, there is a relatively high incidence of deafness due to homozygous individuals (~3% for ABSL and ~2% for KK), and such high frequencies of deafness have persisted for several generations due to high rates of *inbreeding* (within-community marriages). Inbreeding increases the chances that the two parents carry one (heterozygous) or two (homozygous) copies of the mutation, as they descend ultimately from the original carrier of the mutation in the population.

Such a situation in which congenital deafness persists could result in the long-term marginalization of the deaf. However, in both cases discussed here, not only are the deaf well integrated socially, including marriage and reproduction with other members of the wider community, but most *hearing* members of the communities are also able to communicate effectively with the deaf members. This has encouraged the feedback between the two systems: the culturally evolved sign language which allows deaf members to be fully fledged members of the society, and the inbreeding thus facilitated which maintains the strong strand of hereditary deafness. There has been some interesting mathematical and computational work that has addressed the population characteristics and assortative mating required to maintain this positive feedback relation between language and genes (Feldman and Aoki 1992; Nance and Kearsey 2004), but a more general model is needed.

We have focused on village sign languages as a microcosm to illustrate the interplay between genes and culture in the evolution of a language. Of course, the principles here are special, and the genetics of high-incidence deafness

Table 12.1 Probabilities (as percentages) of hearing and deaf children resulting from all possible marriages by parental genotype. N represents the normal allele, M the mutated, deafness-causing allele. Possible children genotypes are given by phenotypes; percent.

		Mother's genotype (phenotype)		
		NN (hearing)	NM (hearing)	MM (deaf)
Father's genotype (phenotype)	NN (hearing)	NN (hearing; 100%)	NN (hearing; 50%) NM (hearing; 50%)	NM (hearing; 100%)
	NM (hearing)	NN (hearing; 50%) NM (hearing; 50%)	NN (hearing; 25%) NM (hearing; 50%) *MM (deaf; 25%)*	NM (hearing; 50%) *MM (deaf; 50%)*
	MM (deaf)	NM (hearing; 100%)	NM (hearing; 50%) *MM (deaf; 50%)*	*MM (deaf; 100%)*

forces a modality change in language. Nevertheless, the examples serve as an important reminder of the power of genetic biases to send cultural evolution down novel paths. In the following, we will focus on much weaker effects, examining first the amount and nature of interindividual genetic variation affecting language and speech, and then looking into the effects of interpopulation differences on language change, diversity, and universal properties.

The Fuel of Evolution: Genetic Variation in the Population as Evidenced in Language Performance

We have mentioned the common assumption in the cognitive and social sciences that human evolution is somehow in abeyance, and that we all have the same essential cognitive endowments, linguistic abilities being a prime example: Fictions (2) and (3) above. In fact, linguistic abilities are clearly conditioned by genetic variation, and this variation can be found within both the clinical and normal population.

A first approach concerns the estimation of the heritability of aspects of speech and language. In essence, heritability is defined as the *proportion* of phenotypic variance explained by variance in the genotype,[3] $h^2 = \mathrm{var}(G)/\mathrm{var}(P)$ (Visscher et al. 2008). Its estimation classically involves *twin* and *adoption studies*, but new methods of estimation from unrelated individuals have been recently developed. In twin studies, the phenotypes of *monozygotic* (identical; MZ) and *dizygotic* (nonidentical; DZ) twins are compared; since MZ twins are genetically identical[4] whereas DZ twins share on average 50% of their genes, the (narrow sense) heritability of the phenotype can be estimated. In its simplest form,[5] $h^2 = 2 \times (r_{MZ} - r_{DZ})$, where r_{MZ} and r_{DZ} are the correlation between the phenotypes of the MZ twins and the correlation between the phenotypes of the DZ twins, respectively. Heritability can vary between 0 (genetic factors do not account at all for variation in the phenotype) and a maximum of 1 (all phenotypic variation is accounted for by genetic variation), and several caveats and limitations must be kept in mind when interpreting it (Visscher et al. 2008; Charney 2012).

There is a wealth of twin studies on various aspects of language (for a good review, see Stromswold 2001), and they suggest that most have a genetic component. For example, various *disorders* of speech and language usually show moderate to high heritabilities ($h^2 > 0.50$), such as liability to *stuttering*, *specific language impairment*, or *dyslexia*. Likewise, an important genetic

3 There is a distinction between *broad sense* heritability (denoted H^2) and *narrow sense* heritability (h^2). Here we focus on the latter, given that it is easier to estimate and it is generally used in the literature. This is also the additive component of genetic variation upon which selection can work.

4 Ignoring *de novo* mutations.

5 Presently, more complex methods based on *structural equation modeling* are increasingly used.

component was found for some aspects of *normal variation* in language and speech such as *vocabulary* size and *second language learning*. In addition, the physical aspects of producing speech are under genetic influence as shown by the heritability of the shape and size of the *hard palate*, *tongue* size and shape and of various *acoustic properties of speech* due to vocal tract anatomy and physiology. Heritability studies establish prima facie reasons to suspect genetic causal correlates of phenotypic differences, but they do not themselves pinpoint the genes or pathways involved, which need a different set of methods.

A very successful and productive research program was initiated by the discovery (Lai et al. 2001) that the *FOXP2* gene is responsible for a complex pathology ("developmental verbal dyspraxia") that affects speech and language (OMIN 602081). This gene is highly conserved across mammals and, against this background, the human form (which we share with our cousins, the Neandertals; Krause et al. 2007) has specific mutations that might be related to the emergence of speech and language in our lineage. It is currently uncertain whether *polymorphisms*[6] in *FOXP2* affect normal variation in speech and language, but there are early indications that they may, at least at the neural level (Pinel et al. 2012). *FOXP2* gave us an entry point for exploring the genetic underpinnings of speech and language because it is a regulatory gene (Fisher and Scharff 2009). The identification of other genes regulated by *FOXP2* proves to be an active research field, leading to the discovery of genes such as *CNTNAP2* (Vernes et al. 2008) which is also involved in language pathologies.

Another syndrome that has offered valuable insights is *dyslexia* (Scerri and Schulte-Körne 2010), a spectrum disorder which manifests as a reading disability but which often seems nevertheless to reflect deeper phonological and linguistic processing problems. Risk factors for dyslexia are association with alleles of the genes *KIAA0319* and *ROBO1* among others. Interestingly, it was recently found that *ROBO1* is involved in normal variation in Non-Word Repetition (Bates et al. 2011), a task thought to measure a component essential for language acquisition. In this sort of way, research into syndromes can reveal variants that have milder effects in the general population.

To summarize, during the last decades we have accumulated a lot of evidence that (a) most normal and pathological aspects of speech and language have a genetic component, (b) these genetic foundations are extremely complex (Fisher 2006), and (c) there is extensive variation between individuals. This last point is fundamental for understanding the relationships between language and genes and needs to be emphasized: close investigation reveals variation between normal individuals in almost all aspects of speech and language one cares to consider. This includes variation in *grammaticality judgments* (e.g., Schütze 1996; Dąbrowska 1997) and *sentence processing* (Farmer et al. 2012), suggesting that the normal, fully competent native speaker's language carries

6 That is, normal variants of the gene, as opposed to the catastrophic mutations, which give rise to developmental verbal dyspraxia.

that individual's genome signature in subtle ways. It must be highlighted that this signature is not direct and deterministic, but modulated by the environment, such that the individual differences result from the complex interaction of the genetic background and the individual's experiences (Lieven, this volume). Such a signature may remain hardly discernible until the individual's language system is under stress or when the cultural and social circumstances change, such as when literacy became a widespread and essential requirement, unmasking hitherto hidden genetic variation in the form of what we now diagnose as dyslexia (see Tomblin and Christiansen 2009). In short, just as we are short or tall, fat or thin, good at running or dexterous with our hands, so all the myriad capabilities that underlie language capacity vary within any population.

We turn now briefly to the differences *across populations*. Cavalli-Sforza et al. (1994) produced a magnum opus that estimated the most likely phylogenetic trees for worldwide human groups based on blood groups and other classical genetic markers, and found close matches to the trees so obtained with the boldest suggestions for the phylogenetics of the world's language families then available. It is now conceded that both kinds of trees were flawed (e.g., Sims-Williams 1998). They presumed successive splitting with minimal hybridization, which is wrong for both language and genes, and used language trees that few linguists now subscribe to. Since then, with the full sequencing of the genome, the simplicity of the phylogenetic story has unraveled, and the many factors from disease to ecology shaping genetic diversity have become apparent (Jobling et al. 2004; Novembre et al. 2008). Even small gene trickle between Pacific populations, for example, has wiped out close correlations with language families in that area. Spectacularly, recovery of archaic DNA shows that some populations have inherited genes from different premodern humans, with out-of-Africa peoples picking up Neandertal genes (Green et al. 2010), and Australians and Papuans further picking up genes from the Denisovans (Reich et al. 2011), a sister branch to the Neandertals. Still, when the dust settles from the intensive work on the human genome now in progress, we can be fairly sure that the migration history and kinship of most human groups will be at least partially recoverable from the genetics (see Novembre et al. 2008; Paschou et al. 2010).

Human groups then tend to have relatively different gene pools, within which the gene variants governing language-related phenotypes will no doubt be shown in due course to vary in kind or more likely proportion. It is, however, important to remember that this kind of variation across human groups is a minor part of the story. Modern humans are genetically a very homogeneous species when compared to other mammals (Barbujani and Colonna 2010), due to our evolutionary history involving small population sizes and repeated bottlenecks, with few differences between populations due to selection, such as in the immune system (Mukherjee et al. 2009) or skin color (Jablonski and Chaplin 2010). This is reflected in the distribution of genetic differences: about 85–90% of the genetic variation is between the individuals of the same population with only 10–15% lying between populations, and even these form smooth clines without sharp boundaries that could be taken as anything like "human races" (Cavalli-Sforza et al. 1994). These small differences between populations, though distinctive, are mostly due to differences in the frequency of ubiquitous ("cosmopolitan") genetic variants and very rarely to group-specific ("private") ones.

To summarize this section:

- We are beginning to discover the genes that build the brain and the vocal tract and make language and speech possible.
- These discoveries rest on significant variation in genes within populations; these govern distinctive phenotypes, which provide a reservoir to be exploited for language evolution.
- There is also lesser but significant variation in gene pools across populations in different parts of the world.

Language Diversity Seeded by Genetic Variation

We wish to put forward a startling hypothesis: Genes may be partly responsible for the cultural diversity of language, and it certainly is the antithesis of Fiction (1). They may play this role in two ways:

1. Genetic diversity within populations harbors a range of potentially culturally reinforcable language phenotypes.
2. Small genetic differences between populations may slightly bias cultural transmission and so tip cultural evolution in particular directions.

We should emphasize right away that this is not a theory of genetic determinism of culture, nor a kind of closet racism. The spectrum of phenotypic possibilities in Hypothesis (1) will be mostly shared across populations right around the world. The biases in Hypothesis (2) just amplify tiny imbalances by cultural processes and are highly contingent on many other factors, just as in chaos theory where small differences in input conditions can lead to wildly different outcomes.

Let us begin with Hypothesis (1), which is largely speculative but plausible. As noted, most human genetic variation is shared across widely spread populations; just as there are tall versus short, lighter- versus darker-skinned, thin versus fatter individuals in every population, so there are potential lispers, stammerers, hyper-multilinguals, gifted poets, singers, or public speakers in every population. In addition, as remarked earlier, the variable genetic substrates may often be masked by adequate linguistic performance. Think of a language as a mapping from sound (or sign) to meaning: there may be multiple possible algorithms that will do the mapping, and there may be multiple ways that any one of those algorithms can be instantiated in wetware. Given the fundamental variation in language types around the world (Evans and Levinson 2009a; see also Evans, this volume), it is likely that different types of language favor different algorithms. Consider, for example, that some languages have contrastive tone on words and some do not. Chinese speakers and even speakers of tonal dialects of European languages show greater activation of right hemisphere neural circuitry, probably because the right hemisphere better handles the 200 ms processing window required. Yet we already know that speakers of nontonal languages show considerable variation in the degree of right hemisphere language processing, that this is reflected in cortical structures and linked to biological factors like sex (Catani et al. 2007). In this way there is a reservoir of potentiality in any population: some individuals will be preadapted for more efficient processing of a tonal language, others for a nontonal language. A particular cultural form of language exploits the special facility of some individuals and makes the rest work harder.[7]

In a similar way, it is likely that languages with surface phrase structure and fixed order must be processed differently than those with case markers and scrambled word order; the former predisposes a more syntax-driven route to parsing, the latter a more semantic (thematic-role) driven route. Shallow, "good enough" processing shows we all probably use both routes, but the differences in grammaticality judgments mentioned earlier probably reflect the graded interindividual differences in the use of various strategies. At present we know little about intrapopulation differences in language processing due to the unfortunate Fiction (3) of cognitive uniformity, which has dominated the cognitive sciences (Levinson 2012a). Variation is avoided by using Western undergraduate participants for experiments (Henrich et al. 2010b)—but even in this very restricted population, variation, of course, exists (Farmer et al. 2012)—and by subsuming variation in group averages, while brain imaging has historically mapped results onto a generalized brain and swept the left-handers and even the "other" sex under the rug.

Let us turn now to Hypothesis (2): genetic differences between populations may slightly bias cultural transmission and intragenerational communication and so tip cultural evolution in particular directions. Hypothesis (1) is the idea that a single population encompasses most of the variation. Hypothesis (2) is based on the assumption that the variation is not equally distributed across all populations, and that this can have consequences for the likely direction of cultural evolution. It is clearly in line with what we know about the distribution of genetic diversity in general across the human species, as discussed above.

7 An application of the same idea to nonlinguistic material may perhaps be found in observations about cultural "techniques du corps" made by Mauss. Some traditional populations throw overhand, others underhand; some swim breaststroke, others prefer crawl; some relax by squatting on their haunches while others stand on one leg, and so forth. Big calf muscles make it hard to squat, thin calves make it easy, and in this way genetic variation may seed cultural norms (Mauss 1973).

Hypothesis (2) has some empirical backing. First, computational and mathematical modeling of language transmission shows that weak biases in individual processing get amplified through cultural transmission, resulting in universal tendencies when the bias is shared by all individuals (Kirby et al. 2007). Experimental studies of iterated learning provide more evidence that the transmission of language across chains of human participants amplifies much weaker biases, such as a bias towards systematicity (Smith and Wonnacott 2010). Further modeling shows that when the bias is unequally distributed across populations, different structural solutions recurrently emerge (Dediu 2008). In short, innate biases in the form of slightly different proportions of alleles in a population are sufficient to breed extensive cultural differences.

Slight differences in the facility with which sounds can be produced would constitute one of these kinds of biases. Consider that the authors of this paper both have their linguistic Achilles' heels: one had a childhood lisp and the other cannot do his alveolar trills to this day, but of course not all languages have sibilants or trills, and thus we could both have won elocution prizes in the appropriate cultures. A population heavily seeded with Levinsons and Dedius would not have evolved a language with sibilants and trills. There is some empirical evidence for the effects of these production biases on linguistic typology. Naturally enough, vocal tract anatomy exhibits structured variation across the populations of the world. It has been noted that some of these differences likely contribute to the probabilities of certain phoneme inventories. For example, Ladefoged (1984) noted differences in the second formant (F2) between the otherwise similar vowel systems of Yoruba and Italian, arguing that it might be due to differences in the anatomy of the upper tract between the typical speakers of these languages, ideas that have been further examined by Dediu (2011b) and Ladd et al. (2008). Another example concerns the shape and dimensions of the *hard palate* (the bony roof of the mouth), which is known to vary across populations in the degree to which it is wide and flat, or narrow and domed (Sugie et al. 1993; Byers et al. 1997), and known to be under genetic influences (Townsend et al. 1990; Dellavia et al. 2007). These differences induce differences in tongue contact, and thus have acoustic correlates. It is probable that a high-domed palate facilitates the production of *apical retroflex consonants* (sounds produced with the tip of the tongue curled back, such as [ʈ] and [ɖ]), even though anyone can learn to produce these sounds. The hypothesis advanced here is that there will be a positive correlation between the proportions of palate shapes in a population and the phoneme inventories likely to arise. This is because palate shape would constitute one of these weak but insistent biases that have been shown in modeling to induce population-level cultural differences.

Probably the best-supported proposal to date for such a genetic bias channeling cultural evolution links the distribution of *linguistic tone* (the use of voice pitch to convey lexical and grammatical distinctions; Yip 2002) and the distribution of alleles of two genes involved in brain growth and development,

ASPM and *Microcephalin* (Dediu and Ladd 2007). The frequency of certain variants of *ASPM* and *Microcephalin* is correlated with the presence or absence of linguistic tone in the language used by those populations, and this correlation survives multiple controls such as shared history and language and genetic contact. It is not entirely clear what the underlying mechanism is which constitutes the learning or production bias, but a recent report (Wong et al. 2012) suggests both behavioral and neurophysiological correlates of *ASPM* variants in pitch processing. Thus the strong suggestion is that the odds of developing a tone language by cultural evolution are slightly weighted by the proportion of these alleles in a population.[8]

To summarize, then, there is both plausibility and prima facie evidence that population genetics provides both a reservoir that would facilitate the huge range of alternate cultural possibilities open to the species and, at the same time, biases the local outcomes of cultural evolution through structured variation in the local population.

Conclusions

Polite fictions of the current independence of cultural and biological evolution, of biological evolution being now in abeyance and the cognitive uniformity of the species, need to be swept away. Instead, our theoretical and empirical investigations into cultural evolution need to be informed by the feedback relations between biological and cultural evolution (see also Lieven, this volume). The example of village sign languages serves as a startling reminder that this feedback is crucial for understanding ongoing cultural evolution.

We emphasize the pervasiveness of genetic variation in any population and its inevitable links to phenotypic (cognitive or behavioral) variation. We outlined the rapidly developing insights into genetic factors in language capacities, and how a single population harbors most of the variation in the species, but nevertheless local populations evidence distinct proportions of genetic variants. These slight local population differences may be sufficient to seed cultural differences and there is some evidence that these processes actually occur in the language domain.

The crucial message for this volume is that we cannot sensibly divorce cultural evolution from biological evolution. The two are intertwined. The distribution of genetic variants can bias the direction of cultural evolution, and cultural evolution can, in principle, channel biological evolution. Theories like twin-track models of coevolution of culture and biology, or the concept of niche construction (Odling-Smee et al. 2003), should always be borne in mind.

8 Of course, it is entirely possible that the reverse process, whereby speaking a tone language generates weak selective pressures on the processing and acquisition of specific cues, is active as well, but this probably cannot explain the observed genetic differences.

Language is par excellence a bio–cultural hybrid: a cultural system that runs on biological infrastructure and offers general insights into cultural evolution.

Acknowledgments

We wish to thank the participants of the Forum for extensive discussions and, in particular, Simon Garrod for highlighting the importance of horizontal processes in amplifying weak genetic biases. We also thank one anonymous reviewer for constructive comments and suggestions.

13

Language Diversity as a Resource for Understanding Cultural Evolution

Nicholas Evans

Abstract

The study of linguistic diversity, and the factors driving change between language states, in different sociocultural contexts, arguably provides the best arena of human culture for the application of evolutionary approaches, as Darwin realized. After a long period in which this potential has been neglected, the scene is now set for a new re-connection of evolutionary approaches to the astonishingly diverse range of languages around the world, many on the verge of extinction without trace.

This chapter outlines the various ways coevolutionary models can be applied to language change, and surveys the many ways diversity manifests itself both in language structure and in the organization of diversity beyond the language unit. Problems of establishing comparability and characterizing the full dimensions of the design space are discussed, including the distribution of characters across it, the correlations between them, and the challenge of establishing diachronic typologies (i.e., establishing the likelihood of different types of transition, including the insights that could be reached through properly focused studies of micro-variation). It concludes by surveying the main types of selection that mold the emergence of linguistic diversity—psychological/physiological, system/semiotic, and genetic/ epidemiological—and spells out seven major challenges that confront further studies of linguistic diversity within an evolutionary framework.

Introduction

Languages occupy a central role in studies of human cultural diversity, whether viewed through the prism of social, cultural, historical, or psychological variability. With something between 6,000 and 7,000 distinct languages spoken today, they offer a kaleidoscope of largely independent natural experiments in evolving complex cultural systems without formal planning. This has led to widely diverse outcomes both in terms of linguistic structures themselves and their articulation with culture and society.

The complexity of linguistic systems, their amenability to rigorous modeling, the assessability of these models against external data,[1] and increasing sophistication and commitment to gathering matched data on all of the world's languages has allowed linguists to accumulate a vast body of information on patterned diversity. With current moves to recast linguistics away from universalizing accounts that relegate diversity to a bit part, and toward coevolutionary models which assign it the central role, the time is now ripe to reconnect linguistics with evolutionary theory in a range of ways.

Linguistics, as the discipline most centrally occupied with language, has had an ambivalent relationship with evolutionary biology, and now with studies of cultural evolution. On one hand, at various moments in its history, linguistics has been particularly interfertile with evolutionary biology (going back to Darwin) and the development of rigorous modeling of evolutionary processes (Atkinson and Gray 2005). The historical linguist Brugman's 1884 distinction between shared retentions and shared innovations in language, for example, anticipated Willi Hennig's comparable distinction between symplesiomorphies and synapomorphies in systematic biology by nearly 70 years (Hennig 1950). On the other hand, for half a century, throughout both the structuralist era of Saussure, Sapir, and Bloomfield and the Chomskyan generativist era which followed, these classical connections of linguistics to biology were allowed to languish. Gains in the elaboration of techniques for "describing each language on its own terms" (structuralists) and for modeling the open-ended nature of linguistic systems in mathematically tractable terms (generativists) were achieved at the expense of developing procedures for comparing large matched data sets that respect languages' individuality while still allowing meaningful comparability.

It is only with the growing sophistication of linguistic typology, now amassing large bodies of cross-linguistically comparable material, that comprehensive ontologies of linguistic design choices, and statistically well-grounded testing of the relations between them, and to nonlinguistic factors such as group size or population genetics, have become possible. Since Labov's pioneering work on linguistic variation (Labov 1972, 1994, 2001, 2010) and the development of techniques by functionalist linguists like Bybee (2007) for studying the impact of use on form, the field has begun to have a better understanding of what promotes micro-variation and how social factors impact upon it.

---

1 Different types of linguists will interpret this as testing the characterizations in a reference grammar (written in disciplined normal language rather than a formalism) or a dictionary against a corpus, or of testing formal syntactic models against speaker intuitions. Though lively debate rages around these differences, a (rare) feature that unites most linguists is the belief that, in principle, the highly complex models we build do permit objective validation of some sort; linguistics has not gone down the postmodern road of seeing these models as simply subjective and untestable.

Coevolutionary Models of Language Change

Before plunging into an examination of language diversity, it is worth sketching a basic coevolutionary model of the language change which engenders it (cf. Evans 2003b; Evans and Levinson 2009a, b; Levinson and Evans 2010). First, languages are what Rudi Keller (1994, 1998) calls "objects of the third kind." They are neither straightforward results of biological evolution, like eyes or kangaroo hops, nor the intentional results of conscious human planning, like suspension bridges or constitutions. Rather, like equal-length supermarket queues or paths worn straight to the front door, they are the "unintended outcomes of intentional behavior," where the intention might be to get through the supermarket checkout or to the front door as quickly as possible, or (in the case of language) to persuade, deceive, sound like or unlike a particular person, or avoid homophony with an obscene sound-alike word. Second, the notion of coevolution applies here at a number of levels, of which I mention the three most important here.

Biology–Culture Coevolution

The biology–culture interaction is familiar in studies of human evolution (Boyd and Richerson 1985; Durham 1991). At the most gross level, this refers to the spiraling interactant effects between the hardware of the human brain plus its physical input-output system (e.g., larynx, tongue, ear) and the cultural software of language. Physiological changes created affordances which supported more complex language, while the cultural evolution of linguistic systems in its turn placed increasing demands for complex cognitive skills and a streamlined input-output system. Changes to the vocal tract that allow more consonant distinctions to be articulated permit the evolution of more complex consonantal phonologies, which favor individuals whose vocal tracts and perceptual acuity can best produce and discriminate the sophisticated new phonologies. In a more interesting guise, to which the field is just returning after many years of dogma-driven neglect, the possibility is opening up that different genes favor different linguistic structures (e.g., pitch perception and tone, Dediu and Ladd 2007) and that some of the distribution in language structures over the design space can be sourced to genetic differences in the populations that have shaped them through history. I will discuss this more later (see section on Mechanisms of Selection), and it is also addressed by Levinson and Dediu (this volume).

Social–Psychological Coevolution

Languages lead a double life, as social institutions and individual representations/dispositions. This gives a double set of mechanisms by which change can occur:

1. At the individual level: imperfect learning or generalization of new forms, streamlining of pronunciation leading to phonological reduction, modification of accent accompanying social reaffiliation.
2. At the societal level: norm resettings weed out or repurpose some variants through such social mechanisms as stigmatization, revaluation (e.g., the recentering of crucial vowel norms on the mercantile class in Shakespeare's time), or identification of variants with a particular social group.

Normally, individuals construct their own linguistic representations from a complex society of people around them. To be sure, some caregivers have disproportionate input in early childhood, but as life proceeds, the set of sources expands—try the thought experiment of going through your own vocabulary, including variants of pronunciation or meaning, and thinking the source from which you learned each item. Societal norms about language, to a large extent, are "out there" in some generalized sense, but there are important subsets of society (e.g., parents, other kindred, neighborhood, clan, professional networks) which may be disproportionately important.

During the structuralist-generativist period, dominant theories focused on just one or the other of these two loci. For Saussure, the most tractable object of study was *langue*, a set of social conventions which he fictively conceived of as a set of identical dictionaries deposited in the brains of all speakers. For Chomsky, the true object of linguistic study was (*individual*) *competence*, the knowledge of an idealized speaker-hearer in a homogeneous speech environment. Each was an abstracting move, with Saussure sidelining *parole* and Chomsky sidelining *performance*, both deemed too chaotic to be analytically tractable.

By focusing on just one linguistic locus, both approaches disfavored the adoption of a coevolutionary approach. More recent work, however, is leading to a view that synthesizes both of these loci, using the sorts of variationist methods developed by Labov to find systematicity in social variation as well as functionalist methods to find systematicity in individual variation in production. The interaction of evolution at these two sites can then amplify small selection biases through evolutionary funneling effects. Christiansen and Chater (2008:507) argue that "language has evolved to be learnable": this works through "C-induction" (Chater and Christiansen 2010) by which "cultural transmission delivers the restricted search space needed to enable language learning, not by constraining the form language takes on an innate basis, but by ensuring that the form in which language is presented to the learner is learnable" (Merker 2009:461).

Culture–Language–Cognition Coevolution

A third bidirectional conception of coevolution is that (a) cultural design choices are an important selector for language (typically over a scale of centuries or millennia), and that (b) linguistic design choices are an important selector for cognition (over the scale of the lifetime or shorter). The latter phenomenon, whereby language is argued to have some influence on thought, is typically labeled neo-Whorfianism (or the Sapir–Whorf hypothesis) after two of its major mid-C20 proponents. It is useful to distinguish the former as Vico–Herder effects, after two early romantic proponents of the view that languages express unique aspects of the history and worldview of their cultures (or peoples, or nations). Though the two effects are frequently conflated loosely under the "Sapir–Whorf" rubric, they operate on very different timescales and require very different evaluation methods, which makes it handy to have different labels for the two types of effect.[2]

It has taken decades to develop suitably operationalized domains of language and experimental paradigms to evaluate the Sapir–Whorf hypothesis, and it remains a topic of fierce debate. Proponents of its "weak form" (i.e., choices in language structure have some impact on habitual thought) have pointed to effects of grammar or lexical structure on such issues as color perception, preferred mode of spatial orientation (absolute vs. egocentric/body-centered), or categorization and representation of information about motion events. This is not simply "thinking for speaking" (Slobin 2003), but more generally "scanning and coding for speaking," given that particular grammars may require their speakers to report on particular aspects of reality (say, absolute compass orientation) at any point in the future, so that attention for future use becomes paramount, and this in turn influences how memories are coded. Methodologically, the focus in neo-Whorfian research has been on the effects of language on individual cognition (see the collection of studies in Gentner and Goldwin-Meadow 2003), whether in adult processing or child development. There have also been interesting studies on the ways these are summed across the society of language speakers to produce statistical effects of a language variable (say, grammatical gender) on a social variable (e.g., female workforce participation, Mavisakalyan 2011).

For Vico–Herder effects, two main variants can be distinguished:

1. "Ethnosyntactic" phenomena, where some "cultural preoccupation" ends up shaping the grammar in some way (for examples, see section on Social and Cultural Selection). Appealing case-study type examples abound, but it is fair to say that the problems of evaluating causal effects in this domain have not been overcome; since they are correlation based across large numbers of cultures, they rest on solving coding problems (see section on The Problem of Comparability) in both language and culture.

2 Indeed, Whorf explicitly denied any causal link of culture on language: "The idea of 'correlation' between language and culture, in the generally accepted sense of correlation, is certainly a mistaken one" (Whorf 1956:139).

2. Social-scale phenomena, where such issues as size of speech community and number of speakers who have acquired the language late in life are argued to impact on factors like the degree of grammatical elaboration and complexity. Once again, coevolutionary accounts of this type are a new game, and the field has yet to develop rigorous methods for testing these interesting claims (see discussion in section on Social and Cultural Selection).

These three nested levels of coevolutionary modeling in linguistics give some indication of the wide range of factors which may give rise to the diversity found across the world's languages. We now turn to the languages themselves, with an eye to indicating the many dimensions on which language diversity can be identified.

Diversity in Language Structure

All languages are made up of a large set of signs: conventionalized associations of form, meaning, and combinatorics. Signs are usefully partitioned into lexical signs (basically words that would go into a dictionary) and grammatical signs (like affixes, or syntactic facts like word order), as well as less straightforward sign types like intonation and other forms of prosody. The fact that English builds noun phrases in the order demonstrative—adjective—noun ($this_1$ big_2 $book_3$) whereas Indonesian does the reverse ($buku_3$ $besar_2$ ini_1) provides one example of a grammatical sign (signaling how units within a noun phrase, NP, will be assembled and have functions like "modifier" or "determiner" assigned).

For all signs, each of their three dimensions is conventional and largely arbitrary. Consider the English word "know." As Saussure emphasized (using other examples), in a tradition going back to the Greeks, it is largely arbitrary what form is employed to designate concepts, as can be illustrated by finding the (rough) equivalents in other languages:[3]

- Russian знать *znat'* (cognate with English *know* but highly divergent in form)
- Japanese 知る *shiru*
- Tamil தெரி• *teri*
- Kayardild *mungurru*
- Dalabon *bengkan*

3 Premodern linguistics was obsessed with written language to the exclusion of the spoken; much modern linguistics has erred on the other extreme, treating writing systems simply as transcription methods. A more useful approach (and interesting from the point of view of incorporating writing systems as cultural innovations potentially impacting on other aspects of language) is to treat sound and written forms as partially independent but partially linked signifying systems, something I emphasize here by giving the Russian, Japanese, and Tamil words in their respective orthographies as well as in a Romanized transcriptional system.

Despite this traditional view of the "arbitrariness of the sign," evidence reclaiming significant levels of non-arbitratiness has been accumulating from a variety of quarters. Onomatopoeic words (e.g., bird names resembling bird calls) are one type of case. A second (see Haiman 1980, 1983) is "grammatical iconicity": if there are two types of possession, e.g., "inalienable" like "my eye" vs. "alienable" like "my yam," the semantically "closer" inalienable type will use a more "direct" syntactic construction, as in the Paamese contrast between *mete-ku* [eye-my] "my eye" vs. *auh aa-k* [yam possessed.edible-my] "my yam" (Crowley 1982). A third is "diagrammatical iconicity" by which the ordering of affixes normally reflects their logical scope.[4] A fourth is the fact that in many languages phonological cues within a word provide probabilistic cues to word-class, helping the child learn the combinatorics of new words on the basis of their forms (Monaghan et al. 2007, 2011). In Japanese, for example, all verbs end in *-u* (and a high proportion in *-ru*) whereas for nouns this proportion is much smaller; thus the fact that the Japanese verb for "(come to) know" is *shiru* rather than *shika*, is not arbitrary. A more balanced view, therefore, is to see linguistic forms as exhibiting a mixture of arbitrariness and nonarbitrariness in their forms across the whole vocabulary and grammar.

The meaning of signs is likewise highly variable; we are not just dealing with pinning different labels on the same conceptual object. To translate English *know* into Russian we need to distinguish (roughly) factual knowledge (*znat'*), procedural knowledge (*umet'*), and acquaintanceship (*byt' znakom s...*). Japanese *shiru* is better translated as "come to know," with "know" a consequence of certain aspectual operators (completed transition) on the basic verb meaning. In Dalabon *bengkan*, even though it is the closest to an equivalent, is better translated as "have in mind on a long-term basis" (Evans 2007); it can include long-term remembering, belief, and lengthy contemplation, and unlike "know" only implicates rather than indefeasibly entails the truth of its cognitive content. Thus, unlike English *know*, you can follow *bengkan* with a word meaning something like "believedly" to indicate skepticism with regard to the accuracy of the mental state.

Finally, the combinatorics can be rather different. Though the corresponding words in some of the above languages (Russian, Japanese, Dalabon) are like English in being transitive verbs, other languages do things differently. The Tamil equivalent *teri* is also a verb, but assigns quite a different case pattern, with the knower in the dative case (literally "to Kumar this place knows" for "Kumar knows this place"). Dalabon *bengkan* is a transitive verb, and its

4 As an example, consider Kayardild *karndi-wala-nurru* [wife-many-having] "polygynous" vs. *karndi-nurru-wala* [wife-having-many] "many married (men)." The semantic scope of *wala* and *nurru* seems "compositional" to English speakers even though the ordering is alien, as is the fact that these categories are expressed by suffixes rather than free words. However, diagrammatic scope is not found in all cases. For a discussion of one such case and the unusual evolutionary pathways which gave rise to it, see Evans (1995b).

syntax is not grossly different from English, but its morphology differs radically. Like other transitive verbs in Dalabon, it takes prefixes indexing both subject and object, so "(s)he knows him/her" is the single word *bûkahbengkan*, where *bûkah-* indicates "(s)he acts upon him/her." The verb can also accrue all sorts of modifying prefixes (e.g., *bûkahkakkûbengkan*, "(s)he really knows him"). Passing to languages which do not use verbs to encode this concept, in Kayardild the corresponding word, *mungurru*, is a predicative adjective (much like the English word "'knowledgeable"), though one still capable of taking an object, so that *ngada mungurru ngumbanji* "I know you" is more like "I [am] knowledgeable of you."

Combinatorics is fundamental to describing grammars of languages because the rules grammarians write to build words from their parts, or phrases and clauses from words, refer to classes of entity (like noun, verb, adjective, determiner, root, verb phrase, and so on). The first four of these (noun, verb, adjective, determiner) exemplify "parts of speech" or "word classes." These are sets of words united by common combinatorics, and these sets are indefinitely large, in the case of the "major word classes." It is their common combinatorics which allow us to formulate syntactic rules. To build an English noun phrase up as determiner + adjective + noun rather than the opposite order as in Indonesian, we need to know which lexical items go in which slots: "the" or "this" can fill the determiner slot, "big" or "red" the adjective slot, and "man"' or "building" the noun slot.

An important dimension of variation across languages lies in the patterning of word classes across languages. It is well established that there is considerable variation here. English, for example, lacks developed word classes that correspond to such classes as expressive or mimetics in languages like Semelai or Japanese, or co-verbs in Jaminjung, whereas there are many languages that do without prepositions (e.g., Kayardild) or adjectives (e.g., Lao). Whether this variability extends to the most fundamental distinction of all—that between nouns and verbs—is a topic of continuing debate. For a language like Straits Salish, Jelinek (1995) has argued that there is just one open class of predicates (as in predicate calculus), with meanings like "eat," "be a man," "be Eloise." On this analysis, in a language like Straits Salish "the men ate the fish"' would be rendered as "the (ones) who are.men ate the (ones) which are.fish."

This does not exhaust the ways signs can vary. Consider polysemy, where what looks like a single sign (same form, same combinatorics) has more than one meaning. Languages differ again here, in terms of which meanings they co-link to the same form. The polysemic ranges stretch from near universal at one end (e.g., association of "high"' with happy/active rather than sad/inactive) to highly idiosyncratic at the other. Consider the extension of "see"' to "understand," as in "I see," which is sufficiently widespread to have led Sweetser (1990) to argue this was based on universal metaphors of visual perception as cognition. Yet this particular polysemy is absent in Australian Aboriginal languages, where it is "hear" rather than "see" which is the modality underlying extension into metaphors of understanding (Evans and Wilkins 2000). In many cases, figurative language draws heavily on rather specific cultural scenarios or presumptions. The celebrated extension from "fly" to "steal" in medieval French, retained in these two meanings of *voler*, was mediated by the specific practice of falconry (*le faucon vole le perdrix*, "the falcon 'flies' the partridge") (Benveniste 1966).

Returning to the form of signs, and now focusing on the syntagmatic[5] dimension (i.e., the dimension of how things combine in sequence), we see once again that what appears superficially to be the same form can result from quite different types of structure. Consider the situation where three languages, A, B and C, all have a sound sequence *kákì*, where ´ represents a high pitch and ` a low pitch (thus a pitch contour like [⁻ _], and where this contrasts with a word *kàkí* [_ ⁻].[6]

We could class all three languages as tone languages, on the grounds that all use pitch to discriminate meaning. However, lurking under this apparent similarity we find that pitch is organized in a very different way. If we expand the number of syllables we are looking at, we find that for A, the number of different pitch combinations is 2^S, where S is the number of syllables; this is what we would expect if we could make an independent two-way choice on each syllable. This is more or less the situation that is found in "classical" tone languages like Mandarin or Vietnamese, except that these have more tone contrasts. Such languages are sometimes called "syllable tone" languages. A second possibility would be that, however many syllables there are in the word, there are just two "melodies"—a rising one and a falling one—which can get squeezed onto just one syllable or stretched out over a long word. On a single syllable, the contrast would now show up as rising versus falling tone as the melody gets compressed. In this case, tonal phonologists generally talk of "word tone" languages.

A third possibility is that the number of contrasts is linearly related to the number of syllables, as s or $s + 1$; the latter is more or less the case for Tokyo

5 This is usually contrasted with the "paradigmatic" dimension, or the dimension of opposition as opposed to the "dimension of combination" which is the syntagmatic dimension. We can illustrate the difference with the *p* in the word *sprite*. It is a syntagmatic fact that *p* can be preceded only by the sound *s* (i.e., *hprite*, *mprite*, etc. are impossible in English), and a paradigmatic fact that it could (in hypothetical but perfectly pronounceable words) be substituted by *k* or *t* (*skrite*, *strite*) but not by *m* or *n* (**smrite*, **snrite*). We can transfer these concepts directly to many other informational systems, such as DNA; for example, the fact that any one site is part of a four-way opposition between bases is paradigmatic, while the groupings of bases into longer sequences of various sizes is syntagmatic.

6 In Japanese the first (meaning "oyster") would be written 蛎, whereas the second could mean either "fence" or "persimmon," and would be written 垣 and 柿, respectively: the character-based writing system distinguishes them totally. It is also worth pointing out that the latter two can, with more phonological finesse, be distinguished, according to whether the high tone is continued onto following postpositions like the topic marker.

Japanese, except that words more complex than the ones shown here would demonstrate that instead of syllables, *S* should be counting slightly smaller units, known as morae. A consequence is that in Tokyo Japanese, just one place has to be marked as the "inflection point," which is the point at which pitch falls.[7]

These possibilities are summarized in Table 13.1. We could think of them as three sets of rules negotiated between a composer and a librettist in terms of how they line up their contributions as they collaborate on an opera. In the first, the composer is allowed to put any note to any syllable. In the second, the composer hands a tune to the librettist and instructs her to stretch it out, once per verse, for verses of any length. In the third, the composer writes tone rows (here, of just two tones) but gives the librettist choice about where in the lyrics she may move from one tone to the next.

I have chosen an example from the realm of tone because it sidesteps some of the gridlocked debates that have led to rather unproductive standoffs in the realm of syntax. What it should illustrate very clearly is that diversity in how languages are organized (in this case, in how they harness melody as part of meaning-signaling form and link it up to segmental elements) may not be immediately obvious, and that benefit is drawn from having abstract representational mechanisms able to capture different deeper patterns operating in what may seem to be the same form. Conversely, more abstract representations may show similar patterns lying under different forms, as when a monosyllabic falling tone and a disyllable with a high-low pattern are shown to both be instantiations of a falling melody in a word tone language. In other words, the use of abstract representations is neutral in terms of increasing or decreasing the level of variation we postulate as underlying the system.

The role of abstract representations of one form or another has been a major problematic in debates about linguistic diversity over the last half a century. There has been a general tendency for those from the generativist tradition to use more abstract representations and those from the descriptive and typological traditions to use more concrete ones. Both are clearly necessary to do justice to many linguistic phenomena, but they also introduce a dangerous possibility of glossing over significant diversity by viewing the "real" phenomena as invariant at some underlying level. Even where they are justified (far from a simple point to determine), their use can still cause two major problems for comparative work: (a) since abstract representations often depend on more sophisticated analysis of the data, there will be fewer data points in terms of languages for whose structures we can vouch at this abstract level; (b) if the abstract representations exhibit elements that are specific to particular languages, this also makes comparison harder.

7 The rising pattern is produced because of a rule that lowers the first syllable just in case the inflection point does not produce a low tone on the second syllable.

Table 13.1 Types of linear patterning of pitch choices in three types of language. N(S) is a function from the number of syllables N to the number of word-tone contrasts maximally available for an N-syllabled word.

Hypothetical Language	Form	Process	Representation	N(S)	Type Language
A	*kàkí* [_ -]	Concatenation of contrasting syllable tones	ka[_] + ki[-]	Power of syllables: 2^s	"Classical" tone languages (e.g., Mandarin, Vietnamese)
B	*kàkí* [_ -]	Association of melody with whole word	kaki + [_ -]	Constant: 2	Word tone languages (e.g., Mian)
C	*kàkí* [_ -]	Change in pitch level at inflection point	ka'ki	Linear: S(+1)	Pitch accent languages (e.g., Japanese)

Further Dimensions of Diversity: Beyond the Language Unit

In the previous section I focused on structural diversity, assuming that each language is relatively standardized and internally variant, and then compared given structures for each such language in terms of form, meaning, or combinatorics. Three other types of diversity should not be neglected and are thus briefly discussed here.

Sociolinguistic Variability within Speech Communities

This concerns the way variation is organized within the unit rather imprecisely designated as the speech community: a unit which according to the situation may be smaller or larger than the units we conventionally and uncritically designate as languages.

This kind of variation is fundamental. It is vital to establish, empirically, how robust linguistic systems are against communicative degradation when norms are not shared, an area where modeling can provide vital insights (Hruschka et al. 2009; McElreath et al. 2003). Experience with our own languages makes it clear that Chomskyan levels of idealized homogeneity are not necessary to assure (largely) successful communication, but how far can they drop without ceasing to function as an efficient shared code? Are there different patterns of internal diversity by domain (grammar vs. lexicon vs. phonology) or by medium (speech, writing, sign vs. spoken language)?

In fact we find that variation is often less a matter of degraded signal than an additional semiotic layer employed for social-signaling purposes, identifying the speaker's regional affiliation, class, caste, religion and so on, as well as aspects of the communicative setting. This may, in turn, have adaptive functions in terms of flagging membership in communities with the same norms of cooperation and coordination (Richerson and Boyd 2010). Once linguistic variability is deployed in the same way, it creates a more complex set of shared norms: not just a <form : meaning : combinatorics> triple, but a link from each of these to some social information. This can drive linguistic change in particular ways. For example, historical linguists have recently become aware of processes like "correspondence mimicry," where speakers are aware of proportionalities between languages and use these to refashion foreign adoptions in their own language to make them look more native-like (Alpher and Nash 1999). Likewise, an increasing number of otherwise inexplicable historical changes are cropping up, for which the best explanation is that a highly unusual variant got promoted in some variety precisely to signal that speech community's distinctness from a neighboring one, in the grammatical equivalent of the *you say tomato, I say tomahto* principle (for an Iwaidja example, see Evans 1998). Such changes can only occur against a background of shared knowledge of both languages.

Not all speech communities are equally diverse internally. Some (e.g., precontact Kayardild) were exceedingly homogeneous, whereas others are highly diverse (e.g., the "dialect chains" of Western Desert in Australia or Numic in the Great Basin). It is sometimes proposed—thus far without clear comparative evidence—that such dialect chains are found among hunter-gatherers in desert regions where the vagaries of rainfall and yield constantly drive small bands of desert-dwellers to recoalesce with others living in areas that got a good recent rainfall. Having a shared grammar then makes it easy to learn whatever new code one needs to blend into the environment (cf. Shaul 1986). We can get situations where there is effectively a common grammar that gets localized by differences in vocabulary (e.g., parts of Northern Vanuatu, or the Western Desert Chain mentioned above). Conversely there are situations with such high levels of shared vocabulary that mutual comprehension is assured despite the grammars being organized in significantly different ways (this was the situation between Kayardild and Yukulta; Evans 1995a).

Once writing is added as medium, of course, we can have speech communities of great diversity in speech (effectively many different languages) united by a common writing system; this is the case for Chinese. The converse case involves a single language (perhaps spread across dialects) divided into two or more "political languages" with attendant different literatures, through the use of different writing systems, as between Serbian (written in Cyrillic) and Croatian (written in Latin script). The Hindi–Urdu case is similar though complicated by greater divergence in learned vocabulary, with Hindi stocking from Sanskrit and Urdu from Persian.

On top of this, there are situations where more than one language is deployed inside a speech community for designated roles—so-called diglossia. These languages may be related as a more archaic/classical and modern/vernacular version, such as classical Arabic and local vernaculars throughout the Arab world, or may be unrelated, as is the case between English and many other languages today (e.g., Swahili, Malay) or, traditionally, between Latin and European vernaculars or Classical Chinese and other languages of the Sinosphere (e.g., Korean, Japanese, Vietnamese).

Responding to diglossic situations simply by treating them as two separate languages misses key insights into how language spans its communicative goals as well as how linguistic systems interact. In particular cases of bilingualism, such as the traditional French–Cree bilingualism of Métis trappers in the Canadian prairies, the severing of ties between the mixed speech community and its two reference groups can produce a peculiar outcome: a so-called mixed language like Michif. In the Métis–Michif case, this happened once racial redefinitions cut off people of mixed descent from both groups. Mixed languages intimately combine features of two languages that would normally not be transmitted by borrowing or contact, in what Bakker (1997) has called language intertwining. It can happen that speakers of mixed languages sometimes know neither of the two source languages, i.e., a group of Michif speakers developed who, unlike their forebears, knew neither French nor Cree but preserved the complexities of both languages in a new mixed code.

A key tenet of linguistics since the structural era has been Meillet's dictum that "language is a system where everything hangs together."[8] This formulation suggests that certain hypothetical points of the design space (characterized by combining elements which would not "hang together") are unpopulated because any system combining them would somehow be dysfunctional. Now that we know more about mixed systems, which often combine elements in highly unexpected ways, an alternative explanation suggests itself; namely that co-transmission of traits is what produces any "Meillet effects" that may be out there, and that once particular social situations are present to engender mixed languages, then we may see the co-occurrence of unexpected traits. At a period when there is growing evidence that traits hitherto believed to be tightly coupled can in fact exhibit more independent evolutionary trajectories (Dunn et al. 2011b), the need to examine the impact of given social deployments of linguistic variation takes on a new significance.

Also highly variable is the way given speech communities harness variation to signal social significance. English speakers are accustomed to variation (phonological, lexical, grammatical) being used to signal regional origins, as with the different pronunciations of *butter* as [bɐtʰə], [bɐdəɹ], [bɐtʰəɹ], [bɐɾə], [bɐʔə], [bʊʔəɹ], and so forth. We also know that speakers slide between more conscious, educated upstyle variants (e.g., *aren't you*) and more casual ones (e.g., *arntcha* or *arntchas*[9]) according to factors like formality. Since Labov (1966), we also know that patterning can emerge across the whole speech community even though no individual has knowledge of the whole

8 Famously formulated by Meillet (1906) as *une langue est un système où tout se tient.*

9 In those parts of Australia and elsewhere in the English-speaking world that use *youse* (~ [jəz]) as a (typically low-style) plural second person pronoun.

pattern. However, this sort of patterning only scratches the surface of what languages can do with variation. For Tamil (Andronov 1962), for example, there are choices between classical/literary and colloquial, regional colloquial varieties, and caste varieties (e.g., the variety used by Brahmins) as well as a range of situationally based choices for using honorifics depending on the speaker-hearer or speaker-referent relationships. For Javanese (Errington 1988), there are three main registers—*krama*, *madya*, and *ngoko* (in order of descending "refinement")—with complex conventions for who can use what to whom, taking into account the social position of both speaker and hearer. These three registers are so different that a given sentence may differ in every word between them (e.g., Krama *Menapa nandalem mundhut sekul semanten* vs. Ngoko *Apa kowé njupuk sega semono* for "did you take that much rice?").

For the Australian language Bininj Gun-wok (Evans 2003a), there is an everyday or default variety known as *gunwokduninj* "real language" (though itself divided into half a dozen main dialects); a special variety known as *gun-gurrng* "in-law language" used with certain high-respect affines like one's wife's mother's brother; an obscene joking variety used between potential (but unactualized) affines; a series of clan lects (with shibboleth terms down to a much finer grain than the main dialects) used to establish clan identity at the beginning of encounters as well as in addressing ancestral clan sites; and many special lexical items found just in poetic or song language. As with Javanese, there may be total lexical differentiation between registers (though grammatical affixes are typically unchanged): "Have you got any meat? No, nothing" would be *Gun-gunj yigarrme? Gayakki* in *gun-wokduninj* but *Gunmulbui yiwalebonghme? Gayagura* in the decorously multisyllabic *gun-gurrng*.

Multivarietal "lect clouds" may exhibit sufficient systematicity to suggest systems of shared norms that transcend what linguists take as the unit of description when writing grammars of a single variety. In Northeastern Arnhem Land, for example (Wilkinson 1991), a substantial part of the variation is factored across two orthogonal dimensions:

1. a geographical one, running broadly west to east and shown by the optional loss of initial ŋa from pronouns in the western area, and
2. a social one based on the assignment of every clan and language variety to one of two patrilineally transmitted moieties.

Languages spoken by clans of the Yirritja moiety have vowel-final phonologies (like Italian) whereas those of the Dhuwa moiety have dropped most final vowels giving a more staccato phonology (like Catalan). The intersection of these two factors (see Table 13.2) creates a four-way matrix in which most stretches of speech in a number of languages in this region can be rapidly located in social and geographical space. This attests to semiotic rules shared across a multilingual region that transcends the boundaries of the single languages (like Djapu or Djamparrpuyngu) which form the normal units for linguistic description.

Table 13.2 Geographic and social patterning of the first-person plural pronoun form in some Yolngu dialects (after Wilkinson 1991:187).

Geographical [initial ŋa-drop]	**Social (patrimoiety) [Final vowel drop]** *Yirritja moiety* (Dhuwala varieties) ŋa		*Dhuwa moiety* (Dhuwal varieties)	
Western	Gupapuyngu:	[ŋa]napuru	Djamparrpuyngu:	[ŋa]napur
Eastern	Gumatj:	ŋanapuru	Djapu:	ŋanapur

A still underexamined area, particularly important for theories in which gesture played at least some part in the original evolution of language, is the semiotic partition between spoken and signed or gestural codes. In Dhuwal (Northeastern Arnhem Land), for example, the conventionalized touching of symbolic body parts instead of verbal reference is a regular practice for some respected types of affine, so that a man might ask "where is (your) wife's mother?" by saying the equivalent of "where is that kind," slapping his knee while saying "that kind" (Heath 1982:255). Alternatively, gesture may be harnessed to clarify thematic roles like instrument. Another Australian language, Iwaidja, is unusual for Australian languages in not having overt marking for instrumental case, but I have recorded Iwaidja speakers saying things like "I went.out bark.torch" while raising their hand in a gesture of holding a torch concurrently with the word for it. In this case, the gesture marks the thematic role in a way that would be carried out by the grammatical device of case-marking in most other Australian languages. There has been so little work on the integration of language and gesture (or spoken and sign language in communities that have both) that we cannot currently compare the functions and degrees of integration that these two modalities have, but it is an important dimension of variation for future study.

Until recently, linguists have kept the study of variation in the various guises outlined above as a distinct domain from the study of structure. Though this decision had temporary heuristic value, bringing together the study of sociolinguistic variation and the study of structural diversity is likely to yield many interesting findings in the coming decades.

Phylogenetic and Typological Diversity

Zooming out from our dialect maps to the level of language families and subgroups, we encounter another level of diversity. Once again, this has only been conceived as a phenomenon in need of explanation in the last two decades or so, prompted by Johanna Nichols' seminal book *Linguistic Diversity in Space and Time* (Nichols 1992) and, more recently, books like Daniel Nettle's (1999) *Linguistic Diversity* and the large compilation of data assembled in the

WALS[10] survey. Work like this has drawn attention to striking disparities in the worldwide distribution of diversity, be it phylogenetic (i.e., in terms of numbers of families, subfamilies etc.) or typological (i.e., in terms of the amount of gross typological variation).

Some of these can be attributed to colonial expansion erasing ancient diversity. This may be modern, as in the spread of English or Spanish—the latter case eliminated all indigenous languages (e.g., Cuba and Uruguay). However, it may also be ancient, as in the spread of Latin and its descendants with the attendant elimination of Etruscan, Umbrian, Gaulish, and many other languages, or the spread of Austronesian languages through islands in southeast Asia (which eliminated, e.g., all non-Austronesian languages from the Philippines and Western Indonesia). There has been a tendency by some authors (e.g., Renfrew and Bellwood 2002) to see "demic expansion," typically agriculture driven, as the only cause of widespread language families; that is, the higher populations of farmers simply squeezed out the hunter-gatherers demographically, so that the farmers' language ended up displacing that of the aboriginal hunter-gatherers. However, there are many large language families in the world (e.g., the earlier levels of Niger-Congo, Pama-Nyungan in Australia) where we currently lack a plausible explanation in terms of agriculture-driven demic expansion. This is one type of phenomenon, then, where linguistic facts (distribution of language families) abut scenarios of cultural evolution more generally, leading to multidisciplinary accounts of prehistory that range from canonical and convincing (e.g., the Austronesian expansion) to still enigmatic (Pama-Nyungan, Trans-New Guinea).

Even if we strip away the effects of colonization and expansion and its presumed steamrollering of prior phylogenetic diversity, we are left with major distributional puzzles. In terms of its number of maximal clades (i.e., phylogenetic groupings, including isolates not currently relatable to any other), New Guinea and its immediate surrounds comes in with around 35—a number greater than that found for the whole of Eurasia. This cannot simply reflect New Guinea's long-standing isolation from the centralizing effects of larger states. Australia, just next door, joined to New Guinea for most of its human history, and with comparable or greater levels of linguistic diversity in terms of number of speakers per language, is covered by what increasingly looks like a single language family, as more and more evidence accumulates (Evans 2003c). The differing levels of diversity in Australia and New Guinea (and comparable asymmetries elsewhere) remain one of the great unanswered puzzles of linguistics. Its solution will most likely need to bring in both (a) proximal examinations of degrees of sociolinguistic dispersal in traditional speech communities of different kinds and (b) distal research on what social configurations accelerate or retard microdiversification within a speech community, and what determines how variants get invested with social meaning to allow

10 WALS = World Atlas of Linguistic Structures, now available online: http://wals.info/

them to survive as transmitted signs rather than being weeded out as mistakes or stigmatized forms.

The worldwide distribution of typological diversity is likewise shaping up as a key area of research, replete with currently puzzling data. To illustrate the distinction between typological and phylogenetic diversity consider the following: both Vanuatu and New Caledonia contain low-level branches of Oceanic within Austronesian, and in fact Vanuatu is somewhat more diverse phylogenetically since it is normally held to contain several low-level branches. Yet New Caledonia appears to be much more diverse typologically: it contains, for example, several tone languages and has much greater diversity in phoneme inventories than those found in Vanuatu. Once we recognize the independence of phylogenetic and typological diversity, two immediate challenges face us: How can we quantify the degree of typological dispersal in an observer-independent way that is comparable across all languages? What types of models can account for why there seems to be much greater typological diversification in some areas than others? Once again, we need to attend both to linguistic microlevels (measuring amounts of typological variation inside speech communities) and to types of social organization or other coevolutionary selectors which may afford or downplay typological diversification.

The Problem of Comparability

> Coding comparative cultural and linguistic data is a task fraught with the difficult reduction decisions familiar to all scientifically minded anthropologists.
> —Fiona Jordan (2013:47)

Finding ways of coding data in cross-linguistically comparable ways is a difficult but essential task if linguistics is to draw the full power from the vast data sets it is beginning to assemble. In the development of the field, both structuralist/descriptive and generative traditions took intellectual positions that hindered cross-linguistic comparison. Somewhat simplistically, we can say that structuralists were guilty of bongobongoism, exaggerating noncomparability, and generativists were guilty of procrusteanism, smoothing away cross-linguistic differences by using the great representational flexibility afforded by deep structure to surface mappings.

For the structuralists, both the Boas-Sapirian program (i.e., each language should be described on its own terms) and the Saussurean dictum (i.e., the significance of any sign was its place in the system) worked to favor descriptions which could be exquisitely finely tuned to linguistic particularity. However, their premises made cross-linguistic comparison impossible.

For the generativists, the quest to look below the surface to a more abstract "underlying structure" had several effects. One was to downplay surface variation as insignificant. The famous Binding Condition A (Chomsky

1981)—anaphors (unlike pronouns) should be "bound in their governing category"—predicts that there should not be languages with a single item *them* whose meaning would range across "disjunct" pronominals like "they saw them" to reflexives like "they saw themselves" and reciprocals like "they saw each other." This is because the first reading would not be "anaphorically bound within the clause," whereas the second and third would be. Put another way, languages should always have distinct anaphors for the bound and nonbound situations. Now when languages turned up which appeared to violate this (Old English, but also many others, such as Mwotlap and Tinrin in the Pacific, or Javanese), the response was either to postulate two distinct entities at some deeper level of analysis that happens to be homophonous on the surface, saving the putative universality of the Binding Condition.[11] Certainly there may be cases where more subtle investigation reveals justification for this, but to assume that such reanalyses of the data can be taken as the default (or to dismiss apparent counterexamples from less-described languages as likely to yield a deep-level distinction on further analysis) had the effect, within the generative program, of ignoring many populated regions of the design space.

This is not to deny the very real difficulties in comparing phenomena across languages. Suppose we are surveying the world's phoneme inventories and are deciding whether particular languages have a /p/ phoneme or not. First we compare English and French: both have a /p/ contrasting with a /b/, so we could decide affirmatively. However, if we are fussier about phonetic detail, we might decide that the English phoneme is really an aspirate /p^h/ whereas the French is really a /p/. Then we bring in Hindi, which (leaving aside the voiced aspirate or murmured /b^h/) has a three-way distinction between a /p/ close to the French sound, a /p^h/ closer to the English one, and a /b/. Then we add Mandarin, which contrasts /p^h/ and /p/ but lacks any real voiced sound,[12] and Korean, which contrasts an aspirate /p^h/ with a lax unaspirated sound (revealingly, sometimes transliterated as *p* and sometimes as *b*) which alternates between these two according to word position. Finally we add Kayardild, which has just one bilabial stop phoneme realized variably as [b], [p] and [p^h]; though the aspirated pronunciation is marginal enough to make this unappealing as

11 An alternative response has been to split the phenomenon into such a large number of factors that it becomes difficult to gather sizeable cross-linguistic data (e.g., Reuland 2008). A third option, however, may be introduced, as Cole et al. (2008) do for Peranakan Javanese; namely, there are words which are unspecified for the anaphoric vs. pronominal contrast and hence compatible with both conditions, thus making it possible for them to conclude that Peranakan Javanese "does not…constitute evidence that there are languages in which the Binding Theory fails to apply. Indeed, Peranakan Javanese provides compelling evidence that the Binding Theory is active in languages containing forms that appear to be exempt from the Binding Theory" (Cole et al. 2008:585).

12 Pinyin, the system now most widely used for Romanizing Mandarin, writes p for /p^h/ and b for /p/. However, earlier systems of Romanization, such as the Wade–Giles system, used a different method, writing these sounds respectively as *p'* and *p*.

the candidate for naming the phoneme, the grounds for choosing between the other two are pretty arbitrary. This gives us the situation shown in Table 13.3.[13]

Depending on how we operationalize the question, we could answer that:

- all six languages have a /p/ (since all have phonemes including this in their allophonic range),
- just one language (Hindi) has a /p/ (since only Hindi in our sample contrasts it with both other bilabials), or
- three languages (Hindi, French, Mandarin) have a /p/ (since just these three languages have a /p/ phoneme that does not stray into the phonetic territory of the other phonemes).

Comparable problems recur in the comparison of every part of the linguistic system. In terms of semantics, does a language have a word for "hand" if the same word extends its denotation to the whole arm? Do we treat this as polysemy (i.e., meaning 1: "hand"; meaning 2: "arm") or do we claim it has just made a different cut on reality that bypasses any concept exactly translating "arm." In terms of syntax, does a language (say, Lao) have adjectives if they are merely a sub-sub-subclass of verbs? Or, comparing word orders and deciding on whether a language has SOV (subject–object–verb) word order, how do we deal with, for example, the following factors?

- Different orders, if one or both elements are pronouns rather than nouns.
- Languages which differ in the degree of rigidity of order, so that in some there is strict SOV order whereas in others it is a mere statistical preference.
- Languages which have different orders in main and subordinate clauses.
- Languages where it is not clear that there is a notion of "subject" in the familiar sense, but which organize their syntax in a way that conflates the patient of a transitive with the sole argument of an intransitive. In this case should the patient count as the subject in assessing order?

The subfield of linguistics known as linguistic typology, which has for its core goals the systematization of cross-linguistic comparison, has tried to find a

Table 13.3 Overview of comparability problems when determining which languages have a /p/ phoneme.

<table>
<tr><th>Phonetic realization</th><th>Hindi</th><th>Korean</th><th>French</th><th>Mandarin</th><th>English</th><th>Kayardild</th></tr>
<tr><td>[p^h]</td><td>/p^h/</td><td>/p^h/</td><td></td><td>/p^h/</td><td rowspan="2">/p/ or /p^h/</td><td rowspan="3">/ p^h~p~b/</td></tr>
<tr><td>[p]</td><td>/p/</td><td rowspan="2">/p~b/</td><td>/p/</td><td>/p/</td></tr>
<tr><td>[b]</td><td>/b/</td><td>/b/</td><td></td><td>/b/</td></tr>
</table>

13 In fact, the problem is much deeper than this, since the "phones" used in the exposition here are themselves idealizations that are not exactly equivalent across languages. For an important discussion of this problem, see Ladd (2011).

way out of these impasses by developing frameworks that suggest replicable ways of getting around some of these problems. The key analytic elements in doing this are:

1. The systematic distinction of language-specific terms (where structuralist methods can be employed so as not to camouflage linguistic particularity) from "comparative concepts" that serve as anchor points (or *tertium comparationis*) between languages.
2. The use of flexible definitions employing either prototypes or *canonical types* to place the anchor points at those points in the design space which make the comparison with actual language material most useful.
3. The progressive factorization of comparative concepts so as to allow for the formulation of a number of dimensions, which can then be assessed independently, such as the breaking down of criteria for "subject" into grammatical elements (again factorizable into such dimensions as position, government of agreement, case choice etc.), discourse elements (e.g., topicality), and semantic elements (e.g., preferential projection of agentive roles).

None of these steps are analytically simple, and there is typically a substantial time lag between primary descriptive materials on many languages and the assembly of data from those into typological surveys, with the result that key definitional facts may be missing because they were not held to be important at the time of description. For this reason, the typological surveys with the highest-level of coding consistency have resulted from "Leipzig School" projects, (e.g., APICS[14] or the Leipzig LoanWord project), where the project design involved a group of typological masterminds calling iteratively on the expertise of a number of language experts who would be trained into the comparison methods. The obvious disadvantages of this approach, however, are the relatively limited number of sample points and the lack of extensibility: when the project ends, there is typically no new data entry. For these reasons, others advocate a much more open approach where any expert is, in principle, free to add their own data to a permanently updatable database. The advantages and disadvantages of this approach are almost the converse of the Leipzig school: greater coverage and extensibility on the one hand, but potentially lower data reliability on the other.

Whatever the problems outlined above, it is clear that the field of typology is very significantly extending our knowledge of the world's linguistic diversity as it lumbers through these various difficulties. We now have large databases, like WALS or APICS, and scores of books synthesizing worldwide linguistic data to give reasonably workable ontologies for many of the categories linguists wish to compare (e.g., the Cambridge series with titles like Number,

[14] Atlas of Pidgin and Creole Structures; see http://wwwstaff.eva.mpg.de/~taylor/apics/

Case, Aspect, Gender etc., each addressing a particular morphosyntactic category). Although we are still a long way from having a workable ontology of linguistic categories, there is substantial and accelerating mutual feedback between typologists and descriptivists, which is breaking down many of the old inconsistencies in descriptive practices. The biggest challenge that remains is the basic one of getting descriptive data (in the form of the classical "Boasian trinity" of grammar, texts, and dictionary) in the first place. Depending on where we set the bar for a reasonable description, we still only have coverage of perhaps 10–20% of the world's linguistic diversity. Without extending this, we are a long way short of having the sorts of data sets we need to study cultural evolution seriously.

Dimensions of the Design Space

Toward a Total Ontology of the Design Space

Perhaps the biggest challenge for linguistic typology is to develop a total ontology of the design space (i.e., a clearly defined ontology for all possible linguistic phenomena). Although this has in fact been a relatively unarticulated goal of typology for several decades, the recent adoption of the term "ontolinguistics" by some linguists (Schalley and Zaefferer 2007) and the so-called GOLD (general ontology for linguistic description) initiative have begun to make the ontological aims of much of typology more explicit. Within phonetics, of course, this goal has been a driving force for a long time—the explicit design goal of the International Phonetic Alphabet is to unambiguously represent and characterize any attested speech sound—and there are explicit procedures for admitting newly discovered sounds to the set.

An initial and deceptively straightforward example of exhaustively characterizing one subset of the design space for the dimension of main clause word order is Greenberg's famous "word-order typology" (Greenberg 1963). Using three elements—subject (S), object (O), and verb (V)—six possible orders are generated: SOV, SVO, VSO, VOS, OSV, OVS.[15] At the time it was initially formulated, some points in the design space were believed to be empty, but subsequent discoveries (e.g., the finding that the Carib language, Hixkaryana, was OVS; Derbyshire 1977) have shown that all combinations in this set are attested, albeit in highly skewed fashion.

15 Among the assumptions needed, for this to be a complete ontology, are: (a) S and O are unproblematically identifiable as units in the language, not at all clear in the case of ergative languages like Dyirbal, for example, or in those where "subject" properties are split between both NPs; (b) the language will have an identifiable basic word order (if a language has free order of elements, with roles signaled by case, it must then be treated as uncategorizable or one of its orders chosen as basic on grounds such as frequency).

Our grand global ontology can be built on any of the unit types discussed earlier, or any combination thereof. For example, one can survey whether nasal segments are in fact more common (yes) or universal (no) as cross-linguistic expressions of negation. Two very important lines of research in typology focus on (a) constructions used for expressing particular meanings (e.g., reciprocals comparable to English "each other") and (b) semantic dimensions that get grammaticalized and hence form core building blocks in the semantic machinery of languages. A brief look at each of these will show some of the challenges linguists still face in making their ontologies comprehensive.

First consider reciprocal constructions, a central preoccupation both because of the importance of reciprocity in human culture and ethics (cf. Gintis, this volume) as well as the syntactic intricacy of reciprocals in many languages (leading "anaphors" like *each other* to play leading roles in syntactic analyses in the generative tradition). An obvious question[16] is: How many ways can languages do reciprocals? In other words, how many basic types of construction types are there? WALS is no use here; it simply tells us whether languages have "dedicated" reciprocals (like *they see each other*) or coerce the means used for expressing reflexives (*ils se voient*, *sie sehen sich*, etc.). However, there are so many ways of implementing dedicated reciprocals grammatically that we need a whole raft of factors to account for the design space of just this one grammatico-semantic category. In an elegant article, König and Kokutani (2006) proposed four types: *quantificational* (like each other), *pronominal*, *affixal,* and *deverbal*. However, further research (Evans 2008) shows that this typology is still far from comprehensive.

Among other means, languages can use special reciprocal auxiliaries, call upon symmetric simultaneous signs (in Indo-Pakistani sign language) such as two fists going toward each other, or develop strange "clause-and-a-half" constructions which have fished a "contrastive subject" pronoun out of a tit-for-tat following clause, giving something like "she-him gave and.he.in.turn money" for "the two of them gave each other money." What a more comprehensive investigation shows, then, is that many linguistic phenomena have a "long tail distribution": a small number of regular structural solutions account for most languages. However, to arrive at a full account, we need a much larger design space. Although for some purposes it is convenient just to work with the common types, there are other goals, such as getting the constructions which represent the full semantics most explicitly, or accounting for seemingly unmotivated characteristics found in the common constructions,[17] or simply

[16] There are also interesting questions about the semantic content of reciprocals: Do we count "the students followed each other onto the stage" or "the cop and the robber chased each other down the road"? These are nontrivial questions but they would take us too far afield here; for discussion, see Evans et al. (2011).

[17] See Evans (2010) for examples. There it is argued that a rare construction shown in some highlands New Guinea languages actually come closest to representing the full semantics of reciprocal constructions, and that by making the presence of both transitive and intransitive

accounting for where languages seem to have the most diverse range of structural solutions. In the case of reciprocals, this is arguably because of the engineering challenge of mapping several propositions—an action going in each direction, plus the coordination or feedback joining the actions together—into a single clause. For phenomena like this, then, it may be the rare constructions that are actually more informative.

Second, consider the problem of factorizing the semantic dimensions of the design space, which I will exemplify with a relatively "new" category known as the mirative. General linguistic theory, and logic, have long been aware of the categories of tense, aspect, and mood, each deeply intertwined with the systems of verb inflection in most familiar languages. (Many analytic problems remain in each of these, but the general dimensions are relatively familiar.) As of the 1950s, at an accelerating pace, a further dimension of so-called "evidentials" began to be explored, as accounts from languages from the Amazon to the Andes to the Caucasus to the New Guinea Highlands have come to indicate what a vital role the marking of evidence source plays in the core grammatical systems of many languages. This led to a basic framework for describing the semantics of evidential systems (e.g., direct participation, direct perception, visual vs. other, inference, hearsay), the degrees of evidentiary weight assigned to each of these, who is taken to be the evidential source, and so on. Following hard on the evidential wave, another interesting type of verbal category has begun to come to light, initially on the basis of languages like Turkish and Macedonian on the one hand and Lhasa Tibetan on the other; this category presents the degree to which the information is new (mirative) or already cognitively integrated. The discovery of a category like this, however, immediately raises a whole host of questions: Is it just the speaker's cognitive state that is at issue or could it be, for example, the hearer's in a question, or even the intersubjective cognitive state of both? Must the relevant cognitive state be in the here-and-now of current conversation or can it be projected back to an earlier moment of realization (perhaps tensed)? Does the engagement with the new information occur just at the level of the proposition or can it be narrowed down to particular entities that participants are for example, pointing to? In other words what is the scope of the cognitive attitude at issue? All of these are currently the subject of lively exploration, and it is not my goal to explore these thoroughly here. I simply want to show that the turning up of some new semantic dimension in one or two languages can sometimes open up a whole new multidimensional space for ontological exploration.

An important trend in typology has been to map the semantic topology of the design space by successively smaller points. Indeed, we could say that every time a language is found that makes a previously unreported distinction,

predicates explicit in the semantic structure these rare types then motivate the widespread mixing of transitive and intransitive features found in the commoner types of reciprocal construction (Evans et al. 2007).

then the need to distinguish two points is made clear, across which the use of common forms in a given language is then mapped. Consider two (among many) senses that are formally conflated by the English perfect: *The dog has eaten the roast* and *John has eaten fugu fish.* When we look at how these get rendered in other languages, we find that many (e.g., Japanese, many Sinitic languages) have a special "experiential perfect" that would be used for the second but not the first, which would be expressed by other means. The use of a single construction to express both in English is thus a contingent fact; we accommodate this within a more general account of aspectual typology by postulating distinct points in semantic space (resultant state, experiential perfect) which some languages (English) conflate while others (Japanese) do not. The technique of "semantic maps," which overlays the semantic range of particular forms in a number of languages over a language-independent semantic grid, is used to explore which points in this space attest conflation of this sort. Results in a number of areas of grammatical semantics, such as mood/modality and indefinite pronouns, suggest that semantic extensions are highly motivated (by semantic closeness and possession of similar elements). This makes it possible to set up a general topology such that language-specific semantic ranges always span contiguous points (see Figure 13.1).

These sallies into the grand realm of what a total linguistic ontology would look like should give some idea of what a vast set of dimensions and multifactorial subspaces would be needed to characterize the design space completely in a way that every human language could be accurately characterized within it. The typological work that goes into constructing this ontology has a double

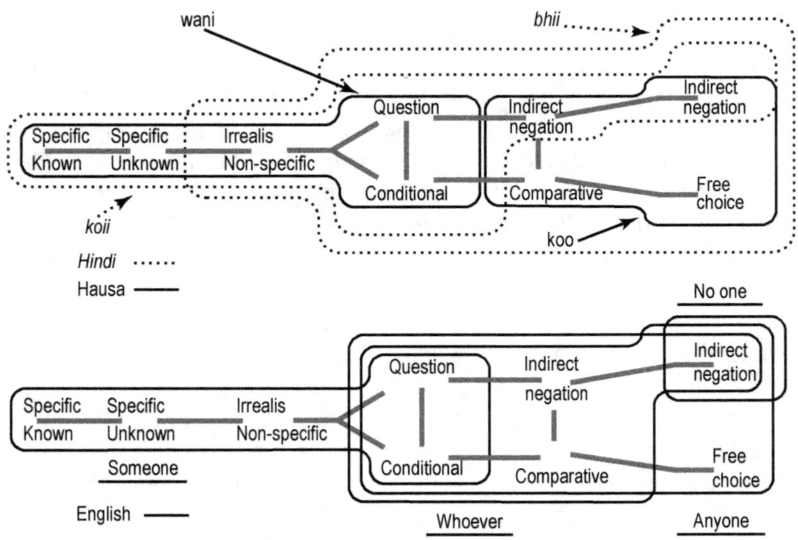

Figure 13.1 Partial semantic map of indefinite pronoun types, comparing English, Hindi, and Hausa (adapted from Haspelmath 1997)

purpose. First, it aims at being able to characterize any linguistic phenomenon, in any language, in a precise way that relates it to what is found in all other languages. Second, it allows any language to be given a comprehensive typological profile, in terms of a clear value (or "character") at each of many hundreds of thousands of dimensions: phonological, combinatoric, semantic, including complex combinations of these such as a given polysemy, or a given construction for expressing a given meaning.[18]

Profiles of this latter type are likely to play an increasingly central role in examining phylogenetic relationships between languages, simply because of the vast amounts of information they include. There have already been a number of interesting attempts to do this (Reesink et al. 2009), and despite some interesting results the field has not yet been willing to put such findings on an equal footing with the gold standard yielded by the comparative method. My personal view is that it is just a matter of time and of experimenting with the right choices of traits before this method really comes into its own. I will close this section by considering three crucial issues which bear on this question.

Distribution of Characters in the Design Space

Most points in the design space have a very skewed distribution of characters. This is the same whether one looks at basic word order (87% of languages are either SOV or SVO), vowel inventories (only about a dozen of the world's vowel inventories have a "vertical" system where only tongue height matters, not frontness or roundedness), the sensory modalities to which "see" and "hear" can extend, and so forth. The great rarity of many theoretically possible characters has given rise to many claims about "universals," but typically what begins as a claimed absolute gap turns into a mere rarity as the sample is extended, another manifestation of the "long tail" phenomenon mentioned above. This is not to say that there may never be gaps which remain absolutely unfilled, but the combination of skewed distributions with discovery lag and sampling issues mean that at any one historical moment in the evolution of typology we will have many "falsely absolute gaps." Instead of stipulating these gaps as cognitively impossible, a more useful universal approach would be to develop general and comprehensive models of selector bias. These models are needed anyway for the rest of the distribution (whether or not absolute gaps are found), and still retain their validity in accounting for rara even where these are "rehabilitated" gaps. As the title of an important article by Dryer (1998) put it: "statistical universals are better than absolute universals" (though personally

18 Here I skirt around the problem of how to convert typical typological characterizations into attribute-value pairs and merely point out that, mathematically, each typological attribute can be defined as having a number of potential values, for which a score can be assigned. Where the choice set is greater than two (as with the Greenbergian word-order characters), it is possible to factorize down the dimensions to yield binary codings (e.g., treating the S|V, the O|V and the S|O orders separately).

I would prefer "statistical trends" to "statistical universals"); cf. Christiansen and Chater's (2008:500) remark that "universals" are more akin to statistical trends tied to patterns of language use.

Many approaches have been taken to explain these asymmetries: generativists in terms of universal grammar stipulating some systems as possible; functionalists and cognitivists appealing to "markedness," "naturalness," frequency, or general cognitive preferences. Regardless of theoretical persuasion, all such skewings ultimately require explanation, with the due caution that some may be due to sampling biases.

Unless one takes the asymmetries themselves to be hardwired, a plausible line of general attack is through selector biases on what can evolve, drawing the selectors from a wide range of types. These range from perceptual or articulatory constraints, to general cognitive constraints, to system constraints favoring combinatoriality and discreteness in the evolution of efficient signaling systems, or conjunctive category characterizations (A and B) over disjunctive ones (A or B).[19] As with other complex systems, such as engineering solutions or complex species adaptations, the number of equivalently adaptive solutions increases with the number of selectors, whose interactions generate large numbers of more-or-less equivalent local optima (in evolutionary biology, cf. Niklas 1994; 2004; for an application of complex adaptive systems perspective to language, see Beckner et al. 2009).[20] Some of these selectors may not be equally distributed across all speech communities (see below), thus opening a chink for genetic, cultural, and sociolinguistic biases to apply.

Coupling of Characters (Implicational Universals)

A rich seam of typological work from the 1960s, originating in the Greenbergian tradition, has explored "implicational universals": statements about the likelihood or possibility of one character in a language, given another (i.e.,

19 For example, many languages have "Eskimo" kin systems like English "father" vs. "uncle" (F ≠ FB = MB) (lineal +1 generation male vs. collateral +1 generation male); many have "Dravidian" kin systems like Kayardild *kanthathu* "F, FB" vs. *kakuju* "MB" (patrilateral +1 generation male vs. nonpatrilateral +1 generation male); many have Hawaiian kin systems which use a single term for all three (F= FB=MB); and many have "Sudanese" systems which distinguish all three (F≠FB≠MB). None, however, are known in which F = MB ≠ FB. Greenberg (1990) attributed this gap to the fact that it would require a disjunctive definition whereas all the others can be characterized conjunctively.

20 "[E]ngineering theory shows that the number of equally efficient designs for an artifact generally is proportional to both the number and the complexity of the tasks that an artifact must perform" (Niklas 1994:6772), and with respect to simulated biological evolution, "morphological diversification became easier on complex as opposed to simple fitness landscapes. Likewise, it is biologically reasonable to suppose that the morphological diversity manifested by extant species occupying similar or identical habitats vouchsafes that very different phenotypes can have equivalent capacities for growth, survival, or reproductive success" (Niklas 2004:65).

pertaining to the coupling or linkage of traits). Three well-known examples, one each from grammar, semantics, and phonology, are:

1. Word-order universals: linking basic clause order (SVO, SOV, etc.) to the order of adpositions or possessive expressions with respect to the NP they are connected to. For example, Greenberg Universal #4 claimed that SOV order implicates postpositions rather than prepositions.
2. Semantic universals: claims about the structure of the color lexicon to the effect that a language would, for example, only possess a distinct word for orange or purple if it also possesses distinct words for green and blue. The implicational hierarchy for the elaboration of color term terminologies, proposed by Berlin and Kay (1969), generates a large number of specific implicational universals of this type.
3. Phonological universals: if a language has N distinct oral vowel phonemes, the number of distinct nasal vowel phonemes will not exceed N.

Like unconditional[21] universals, implicational statements generally tend to turn out to be statistical tendencies as research proceeds. Often this takes longer to discover because implicational statements generate a greater number of logical cells that require independent testing—the product of possibilities in each dimension under examination.

As with absolute universals, each statistical association needs explanation. In the case of word-order associations, for example, explanations have been advanced in terms of consistent parsing or processing orders across units of different types; in the case of color terms, explanations appeal to the differential sensitivity of color receptors at different wavelengths.[22] Each of these has its own trajectory of unfolding debate. The point is that implicational universals introduce additional phenomena (i.e., those defined by trait linkage), which need external explanation above and beyond the selector biases operating on individual traits.

Implicational universals thus relate to the correlated evolution or "coupling" of characters in more general evolutionary terms. We can view correlations as a static result (i.e., languages are more learnable or processable if particular traits cohere) or as the result of correlated evolution (i.e., the evolution of one trait favors or disfavors another). In either case, because trait correlation must result from correlated evolutionary processes, any claims about coupling are particularly susceptible to Galton's problem (i.e., apparent statistical correlations may merely reflect oversampled inheritances from a clade in which they happen to be correlated). In addition, recent work (e.g., Dunn et al. 2011b) has

---

21 I avoid the term "absolute universal" here, since it incurs an ambiguous double opposition: to statistical (tendency) and to conditional.

22 See Loreto et al. (2012) for a recent evolutionary multiagent simulation producing the expected emergence of color vocabularies based on the attested uneven distribution of just noticeable-difference across the wavelength.

turned up evidence for what might be called "pseudo-coupling" (i.e., apparent universal correlations which turn out to be lineage specific, such as a word-order coupling which holds in Austronesian but not in Bantu or vice versa).

Among the interesting reasons for this sort of finding are (a) possible noncomparability of units (e.g., that what gets counted as an adposition is not the same, functionally, from one group to another), (b) possible noncomparability of syntactic environments which act as selectors on units, and (c) the possibility that extrinsic selectors (e.g., discourse styles) are the real drivers of directed evolution, rather than the co-occurring features under examination. The latter is an example of the phenomenon Sapir called "drift," which has rather a different meaning in linguistics than in evolutionary biology, and denotes the mysterious phenomenon of related languages following parallel evolutionary pathways for no evident reason.

A more precise picture of character coupling than we now possess would bring many advantages. In terms of harnessing character profiles to phylogenetic inferencing, the more independent (uncoupled) or weakly coupled characters we can find, the more likely we are to detect signals of deep relatedness. It can help us evaluate the degree to which linguistic systems really are tightly integrated from a functional point of view (which tends to be the default assumption by linguists), as opposed to just looking like this because hundreds of thousands of traits usually get transmitted together. On the other hand, because our shallow time barrier for demonstrating linguistic relatedness will fail to detect cases where many tips grow out of the same iceberg, there is enormous potential for Galton's problem to go undetected, and for co-inherited traits to appear to result from correlated evolution.

Diachronic Typology

The grand ontology described above conceives of a design space of language states: what is their phoneme inventory, what is the meaning and syntax of "know," what is the basic clause order, and thousands upon thousands of other questions. Equally, we can turn our attention to the transitions between states. For example, how does a language develop tone (tonogenesis), or how does it get from one basic word order to another? Are all logical types of transition equally probable? What preconditions need to be in place for a particular evolutionary step to occur? These are the concerns of the field of diachronic typology.

A general evolutionary postulate here is that every synchronic phenomenon has a diachronically understandable pathway. To understand tonogenesis, for example, we need to understand how particular features of the consonants impact on the pitch of the vowels they adjoin. To understand word-order change we need to understand how other processes either move around some elements (topicalization or focalization) or reduce them to the point where they are out of the word-order game (e.g., the reduction of unstressed pronouns to affixes).

Conversely, at least some synchronic gaps or rara can be explained by the lack of ready diachronic pathways for evolving them (Evans 1995b) or the need to combine a number of rare features in the springboard state (Harris 2008), rather than any intrinsic lack of processability or learnability of the phenomenon itself.

For only a tiny fraction of the world's languages do we possess actual records of how the language has changed through millennia. Languages like Chinese, Japanese, Greek, Latin, or Sanskrit and their daughters are invaluable because they allow us to see what actually happened down the centuries and what features went together at any one moment. However, they represent such a small sample of the world's languages that for most diachronic changes we need to draw on other types of evidence, typically by comparing closely related languages. This is particularly the case when dealing with rara, where the diachronic evidence may, like the phenomenon itself, be confined to just one family. Synchronic rara thus create a particular aura of urgency around documenting the closely related languages that can help us understand them.

Changes, like states, tend to display stochastic distributions, and diachronic typological databases aim to characterize frequency as well as possibility, both in terms of topology (between A and B) and direction (from A to B, or B to A?). Important new developments here include the synthetic studies of attested and unattested patterns of sound change and their mechanisms, under the banner of "evolutionary phonology" (Blevins 2004), and growing databases of semantic shifts.

Macro- and Micro-Variation

Another postulate of diachronic typology is that categorical change is typically preceded by variation: within an individual, across a speech community, or both. As Labov has frequently argued (e.g., Labov 1994), this makes micro-variation a crucial part of understanding linguistic change, since it gives a way of detecting the seeds of large shifts which may themselves be unobservable because of the lengthy time spans involved.

Two types of methods have, since the 1960s, given us a very detailed picture of a number of changes and how they proceed. Functionalist studies, such as those by Bybee and Scheibman (1999), have carried out very fine-grained studies of how particular tokens are pronounced (e.g., *don't* and its various reductions) with regard to the frequency with which they occur in particular environments (e.g., before *know*, as in *dunno*). These studies tend to focus on the effects of repetition and frequency on form and emergent structure. Another approach has come from the Labovian variationist school, which focuses on the dynamic distribution of variants through speech communities. In general, variationist models of change involve three steps: generation of the change itself (e.g., within an individual, during learning, the streamlining of production or social positioning), propagation (e.g., through social networks), and valuation (i.e., the valorization or stigmatization of a variant, or its categorization as associated with a particular social group). In terms of the types of linguistic phenomenon covered, there has been a strong bias toward the form end of language, often with detailed studies of micro-variation in pronunciation, with less on grammar and very little at all on semantics.

A key question here is: Do mechanisms of social selection interact with the sorts of structural preferences outlined above? For example, will two variants, A and B, one of which is cross-linguistically highly preferred, be treated equally by processes of social selection, and will these processes apply equally in all types of speech community? I have already suggested above (see section, Diversity in Language Structure) that some processes may depend on high levels of shared bilingualism, such as the promotion of one highly marked gender form to shibboleth status by Iwaidja speakers and its subsequent generalization. More generally, it has been argued that a process of "esoterogeny" (i.e., the elaboration of difficult difference) operates in certain types of speech communities, such as the multilingual and highly metalinguistically aware communities of much of New Guinea (Thurston 1992). In such cases, the "expensive" options may actually be selected, due to interests of signaling group affiliation from childhood. If, further, bilingual-awareness-driven change of this type is particularly characteristic of small speech communities, then we would have a situation where the sociolinguistic characteristics of the speech community impacts upon the type of change and, ultimately, on the type of language found in particular communities. I will return to this particular variant of the society–language structure coevolution hypothesis below (see section on Social and Cultural Selection).

These examples should make clear that studies of micro-variation need to be connected to studies of macro-variation. Unfortunately, micro-variation in small-scale speech communities is one of the most neglected aspects of current linguistic research, and the last two decades of intensive language documentation have by and large failed to treat the organization of micro-variation as one of the dimensions in which we will discover significant cross-linguistic diversity.

Mechanisms of Selection

In the preceding sections I have argued that a central goal of linguistics is to give a full account of how and why language diversity is skewed across the design space—whether of synchronic structures or of attested diachronic pathways. In evolutionary terms, this requires us to identify the various types of selectors involved. When they are postulated, we must ultimately provide experimental or other detailed replicable evidence of how they work, such as studies showing acoustic similarities or articulatory adjustments as responsible for certain types of sound change. Here are a number of broad types of selectors.

Psychological and Physiological

These are the traditional staple of selection-based accounts. They include cognitive biases in cognition/processing of various types, the acquisition process, and the physiology of perception and production. Data on specific conditions facilitating selection have come from laboratory studies (especially in phonetics), quantitative studies of linguistic corpora (e.g., looking at the frequency with which new vs. established mentions occur in different grammatical roles), and, increasingly, computer simulations. The general logic of these accounts is to attribute some skewing in the cross-linguistic data to the greater ease or frequency with which certain traits co-occur, leading to correlations via Zipfian effects (Zipf 1935) or amplified cross-generational learnability. Well-known examples are the complementarity of palatal versus velar articulations before high front versus other vowels; the frequency of "up" polysemies with health, dynamism, activity, and happiness; or the association of ergative/absolutive case patterning with nouns and of nominative/accusative patterning with pronouns (on this last point, see DuBois 1987).

System Selection

This includes properties which favor the overall efficiency of languages as intricately interconnected semiotic systems. An interest in system architecture goes back to the grammatical tradition of Paṇini in ancient India which already had a deep concern with informational parsimony and developed the first data compression algorithms for representing complex and variable linguistic forms with extreme economy This has resurfaced over and over in the history of linguistics through both structuralist phases (e.g., the importance of structured sets of orthogonal feature oppositions) and generative phases (including the overt reconnection of generative methods with the Paṇinian tradition, such as rules deriving variable realizations of single underlying forms).

In his seminal identification of various "design features" of language, Hockett (1960) pinpointed a number of key properties of languages as systems. These include double articulation (i.e., phonological units are inherently meaningless but then combine into meaningful morphemic units which are the level at which semantic composition begins) and arbitrariness (freeing sign form from referent properties, which is useful, e.g., in coining short names for giant numbers or creatures). Though some linguists in the generative tradition have wanted to attribute these properties to the human mind, in the form of a putative "Universal Grammar," computer simulations (Kirby 2002; Zuidema and De Boer 2009; Reali and Christiansen 2009) have shown that properties like arbitrariness, discreteness, compositionality, and consistency in branching can emerge through recursive transgenerational selection.

Another type of system selection concerns linked typological traits. Traditions of doing this go back a long way in linguistics; for example, to Wilhelm von Humboldt in the nineteenth century, with his postulations of agglutinating, fusional, incorporating, and isolating types. More recently they have recurred through claims about implicational universals in the typological tradition or as flow-ons from "parameter settings" in the generative tradition. An example of the latter approach was taken by Baker (1996) to link a number of traits of polysynthetic languages (e.g., incorporation, lack of nonfinite forms). I have already outlined the difficulties in evaluating such claims empirically, because of the large numbers of traits that need to be checked across the sample, but typology is increasingly able to do this. Although many linked-trait hypotheses may be wobbling or tumbling (Dunn et al. 2011b), the precarious complexity of natural languages makes it likely that at least some will survive, at least as statistical correlations. A compelling example concerns the way modality (spoken vs. signed) impacts on a range of language structures, from word classes to semantic structures: Sign language typically includes a type of "classifier" word class not found in spoken language, and many classifiers lack a defined class comparable to pronouns in spoken language. The structure and semantics of reciprocal constructions may also show significant differences from that found in spoken language (Zeshan and Panda 2011).

Social and Cultural Selection

Explanations of this type appeal to properties of the social setting or of the culture in which a language is spoken. Many linguists, such as Perkins (1995), and most recently Bentz and Christiansen (2010), Lupyan and Dale (2010), and Trudgill (2011), have suggested that the most complex and unusual linguistic systems are most likely to evolve in such small communities (see also Bentz and Christiansen 2010). This is partly because speakers in small face-to-face communities can draw on a wider range of mutual knowledge, facilitating the grammaticalization of, for example, detailed kinship information, and partly because widespread multilingualism produces a different semiotic in which sounds, words, and grammatical items are positioned in a complex multilingual space—establishing linguistic identity, in this complex semiotic space, may involve promoting forms which are "marked"—and hence unlikely to arise if psychological and perceptual/articulatory selectors were given a free rein.

The parallels with sexual selection in biology are intriguing here. For certain social-signaling purposes (e.g., signaling lifelong group membership), the most "expensive" structures (i.e., those that are hard to learn or process) may be the most suitable.

Cultural selection is most likely to operate on the semantic dimension of language organization, in the realm sometimes known as "ethnosyntax"

(Enfield 2002). Examples are the impact of particular marriage rules and family structures on kinship semantics, of social categories on honorific behavior, of certain ways of talking on the development of "kintax" in Australian languages (Evans 2003c; Blyth 2013), or of patterns of journey discussion on the emergence of "associated motion" categories (Simpson 2002).

There may be interesting and complex feedbacks between social and other selectors. An interesting example comes from Alipur sign language in southern India (Panda 2013). There, an Urdu-speaking Muslim community transplanted from Andhra Pradesh to a predominantly Hindu area in Karnataka became reproductively isolated for religious and linguistic reasons, and consequently developed high rates of congenital deafness, leading to the evolution of a village-level sign language. Here cultural features (religion) impacted on social unit size, affected some genetic traits in the population, and selected for the development of a particular linguistic modality. Given what we know of other sign languages, it is likely that the evolution of Alipur sign language would be accompanied by the selection of particular linguistic characteristics, but so far we lack detailed descriptions of this sign language.

A further example, also involving sign, comes from Warlpiri, where cultural traditions enforce a speech ban on widows for around a year after their husbands' deaths. During this time, widows dwell in women's camps where the use of sign is widespread owing to the large number of widows. This situation makes it one of the few sign languages to be primarily used by speaking people (though, of course, its existence is also a bonus to deaf Warlpiri) and, most likely as a consequence of this, it displays far greater parallelism to the structures of spoken language than is normal in sign languages (Kendon 1988).

Genetic and Epidemiological Selection

For more than a century, since Boas's forceful denials of any link between language and race, linguists and anthropologists have placed the various mechanisms outlined above beyond the effects of genetic variability. This was based on the "self-evident" fact that children raised in any culture appear to learn that culture's language flawlessly. However, it neglects to consider the possibility that iterated selector effects, minor in any one generation but amplified through coevolutionary bottlenecks over many generations, can produce significant biases over time in the emergence of particular types of system. The genie was let out of this particular bottle by Dediu and Ladd's (2007) finding that the distribution of tonal phoneme systems correlates significantly with genes (*ASPM* and *Microcephalin*) that code pitch discrimination; other vocal tract differences such as lip and palate configurations are now emerging as candidates for genetic biases that may have loaded the dice in selecting for the emergence of particular types of phoneme system (or phonetic realization of particular phonemes). We now need to contemplate, and thoroughly investigate, the possibility that a significant part of the world's linguistic variability is due to genetic differences in the populations of speakers (for further details, see Levinson and Dediu, this volume).

Biological features of speaker populations are not confined to genes. An intriguing series of studies by Butcher (2006) and his colleagues (Stoakes et al. 2011) raises the possibility that particular properties of the phonologies of Australian languages may have been shaped by longstanding epidemiological particularities of the Australian indigenous population (in the original, medical sense of "epidemiological"), notably chronic otitis media. Butcher suggests that a number of phonological features may be adaptations to systematic gaps in frequency perception in the high spectral range in much of the population as a result of chronic otitis media. These include "long flat" phoneme inventories lacking fricatives and phonotactics that permit large numbers of "heterorganic" nasal + stop clusters; for example, the Kayardild triplet ŋa**nk**i "temple," ŋa**ɳk**i "beach (LOC)" and kaŋki "word (LOC)."

In other words, against a background of prevalent chronic otitis media knocking out perception in the high-frequency range, Aboriginal languages could have evolved (or maintained) "long flat," otitis-robust phonologies that concentrate the bulk of the perceptual discriminations in the mid-frequency range. Evaluating this hypothesis is a fraught enterprise. To be fully convincing it would need to correlate detailed data on prevalence of otitis media across large numbers of language groups, and deal with the difficulty of determining prevalence in past populations whose health status may have been different.[23] Thus, in principle, epidemiological as well as genetic factors may play a role in weighting selective processes in a particular direction.

Prospects and Challenges

The extraordinary diversity of the world's languages is being reconceptualized from noise to signal. As a result, linguistics is now entering an exciting phase where language diversity and the evolutionary processes which shape it are assuming center stage. A steadily expanding set of descriptions from languages across the globe, coupled with increasingly successful methods of setting up cross-linguistic comparison, is coming to furnish the most detailed, far-reaching, falsifiable and interrogable data sets we have for the world's many thousands of culture-defined groupings. Still, as the field stands on this threshold, it faces seven great challenges: the first two empirical in nature, followed by five theoretical issues:

1. *Basic descriptive coverage*. Despite recent advances, we are a long way from having anything like comprehensive descriptive coverage of

[23] Interestingly, the only existing paleopathological study, by Roche (1964), finds extremely high rates of aural exostoses in Aboriginal skulls, attributing much of the effect to prevalent otitis media in earlier populations.

the world's 6,000–7,000 languages. Getting beyond our current levels of around 20% coverage is a giant challenge, particularly at a time when major international programs (DoBeS, HRELP) and departments within institutions (e.g., Max Planck Institutes for psycholinguistics and evolutionary anthropology) are coming to the end of their funding lifetimes.

2. *Sociolinguistic studies of micro-variation in small-scale speech communities*. This represents a huge gap, not just in terms of coverage but also at the level of theorization and investigative tools.[24] But without studies of micro-variation in communities of the type that have shaped most of our human past, many central questions cannot be evaluated rigorously: these centrally include those concerning diversification, or the impact of population size on structural options.
3. *A complete characterization of the design space*. This is intimately tied up with developing an ontology for calibrating coding (and for quantifying uncertainty/ambiguity of analysis), and functional accounts of distributions. Typologists have been steadily working away at this, but this remains a work-in-progress, leaving linguists "like chemists without a list of the elements, or physicists with no account of particles" (Corbett 2012).
4. *A comprehensive categorization of transitions* compiled by induction from known changes across the world's languages and supplemented by experimental and computational studies of what leads to change. More extensive knowledge of which changes are common and which are rare will be of great help in weighting the likelihoods of alternative phylogenies. Historical linguists already do this implicitly and intuitively; it is a central part of their unquantified art and craft. However, explicit measures need to be developed and tested.
5. *A complete phylogenetic tree of the world's languages*. Linguistics is in the sad state, unthinkable to geneticists, of having hundreds of unconnected families with no methods for joining them up at deep levels. This is analogous to a vast collection of twigs and small branches without any larger tree that joins them together. The ±10 ky time barrier set by the comparative method, just like the time barrier set by radiocarbon dating, can only be overcome through the development of new methods capable of picking up heritable information in more sensitive ways. Here, the most promising techniques are new methods that apply phylogenetic algorithms to huge numbers of traits. As we get cross-linguistic data on an ever-growing number of characters, this will become increasingly informative.

24 An honorable exception is the series of studies by James Stanford (2008a, b) on how Sui speakers in Southern China organize tonal variation around patrilines rather than age- or gender-based signaling.

6. *Understanding of microprocesses generating diversification* and why it appears to proceed at different rates (and in different ways) in different settings. This is the theoretical counterpart to the empirical gaps delineated in Pt. 2. Labovian sociolinguistics has developed a sophisticated model of microevolution, but it is unlikely that this covers all types of microprocesses in small-scale societies.
7. *A revisiting of external selection.* The last sixty years have been characterized by a largely unexamined consensus between generativists, functionalists, cognitivists, and typologists: that "external" factors (genetic, epidemiological, sociolinguistic group size, and cultural) can be set aside when it comes to explaining the distribution of linguistic phenomena across the design space, in favor of properties of the human cognitive, perceptual and articulatory systems (all presumed to be invariant across human populations) along with universal discourse properties. Coming into the twenty-first century, it is looking increasingly plausible that at least some of the cross-linguistic variability may reflect these external factors. Testing hypotheses of this type will be an intricate affair, requiring much more attention to the gathering of matched nonlinguistic data sets to go with our linguistic information. If they turn out to have some currency, it will show the interplay of language, culture, and biology in human populations to be a much more dynamic and interactive process than we have imagined so far. In the process, it will give quite a different sort of answer to the basic question of why languages differ so much.

Acknowledgments

I gratefully acknowledge the support of the Australian Research Council (Grant DP878126: Language and Social Cognition – the Design Resources of Grammatical Diversity) and the Humboldt-Stiftung (Anneliese-Mayer Forschungspreis) for their material support which contributed to some of the work reported on here; Fiona Jordan for presenting many relevant ideas in a Master class on Cultural Phylogenetics taught at the Australian National University in December, 2011, and the audience at my 2011 Nijmegen lectures, as well as fellow participants in the 2012 Strüngmann Forum, for feedback on some of the ideas presented here; and Morten Christiansen and Pete Richerson for useful comments on the original manuscript. All errors of fact or interpretation are of course my responsibility.

14

Language Acquisition as a Cultural Process

Elena Lieven

Abstract

It is possible to identify three elements involved in the phylogenetic and ontogenetic development of the human language ability: (a) specific speech-related abilities, (b) cognitive abilities related to signification, symbol manipulation, and categorization, and (c) communicative or interactive abilities. This chapter suggests that although all three elements have a universal basis in evolution and development, they are affected at a rapid pace by the culture within which the infant develops. During the first year of life, infants become increasingly sensitive to the particularities of the ambient language(s). Their babbling shows both the restrictions on vocal production caused by the slow maturation of the speech apparatus and the influence of the language they are hearing. By 12 months of age, potentially primate-wide discriminations between types of events have given way to categorizations that reflect those of the language that the child is learning. Despite differences across cultures in both the ideologies and practices of child rearing, the onset of shared intentionality occurs at around nine months of age and does not seem to be affected by cultural differences in ways of interacting with babies. Once infants start to comprehend and produce language, there is a great deal of evidence, mainly from research in modern industrial cultures, that language development is influenced by quantitative and qualitative aspects of the ways in which children are spoken to. The chapter concludes with questions concerning the role of language in socialization and the relationship between concepts of socialization and culture.

Introduction

Children the world over learn to talk on a roughly equivalent timetable. This suggests the presence of biological and/or environmental constraints on language development. The claims for specifically linguistic, biological constraints have been made very strongly. These claims were based in part on the hypothesized "unlearnability" of language: that both word-to-world mappings and syntax cannot be learned without some innate, specifically linguistic, biases. Wide differences in the amount of talk to young children and the absence of explicit corrections to syntactic errors were also claimed to support the idea that children could learn language from relatively minimal input. On the other hand, we know that children must learn language from what they hear. Cultures differ widely in their socialization ideologies and practices as well as in their languages; they "promote and sensitize attention and representation of quite different aspects of social reality" (Evans, pers. comm., see also Evans, this volume). This leads to two possible conclusions: learning syntax is completely separate from learning the semantic contents of a particular language or, alternatively, children learn language "as a whole" and the biological constraints for doing so lie elsewhere. I begin by reviewing the evidence for differences between cultures in the communicative socialization of their children, before discussing the linguistic and sociocognitive developmental underpinnings to language learning that are present, or developing, in the first year of life and the possible cultural influences on them. I conclude by addressing the evidence for effects on children's language, thought, and socialization, respectively, of differences in their language environments.

Cultural Differences in Communicative Interaction with Infants

A number of studies maintain that there are cultures in which there are much lower levels of communicative interaction and child-directed speech with infants than has typically been reported for technologically complex societies (see Lieven 1994). These studies were usually conducted by linguistic anthropologists and tended to be qualitative rather than quantitative; for example, the study of the Kaluli of the New Guinea highlands by Schieffelin (1985) and the study of a Samoan community by Ochs (1982). In their studies of children's linguistic development, the researchers reported that children were spoken to rather little by adults and that this was related to ideologies of child rearing in the communities. They suggested that the child-centered style typical of middle-class families in modern industrial societies was completely at odds with the cultural ideology of the people they were studying (Ochs and Schieffelin 1983). The Kaluli believed that, as they developed, babies had to be weaned away from the animal world by being taught to talk. According to Schieffelin, little interaction took place until a child started to use its first words (often "milk/breast"), and then adults employed a style in which they told the child directly what to say. Ochs argued that the child-directed style, in which middle-class English-speaking adults followed in on their children's utterances, and which was regarded as important for good language development, was highly inappropriate in the Samoan context, in which interactions were governed by status: adults, who have higher status than babies, did not adapt their speech to the baby nor even talk to them very much, but tended to instruct older children to interact with them. Further examples come from Heath's (1983) study of the "Trackton" community in the Piedmont Carolinas of the United States and Ratner and Pye's (1984) study of language development in Quiche Mayan children, both of which reported low levels of speech to children. Interestingly, the ideologies behind these low levels of talk were very different. Heath describes how Trackton children were expected to "fight for the floor" in interaction. She describes the ways in which children started to talk by imitating parts of utterances that they heard exchanged between others and used this to try to break into the interaction. Ratner and Pye report that the Mayan group thought that young infants were very fragile and needed to be protected from too much stimulation and excitement, leading to little intense interaction and talk. The great advantage of these studies is that they were conducted by researchers who lived with the people they were studying for considerable lengths of time; this allowed them to gather in-depth knowledge of the culture, language, and ideology. Nonetheless, it is difficult to make comparisons between the cultures since these studies were largely qualitative rather than quantitative. It is therefore hard to know what constitutes "a lot" or "a little" talk to children.

A more recent series of studies by Keller and colleagues (Keller 2007) provides extensive, quantitative, cross-cultural comparisons of behavioral interactions with babies and the accompanying ideologies. Studies were conducted in rural and urban contexts in Costa Rica, China, India, Cameroon, Greece, the United States, and Germany and focused primarily on parental attitudes (obtained through extensive parental interviews and questionnaires) and their behavioral interactions (from systematic observations) mainly with very young infants of about three months. Clearly, there is a biological framework to infant development and the resulting care that is required. Keller (2007:22) identifies four "parenting systems" that were found across all the cultures studied: body contact, body stimulation, object stimulation, and face-to-face contexts. These were accompanied, however, by very different ideologies, and the relative frequency with which they were used also differed considerably. From this research, Keller abstracts two prototype models of parenting: interdependent and autonomous. The first is more typical of small-scale, rural, and nonindustrial cultures in which the emphasis is on the child's membership of a community: babies never sleep alone, they are breast-fed on demand, with almost continuous bodily contact, but they are talked to less. On the other end of the continuum, parents' emphasis is on the child as an autonomous agent, treated as intentional in interaction from the outset, with much more talk to the infant but the child spends more time sleeping and playing alone. Keller's results do indeed show that differences in the amount of talk to very young babies are roughly correlated along the dimensions she outlines, but she also points out that this is indeed a continuum and that cultures which might be placed close together on the continuum can vary widely on other dimensions. Keller recognizes, however, that there are inevitable problems in placing cultures on these types of dimensional continua. As we shall see below, two traditional societies can vary wildly in the amount of talk to children and there can be major differences in the amount of talk to children in the subcultures of modern industrial societies.

Developing Linguistic Sensitivity to the Ambient Language

All the evidence suggests that, not surprisingly, human infants start with universal sensitivities and cognitive skills and that these become tuned to their specific linguistic and social environment during the first year of life. For some language-related characteristics, this process begins very early. For example, sensitivity to the mother's voice can be shown in the last trimester of fetal development, and newborns can already discriminate between the rhythmic features of their ambient language and that of rhythmically dissimilar languages. Very young infants (between one and four months of age) can show categorical perception for consonants and the ability to discriminate vowel contrasts. Infants become increasingly sensitive to the prosody, phonemes, and vowels of their ambient language, and the capacity to discriminate the sounds of other languages attenuates in the six- to twelve-month period of life (for a summary, see Ambridge and Lieven 2011, chap. 2). Neuropsychological evidence largely confirms this picture. Functional magnetic resonance imaging (fMRI) of neonates and three-month-olds while they are asleep shows left lateralization for speech (Dehaene-Lambertz et al. 2002), though there is obviously considerable flexibility in the infant brain, as shown by the ability of children with severe left-hemisphere damage to learn language (Feldman et al. 2002). Studies using event-related potentials recorded from infant scalps show discrimination of phonetic features by one to four months and of phonemic features by six to seven months, irrespective of the child's ambient language, whereas older infants of eleven to twelve months only show discrimination of contrasts that are relevant in their own language (Friederici 2009). Infants are also capable of domain-general, statistical learning based on transitional probabilities from an early age (Aslin and Newport 2008). A number of experiments have shown that infants can induce underlying patterns in strings of artificial syllables (e.g., at seven months, Marcus et al. 1999; for evidence that nine-month-old infants can extract "rules" of different levels of abstractness depending on the precise structure of the stimuli, see Gerken 2006). In terms of what they actually hear, four- to six-month-olds show listening preferences to some highly frequent words (e.g., their own name, *baby*) and at six months can link the words *Mummy* and *Daddy* to pictures of their own parents in preference to unfamiliar adults. In the last three months of the first year, infants become sensitive to the specific cues that characterize the words of the language they are hearing. Infant babbling shows universal characteristics that result from the immaturity of the speech organs but also reflect certain aspects of the phonological structure of the language that the infant is hearing (Oller et al. 1975; Boysson-Bardies and Vihman 1991). An important question that arises from these studies is the extent to which these findings are specific to humans and to language learning. Categorical perception has been shown for some species of monkeys (Kuhl 1991) and for nonspeech and visual stimuli (Aslin and Newport 2008). It is possible that some species also show statistical learning in the auditory as well as the visual domain (Newport et al. 2004; Heimbauer et al. 2010). In addition, studies suggest that, initially, cognition of the physical world is very similar in human infants and nonhuman primates, for example, Michottian experiments, object permanence, rapid discrimination of small numbers (Gómez 2004). In terms of word learning, we know from studies of human-trained and -encultured apes that some can learn upward of 50–100 words. This suggests that while human language development may well be based on primate-wide skills of distributional learning, categorical perception, and more general learning principles, something more is required. One obvious answer to the question of what this "something more" might be is the mechanisms that underpin the complex motor control required to produce speech. In ontogeny, this takes considerable developmental time, with early speech being very slow and showing major "distortions" from the adult system. However, language is more than speech: it involves meaning and structure and, at this level, can be manifested in different media (e.g., speech, signing, and writing). There have been three general approaches to the question of what underpins the development of meaning and structure:

1. specifically linguistic modularity,
2. cognitive capacities that involve higher-order operation with symbols, and
3. sociocognitive capacities.

In terms of the first the evidence strongly suggests to me that linguistic modularity is the outcome of language development rather than its cause. In terms of the second humans undoubtedly have a cognitive ability to manipulate symbols mentally. The issue, however, is when this ability emerges in development as being clearly distinct from nonhuman primates. From three to four months infants can form prototypes of basic shapes, and at around six months they show evidence of organizing vowels into a prototypical category structure; rhesus monkeys, in contrast, are only sensitive to the distances represented by the absolute values of the stimuli (Kuhl 1991). There is a huge literature on infant categorization abilities that definitely develop over the first year of life (Mandler 2000), but it comes with a fierce debate about the extent to which these abilities are domain general or specific, and whether they are generated by perceptual phenomena or by higher-order representations (Rakison and Yermolayeva 2010). Most of the evidence for specifically human aspects to cognition stems from work on young children, rather than infants. For instance, analogical reasoning has been cited as an example par excellence of the human capacity to manipulate symbols, but the youngest children tested in these types of experiments were between 2;6 (two years, six months) to 3;0 (Gentner et al. 2011).

Metacognitive abilities, such as response inhibition linked to the development of the neocortex, have also been used to explain changes in children's ability to perform on tasks, including theory of mind tasks, but here we are also talking about children between three to five years old. Thus, although human cognition is both qualitatively and quantitatively different from that of other species, it is an open question whether this can be demonstrated in human infants or whether it is evident only later in development.

The third category of what might underpin language acquisition concerns the development of sociocognitive skills from around nine months of age.

Sociocognitive Development across Cultures

A strong alternative to positing language development as being on a separate, encapsulated, developmental pathway is to suggest that the critical human-specific factor underpinning both the phylogenetic and ontogenetic development of language is the development of "shared intentionality" (Tomasello 2008). Shared intentionality is not just the understanding that others have intentions and goals, which nonhuman primates also show, but the ability to participate with others in shared goals; that is, to coordinate joint action with other minds. From about nine months of age, infants show the development of a series of abilities, which, it is argued, are not found in nonhuman primates and reflect this coordination with other minds (Herrmann et al. 2007). Thus when eight- to nine-month-old infants start taking part in episodes of "triadic joint attention," they become able to draw the attention of others to objects and to follow the other's attention, not just with gaze following, which has been in place for many months, but with gaze checking, indicating an understanding of the "sharedness" of the event. They behave as if they expect their partner to be a communicative partner, persisting in attempts to establish joint attention. Around this age infants also begin to point informatively for others: to attempt to coordinate joint interest and to help the other (e.g., to locate a missing object). A short time later they will imitate not just the physical actions of others but also the intentional structure of the other's action (Lyons et al. 2007).

Do major differences in child rearing outlined above have an important impact on when children in different cultures start to show these behaviors? Callaghan et al. (2011) investigated this in interviews and experiments with parents and their children (aged between eight months to three and a half years) from two rural communities (in India and Peru) and one modern, urban community (in Canada). For joint attention skills, imitation, pointing, and helping behavior, as well as language comprehension, similar ages of onset were found both from parental report and in the experiments, strongly suggesting a universal timetable. However, language production was reported to be on average three months earlier in the Canadian children; these children also demonstrated an earlier beginning to pretend play, book reading, and drawing, presumably as a result of cultural differences.

Other studies also suggest that the onset of these early sociocognitive skills is more dependent on child-intrinsic factors, and therefore independent of culture, although the frequency with which they occur may depend on within-culture differences. Liszkowski et al. (2012) used a seminaturalistic procedure to elicit pointing by preverbal infants and their caregivers in seven cultures. By 10–14 months of age, infants across cultures were all pointing with similar frequencies. Liszkowski et al. found that infant pointing was best predicted by the child's age and caregiver pointing—not by culture. Thus, although the caregivers on Rossel Island (Papua New Guinea) pointed at significantly higher rates than those of the other cultures, infant frequencies on Rossel Island did not differ from the other cultural settings. On the other hand, frequencies of caregiver pointing within each culture did relate to the frequency of infant pointing, although not its onset. Liszkowski et al. (2012) identified strong relationships between the interactional timing of infant and caregiver pointing, which they argue indicates a universal prelinguistic structure of proto-conversation. Two naturalistic studies tend to support the conclusions of the Callaghan et al. (2011) study. Lieven and Stoll (2013) compared the onset and frequency of imitation and pointing in children from a community in eastern Nepal who speak Chintang (a Tibeto-Burman language) with children matched for age and gender in a German community. They found that there were no differences in ages of onset, and that differences between individuals were as wide within the two cultures as between them. Brown (2011) compared rates of interactional initiation on Rossel Island, Papua New Guinea, and in Tzeltal, a Mayan culture of Mexico. Concurrent with reports by both Ratner and Pye (1984) and Gaskins (2006) for other Mayan cultures, in Tzeltal there is a relatively low level of interaction and conversational initiation with young children whereas on Rossel Island, adults are constantly engaging small children with high rates of interaction. The interesting finding is that as young children began to initiate interactions themselves, there was no difference in the rates between the Rossel Island and Tzeltal children. This again suggests that entering into the communicative world is driven by child-internal factors which are then shaped by the culture rather than there being a direct influence of styles of interacting with babies on the initial development of communicative skills. A training study by Matthews et al. (2012a) also suggests separate contributions of infants' own sociocognitive developmental timetable and their caregivers' interactions. Mothers were asked to spend 15 minutes per day over four weeks engaging in enhanced pointing with their infants. As in the studies cited above, Matthews et al. found no influence from either the caregivers' pointing in free play or the training on the age at which infants started to point; however, the frequency with which mothers pointed in free play did influence the frequency of their children's pointing. Thus, although infant sociocognitive development is the prerequisite, socialization processes start to affect how this unfolds. Callaghan et al. (2011) found that parents in all three societies reported language comprehension as starting at around nine to twelve months and this fits well with many other studies. Following Bruner (1975) and others, Tomasello (2008:155) argues that the best explanation for the fact that word learning starts at this age is that it depends on the development of the "cooperative infrastructure of shared intentionality." As noted above, six-month-old infants are already able to identify some words and associate them with their referents. But word learning cannot be a simple matter of association between word and referent, as we all know from Quine. The best candidate for solving the Quinean problem is children's ability to infer the intentions of their communication partners; this acts as the common ground through which they can map the sound (or gesture or sign) to some child-extracted meaning. There is plenty of evidence that children can do this: they can learn words for nonvisible referents, for the "intended" referent when there are multiple possible referents, as well as for actions which cannot be pointed to in the way that objects can (Tomasello 2003). The suggestion is that this is only possible because the interactional frame constrains the interpretation of the utterance. Once this fundamental development has taken place, cultural differences in children's communicative environments can start to make themselves felt in language learning.

Learning the Meanings Encoded by a Language

How do children learn meanings that are encoded by a language? One possibility is that children, when they start learning language, initially map universal cognitive categories onto the words that they are learning. This would mean that young children would make similar distinctions between events, whatever the language, and then gradually develop language-specific categories. As far as semantic categories are concerned, this is a debate that has a long history in research on semantic development. As children are learning words, are they also learning the concepts that these words relate to with all their cultural connotations? Conceptual categories develop over the first year of life: the ability to discriminate between patterns of motion, to discriminate nonbiological from biological motion, to segment action events, and to understand the physical relationship between objects (e.g., containment and support) develops earlier, whereas the categorization of path and manner changes in objects develops somewhat later (10 and 13 months, respectively). There is evidence that prelinguistic children start with, or develop, similar concepts about the physical world. A recent review by Göksun et al. (2010) on infants' nonlinguistic conceptualization of event constructs suggests that prelinguistic infants are sensitive to a number of the event categories coded by the languages of the world (e.g., containment by six months, path by ten months). There are clear developments in infants' sensitivity to different event categories over the first year of life, and Göksun et al. (2010:36) conclude that "infants possess a set of nonlinguistic constructs that form the basis of learning relational language" and that, insofar as this has been tested, these encompass the range of distinctions made by languages of the world.

However, it also seems to be the case that as soon as children start to learn language, they begin to develop language-specific categories, and this rapidly starts to influence their attentional behavior and their responses in both linguistic and nonlinguistic tasks. Bowerman and Choi (2001) show that even in preferential looking tasks with children aged 18 and 24 months, who are only just starting to produce language, Dutch and Korean children categorize spatial relations in ways related to the language they are learning (e.g., "tight" vs. "loose" fit in Korean) rather than in terms of any universal categories (e.g., relations of containment and support denoted by "in" and "on" in English). Brown's (2001) data on children learning the absolute spatial system of Tzeltal confirms this: from their earliest productions, children are reflecting the spatial distinctions made by their language rather than any more universal categories. Studies on the expression of containment (Narasimhan and Brown 2009) and verbs of cutting, breaking, and tearing verbs (Narasimhan 2007) in Hindi and Tamil show that children largely observe the regularities of the languages they are learning. They do not start with the most general meaning and the means for expressing it, and only later use more specific forms, and they do not overgeneralize from one type of event to another. Göksun et al. (2010) also conclude that, depending on the precise components of event structure being interpreted, these become increasingly language specific between two and three years of age. Although this evidence on the learning of semantic categories suggests early sensitivity to the categories encoded by the language, it is also clear that there is a long process of development toward the adult system. Thus Brown (2001) says that Tzeltal children's development of the absolute spatial system shows limited productivity by 3;6 and mastery for small objects by 4;0. It takes, however, from between 5;0 to 8;0 for them to acquire the geographical knowledge that allows them to use the system fully abstractly. Lucy and Gaskins (2001) came to a similar conclusion in a series of studies that compared seven- and nine-year-old English and Yucatec Mayan children's preferences for classifying objects by shape or material. In many ways, the finding of an influence of language on categorization does not come as a surprise, although it did initially, given the very universalist theoretical zeitgeist of the 1960s to the 1980s. From a position which argues that children learn form-meaning mappings as a unit and from the input, one would expect that first use of words and phrases would follow the meanings of the language, but without necessarily encompassing the full range of semantic constructs expressed in the language. Much of my own research and that of colleagues has shown the extremely close linkage between distributional aspects of the input and children's own language (Lieven 2010). In addition, although there is little explicit correction of children's linguistic errors, Chouniard and Clark (2003) showed that a group of middle-class U.S. mothers tended to respond to errors in ways that provided corrective models, and that their children quite frequently picked up on these responses by producing corrected repetitions. Obviously, children go beyond what they hear throughout development. However, at any one time, this creativity and productivity is particular to specific aspects of the language and the level of schematicity and abstractness to which these have developed in the child's system (Karmiloff-Smith 1994).

Within-Culture Effects of Input on Language Learning and Developmental Outcomes

The Callaghan et al. (2011) study found that language production was reported to start three to four months earlier for the Canadian children than for the Peruvian and Indian children. This may well be due to differences in the amount of talk experienced by children of the middle-class Canadian parents in contrast with those from the other two cultures. However, it is obviously difficult to compare outcomes in terms of children's language development across cultures that vary on such a wide range of measures. In an effort to control this better, we can look at differences in the language development of children growing up in different circumstances within the same "overarching" culture. For instance, many studies have compared the language development of children from different socioeconomic, ethnic, and linguistic backgrounds growing up in technological societies such as the United States (including the Heath 1983 study mentioned above). Many of these show an effect of the quantity of talk to children and/or its lexical diversity on the growth of children's vocabularies[1] (Weizman and Snow 2001) and that this is related to social economic status (SES) and the educational level of the parents (e.g., Hart and Risley 1995; Hoff-Ginsberg 1991). There are also many studies that relate children's early language development to later reading and educational success (e.g., Snow et al. 1998). These studies suggest that language development is closely related to the communicative environment in which children are raised. This environment, in turn, may reflect a particular "culture" or "subculture." This raises the problem of how to define "culture," and whether culture can really be related to SES in any straightforward way. Importantly, these relationships between the language environment and children's own language development are found at the level of individual differences within a particular SES. For instance, in a

[1] It is important to note that there are methodological problems in using vocabulary size as a measure of children's language development. Parental check-sheets for children's vocabularies suffer from design problems, particularly when cross-linguistic comparisons are made (e.g., they contain far more nouns to be checked than verbs). Estimating lexicon size from naturalistic recording depends critically on how much talk there is (Malvern et al. 2009), which may differ between dyads in a recording situation without necessarily being related to the quality of interaction or richness of talk in other contexts (as noted by Labov 1969).

study of children's development of complex syntax, Huttenlocher et al. (2002) showed that parental use of complex syntax accounted for more of the variance in children's development of complex syntax than did SES, which was only marginally significant. In the same study, after controlling for the children's language skills at the beginning of the school year, the researchers also showed that the use of complex syntax by classroom teachers had an effect on the syntactic development of the children. Pan et al. (2005) found a relationship between the diversity of the maternal lexicon, language and literacy skills, and the growth in children's vocabulary between twelve months and three years in a group of 112 low income families. Marchman and Fernald (2008) found relationships between children's processing speed (in tasks in which they had to look toward a picture of an object when hearing the name of that object) and their vocabulary size, both at the same age and at later ages. When these same children were followed up at eight years of age, processing speed predicted differences in their linguistic and cognitive skills. These children all came from high SES families. In a subsequent study, however, Hurtado et al. (2008) found that the type of input that low SES, Spanish-speaking infants were hearing from their mothers was related (a) to their vocabulary size and (b) to their processing speed six months later.

One important direction for future research will be to look at the ways in which conversational turn taking with language-learning children may provide them with linguistic information geared to their own production and, in particular, errors (Chouniard and Clark 2003) and how this may differ between individual dyads and social groups. Although it is unlikely that this type of implicit correction in vertical turn-taking sequences can provide a full account of how children's language develops, it may well play a role in some, if not all, social groups.

Finally, studies by Street and Dąbrowska (2010) have shown that differences in linguistic competence in adulthood can be related to measures of the amount that people read, to their general cognitive skills, and to their educational attainment. Even among the fairly homogeneous group of college-aged students, there are significant differences in language processing ability that can be tied to their experience with language (Wells et al. 2009). This raises the issue of the influence of reading and writing on children's language development.

Written symbol systems have existed for only about 5,000 years and thus are clear examples of cultural evolution. Children have to be taught to use them, and this can be a very protracted process. There is clear evidence for relationships between various processing skills and reading ability, although whether this is related to oral language ability is the subject of much research. In addition, there is a genetic contribution to dyslexia which probably operates through an influence on these same processing skills. However, there is also considerable evidence for a relation between complexity of language use (and, arguably, competence) and the extent to which, in literate cultures, people engage in literacy skills, as measured, for instance, by level of education (Street and Dąbrowska 2010). This is almost certainly because complex syntactic structures and meanings are more likely to be encountered through reading than through everyday language use (Miller and Weinert 1998). Similar advanced competence in the ability to use and comprehend the more complex aspects of language may be found in experienced storytellers and/or public orators in nonliterate cultures (Goody 1987). These highly developed skills require an ability to reflect on language: it seems probable that both writing and the development of complex oral skills can provide the basis for individual differences in the domain of complex syntax. Thus, it would appear that the way in which children are spoken to affects the manner in which they themselves develop language, and that this may also be implicated in developing cognitive skills such as processing speed and working memory. In turn, these may continue to show effects into adulthood, particularly in the realm of the more sophisticated uses of language. The implicational leap, which may well not be justified, is that if this is true for differences between individuals and groups within a culture, it may also be true for children learning languages that differ much more extensively.

Language and Sociocognitive Socialization

Although there is much less discussion about the relationship between language learning and sociocognitive categories, what there is suggests a similar process to that outlined above. The universal development of intention reading during the last few months of the first year of life is followed by an immediate impact of language learning, which is initially reflected in set formulae and relatively ritualized contexts but gradually develops toward the adult system: the latter is affected by the complexities of the language itself as well as by what is being conveyed. There is also some evidence for more explicit teaching of these sociocognitive categories, presumably because they are extremely important in managing social relations. Clancy's (1985) paper on Japanese acquisition illustrates all three of these processes. First, she reports that children initially learn the use of the plain and polite registers of Japanese, which are used in relation to in- and out-groups as set formulae, and that they mostly hear the polite forms used by their mothers in pretend play, perhaps as an implicit or explicit teaching measure. Second, an example of early learning is the conveying of speaker attitude when providing information to a listener. There are three sentence-final particles: one neutral, one used when "encountering resistance or lack of mutuality," and one that expresses "rapport with the addressee." These are extremely difficult for non-Japanese learners to learn but they appear early on for Japanese-learning children, presumably because they are affective, in a salient position, and highly frequent. Third, the children's use of honorifics happens quite late, presumably because of their extreme linguistic complexity which involves their use not only in addressing others but also in speaking about them.

One area in which there has been explicit discussion of the influence of language on social cognition is children's development of "theory of mind" understanding, as measured by their performance in false belief tasks. Brandt, Buttelman, Lieven, and Tomasello (in preparation) have shown that German-speaking children, age 3;0 and 4;0, are able to distinguish successfully between "believe (*glauben*)" and "know (*wissen*)" when used in first-person matrix clauses in sentential complements: *I think X* versus *I know X*. However, only children, age 4;0, are able to do this in third-person contexts: *A thinks X, B knows X* . Following Diessel and Tomasello (2001), and Brandt et al. (2011), they argued that this is because children first learn these mental state verbs from their frequent, formulaic occurrences in the input—*I think it's raining*; *Know what Daddy said?*—where they act as discourse markers rather than as indicators of the contents of other minds. The authors also found that success by the four-year-old children in the third-person task was correlated with their ability to pass the false belief task. This fits well with other studies which show that children's ability to pass theory of mind tasks, in which they understand that it is possible for others to hold a "false belief," is closely related to specific aspects of their language development. Thus de Villiers and de Villiers (2000) showed a correlation between passing the false belief test and the use of mental state verbs such as "think" and "know" in complement structures, and this finding was confirmed by Lohmann and Tomasello (2003). However, Lohmann and Tomasello found that other forms of training, involving perspective-shifting discourse but not sentential complements or mental state verbs, were also correlated with success, and they suggest that there might be both a direct effect of learning the meanings of the particular verbs but that discourse about deception could have an independent effect.

This leads us into complex territory with very differing theoretical emphases on the role of language in socialization. One approach could be called "psychologizing," in which aspects of language used to, and by, children are correlated with children's behavior; language is thus seen as either reflecting or having a direct influence on underlying psychological states. A good example of this is work by Dunn and Kendrick (1982), and Zahn-Waxler et al. (1979), who show correlations between mothers' tendency to talk about the feelings of others in conflict situations and children's own prosocial behavior. Capps and Ochs (1995:186) characterize this approach as a "tendency to look through language rather than at its forms." Reflecting a rather different theoretical framework is the position that in learning language, children become "cultural subjects." Kulick and Schieffelin (2004:350) state:

> Language is not just one dimension of the socialization process; it is the most central and crucial dimension of that process.... any study that does not document

> the role of language in the acquisition of cultural practices is not only incomplete. It is fundamentally flawed.

Let me give a few examples that seem to exemplify the importance of paying attention to this approach. In a study that compared Italian and American families' ways of talking about food, Ochs et al. (1996) showed that likes and dislikes of food are socialized at the dinner table, with food conceived of as "oppositional" between children and adults in U.S. families and as "pleasure" in Italian families. The importance of what is made explicit and what not is referred to by Kulick and Schieffelin (2004:357) as "what must remain unspoken and unspeakable" and shown to be reflected in avoidance and topic change as well as in direct commands. They point out that "the ability to display culturally intelligible affective stances [e.g., desire and fear] is a crucial dimension of becoming a recognizable subject in any social group" (Kulick and Schieffelin 2004:352–353). Thus, in a comparison of U.S. and Japanese mother–child interactions, Clancy (1985) argues that Japanese children are socialized to command the strategies of indirection and intuitive understanding through early socialization routines. In an analysis of the different linguistic forms used to control children's behavior in Japanese, Korean, and English, Clancy et al. (1997) argue for a potentially direct link between linguistic form and the ways that power is channeled in the different cultures. They show that Japanese and Korean mothers use deontic conditionals (e.g., *If you do this, it's bad*) which have the effect of providing an advance evaluation of the behavior, whereas U.S. mothers, using modals like *can* or *should*, often used an explicit reason (*If you do this, you can get hurt*).

These two positions derive from different disciplinary backgrounds: psychology in one case, discourse theory and ethnomethodology in the other. The first framework aims at showing effects on children's behavior of parental attitudes to socialization as reflected through what parents say to children. Here the focus is on the underlying attitudes of the parents and behavior of the children: language is more of a "measurement tool." The second views language interaction as creating "culturally specific subjectivities" (Kulick and Schieffelin 2004:351) through what is said and what is not said. The first is more focused on individual differences; the second on cultural differences. It is not clear to me whether there is an irremediable theoretical conflict here. My own view is that, in principle, the socialization of individual differences and the creation of cultural subjectivities must involve the same psychological processes, and that children's language development and the ways that language is used in interaction with children are both central to the operation of these processes.

Conclusions

Despite major differences in child rearing and socialization practices, infants seem to adhere to the same developmental timetable for sociocognitive development during the first year of life. As soon as they begin to learn language, however, children show language-specific effects on semantic and syntactic development. The extent to which human infants in the first year of life show clear differences in cognition (e.g., categorization and symbol manipulation) from nonhuman primates remains an open question. Within cultures, individual differences in caretaker language are related to children's own language development as well as to literacy and educational outcomes. Not only typological differences in syntax and semantics but also the different ways that languages are used in interaction with children reflect and, arguably, create differences in both language competence and cultural subjectivities. Much more research is needed on peer–peer communication and its influence, particularly in contexts where, as children move from infancy to toddlerhood, they increasingly spend most of their time with peers rather than adults. Finally, many open questions remain about how to theorize the relationship between language, socialization, subjectivity, and culture.

Acknowledgments

I am grateful to Michael Tomasello, Paul Ibbotson, Simon Garrod, and the editors for very helpful comments on an earlier draft.

15

Phylogenetic Models of Language Change

Three New Questions

Russell D. Gray, Simon J. Greenhill, and Quentin D. Atkinson

Abstract

Computational methods derived from evolutionary biology are increasingly being applied to the study of cultural evolution. This is particularly the case in studies of language evolution, where phylogenetic methods have recently been used to test hypotheses about divergence dates, rates of lexical change, borrowing, and putative language universals. This chapter outlines three new and related questions that could be productively tackled with computational phylogenetic methods: What drives language diversification? What drives differences in the rate of linguistic change (disparity)? Can we identify cultural and linguistic homelands?

Introduction

Evolutionary biology has changed remarkably over the last thirty years. Phylogenies have sprung from the margins to center stage. Open any evolutionary journal, or go to any evolutionary meeting, and you will find wall-to-wall phylogenetic trees. Tree thinking (O'Hara 1997) is now the dominant way of making inferences in evolutionary biology (see Figure 15.1). The phylogenetic revolution in biology has been driven by two main events: the development of computational methods and the deluge of molecular sequence data. Today, molecular phylogenies are used to analyze everything from Aardvarks (Seiffert 2007) to Zoogloea (Kalia et al. 2007).

Despite its apparent position on the other side of the arts/science divide, linguistics is also a discipline that requires making complex inferences from a wealth of comparative data. Moreover, as scholars dating back to at least Darwin (1871) have noted, there are numerous "curious parallels" between

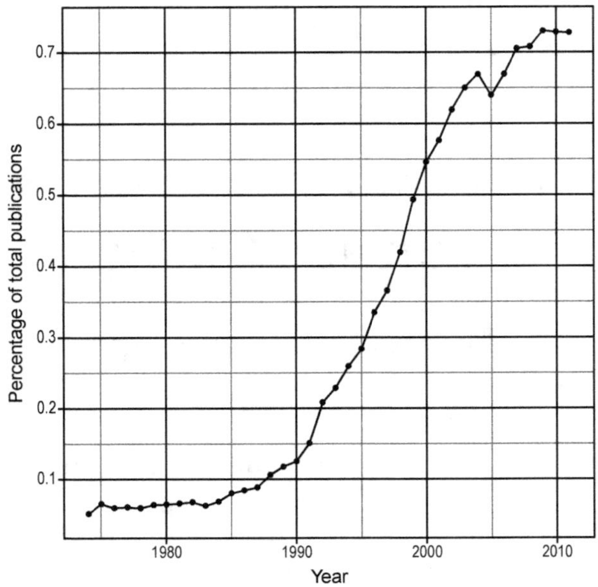

Figure 15.1 A plot showing the percentage increase in papers mentioning the keyword "phylogen*" in the Scopus publication database by year.

the processes of language change and biological evolution (see Atkinson and Gray 2005). Although early attempts to turn historical linguistics into a computational science were far from successful (Swadesh 1952; Bergsland and Vogt 1962; Greenhill and Gray 2009), we are currently witnessing a steady growth in both the use of computational methods and the development of large comparative databases (Greenhill et al. 2008; Dryer and Haspelmath 2011). Computational methods derived from evolutionary biology have been used to construct phylogenetic trees for language families including Aslian (Dunn et al. 2011a), Austronesian (Gray et al. 2009, 2011; Greenhill and Gray 2009, 2010), Bantu (Holden 2002; Holden and Gray 2006), Indo-European (Gray and Atkinson 2003), Japonic (Lee and Hasegawa 2011), Pama-Nyungan (Bowern and Atkinson 2012), Semitic (Kitchen et al. 2009), and even creoles (Bakker et al. 2011). They have been used to:

- Date language divergences and thus test hypotheses about human prehistory (e.g., Gray and Atkinson 2003; Gray et al. 2009).
- Investigate the rates of change in aspects of language (Pagel et al. 2007; Greenhill et al. 2010).
- Quantify patterns of borrowing in languages (Greenhill et al. 2010; Nelson-Sathi et al. 2011; Gray et al. 2010).
- Identify functional dependencies in language and thus test claims about language universals (Dunn et al. 2011b; Levinson et al. 2011).

As these approaches have recently been reviewed by Gray et al. (2011) and Levinson and Gray (2012), we will not cover the same ground here. Instead, we outline three new and related questions about language evolution that could be productively tackled with computational phylogenetic methods: What drives language diversification (cladogenesis)? What drives linguistic disparity (anagenesis)? Can we identify cultural and linguistic homelands?

What Drives Language Diversification?

Vast amounts of ink have been spilt, and millions of computer keys pressed, addressing detailed linguistic questions such as the development of Proto-Indo-European laryngeals.[1] We certainly do not wish to diminish the importance of these endeavors; however, we are surprised at how little attention linguists have given to the question of language diversity. Explaining why the human species currently has around 7,000 languages (Lewis 2009) should be a fundamental task for both linguists and theorists of cultural evolution. Moreover, the patchy distribution of this diversity cries out for explanation. According to Lewis (2009), there are 194 language families. Most of these families, 74, have a single member (i.e., are isolates). At the other extreme, Niger-Congo and Austronesian contain over one-third of the total between them (1,495 and 1,246 languages, respectively). This massive disparity between language families suggests that there has been substantial variation in the rates at which languages diversify and go extinct. The large number of isolates suggests that uneven patterns of extinction have had a major role (Nichols 1997). However, diversification rates vary strikingly as well. For example, both Mayan and Malayo-Polynesian are estimated to be around 4,000 years old (Gray et al. 2009; Atkinson et al., in preparation) and yet there are 69 Mayan and 1,226 Malayo-Polynesian languages. Thus, if we assume no extinction, Mayan gave birth to approximately one language every 58 years, whereas Malayo-Polynesian spawned one language every 40 months or so. Patterns of language diversity also vary strikingly in space. For example, the island of New Guinea, despite covering less than 0.5% of Earth's land area, supports over 900 languages (13% of all languages). Comparatively, Russia is over 20 times the size of New Guinea, but only has 105 languages.

Characterizing language diversity is not straightforward (see also Evans, this volume). Following the literature on biodiversity (see MacLaurin and Sterelny 2008), we will distinguish between three types of language diversity: alpha diversity (the number of languages at a location), phylogenetic language diversity (the sum of the path lengths between a set of languages on a phylogenetic tree), and language disparity (the overall amount of variation between

1 This example is actually one of the triumphs of the comparative method. The brilliant reasoning involved was subsequently confirmed by the discovery of ancient Anatolian languages with two laryngeals.

languages). Note that alpha diversity is only the product of language-splitting events (cladogenesis), whereas phylogenetic diversity and disparity are produced by both change within lineages (anagenesis) and cladogenesis (see Figure 15.2). As Nettle (1999) pointed out, language families are not really ideal units for comparative quantitative analyses because the differing time depths of language families means they are not equivalent evolutionary units.

Our focus here is on ways in which phylogenetic methods can help us explore the causes of the drivers of alpha language diversity (the following section will focus on drivers of language disparity). First, biologists have noted that the *shape* of the tree alone provides clues to the diversification dynamics that gave rise to a phylogeny. If a set of languages are diversifying at fairly constant rate, then the tree will be *balanced*; that is, each node (protolanguage) at a given time depth on the tree will tend to have the same number of descendants in each of its daughter lineages. If, however, there are substantial differences in the rate at which some subgroup diverged, then the tree will be *unbalanced* so that one branch will have more descendants than the other (Figure 15.3). For language families that have undergone large expansions, we would expect them to be highly unbalanced.

There is a suite of tools for quantifying the shape of a tree to identify the signature of variation in diversification rates (e.g., Agapow and Purvis 2002; Fusco and Cronk 1995). To date, Holman (2010) conducted the only study to apply these tools to language trees. Holman calculated the imbalance, *Iw* (Fusco and Cronk 1995), from the trees of 19 large language families from the *Ethnologue* database (Lewis 2009), published language phylogenies, and found that almost all of the language trees were significantly more unbalanced

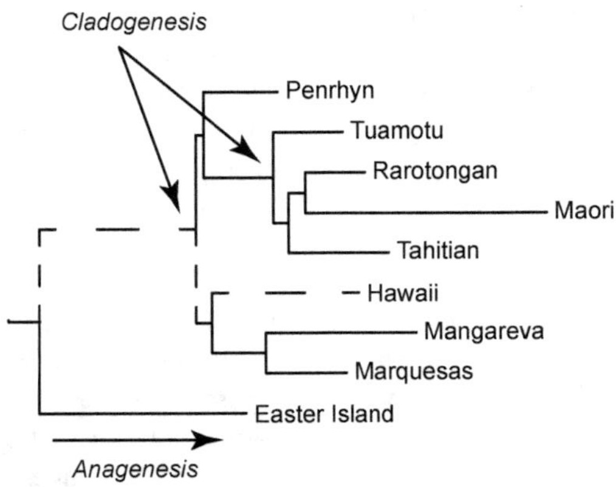

Figure 15.2 A phylogenetic tree for Polynesian languages showing cladogenesis (lineage splitting) and anagenesis (change in a lineage). In this tree the branch lengths are scaled to be proportional to the amount of change in a lineage. The dotted line shows a path from the ancestral language (root of the tree) to a tip (Hawaiian). The length of this path measures the amount of change from the root to the Hawaiian tip.

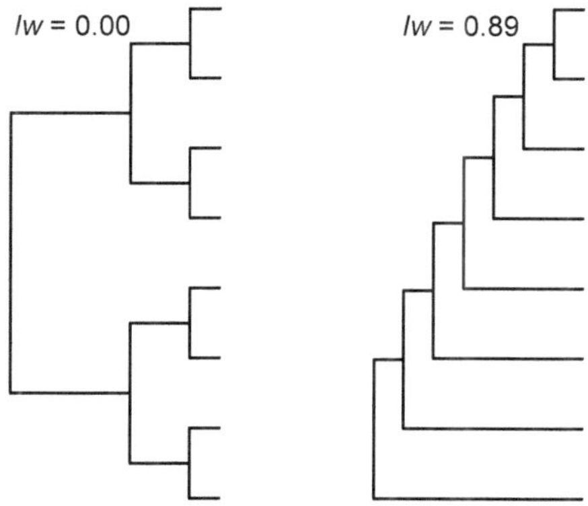

Figure 15.3 Depiction of (a) a perfectly balanced tree versus (b) an extremely unbalanced tree; *Iw* represents imbalance.

than expected by chance. These results indicate that there is substantial variation in diversification rates within families and support the notion that much of the world's linguistic diversity is a result of large-scale expansion events. This statistic, *Iw*, is robust, comparable across trees, and can accommodate unresolved subgroupings (polytomies) and incomplete phylogenies (Fusco and Cronk 1995). We have calculated the same statistic, *Iw*, on some of the major language family trees (see Table 15.1). The *Iw* score varies from 0 for balanced trees to 1 for completely unbalanced trees.

If the tree shows no evidence for differences in rates of language diversification, then the expected value of *Iw* will be 0.5. We can therefore test if the observed tree differs from 0.5 by using a null model of branching that assumes a simple Markov process, where all languages share the same birth rate (Fusco and Cronk 1995).[2] Table 15.1 shows that the most balanced families are Mayan and Austroasiatic. At the other extreme, Austronesian and Semitic are moderately imbalanced.

What factors could have caused this imbalance? One possible explanation is that imbalance is caused by the pruning of branches due to language extinction. Phylogenetic methods can help uncover periods of extinction using birth–death models (Nee 2006). If we were to plot the number of languages over time on a semilog plot, we would then recover a line with a slope proportional to the diversification rate. If the trees grew without any major extinctions (i.e., a pure birth model), we would expect this line to be straight. However, if there is extinction, this line is expected to show an uptick toward the present, as the most recently born languages have not yet had a chance to become extinct. The difference between the diversification rate slope and the rate on this uptick is the extinction rate. Using this logic, we test whether a given phylogeny is best explained by a pure birth model that assumes no extinction or a birth–death

2 See also http://R-Forge.R-project.org/projects/caper/

Table 15.1 Mean *Iw* scores for various language families. Languages are sorted from most balanced to least balanced.

Family	Languages	*Iw*	Source
Mayan	53	0.33	Atkinson et al., in prep.
Austroasiatic	54	0.39	Sidwell et al., in prep.
Pama-Nyungan	194	0.44	Bowern and Atkinson (2012)
Indo-European	103	0.45	Bouckaert et al. (2012)
Japonic	59	0.47	Lee and Hasegawa (2011)
Semitic	25	0.51	Kitchen et al. (2009)
Austronesian	400	0.59	Gray et al. (2009)

process that allows extinction. In both the Austronesian (Gray et al. 2009) and Mayan families, the pure birth model fits the tree significantly better than a birth–death model ($p < 0.001$), suggesting that extinction has played a relatively minor role.

If the observed differences in tree topology are not caused by extinction, then they must be caused by differences in the rates at which the languages diversify. In Gray et al. (2009), we developed a Bayesian method for modeling diversification as a change-point process along a phylogeny. We applied this method to the Austronesian language phylogeny (Figure 15.4) and identified four regions with significant evidence of increases in diversification rate (i.e., expansion pulses). Our results showed that significant pulses occurred prior to the proto-Malayo-Polynesian branch, before the breakup of the Philippines languages, before the diversification of the Micronesian languages, and the branch leading to the Micronesian and Central Pacific subgroups. We suggest that these pulses could be linked to technological advances, such as the development of the outrigger canoe enabling the Austronesian peoples to cross the channel into the Philippines, and the invention of the double-hulled canoe enabling the expansion into Eastern Polynesia (cf. Pawley and Pawley 1994).

However, although there was evidence to suggest that the pulses were linked to advances in canoe technology, we did not directly test this. A new set of methods, BiSSE and QuaSSE, can directly test the effect of a binary trait (e.g., presence or absence of double-hulled canoes) or a quantitative variable on the rates of diversification (Maddison et al. 2007; FitzJohn 2010). These new methods open up exciting possibilities for comparative analyses as they provide a powerful way of testing hypotheses about the causes of cultural evolution and diversification. Many such factors have been proposed, ranging from the simple acquisition of new technological items like canoes, to social factors such as the level of political complexity (Currie and Mace 2009). One of the most prominent suggestions links the advent of farming to the expansion of language families around the world (Diamond and Bellwood 2003).

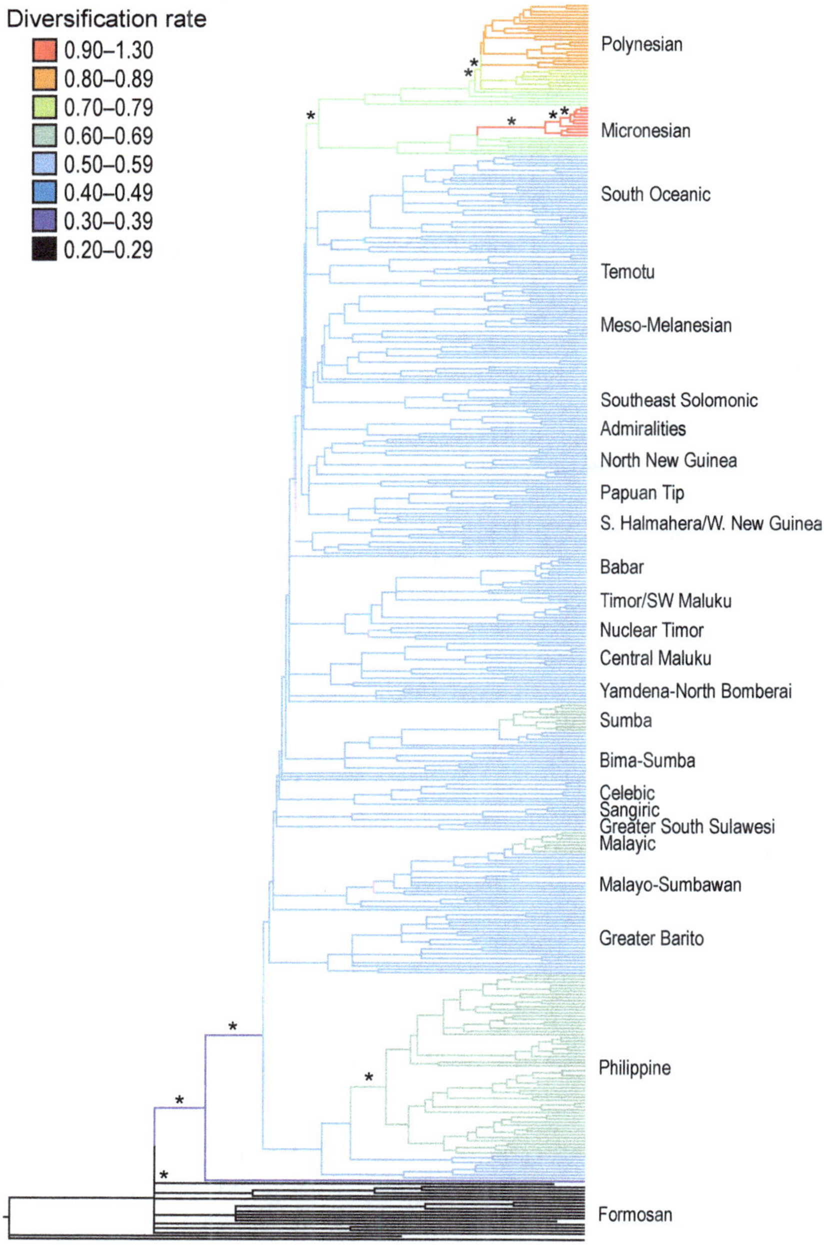

Figure 15.4 Austronesian language phylogeny showing changes in diversification rate due to expansion pulses. Branches with significant shifts in rate are marked with an asterisk. Reprinted with permission from the supplementary material in Gray et al. (2009).

This theory suggests that the invention of agriculture enabled the new farmers to obtain higher yields of food and reach much higher population densities. This advantage allowed farmers to outcompete existing hunter-gatherer populations and led to major population expansions out of agricultural homelands. Diamond and Bellwood (2003) claim that the signature of these farming-driven expansions is evident in the distribution of no less than 13 of the major language families: Afro-Asiatic, Austro-Asiatic, Austronesian, Bantu, Dravidian, Indo-European, Japanese, Nilo-Saharan, Sino-Tibetan, Tai, Trans New Guinea, and Turkic. The new BiSSE and QuaSSE methods provide the means to test these prominent and long-standing hypotheses about the factors that have shaped our modern-day language diversity.

In biological evolution, diversification rates are relatively constant over time after a burst of diversification. It has been suggested that the burst occurs as the species diversifies into new niches. After this initial burst these niches become filled and therefore constrain further diversification (Etienne et al. 2012). Evidence for this "density dependence" comes from many molecular studies showing a slowing down of diversification rates in many species. For example, a meta-analysis of bird families showed significant decreases in diversification in 23 out of 45 families with bigger decreases in larger families consistent with density-dependent constraints on diversification (Phillimore and Price 2008). To date, this idea has not been applied to cultural evolution. This omission is striking as there are strong hints that density dependence operates on cultural diversity. For example, a study of 264 islands in the Pacific found that 195 (74%) had only one language (Gavin and Sibanda 2012). This suggests that once a language or culture fills a niche, it heavily restricts the birth of new languages or cultures. Thus one possible explanation for the immense diversity of the Austronesian language family might be that the invention of better canoe technology combined with a shift to agriculture opened a range of new niches in the Pacific that facilitated the diversification of these cultures. In contrast, the substantially less diverse Mayan family had to compete for niches with hunter-gatherer groups and other agriculturalist populations belonging to the Mixe-Zoquean, Oto-Manguean, and Uto-Aztecan language families.

What Drives Linguistic Disparity (and What Constrains It)?

In his influential book, *Wonderful Life*, Stephen Jay Gould (1989) distinguished between diversity and disparity. He argued that the number of species was a poor measure of the overall amount of phenotypic variation. Diversity in overall body plan does not necessarily correlate well with the number of species in a clade. Whereas there might be millions of species of beetles, they are all still beetles. A similar distinction could be made in linguistics. With over 100 languages spoken across its islands, Vanuatu has one of the highest densities of languages in the world (Lewis 2009). However, all these languages belong to just two subgroups (North/Central Vanuatu and South Vanuatu) of the Oceanic group, which is itself a subgroup of Austronesian. How, then, should we measure language disparity?

Languages differ not only in their lexicon but also on numerous structural levels, including the organization of the sound system (phonology), systems for the combination of meaningful elements into words (morphology) and phrases (syntax), as well as systems for indicating spatial and temporal relationships, speaker attitude, and epistemological status (see Evans, this volume). It is not possible to combine these variables into a global measure of linguistic disparity, just as it is not possible to come up with a global measure of biological disparity (see MacLaurin and Sterelny 2008). We are thus skeptical whether it is possible to conceptualize the "absolute design space" for all possible languages. It is, however, possible to develop measures of disparity relative to particular traits and for specific questions, just as David Raup did in his famous diagram of possible and actual ammonoid shell morphologies (Raup 1967).

These local representations of morphospace have provided theoretical morphologists and evolutionary biologists with powerful tools for analyzing both the drivers and constraints on morphological evolution. Phylogenies can be used to trace phenotypic evolution through these spaces and infer factors that accelerate or constrain the evolution of disparity. A similar approach could be adopted in studies of linguistic and cultural evolution (see Hauser 2009; Levinson 2012b). Just as Kemp and Regier (2012) constructed a design space of possible kinship systems, linguists could construct phonological and typological spaces. For example, we could classify the world's languages based on primary word order (i.e., the order of the Subject, Object, or Verb in a sentence). There are six possible ways of structuring this information, however, not all combinations are as likely (Dryer 1992, 2011):

- 41% order the elements as SOV, while 35% use SVO.
- 13% use no dominant order.
- Less frequent are VSO (7%), VOS (2%), and OVS (0.8%).
- The least common is OSV, with only 0.2% of the world's languages choosing this ordering.

This difference in the frequency of word orders requires explanation. Whereas linguists often claim that these patterns reflect cognitive and functional constraints, the role of historical contingencies needs to be evaluated as well (see Levinson and Gray 2012).

Let us extend this idea of word-order space to many aspects of language typology. If we take the World Atlas of Language Structures (WALS) as an example, then there are 140 different traits that characterize language. Each of these traits has on average 4.6 states. If we trace all possible combinations of these traits, then there are 2.5×10^{89} possible ways of constructing a language. However, in this "WALS space" of possible languages, not all regions will be equally likely. Phylogenetic methods could be used to map the movement of language lineages through this space and thus evaluate the roles of cognition, function, and history in explaining the patterns of disparity that we see among the languages of the world today.

If the rates of both language cladogenesis and anagenesis are constant, then measures of language diversity and disparity will be congruent. Language diversification is unlikely to be constant (see above), and the rate of change in lineages is known to vary markedly (Blust 2000). Phylogenetic methods can be used both to estimate rates of change and to test hypotheses about the factors that influence them. Thus, rather than rate variation being a nuisance, it can become an object of study (just as it is in evolutionary biology). There are numerous hypotheses about the factors that might affect rates of linguistic change. Trudgill (2011), for example, lists five major factors: group size, density of social networks, amount of shared information, social stability, and levels of contact with other speech communities. There is, however, no consensus on which factors most influence rates of change, and little has been done to quantify the relative roles and interaction between these factors. Bayesian phylogenetic model comparison offers a way forward. Rather than fitting a model with a single rate, multiple rates can be estimated for different branches on the tree. Where there is a prior hypothesis about a factor that might affect the rate of linguistic change, the posterior probability of a single rate model can be compared with one that fits different rates for branches with different values of that parameter. For example, if the hypothesis suggested that hunter-gatherer languages had higher or lower rates of lexical replacement than agricultural ones, the hypothesis could be tested by constructing a language phylogeny from lexical data and comparing the posterior probability of a single rate model with one that allowed different rates for hunter-gatherer versus agricultural languages. Alternatively, where there are no prior hypotheses, the analysis could be done in an exploratory fashion using the local random clock approach proposed by Drummond and Suchard (2010), where a Bayes factor is estimated for the probability of a multiple local rates versus a single rate model.

How might social processes affect historical patterns of language change? An important insight from sociolinguistics is that language functions as a mechanism for marking social boundaries (Labov 1963). Human groups under pressure often exaggerate the language differences to make ethnic barriers—a process Bateson (1935) dubbed *schismogenesis* and Thurston (1987) labeled *esoterogeny*. The effect of this process is likely to be particularly marked when speech communities split. If speech communities exaggerate differences at the time when they are drifting apart, then lineages that have been through more splitting events will undergo more change (see Figure 15.5). Atkinson et al. (2008) used phylogenetic methods to quantify the impact of this effect. They used basic vocabulary data to construct phylogenies for the Austronesian, Bantu, and Indo-European language families. Their results revealed that between 10–33% of the vocabulary differences in these families arose during

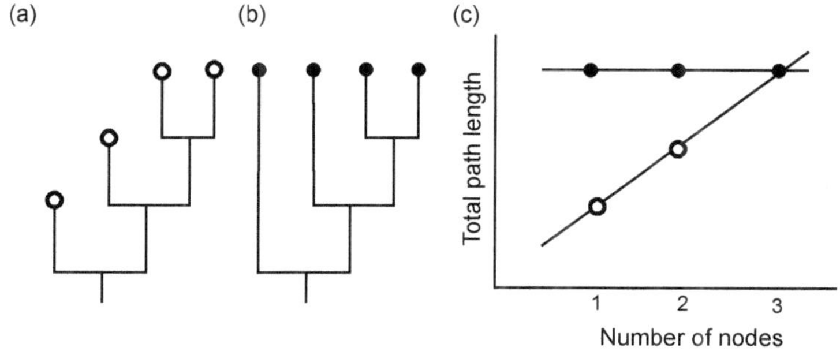

Figure 15.5 The phylogenetic tree in (a) shows the pattern produced by an increase in the rate of linguistic change at the splitting of speech communities; the branch lengths are longer in lineages that have been through more splitting events. (b) In contrast, if the rates of change are not affected by the number of splitting events, then all of the tips of all the branches will be equal irrespective of the number of splitting events the lineage has been through. (c) The size of any schismogenic/esoterogenic effect can be quantified by plotting the path length from the root of the tree to each of its tips against the number of nodes (splitting events) through which the path goes. The slope of the resulting graph estimates the magnitude of the effect.

rapid bursts of change associated with language-splitting events. One interesting extension of this approach would be to see if it holds equally for all aspects of language. To the extent that closely related speech communities differ more in accent than they do in vocabulary, and more in vocabulary than in language structure, it might be predicted that the schismogenesis effect would be most pronounced in phonetics and least in structural features of language.

Can We Infer Cultural and Linguistic Homelands? More Generally, How Do Language Expansions Unfold across a Landscape?

Questions about the origins of human groups and the languages they speak have an enduring fascination. The early European explorers in the Pacific speculated on the origins of the Polynesians after noticing that many words were shared across remote Oceania (Andrews 1836), and for over two hundred years scholars have debated the origins of the Indo-European languages (Jones 1786/2013). Diamond and Bellwood (2003) dub Indo-European the "most recalcitrant problem in historical linguistics." Linguists typically attempt to make inferences about possible homelands by using arguments based on either linguistic palaeontology or area-of-maximum-diversity. The diversity argument postulates that the most likely point of origin of a language family is the area of greatest diversity (Sapir 1916/1949). Linguistic paleontology arguments rely on reconstructions of words tied to specific locations, such as animal and plant names, to locate the homeland. Both arguments are far from infallible. How language family diversity is measured in linguistics is often subjective, and the apparent center of diversity can move as language families expand (Nichols 1997). Reconstruction of the form of ancestral words is a rigorous process based on inferences about sound change. However, the reconstruction of the semantics of these forms is much more speculative; for example, does the proto-Indo-European reconstruction for horse (PIE *h₁éḱwos) actually refer to domesticated horses, wild horses, or some more generic four-legged mammal (see Heggarty and Renfrew 2013)?

Linguists are hardly alone when it comes to rather loose inferences about geography. In biology, studies of phylogeography often consist of a rigorously derived phylogenetic tree and a geographical just-so story. The recent advent of stochastic models have, however, enabled more rigorous phylogeographic inferences (Lemey et al. 2009, 2010). These models have proved particularly adept at tracing the spread of human viruses such as the H1N1 outbreak (Lemey et al. 2009) and the yellow fever virus (Auguste et al. 2010). Virus evolution is perhaps a closer analog to language evolution than is vertebrate evolution (Gray et al. 2007). The obvious question that arises is: Could these phylogeographic methods be adapted to make inferences about linguistic geography?" Walker and Ribeiro (2011) used a relaxed random walk (RRW) model in the Bayesian phylogenetic program BEAST (Drummond and Rambaut 2007) to make inferences about the expansion of the Arawak language family. The RRW model is essentially a Brownian diffusion model in which the rate of diffusion can vary along branches of a tree. Rather than assuming a constant rate of diffusion, rate heterogeneity among branches is accommodated via a single additional rate distribution parameter, $P(r)$, allowing support for rate variation and the degree of rate variation (or "relaxation") to be estimated from the data itself. This approach treats language location as a continuous vector (longitude and latitude) which evolves through time along the branches of a tree. It seeks to infer ancestral locations at internal nodes on the tree, simultaneously accounting for uncertainty in the tree. Thus, the phylogeny and the geographic diffusion are co-estimated. Although there was considerable spread in the posterior distribution of ancestral root locations, Walker and Ribeiro found that the most likely origin of Arawak was in Western Amazonia, with subsequent expansion into the Caribbean and across the lowlands. Interestingly, although Northwest Amazonia has the largest number of Arawak languages, the phylogeographic models did not support the region as a potential homeland.

Could the same approach be used to shed light on the "recalcitrant problem" of the Indo-European homeland? We think so. As part of a large team of mathematical biologists and linguists we have recently assembled a large data set of cognate-coded basic vocabulary for 103 ancient and contemporary Indo-European languages (Bouckaert et al. 2012). To increase the realism of the spatial diffusion modeling, we extended the RRW process in two novel ways. First, to reduce potential bias associated with assigning point locations to sampled languages, we used geographic ranges of the languages to specify uncertainty in the location assignments. Second, to account for geographic heterogeneity we accommodated spatial prior distributions on the root and internal node locations. By assigning zero probability to node locations over water, we incorporated prior information about the shape of the Eurasian landmass into the analysis. Although we do not allow for different rates of movement across specified land types, this approach could, in principle, be extended to incorporate other geographic features such as mountains, rivers, or deserts.

Although there are numerous hypotheses about the origins of the Indo-Europeans, most of the current debate revolves around two theories. The "Steppe hypothesis" proposes an Indo-European origin in the Pontic steppe region north of the Caspian Sea, perhaps linked to an expansion into Europe and the Near East by "Kurgan" seminomadic pastoralists, beginning 5–6 KYA. Evidence from "linguistic palaeontology" and putative early borrowings between Indo-European and the Uralic language family of northern Eurasia (Koivulehto 2001) are argued to support a steppe homeland (Anthony 2007). However, the reliability of inferences derived from linguistic palaeontology and claimed borrowings remain controversial (Heggarty and Renfrew 2013). The "Anatolian hypothesis" holds that Indo-European languages spread out of Anatolia (in present-day Turkey) with the expansion of agriculture, beginning 8–9.5 KYA. Our results unambiguously support an Anatolian origin (see Figure 15.6). To quantify the strength of support for an Anatolian origin, we calculated the Bayes factors comparing the posterior to prior odds ratio of a root location within the hypothesized Anatolian homeland (yellow polygon, Figure 15.6) with two versions of the Steppe hypothesis (blue polygons). The Anatolian homeland was over 150 times more likely in both these analyses. Note that the relaxed diffusion model supports substantial variation in rates of diffusion through time and fits the data significantly better than a model which assumes a constant rate of diffusion, even accounting for the extra rate variation parameter. Nevertheless, there is enough regularity in the inferred rates to allow substantial support to emerge for one hypothesis over another. Additionally, it is not simply the case that these methods return the geographic midpoint of the language distributions. The geographic centroid of the languages we analyzed falls within the broader Steppe hypothesis (green star, Figure 15.6); this indicates that our model is not simply returning the center of mass of the sampled locations, as would be predicted under a simple diffusion process that ignores phylogenetic information and geographic barriers.

The RRW approach avoids internal node assignments over water but assumes the same underlying migration rate across water as land. To investigate the robustness of our results to heterogeneity in rates of spatial diffusion, we developed a second inference procedure that allows migration rates to vary over land and water. We examined the effect of varying relative rate parameters

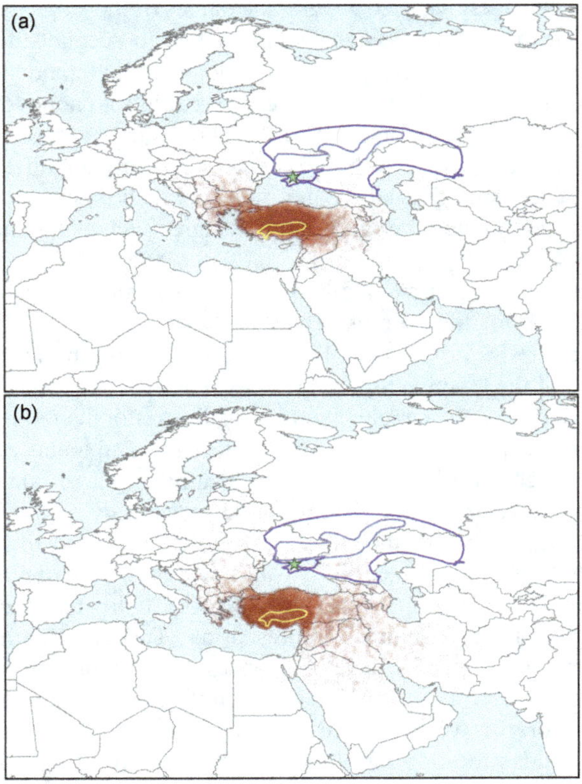

Figure 15.6 (a) Map showing the estimated posterior distribution for the location of the root of the Indo-European language tree. Each point sampled in the posterior is plotted in translucent red such that darker areas correspond to increased probability mass. (b) The same distribution under a landscape-based analysis in which movement into water is 100 times less likely than movement into land. The blue polygons delineate the proposed origin area under the Steppe hypothesis: dark blue shows the initial suggested homeland whereas light blue shows a later version of the Steppe hypothesis. The yellow polygon delineates the proposed origin under the Anatolian hypothesis. A green star in the steppe region shows the location of the centroid of the sampled languages. Reprinted with permission from Bouckaert et al. (2012).

to represent a range of different migration patterns. Figure 15.6b shows the inferred Indo-European homeland under a model in which migration from land into water is 100 times less likely than from land to land. Once again the Anatolian origin is overwhelmingly more likely.

Thus, phylogeographic modeling not only enables us to make probabilistic inferences about ancestral homelands, it also enables us to investigate the robustness of these inferences to a range of assumptions about the spread of languages. Figure 15.7 shows how these phylogeographic models can even be used to plot the spread of an entire language family in space and in time. This

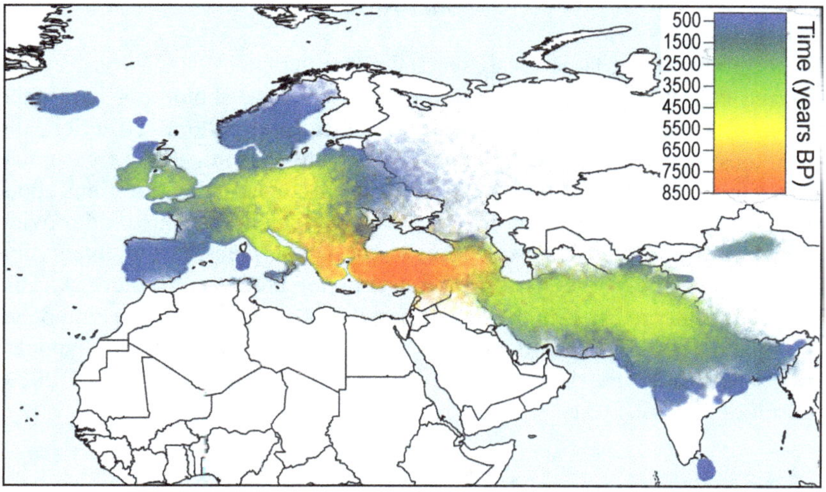

Figure 15.7 Spatial and temporal reconstruction of the expansion of Indo-European languages. The posterior distribution of node location estimates through time is plotted as opaque points with a color that indicates their corresponding age estimate. Older nodes are shown on the foreground to depict clearly the temporal diffusion pattern. Reprinted with permission from the supplementary material of Bouckaert et al. (2012).

figure needs to be interpreted with the caveat that we can only represent nodes corresponding to divergence events between languages that are in our sample. Nodes that are associated with branches not represented in our sample will not be reflected in this figure. For example, the lack of Continental Celtic variants in our sample means we miss the Celtic incursion into Iberia, and instead infer a late arrival into the Iberian Peninsula associated with the Romance languages. The chronology represented here, therefore, offers a minimum age for expansion into an area. Expanding and enhancing these methods to accommodate other aspects of geographic heterogeneity and other language expansions will allow us to test increasingly detailed hypotheses about human prehistory and the processes that drive language diversity and disparity in space and time.

It may even be possible to infer population migration events on a global scale. Atkinson (2011) highlights a global trend of decreasing phoneme diversity with distance from Africa, which is consistent with a serial founder effect in phoneme diversity following the human expansion from Africa. The observed relationship fits with theoretical models of cultural and linguistic transmission (De Boer 2001; Henrich 2004b) and holds after controlling for modern population size, density, and language relatedness. While the finding is, of course, only correlational and remains controversial (e.g., Wang et al. 2012), there are clear geographic trends in language variation across the globe that require explanation.

Conclusion

The combination of large databases and computational methods has revolutionized inferences in evolutionary biology. While we should not ignore the numerous subtle differences between biological and cultural evolution, the three questions we have framed in phylogenetic terms show that there is much to be gained from the nuanced application of this approach to questions about the evolution of languages across the globe. Such an approach would provide a powerful way of resolving questions about human prehistory by integrating genetic, linguistic, and cultural data in a common analytical framework. This ambitious undertaking is not without obstacles, such as the rigorous inference of cognate vocabulary and the detection of borrowing, but already computational approaches are rising to these challenges (Bouchard-Côté et al. 2013; Nelson-Sathi et al. 2011).

Acknowledgments

Some of the ideas in this manuscript were developed at the NESCent-funded workshop on "Modeling the diversification of languages." We thank Claire Bowern, Michael Dunn, and Michael Gavin for useful discussion and the editors and an anonymous referee for comments.

First column (top to bottom): Dan Dediu, Morten Christiansen, Simon Garrod, Stephen Levinson, William Croft, Michael Cysouw, Morten Christiansen
Second column: Stephen Levinson, Andrea Baronchelli, Anne Kandler, Andrea Baronchelli, Group discussion, Elena Lieven, Simon Garrod
Third column: Elena Lieven, Nick Evans and Michael Cysouw, Russell Gray, William Croft, Dan Dediu, Nick Evans, Anne Kandler

16

Cultural Evolution of Language

Dan Dediu, Michael Cysouw, Stephen C. Levinson,
Andrea Baronchelli, Morten H. Christiansen,
William Croft, Nicholas Evans, Simon Garrod,
Russell D. Gray, Anne Kandler, and Elena Lieven

Abstract

This chapter argues that an evolutionary cultural approach to language not only has already proven fruitful, but it probably holds the key to understand many puzzling aspects of language, its change and origins. The chapter begins by highlighting several still common misconceptions about language that might seem to call into question a cultural evolutionary approach. It explores the antiquity of language and sketches a general evolutionary approach discussing the aspects of function, fitness, replication, and selection, as well the relevant units of linguistic evolution. In this context, the chapter looks at some fundamental aspects of linguistic diversity such as the nature of the design space, the mechanisms generating it, and the shape and fabric of language. Given that biology is another evolutionary system, its complex coevolution with language needs to be understood in order to have a proper theory of language. Throughout the chapter, various challenges are identified and discussed, sketching promising directions for future research. The chapter ends by listing the necessary data, methods, and theoretical developments required for a grounded evolutionary approach to language.

Language from an Evolutionary Perspective

Language plays a central role in human cultural life, with thousands of languages being spoken, showing extensive and (from the point of view of other animal communication systems) unexpected variation throughout the world (Fitch 2011). From an evolutionary perspective, a central question posed by this variety of languages is how this diversity arose. Likewise, it is of central importance to elucidate the processes that shaped this variation, and to discover whether these processes differed in the past from what they are today (see, e.g., Baronchelli et al. 2012). Understanding these processes is not only relevant for understanding the history and evolution of languages, it can also function as a unifying force to draw together the widely dispersed and largely unconnected subfields of contemporary linguistics.

Taking an evolutionary perspective on language raises many questions: Does embedding language evolution within a general theory of cultural evolution produce elegant and effective explanations of linguistic phenomena? Are there rich and detailed laws of cultural evolution that apply universally, and thus also to language, and if so what are they? For example, are there general predictions about population size and rates of change or degrees of complexity? Are there predictions about the minimal size of a population still capable of carrying a quantum of cultural information across time? Or are there implications for competition between human groups and rates of change? At this stage we are not in a position to answer such questions, which constitute part of the ongoing challenge in this domain. However, we would like to sketch the main challenges for an evolutionary perspective to language.

This chapter summarizes a wide range of issues for an evolutionary perspective on language. We begin by sketching a few positions—widespread in linguistics—that are at odds with an evolutionary perspective (see section, The Misconception of Language Particularism). These positions all argue for language being different from other kinds of human culture. In contrast, we will argue that these differences are all just a matter of degree and thus not a barrier to evolutionary thinking. Next, to set the scene, we briefly sketch the current (rather limited) state-of-the-art evidence about the biological origin of language (see section, The Antiquity of Language). Thereafter we turn to the central questions for an evolutionary perspective (see section, Function, Fitness, Selection): What are the functions of language that are under evolutionary pressure? How can we operationalize a suitable notion of fitness? What are the entities of replication, and what are the processes leading to differential replication?

Focusing on more empirical questions, we investigate the current diversity among the world's languages and raise central questions about how to proceed with worldwide linguistic comparisons. Why are there about 7,000 languages in this world, and can we say more about their phylogenetic relationship? Can we develop methods to compare traits of languages, given the wide variety of linguistic structures attested? In addition, given a design space of such traits, what are the preferred routes of change through this space?

On a slightly more abstract level, we proceed to a discussion of the shape (the extent to which human language is tree-like) and fabric (the extent to which traits of language develop as bundles) of human language evolution. In this context we also briefly discuss the influence from language ontogeny on language phylogeny (for further discussion of developmental issues in cultural evolution, see Appendix 1).

A central tenet of the evolutionary perspective is that language is not isolated from biological and social evolution. Therefore, we discuss the ongoing processes of biological, social, and linguistic coevolution. Finally, we highlight some of the most pressing further developments that are necessary to bring the field forward. Throughout the chapter, we identify challenges for future research that succinctly summarize some of the main issues that came up during our myriad discussions.

The Misconception of Language Particularism

An alternative nonevolutionary position, still widespread in linguistics, holds that language is a domain that has such special properties that there is no reasonable expectation that general laws of cultural evolution would apply in a uniform way across language and other contrastive domains like religion, technology, or group organization (Pinker 1994). Below we list a number of apparent reasons for thinking that language is apart and special, together with the reasons why these can be discounted. Note that it is not so much the following assumptions themselves which are misconceptions, but it is a misconception that they differentiate language from other aspects of human culture. Also note that the following misconceptions, when accepted, suggest that an evolutionary approach to language is useless. Through this lens, language would be a uniform phenomenon through time and space, and the only way it could have arisen is through a catastrophic change (e.g., Chomsky 2010).

Misconception 1: Language Is Biologically Fixed

Although we agree that language has a deep biological evolutionary background (see sections, The Antiquity of Language and Coevolution of Biology, Culture, and Language), reflected superficially in the specific physiology of the vocal apparatus and potentially in certain brain specializations, we do not think that this deep biological background is special for language. Human technology can equally be seen as reflecting the anatomy of the hand (Shennan, this volume; Stout, this volume). Likewise, group organization and religion are guided by innate behavioral dispositions (Haun and Over, this volume; Whitehouse, this volume).

Misconception 2: Language Changes Constantly

There is a recurrent assumption in linguistics that language changes incessantly, without any direction or advancement (Battye and Roberts 1995; Lightfoot 1999). Indeed, rates of change in language are relatively uniform and constant compared to, say, technology, where technological change shows rapidly accelerating rates of advance. Again, this apparent difference can be eroded, with the caveat that we have no precise way to equate rates of change across domains. Some aspects of language, specifically phonetics/phonology and the lexicon, can change at very fast rates, and it seems better to think of language as structured networks of relations where deep hubs may be resistant to change, while surface nodes can react swiftly to surrounding forces (e.g., Dediu 2011a).

Misconception 3: Every Human Has Full Command of Language

Language exhibits population saturation: every normal person acquires languane over the initial childhood years such that they appear to command language in a way that is not true for technological or religious expertise, for example. Here we are caught in two fictions, widespread in linguistics and psychology: (a) children have full command of their language by three (or seven, or eleven; Crain and Pietroski 2002), and (b) all adults have comparable language competence.[1] In fact, language acquisition continues through the active adult years, as new rhetorical skills and new social arenas are mastered, such as the acquisition of the "triangular" kin-term register in Bininj Gun-wok, which continues into a person's twenties (Evans 2003a). In addition, there are substantial individual differences in almost all aspects of language (e.g., Street and Dąbrowska 2010; Farmer et al. 2012).

Misconception 4: All Languages Are Equally Complex

This misconception arose with good reason at the start of the twentieth century to counter Eurocentric notions of language structure, but it is currently widely accepted that there are differences in structural complexity between languages (e.g., Sampson et al. 2009; McWhorter 2011). However, when there are differences in the complexity between languages, language complexity is often assumed to show reverse patterning to, for example, technological complexity in response to demographic variables: high degrees of morphological complexity in language are mainly found in small-scale societies (Lupyan and Dale 2010), whereas high degrees of technological complexity are primarily found in large-scale societies (Trudgill 2011; Shennan, this volume). Still, once again, careful consideration shows the mismatch to be ill conceived, because languages spoken in technologically complex societies have substantially larger lexica, owing to such factors as occupational specialization (allowing different individuals to have differently elaborated specialized vocabularies) and the elaboration of vocabulary needed to describe the attendant technological complexities. Pawley (2006), who attempts to establish ballpark figures for the size of the lexicon of unwritten languages, finds a range from 5,200

1 As an example, see Pinker (1994:18): "Language is a complex, specialized skill, which develops in the child spontaneously, without conscious effort of formal instruction, is deployed without awareness of its underlying logic, is qualitatively the same in every individual, and is distinct from more general abilities to process information or behave intelligently"; see also Nowak et al. (2001).

(Nunggubuyu) to 31,000 (Wayan Fijian) for 16 languages of Australia, New Guinea, and the Pacific. These are one to two orders of magnitude smaller than the English figures of 460,000 as represented by Webster's Third New International Dictionary or, for the 1933 Oxford English Dictionary, 252,259 entries and 414,825 defined words (headwords and subentries).[2] Once we take into account all the linguistic levels, and use more sophisticated measures of complexity (Bane 2008), it is unlikely that the patterns in language evolution are the inverse to technology evolution.

Misconception 5: Biological Evolution Is Independent of Language Change

When thinking about language evolution, there is a recurrent assumption that there is a clear-cut difference between emergence of language and subsequent language change (e.g., Pinker 1994; Berwick et al. 2013). Confusingly, these two phases are both referred to as "evolution" in the literature. However, this perspective assumes that there are different processes at work in the phase when modern humans separated from other primates compared to more recent processes leading to language change over the last few thousand years. Now, given the enormous difference in the amount of time that these phases encompass (in the order of millions of years in the first case, compared to an order of thousands of years in the second phase), it is not surprising that there are differences between them. However, as a general approach within the evolutionary perspective, it seems much more profitable to assume a general continuous development throughout the whole history of language with differences in degree, but not in quality (while remaining well aware that "more is different" in complex systems; Anderson 1972). An important correlate of this assumption of continuity is that biological evolution and cultural evolution are both ongoing, and there is interaction between the two (see Levinson and Dediu, this volume). It is not the case that biology is an invariant on the basis of which cultural variance can develop. Both biology and culture are sources of variation and constancies, and change is ongoing in both. Biology might be more "stable" than culture in the sense that changes normally proceed more slowly in biology than in culture, but this is likewise a difference in degree, not in quality.

---

2 Many factors complicate the interpretation of these figures, so they need to be taken with caution. Such factors include (a) likely less substantial documentation with some of the languages in Pawley's sample, as compared to English, (b) the effects of morphological type—each Nunggubuyu verb can have hundreds of inflected forms, compared to just four for most English verbs (e.g., *kiss, kisses, kissed, kissing*), (c) the question of whether we are comparing the vocabularies of individual speakers (which may display much less variation) or the summed vocabularies across a whole society (which could differ much more once there is occupational specialization), and (d) the fact that dictionaries of written languages will preserve many words no longer in active use.

The Antiquity of Language

Human languages present a special opportunity for the study of cultural evolution. There are many of them—around 7,000—allowing (in principle) the testing of hypotheses over large populations of sample points, if we compare this to the slim set of data points in lots of archaeological settings, for example. Although our coverage is still nowhere near enough—with something like reasonably sophisticated coverage for only around 20%, and languages becoming extinct at an accelerating rate—we have an increasingly convergent idea about what needs to be done to get good coverage (although the bar for what gets counted as good documentation is constantly being raised). From a data-collection point of view, the complexity of languages—unlike, say, technologies—is of a comparable order of magnitude for all the world's cultures (although the distribution of complexity may vary with community size; Lupyan and Dale 2010).

We know incredibly little about the emergence and early evolution of language in our lineage. Modern humans seem to be at least 200,000 years old (Klein 2009). The split between modern humans and the Neandertal/Denisovan lineage can be dated around 500 KYA (Hublin 2009; Green et al. 2010), and the even earlier *Homo erectus* dates to about 1.8 MYA. The evolution in *FOXP2* (Krause et al. 2007; Green et al. 2010) and breathing control (MacLarnon and Hewitt 1999) may be positioned in the transition between 1.8 and 0.5 MYA. The premodern human audiogram from about 600 KYA appears to be similar to ones from modern humans (Martínez et al. 2004, 2008b). It seems that the Neandertals adopted modern human technology (Floss 2003), and that several modern human cultures left archaeological records strikingly similar (or even simpler) to those of the Neandertals (Roebroeks and Verpoorte 2009). So, there is circumstantial evidence that modern-like language could have been around half a million years ago, and that the Neandertal/Denisovan lineage would have been using some form of language (Dediu and Levinson 2013). Further, when we strictly distinguish between language and communication, it seems to be quite possible that successful communication was available millions of years ago.

Linguists generally take the time barrier for reconstructable clades (language families) to be around 8–10 KYA (e.g., Renfrew et al. 2000). This limitation leaves us with lots of unconnected language families (a couple of hundred) and no deeper phylogeny. It also makes it difficult to harness data from the number of maximal clades and reason back from this to the antiquity of language as a whole. However, there are parts of the world which, when plugged into the overall picture, suggest considerable antiquity. In the Australian case, we have a single continent, inhabited for at least 40,000 years without any major discontinuities suggesting more recent immigration (see, however, the recent finding of an Indian genetic connection about 4,000 years ago; Pugach et al.

2013), and for which all the languages appear to be related at a deep level (Evans 2005). Do we have a 40,000-year-old language family in this case? It may, of course, be that the oldest common ancestor to all modern Australian languages is more recent than this as a result of lineage death for higher-order branches. Nonetheless, it is not implausible to see Australian languages as descended from whatever language was spoken by the first humans to arrive in Australia. Australian languages represent, however, only a tiny fraction of the world's total phylogenetic language diversity. This seems more consistent with an earlier (250,000 years or older) than a later (80–100 thousand years) date for the origins of human language.

There are various proposals in the literature for higher-level groupings such as "Nostratic" or "Amerind," but these are currently not well supported by the available evidence (Renfrew and Nettle 1999). A way forward could be represented by methods which combine multiple sources of linguistic evidence so as to estimate abstract properties of language change, which may be able to break the 10,000-year barrier (Dediu and Levinson 2012).

Challenge 1: To What Extent Do Communication, Language, and Speech Have Separate Evolutionary Histories?

In normal language production, the face, the hands, and the nonverbal paralanguage are all coordinated to produce an overall message. In human development, however, these are dissociated, with early nonverbal communication (smiling, vocalizing) preceding pointing, which itself precedes the first words. It seems not unlikely that during the phylogenetic development of human communication there was a similar dissociation, with general communicative abilities preceding language proper. Nevertheless, in modern human language production, hand and mouth seem coupled, with the emphasis reversible (as in sign languages), suggesting that change may have occurred more in the general weighting of the modalities than in their incremental addition. There are suggestions that speech was essentially modern at least 0.5 MYA (Dediu and Levinson 2013).

Function, Fitness, Selection

Here we investigate some of the central questions for an evolutionary perspective. This involves the functions of language that are under evolutionary pressure, the notion of fitness, and the entities and processes involved in selection.

Function

From an evolutionary perspective, it is necessary to clarify the function of an entity that is under evolutionary pressure. Language is generally considered to be a device for exchanging information, or for the expression of meaning. However, the concepts of "information" and "meaning" are both ill defined. Information is such a broad concept that almost everything can be considered as information. Meaning is a quality that we obviously share with all of humankind, but it is intrinsically impossible to draw out of our minds other than by using language itself.

The different functions of language have been much debated (for a classic account, see Jakobson 1960). Here we emphasize two functions of special relevance to the discussion of language evolution: (a) the role of language to coordinate joint action (including acts of transmission of knowledge) and (b) the role of indicating social relationships. The coordinative role places language use firmly in the class of coordinative social phenomena, with consequences for its temporal stability and resistance to unilateral innovation (e.g., Tomasello 2008). The role of expressing and manipulating social relationships is another possible force for language differentiation, especially as ethnic markers (Boyd and Richerson 1987a) or with a "shibboleth" function (Cohen 2012) as costly signals of in-group membership. People are masters of picking up such linguistic cues rapidly and consistently, allowing language to affect social structuring.

Being a form of coordinative technology, language has a "parity problem"; that is, it works only if we agree on the joint code. However, this problem is not special to language, but is shared by all systems of cultural evolution that have a significant "coordinative" element, which are essentially games of pure coordination. All social rules and norms are of this kind: consider driving on the left- versus right-hand side of the road, or wearing your shield on your left hand in phalanx formation, or singing in unison in a religious ritual. The cultural evolution of language thus falls under all the general laws that apply to coordinative social domains (Chater and Christiansen 2010). Such domains do not include, for example, competitive arenas, where doing precisely what the others do not do may confer significant advantage (e.g., betting against the market, playing tennis or bowling with the left hand, inventing a new kind of fishing fly). A question for future research is whether there are common properties of coordination systems that hold across language, group formation, and religion, but also whether there might be properties unique to language as a cultural system for coordination, due perhaps to the complex multileveled structure of the system.

A central question that emerges from these two main functions (coordinative transmission of knowledge vs. marking social relations) is whether language was evolutionarily designed for one of these two functions, and whether the other was originally an exaptation with only secondary selective pressure. Such questions might sound highly speculative, but they may be profitably transformed into testable hypotheses. Given the approach taken by Cohen (2012), which examines the role of accent as a group marker, all features that mark group membership might be expected to be less strongly expressed in small communities. Another hypothesis is that groups under demographic pressure should show more schismogenetic effects (i.e., traits that are marking differences should proliferate). For example, variation in phonetic realization and intonation ("accent") seems to be a highly prominent social signal of language, as opposed to grammatical structure which stays much more "under the radar"' as far as conscious manipulation by speakers is concerned. This leads to the hypothesis that groups under demographic pressure should show high rates of phonetic change as opposed to grammatical change.

Fitness

Any mention of language function immediately raises the question of what the metric of fitness should be for an evolutionary model of language. Here there are no simple answers, since there are at least three different levels of language fitness to be distinguished: user fitness, group (and whole language) fitness, and item fitness. The relation between these different kinds of fitness (i.e., whether some of them are more fundamental than the others) is an important question for further debate (see next section).

First, from the perspective of a language user, language skills—like eloquence in speaking or writing—may translate directly into biological fitness (but this is an area ripe for further investigation). Rhetoric and persuasion play an important role in leadership qualities, and leadership may in specific social circumstances correlate with potential for increased offspring. A speaker that is able to master more languages or "accents" (something that is difficult for most people to achieve) will be able to reach more people. In addition, the cooperative exchange of information in undertakings important for survival, and in linguistic exogamy, may likewise confer biological fitness advantages, making traits like multilingualism and the ability to express oneself concisely in a biologically relevant way.

Second, from the perspective of group fitness, particular linguistic design choices may impact on issues like the scale of the unified speech community, translating into scale of polity. A classical example is the way written Chinese, by transcending phonetic particularity, has allowed communication across the world's largest polity and its cultural outliers, as well as across time through ancient texts over millennia, contributing to the extraordinary cultural continuity of Chinese civilization. This works because the same character can express the same meaning despite different sounds, both in different Sinitic languages like Mandarin or Cantonese as well as in other unrelated languages of the Sinosphere like Korean or Japanese.[3] As Ostler (2005:157) states: "No

[3] For example, the character 山 for mountain, is pronounced *shān* in Mandarin, *saan1* in Cantonese, *san* in Korean and *san* or *yama* in Japanese, or 中 for middle, is pronounced *zhōng* or *zhòng* in Mandarin, *zung1* or *zung3* in Cantonese, *chung* in Korean and *chuu* or *naka* in Japanese.

alphabetic script, based perforce on the sounds of a language, could now be so conveniently neutral in terms of all the different Chinese dialects."

Phonetic writing systems, on the other hand, need to track the inexorable effects of sound change, so that the written forms of the various Romance languages, for example, despite having a common ancestor around the same time as the Sinitic languages, are not mutually comprehensible across speakers of different Romance languages in the same way as in Chinese. In addition, languages get associated with ethnic identities, and where the viability and desirability of such group identification is lost, language shift typically takes place.

Third, when we take the perspective of the linguistic item, its chances in differential reproduction may depend on its value by association with a prestige group (as in the adoption of accent or fashionable phrases) and the degree to which it is unencumbered by systemic constraints (thus a noun is easier to borrow than a verb, a content word easier than a function word or affix). This perspective is elaborated in more detail below.

Replication and Selection in Language Evolution

A cultural evolutionary approach to language change recognizes that there are two interconnected processes: (a) the generation of variation via descent with modification and (b) selection operating on that variation. Indeed, this is a basic prerequisite to describing cultural change as evolutionary (Darwin 1859). In the case of language change, cultural evolutionary processes give rise to lineages of languages and of linguistic structures (such as sounds, words, and constructions), and methods used in biology to reconstruct lineages can be applied to language change (e.g., Bouckaert et al. 2012). Several proposals have been made for a more specific model of causal relations between variation and selection, and the entities that are involved in biological and cultural evolutionary processes.

One influential model that has been applied to cultural evolution is Dawkins's "selfish gene" (Dawkins 1976, 1982). Dawkins generalizes the role of a gene to a replicator, focusing on the copying process involved in descent with modification. He also proposes that cultural replicators (the so-called "memes") exist and that they evolve in the same way that genes do. Dawkins's theory of replicators is part of a general theory of "memetics" (Blackmore 1999), which proposes that only genes are replicators in biological evolution, that they are the units of selection, that organisms are mere "vehicles" for genes/memes, and that cultural meme replication and selection is analog to biological parasitism. This more general theory, in our opinion, has generally not been successful in providing novel insights into cultural evolution (for an application of memetics to language change, see Ritt 2004).

Beginning with Dawkins's concept of replicator and the lineages that replicators form, Hull (1988, 2001) constructs quite a different general model of biological and cultural evolution. Hull treats the replicator as a role that can be filled by any entity that fits his definition of replication and introduces another role (the interactor), also potentially filled by many entities. Unlike Dawkins's vehicle, Hull's interactor plays a significant role in selection. Croft (2000) uses Hull's model as the starting point for a cultural evolutionary model of language change in which speakers and the utterances they produce each play significant roles in the linguistic evolutionary process. Baxter et al. (2006) developed an agent-based model based on Croft's theory that has been applied to issues regarding propagation (selection) of linguistic variants in sociolinguistics (Baxter et al. 2009; Blythe and Croft 2012). A number of important issues remain in treating a model of cultural evolution such as Croft's as an instance of a general evolutionary model.

Replicators are extremely difficult to identify in biological evolution. There is no simple relationship between a "gene" and the DNA molecules that actually undergo the physical copying process in biological evolution. A single "gene" may be distributed across multiple discontinuous sites in the genome; the sites for two "genes" may overlap or even be identical (the different "genes" being read differently); "genes" may interact with each other in complex ways such that they form a network functioning as a unit; and so on. Thus, the status of a "gene" as a unit independent of the DNA sequences that contribute to it is unclear. "Genes" are often individuated in terms of their relationship to the phenotype, namely as an instrumental (Griffiths and Stotz 2006) concept in modeling the transmission of a heritable phenotype. However, the past few decades have demonstrated that this relationship between genotype and phenotype is very indirect and complex. The phenotype is strongly influenced by developmental processes, and the interaction of the developing organism with its environment is essential. These new insights in understanding biological evolution are leading to dramatic changes in evolutionary theory, with various proposals of extension such as the so-called ecological evolutionary developmental biology, or "eco-evo-devo" (e.g., Jablonka and Lamb 2005; Gilbert and Epel 2008; Jablonka and Raz 2009; West-Eberhard 2003; Koonin 2012).

It is unclear what the consequences of these developments will be for replicator-based models of cultural evolution. Intriguingly, there are analogs in language to these more complex phenomena in biological evolution. For example, linguistic units which form lineages include other lineage-forming units (e.g., constructions include words, and words include sounds); constructions may form discontinuous parts of utterances; linguistic units interact with one another in such a way that they form a unit. Linguistic units are also individuated instrumentally, in terms of their function in the linguistic system and in communication, and there are many complex issues as to how the linguistic system is constituted and how linguistic communication takes place. These phenomena suggest that linguistic evolution and biological evolution do share basic features.

In addition to these issues in biological and cultural (including linguistic) evolution, a second problem in cultural evolution is what sort of cultural entity replicates. Various scholars have proposed that concepts, cultural behaviors, or artifacts may function as replicators. It remains to be seen whether any, all, or some combination of these entities are reasonable candidates for cultural replicators.

Despite these theoretical problems, research that applies phylogeny reconstruction techniques from biology to language change have been productive, as have evolutionary agent-based models of language change to issues in language origins and the propagation of linguistic variants in a speech community (for a discussion, see Hruschka et al. 2009). How much further such methods and models from evolutionary biology can be applied to language change, and to what extent such applications will help us in understanding cultural evolution as a general evolutionary process, is a major question for the future.

Challenge 2: Can We Identify the Signature of Mechanisms for Generation of Variation and Selection in Language Evolution?

It is generally agreed that evolution involves two processes: the generation of variation and selection operating over that variation. A wide variety of mechanisms of language change have been proposed by linguists for these two processes. One set of factors are cognitive, including the phonetic (articulatory and auditory) motivation of sound change, analogy, frequency-driven factors (e.g., the shortening of linguistic forms: "cellphone" to "cell" or "going to" to "gonna"), meaning, pragmatics, discourse interaction, and relations between units in the linguistic system (see Keller 1994; Croft 2000). A second set of factors are social, including network structure, the structure of adopter groups, and social valuation of linguistic variants. Given the wide variety of factors that have been proposed for both the generation of variation and selection, how can one distinguish them in terms of effects on language evolution? Is it possible that different types of factors leave an identifiable signature in language change?

Diversity

Cladogenesis: Why Are There 7,000 Languages?

Why is there not just one single world language, or alternatively, why are there not 2.8 million different languages (the figure we would get if we extrapolated the ratio of languages to speakers in Vanuatu to the rest of the world)? Why is there a high-language density in some regions of the world, whereas in other areas only a very few languages are spoken?

To address such questions, we need a better understanding of cladogenic processes (i.e., processes that lead to the split up of languages into various daughter languages). Many factors appear to be important for the process of cladogenesis in linguistics: First, the density of the population itself might influence the process, in the sense that the availability of empty niches allows for the rise of separate groups. Second, the general tendency of humans to divide humanity into in-group–out-group oppositions is a force to develop different languages. The widespread practice of exogamy is a special case of this, because for exogamy to be feasible, group opposition is necessary, and linguistic diversity seems to be a primary way to enhance such an opposition (Sorensen 1967; White 1997). Third, the human capacity to maintain cohesive groups appears to be limited (e.g., Dunbar 1993), so split-ups are inevitable to some degree. Fourth, Nettle's (1999) proposal for the latitudinal asymmetry (there are more languages around the equator) is that there is a longer mean growing season in this area, leading to the possibility for smaller self-sufficient group sizes, and consequently for the possibility of smaller groups and more languages. Finally, multilingualism is important in altering the selective process in two key ways: (a) it acts as a conduit for replicators to pass between speakers of different languages and (b) it extends the range of interactors for whom signs carry social-affiliation information to a broader speech community.

Still, most potential factors and explanations for language diversity and skewed patterns of language cladogenesis are strongly under-investigated. It is unclear why there has been reluctance in linguistics to investigate such pressing questions further. Nevertheless, there are some early indications that language splits are somehow special, in the sense that both the basic vocabulary (Atkinson et al. 2008) and structural features (Dediu and Levinson 2012) show punctuated evolution, change in both being accelerated around language splits.

For the future, we see two main desiderata for the study of these issues. First, at the macro level there are questions regarding the global prediction of linguistic diversity based on political, ecological, cultural and social structure. For such research, we need large global databases, along the lines of the World Atlas of Language Structures (WALS),[4] and the Human Relations Area Files (HRAF).[5] However, the data currently available is far from ideal and can only be taken as provisional. In contrast, at the micro level, there is a need for detailed sociopolitical studies of multilingual situations to establish models of the processes that are happening in interaction. We lack, for instance, field studies of the processes involved in language differentiation (e.g., during the breakup of Yugoslavia or the tribal conflicts of the Sudan or Somalia). Such studies could then be used to inform simulations, which need detailed knowledge of the relevant variables to be successful.

4 http://wals.info/

5 http://www.yale.edu/hraf/

Challenge 3: How Does a Theory of Language Diversity Look?

There are at least two aspects to language diversity: the number of languages and the overall amount of variation between languages (disparity in biological terms). To date there is no formal theory about what drives either the rates of linguistic diversification (cladogenesis) or the rates of linguistic disparity (anagenesis). Linguistic cladogenesis is produced when speech communities split, and thus factors which promote group boundary formation are also likely to produce more languages. Possible factors include migration, environmental heterogeneity, increases in population size, and selection that favors some groups over others. Possible factors that drive disparity include group size, social networks density, contact with other languages, and the social processes of schismogenesis (esoterogeny). What we need now are formal theories of the relationships between these variables.

Challenge 4: Can We Build a Global Tree of All Known Languages?

A complete understanding of the complex historical relationships between the world's languages would be a major scientific advance. More importantly, it would provide a backbone on which many more specific questions could be meaningfully asked, such as the relationship between various components of language and culture and their rates of change. Although such a project is conceptually simple, it faces two main problems: First, the rate of linguistic change is such that most historical linguists believe that any genealogical signal is obscured by chance and borrowing beyond a 10,000-year time depth. Although there are some suggestions that highly conserved items of basic vocabulary and some structural features might retain historical signal beyond this, the prospect of rigorously inferring language relationships right back to "protoworld" or the African diaspora of modern humans seems remote. Second, it is not a priori clear whether a tree-like model would be sufficient, given the extent of horizontal processes in language. The hunt for a global tree of languages might end up as tangled as the hunt for the tree of life.

How Does Cultural Evolution Explore the Design Space of Languages?

When studying human language, one of the central questions concerns the possible structural variation: How large can it be, and what constraints are acting on it? It is useful to introduce the notion of a "design space," which can be characterized in terms of the ranges of variables, bounded by the parameters of a domain, within which a design solution must be found. Two different approaches can be taken to address this question about the nature of the design space for human language (see Evans and Levinson 2009a; Levinson and Evans 2010). First, the a priori question (necessary to be able to study linguistic variation in the first place) is what kind of design space can we use to describe language variation? Here, the problem is that there is no easy external "measuring stick"' with which to compare languages. Second, given an a priori grid of possible variation, how does the actual a posteriori variation of the world's languages look?

The large set of data points afforded by thousands of languages, as well as the possibility of setting up frameworks of comparability able to make unified categorizations despite very significant differences in how languages work, opens up possibilities for many types of hypothesis testing. A key part of doing this is setting up a comprehensive, precise, and operational ontology of the design space—or, perhaps more realistically and slightly less ambitiously, of the multiple design spaces found for each relevant variable in phonology, morphology, syntax, semantics, and so on (as well as in such other areas as register and other sociolinguistic distinctions). The subdiscipline of linguistic typology has been steadily doing this over the past half-century, although the number of dimensions that well-known databases such as WALS cover is only a tiny fraction of the full set, and new dimensions continue to be discovered (e.g., grammatical encoding of speaker and hearer attentional phenomena, in such subsystems as demonstratives and verbal inflections). Categorization decisions continue to dog this enterprise. We see three main ways for the field to break out of the current impasse of the arbitrariness of cross-linguistic categorizations:

1. Switch to continuous rather than discrete variables (e.g., time measurements of voice onset time as opposed to a simple ± voicing contrast).
2. Break down higher-level categories (e.g., "subject of") to lower-level ones (e.g., "argument triggering verb agreement").
3. Use direct comparisons of (parallel) texts (allowing multiple values to surface in the one language, measured with respect to the statistical occurrence of different choices) as opposed to grammatical descriptions in which structures tend to be essentialized.

Thinking about the structure of this design space, various issues must be considered. First, is the design space tractable; that is, when looking at extant languages, can we determine the design space of language? The design space of the whole of language may not be tractable, given the number of variables and parameters which must be modeled; however, restricted domains seem to be within our grasp. For example, the tradition of linguistic typology (as exemplified by the WALS database) attempts to survey specific parts of linguistic structure and classify the variation attested in these domains. Although such investigations are far from unproblematic or conclusive, it would appear profitable to map out the possibilities of linguistic structures within restricted domains.

Second, is the design space immutable or changing? Namely, can we assume that the design space of our earlier ancestors was the same as ours and, if not, in what respects did it change? Can, for example, advances in other cultural systems (e.g., the invention of writing or of information technology) change the design space by making whole new dimensions possible or altering existing ones (e.g., literacy affords increased complexity in both the lexicon and in complex syntax; Karlsson 2007)? Would such changes simply extend existing dimensions, or add to them, or might the topology and metric properties of the space also change? The latter would imply that the "closeness" between language states could change, allowing different evolutionary pathways for language change. Thus, if the design space has this dynamic aspect, it could be possible that changes which did take place in the past might seem implausible today, or more likely that changes today (spurred by literacy and telecommunications) would not have taken place in the past. If the design space is indeed dynamic, a new dimension of complexity to language comparability and language evolution will be added, and computational modeling will have to play a major role in understanding it.

A third, and related, question concerns the reconstruction of the path a particular language has taken through the design space and of the possible paths and associated probabilities that it could have taken. This requires an understanding of the constraints and metric properties of design space acting at each point in this space and possibly their dynamics (see above). To achieve this, we will need much more data on the actual paths languages have taken (e.g., using phylogenetic methods) and of the properties of language learners presented with certain constructions (e.g., using natural and artificial language learning paradigms combined with computational modeling).

Modeling a possibly correlated random walk through the design space could lead us to theoretical expectations of the distribution of languages under different biases to which we then can compare the observed distribution and infer how likely it is that a certain evolutionary hypothesis could have produced this distribution. Similarly, such a theoretical approach could be informative about the fraction of the design space covered under different hypotheses, and could fruitfully be used to explore the dynamic nature of the design space by altering its metric properties following cultural or biological innovations. Moreover, even the dynamics of the landscape exploration could depend upon time and be heavily influenced by the previous history of each language. Computational modeling, for example, suggests that linguistic categorization in isolated populations might correspond to a metastable state where global shifts are always possible but progressively more unlikely, and the response properties depend on the age of the system itself (Mukherjee et al. 2011). The system actually "freezes," spending progressively more and more time in local minima of the landscape (Mézard et al. 1987). In this general scenario, shared linguistic conventions would not emerge as attractors, but rather as metastable states.

Setting all these complexities aside, and assuming we can in fact set up a maximal design space, what can we do with it? First, we can observe the distribution of data points across it. What we tend to find is a crowding in one corner—vast tracts of the design space are empty (e.g., words which mean father or mother's brother, but not father's sister) or only sparsely populated: of the six basic orderings of subject, verb, and object, three (SOV, SVO, and VSO) account for 96% of the world's word orders; the object-initial order (OVS and OSV) are rare enough that they were believed not to occur at the time Greenberg wrote his seminal paper on word-order typology (Greenberg 1966), though the latest WALS survey gives summed figures for these two orders of just over 1% (Dryer 2011). This has driven many sorts of investigation of why particular options are rare. A maximal view of the explanatory challenge would be that for every asymmetry in populating the design space, some explanation needs to be sought, whether in some form of selector bias or as a fossilization of particular design choices in the past (i.e., as attesting to inheritance from some deep ancestor) or accidents of history. It is also important to remember that given the short timescale in which modern humans have spread across the world, there simply has not been sufficient time for cultural evolution to explore the space: languages may fall into a corner for no greater reason than that is where the space began to be explored (Evans and Levinson 2009a).

Second, we can also ask whether the population of the design space has always exhibited the same distribution; for example, was it different in the early phases of language evolution (e.g., the "non-doubly articulated" portion is currently empty except for recently evolved village sign languages [Sandler et al. 2011] but may have been populated in early phases of language evolution)? All modern languages include a number of design elements, but it is entirely conceivable that in an early phase of linguistic evolution, different human groups developed different elements from this list, and these were transferred horizontally between groups to form an integrated "language package." For some of these elements there are temporal dependencies; for example, well-developed intention-attribution (driving pragmatic enrichment of what a sign means) must have preceded the conventionalization of code. For other elements it is quite plausible that they were produced independently, in different groups; for example, developing the notion of abstract property concepts (big, green) independently of what they are applied to (big elephant, green leaf) is logically independent of developing a pronoun system. Thus a plausible coevolutionary scenario for language origins is that different groups made different "technological" breakthroughs in evolving early language systems, these were then picked up by other groups, and the resultant package was so efficient and advantageous that it fed back into biological selection (e.g., vocal tracts favoring fine articulatory movements would have been more and more favored as phonologies became more complex). At the level of more specific properties (say, particular patterns of case marking or types of consonant inventory), was their distribution different when all humans were hunter-gatherers?

Finally, we can seek other "external" factors—genetic, cognitive, social/demographic (e.g., group size)—and ask if they correlate with particular design choices (e.g., Ladd et al. 2008).

Challenge 5: Can We Formulate a Total Design Space That Can Serve as a Basis for Worldwide Language Comparison?

What are the units of comparison across languages and how do we best deal with the fact that phenomena do not match up exactly? This is part of the ongoing task of linguistic typology (see Evans, this volume). An important question is whether the design space has remained constant through time or gradually developed new properties. Further, what is the shape of the design space? Another issue is what constraints exist on pathways of movement through the design space: Can individual design states simply move to any other state (unlikely), or are there particular pathways between states? Some progress has been made on doing this systematically with approaches like evolutionary phonology (Blevins 2004). Extending this to a wider range of phenomena is one promising way of developing new methods for obtaining deep-time phylogenies.

The Shape and Fabric of Language

Two important questions in understanding cultural evolution concern the shape of cultural history (the extent to which it is tree-like) and its fabric (the unity of that history). Proponents of cultural phylogenetics are often accused of assuming that human history has been both highly tree-like and consisting of tightly linked lineages, but there are obvious exceptions to these assumptions. We suggest, however, that such highly polarized discussions distort a much more complex reality better conceptualized as involving positions along continuous dimensions. The key challenge is to quantify empirically where particular aspects of culture and language lie on these dimensions, and we believe that current computational methods derived from evolutionary biology coupled with computer simulations are able to address these questions meaningfully. A consequence of this approach is that various components and subsystems of language and culture (such as the basic vocabulary or structural features) might show differing amounts of tree-like evolution and degrees of coherence in different parts of the world and language families.

In this vein, another intriguing parallel can be drawn with evolutionary biology, this time with the unicellular "prokaryotes" and the viruses. The evolutionary history of multicellular organisms can be quite accurately represented by species trees, as the histories of their individual genes[6] tend to

[6] We will not go here into the details of what a gene is, but just use this as shorthand.

coincide. There are, however, cases of nonvertical inheritance as well, where the history of some genes is decoupled from that of the containing organisms (Arnold 2008; Keeling and Palmer 2008). In the world of microorganisms, these so-called *horizontal genetic transfer* (HGT) phenomena are very important, as they have the capacity to incorporate foreign pieces of DNA and there are mechanisms adapted for transferring genetic material between organisms (Harrison and Brockhurst 2012; McDaniel et al. 2010).

This "rampant" HGT has led some researchers to propose that the metaphor of the "tree of life" might not reflect the biological reality (e.g., Doolittle and Bapteste 2007; Koonin 2009). The fundamental issue is that while each individual replicator (e.g., gene, lingueme) has a tree-like, vertical history, these histories might fail to coincide. In extreme cases of widespread disagreement, one could still reconstruct an agreement tree, but this "tree of one percent" (Dagan and Martin 2006) might not represent anything real. Other methods propose to reconstruct a "forest of life" and try to identify major trends within it (e.g., Puigbò et al. 2009), or use various types of phylogenetic networks (for an application to language, see Nelson-Sathi et al. 2011). However, despite this, there are coherent systems of genes which probably represent stable islands of fitness maxima, and not all genes are equally prone to HGT—those that are hubs in complex gene networks or are involved in the "informational" aspects of cell functioning are more resilient (the "complexity hypothesis," Jain et al. 1999). Thus, the potentially enormous sea of combinations due to HGT is in fact sculpted by natural selection, resulting in stable "bundles" of genes that are optimally integrated and stable through time, forming coherent lineages.

Similar processes might be at work in language: despite the maximally diffusionist position (e.g., Thomason and Kaufman 1988) that virtually anything can be borrowed between languages, there nonetheless appear to be stable lineages of traits, such as morphological paradigms, that we can use as coherent and stable subsystems. Moreover, recent phylogenetic work strongly suggests that at least the basic vocabulary tends to be inherited as a coherent unit (Pagel 2009; Gray et al. 2010; Bouckaert et al. 2012), and even important amounts of borrowing among languages can be detected by such methods (Currie et al. 2010b).

Pulling in the opposite direction, a widely held belief among linguists is that language is a system where "everything hangs together," whose system coherence means that changes in some feature (e.g., order of basic clause constituents, or the height and frontness of one vowel in the space) will pull along changes in some other features (e.g., order of adpositions with respect to nouns and of relative clauses to their heads, or of the realizations of other vowels). As more evidence from a greater range of languages has accumulated, an increasing number of these correlations turn out to be probabilistic rather than absolute. This reflects common preferences for processing (Hawkins 1994) or historical links between how some categories (e.g., adpositions) derive from others (e.g., verbs or nouns), which means that some of the claimed word-order correlations may be lineage specific rather than universal (Dunn et al. 2011b; cf. Croft et al. 2011).

Challenge 6: What Are the Relative Contributions of Vertical and Horizontal Processes in Language Evolution?

Much recent work on the cultural evolution of language—especially, but not exclusively, in terms of modeling (for a review, see Jäger et al. 2009)—has focused on either vertical transmission of linguistic structure across generations of language learners or horizontal transmission of linguistic elements through interactions between language users. Both lines of work have suggested that biases in cultural transmission can lead to the emergence of language-like structure from a starting state without such structure. However, relatively little work has sought to investigate the two types of transmission within a single framework. Further, we know relatively little about what the relative contribution of vertical and horizontal transmission is in language evolution from an empirical perspective (Gray et al. 2010). More generally, we lack a theory about the interplay between horizontal and vertical transmission in the cultural evolution of language, and the degree to which this interplay may vary for different aspects of language and across different points in time. That is, a key outstanding question pertains to whether we can formulate a theory about the cohesion of transmission of traits vertically and/or horizontally. It is difficult to differentiate the underlying differences between horizontal and vertical transmission, as in one sense there are just traits being transmitted. Thus, the deeper question is how cohesively these traits behave in transmission.

Challenge 7: How Much of Language Consists of Subsystems of Tightly Interlinked Traits?

This problem is about networks of traits in languages (i.e., the systemic view of language): How are traits interlinked within languages, and how lineage specific are these trait linkages? As in biology, where genes interact with each other in complex networks, we can view the various aspects of language as connected in similarly complex networks. Interestingly, in biology these networks tend to be highly structured, with identifiable subsystems of tightly linked genes and various genes constituting "hubs" due to their importance in interacting with other genes (Caldarelli 2007). Moreover, it seems that the resistance of genes to change and horizontal transfer depends on their network properties (Jain et al. 1999; Aris-Brosou 2005), and a similar question arises in language change (Dediu 2011a). These network properties might differ between languages and language families and might influence the trajectory of language change.

Although there are good reasons for extreme caution in the suggestion that ontogeny repeats phylogeny, it is possible that aspects of ontogeny might be informative as to how the evolution of various aspects of language might be separated. For example, in development, early turn taking provides a framework within which joint attention formats develop. Joint attention is one of the frameworks within which children's "mind reading" and intention reading skills start to manifest themselves before there is any language. Comprehension often precedes production (probably through the use of heuristics). Finally, arriving at the ability to produce all the sounds and structures of one's language is an extremely long-drawn-out process.

Challenge 8: How Does the Study of Language Development and Language Evolution Inform Each Other?

First, in development, different abilities appear at different times and, to some extent, may have different developmental trajectories: turn taking, intention reading, coordinated action, comprehension, simple syntax, fully accurate phonology, complex syntax (see Lieven, this volume). Can this inform the processes involved in the evolution of language? Second, language has been shaped by cultural evolution to be as learnable as possible by children given their cognitive and other limitations (and the way these may change across development). That is, language has been shaped by previous generations of language learners (and users) to fit those biases that children bring to bear on language acquisition (Chater and Christiansen 2010). We may further speculate that gradual changes across development could further result in developmental scaffolding in the cultural evolution of language, in which certain aspects of language are acquired before others, as development unfolds. This may place constraints on the nature and the kind of language systems that can emerge.

Coevolution of Biology, Culture, and Language

Biology and Language

It is undeniable that there has been some coevolution of language and biology in the early phase of language evolution, involving evolution of the vocal tract and possibly the brain. However, it is often assumed that since then biological evolution has become "frozen" relative to language evolution, as if language variation and change works on a "fixed"' biological background (e.g., Chomsky 2010; Hauser et al. 2002). Several reasons suggest, however, that this biological basis is far from "universal" and fixed among individuals (Levinson and Dediu, in this volume). It seems also obvious that language adapts to the brains, the vocal tracts, and the hands of speakers (Christiansen and Chater 2008), but it is also important to recognize that language feeds back on cultural evolution, and thereby potentially influences our biological evolution (Laland et al. 2010).

A number of examples suggest, for example, that differences in vocal tract anatomy might influence language structure, such as the correlation between the Yoruba/Italian vowel systems and the anatomy of the upper vocal tracts of their speakers (Ladefoged 1984). In turn, speech will generate selective pressure on the biological mechanisms used to produce and perceive it, as suggested by the various features of the vocal tract that seem designed for speech (Lieberman 2007; MacNeilage 2008). Thus, the gene–culture coevolution might be very profitably investigated by looking at the evolution of the vocal tract in the human lineage, including modern variation in its physiology and anatomy.

It is thus clear that genetic differences between modern populations might affect language, but it is important to highlight that in most cases these genetic differences exist not because of feedback selective pressures generated by language, but rather as a result of neutral evolutionary processes such as genetic drift and founder effects. In fact, we are a quite genetically uniform species, and the amount of genetic variation present between humans is mostly distributed within populations. Nevertheless, there is genetic variation between populations (Barbujani and Colonna 2010; Novembre et al. 2008), some of which might be due to natural selection (e.g., skin color, resistance to infectious diseases), but the vast majority is probably the result of random sampling.

Are there other aspects of language (e.g., morphology, syntax) that might be influenced by genetic biases? Many aspects of language and speech show moderate to large heritability (Stromswold 2001), which, despite the rather substantial inherent problems of such estimates (e.g., Charney 2012), seem to suggest genetic influences. For example, vocabulary size is somewhat heritable (Stromswold 2001), as is short-term memory buffer size (recently associated with the *ROBO1* gene; Bates et al. 2011); for recent reviews, see Graham and Fisher (2013) and Bishop (2009).

Language is a socially shared system, which is constantly (re)shaped by its users. During the process of social agreement, cultural evolution may introduce accidents which, once emerged, "freeze" and act in their turn as sources of bias for the further evolution of that specific language. These cultural biases compete with genetic predispositions and will in many cases mask them. It is precisely in this sense (i.e., in contrast to cultural biases) that a genetic bias can be defined as "weak" or "strong." Moreover, this tension between culture and biology accounts for the fact that while some properties of language are shared by all languages, other language "universals" (or better termed as "trends") are statistical in nature (Baronchelli et al. 2010).

A model for the emergence of color-naming systems (Puglisi et al. 2008), capable of capturing the statistical properties of the World Color Survey, clarifies this picture (Baronchelli et al. 2010). Whereas a psychophysiological bias (namely, the human "just noticeable difference" bias for hues) acts as a cross-population unifying force in shaping the color categorization of different groups, cultural evolution operates as a source of random yet history-dependent bias at the level of the single populations. It is only through a statistical analysis performed over many populations that the presence of the genetic bias, or equivalently the universal properties shared by the different naming systems, can be revealed (Baronchelli et al. 2010). Interestingly, such biases could be amplified not only by vertical transmission across generations, but also by interactions within generations (horizontal processes), as suggested by Nicaraguan sign language or the emergence of the various village sign languages (Levinson and Dediu, this volume).

When discussing the evolution of cognition, it is often assumed that the growth of specific brain areas happened in response to adaptation to specific cognitive niches (e.g., Pinker 1997; Tooby and Cosmides 2005). However, analyses of allometric data from mammals (Finlay and Darlington 1995) to sharks (Yopak et al. 2010) suggest a scenario more in line with predictions from cultural evolution (Finlay et al. 2001). Specifically, these data suggest that as brains grow bigger, some areas grow proportionally bigger compared to others due to the highly conserved order of neurogenesis following a basic axial structure in development. That is, there is no specific selective pressure required for specific brain areas, only a general pressure for larger brains (though, in principle, selection for a specific brain area could lead to the enlargement of the whole brain). Having a larger brain may have resulted in the availability of more neural hardware which, in turn, could be recruited into brain networks to accomplish specific tasks, without specific adaptation for those tasks. One such example is our ability to read, which is clearly an ability to which we have not been adapted but where brain networks are recruited and emerge during development, specifically, the left occipitotemporal sulcus (Dehaene and Cohen 2007). Similarly, it is possible that other brain networks are recruited during development to support various language functions (Christiansen and Chater 2008)—an evolutionary scenario consistent with recent meta-analyses of the emergence of brain networks as reflected by neuroimaging studies (Anderson 2010).

Challenge 9: What Is the Evidence for Gene–Language Coevolution? Is There Any Other Evidence besides Speech?

Gene–language coevolutionary processes have the potential to broaden our understanding dramatically, but it is currently unclear to what extent these processes actually affect language. There are a few sources of evidence, especially concerning the complex apparatus used to generate speech, where it is hard to find alternative explanations for its apparent design. Other sources of evidence concern larger-scale phenomena where language may have played a major but indirect role in shaping various cultural niches, such as, arguably, in the domestication of plants and animals, which in turn feeds back on our immune and digestive systems. A focused research program aimed at identifying any such putative cases and testing them is needed.

Social Structure and Language

An important aspect of linguistic interaction is the ability to replicate linguistic forms and meanings with a high degree of fidelity. Error correction allows interlocutors to coordinate on the intended meaning of utterances. However, social networks vary considerably in the possibilities for error correction of this kind. For example, more interactive network links, such as conversational interaction (e.g., phone calls, face-to-face meetings), allow for greater error correction, whereas more broadcast sorts of communication (e.g., speeches, written communication, television) allow for fewer possibilities of error correction and hence greater likelihood of modification of the interpretation of the language (Garrod and Anderson 1987; Garrod and Doherty 1994; Fay et al. 2000). These differences in social network structure may influence the nature and tempo of language evolution, especially given the accelerating pace of modern telecommunication systems.

Several distinct mechanisms of propagation of linguistic variants through a social network have been proposed (for further discussion, see Hruschka et al. 2009). The first is social valuation of one linguistic variant over another. This is the classic Labovian model, although Labov (2001), like other sociolinguists, allows for other mechanisms as well. A second mechanism is that differences in frequency of conversational interaction and/or tie strength between speakers may result in the differential replication of the variants used by the more talked-with/stronger-tied speakers (Milroy and Milroy 1985). This mechanism presupposes that changes in use of variants proceeds by some sort of accommodation (Trudgill 1986, 2008). A third mechanism proposed by sociolinguists is an adopter group model (generally inspired by Rogers 1995): a community can be divided into groups based on each group's role in adopting an innovative variant, such that some are leaders, others early adopters, others later adopters, and so on (also influenced by Labov 2001 as well as Milroy and Milroy 1985; recent studies utilizing adopter groups in some detail are Sankoff and Blondeau 2007 as well as Nevalainen et al. 2011).

Empirical studies, however, have been unable to distinguish between the operations of one mechanism over another. Thus, a major open issue is the construction of models able to identify the signatures of different sorts of selection mechanisms, such as those proposed by sociolinguistics. For example, the naming-game model shows that simple horizontal pairwise interactions between peers are able to trigger the emergence of shared linguistic conventions in a population of individuals (Steels 1995; Baronchelli et al. 2006). Thus, more complicated processes might have played a role in the rise of a shared (proto) language, but they cannot be considered as necessary. In the context of the naming game, moreover, the role of the social network of communications has been extensively studied. The time needed to reach a global consensus, along with the individual cognitive burden during the process, turns out to depend dramatically on the properties of the underlying interaction patterns, from fully connected graphs to (spatial) lattices and complex networks (Dall'Asta et al. 2006). Adopting a different perspective, Blythe and Croft (2012) propose distinct mathematical formalizations for social network structure, adopter group models, and social valuation of linguistic variants. The latter is modeled in the same way as fitness in biological evolution, whereas the first two require different mathematical models involving the interaction frequencies and weights of the speakers/agents rather than fitness values on the linguistic variants. Mathematical models such as the one proposed by Blythe and Croft need, however, to be enriched to identify signatures of different types of selection mechanisms.

Another issue is that when multiple mechanisms are in operation, which is highly likely in the case of selection of linguistic variants in speech communities, it is even more difficult to identify the operation of all the mechanisms involved. For example, any selection bias in a small population (speech community) may be swamped by random processes; alternatively, the interaction of multiple processes may lead to significantly different change trajectories compared to each process operating independently. Modeling the change trajectories that result from different selection mechanisms (and their interactions) can provide examples of expected patterns that can be compared to observed data on the trajectories of language changes.

Challenge 10: What Are the Effects of Demography on Language Change and Dispersal?

Demography seems to have fundamental effects on many aspects of cultural evolution; for example, large populations correlate positively with the complexity of technology (Kline and Boyd 2010), but inversely with the complexity of demonstrative systems (Perkins 1995) and language morphology (Lupyan and Dale 2010). What are the generalizations here? Are there principles that would tell us what the largest sustainable speech community would be or, conversely, what is the smallest speech community with long-term viability? For example, generally it seems that a couple of hundred individuals are needed to sustain a distinctive language, but work by Green (2003) demonstrates apparent stability and long-term viability for a speech community of only 70 people in Central Arnhem Land.

Future Considerations for the Study of Language Evolution

To make substantial new progress, we need new kinds of data, new methods, and new integrative theories that will bind together the many different levels and ontologies for language and its use in communication.

Necessary Data

Comparable Data across Languages Documenting Child Language Development

Dense developmental data are required from different cultures. Theories of language development are largely based on what we know from a very small number of languages. Naturalistic corpus data of children's language development and the language they hear from a wider range of typologically contrasting languages are urgently needed. For instance, take the contrast between learning English, with its fairly rigid syntactic word order and almost total absence of inflectional morphology, and learning most other languages of the world. Corpus data needs to be as dense as possible to be able to determine when children have productive control of a system, rather than repeating rote-learned strings or using low-scope formulae.

Dense Sociolinguistic Data from Different Linguistic Situations

At present we have a large number of studies of sociolinguistic variation from social groups in large-scale, urban, industrial societies. However, next to no variationist sociolinguistic studies have been conducted on small-scale multilingual societies that have marked most of human history. In fact, hardly any variationist sociolinguistic studies have been carried out even on rural social groups in small settlements within large-scale industrial societies. As a result, we know little about within-society linguistic variation and dynamics for the sorts of societies that existed in most of the (pre)history of modern humans.

Data on Variation in Biological Parameters Relevant to Language

It is becoming clearer that variation in diverse aspects of our biology (including vocal tract anatomy, hearing, and associated genes) might affect patterns of linguistic diversity within and across languages (Ladefoged 1984; Butcher 2006), making the construction of a database of such variation an important goal. Such a database will need to contain, for example, information concerning variation in diverse parameters of the vocal tract within and across populations, color perception and naming data, standardized psycholinguistic tasks or brain imaging protocols, and links to databases that contain genetic polymorphisms and fully sequenced genomes.

Language Documentation, Description, and the Need for Large Corpora

Getting large enough corpora to detect the effects of frequency is a challenge, particularly if we want to get this for the full diversity of the world's languages. Although there has been a big push to build corpora for little-known languages through projects, such as the DoBeS project, that aim to document endangered languages, these rarely contain more than 40–50 hours of data, and usually only a subset of this is transcribed. Psycholinguists know that word frequency has a vital bearing on many aspects of language processing. To get the sort of million-word corpus needed to provide the foundation for processing investigations we would need closer to 1000 hours of data. Much recent work in the psychological modeling of language acquisition and processing has come to rely on dense databases of language consisting of a million words or more. In this context, large data sets are needed to capture the full diversity and idiosyncrasy of language learning and use. Thus, to explore more psycholinguistically motivated models of language evolution, larger databases are needed and they must include crucial use information.

Methods

Computer-Assisted Comparative Historical Linguistics

Comparative historical linguistics is a great method to uncover historical relationships between languages. However, it was conceptualized in the nineteenth century, long before the power of computer-assisted methods was known. We need to reformulate the methods of the "comparative method" so that computer power can profitably be used to reconstruct languages (for an initial step in this direction, see Bouchard-Côté et al. 2013). Basically, the desideratum is to produce not just language trees, but to identify the actual cognate sets, sound correspondences, sound changes, loan words, calques, grammatical borrowing, etc., that are used to infer the trees.

Automated Grammar Extraction, Transcription, Alignment

To transcribe and gloss foreign language text, it is estimated that roughly 100 hours are needed for each hour of recording. Much of this could be semi-automated, together with the temporal alignment of orthography with recordings. Moreover even from small transcribed samples, parts of speech can be automatically extracted by collocation, and using frequency data it should be possible to extract some kind of skeleton grammar. If parallel texts are available, much further automatic grammar extraction becomes feasible. These methods would allow us to work toward the grammatical analysis of all the languages in the world—the exhaustive database we need for understanding the full spectrum of linguistic variation.

Phylogenetic Modeling

Although considerable progress has been made using phylogenetic models derived from evolutionary biology to model the relatively tree-like evolution of basic vocabulary, computational methods that address more complex histories are likely to be needed to model accurately other less cohesive aspects of language. The new multilocus models in *BEAST, which directly model intraspecies polymorphism and incomplete lineage sorting (Heled and Drummond 2010), might prove a promising place to start. Moreover, recent developments in bioinformatics used to model the evolution of microorganisms that transfer genetic material both vertically and horizontally might be usefully exploited by linguists (e.g., Nelson-Sathi et al. 2011).

Experimental Semiotics

Experimental semiotic studies present living participants with communication challenges in mini-experimental situations, especially in the form of web-based studies with large and structured communities (Galantucci et al. 2012). This may offer a way of testing hypotheses generated by computational studies of language evolution on populations. It may also offer ways of studying the consequences of different kinds of social/communicative networks.

Agent-Based Modeling

There is a strong need for linguists, psychologists, and (agent-based) modelers to work closely together so that complex reality can be matched to modelability, thus permitting the development of useful models to address empirical questions in language change. Much work in agent-based modeling has proceeded in the absence of empirical linguistic data, input from linguists, or psychological considerations regarding learning, memory, and processing. However, it is very important for linguists to specify a priori what linguistic questions they want answered in a model, or to specify what empirical linguistic patterns they would like to see modeled. Given that linguistic reality is very complex, these models are very limited, at least at the present time: that is, there is not a very close match between what models can model and what data linguists have in detail. Moreover, it is also imperative that developmental and cognitive constraints are taken into account to ensure that the models involve psychologically, neurocognitively, physiologically, anatomically, and physically plausible computational constraints. Progress in modeling language change processes can be achieved only through close collaboration between linguists, psychologists, and modelers. Such collaboration requires time for each to understand the other's aims and methods of analysis, as well as to develop a collaborative understanding of the sorts of models and types of empirical questions that can be fruitfully combined to yield linguistically interesting results.

Building Methodological Bridges to Other Disciplines

Given that some of the properties of data concerning language are (partially) shared by data from other domains, it is crucial to build bridges toward these other domains in what concerns methods for representing, visualizing, and analyzing such data. One example could be the spatial nature of some of our data, and useful parallels might be drawn with geostatistics, epidemiology, and Geographic Information System (GIS), or the historical aspects of language and parallels with evolutionary biology. Many more of these types of bridges need to be identified and constructed in the near future.

Theories

Language is central to human social interaction and is, at several levels, a fundamental question for the social and biological sciences. How does language evolve in response to social and biological forces? How is language acquired by each new generation? How is language processed "on-line" in social interactions? These questions have frequently been treated as separate topics, to be addressed more or less independently. As such, studies of the evolution of language typically downplay issues related to language acquisition and processing. Similarly, work on language acquisition tends not to address questions pertaining to the processing and evolution of language, and studies of language processing usually pay little attention to research on language acquisition and evolution. We believe that this tendency is misguided, as there are strong constraints between each domain, allowing each to throw light on the others.

We think that the evolutionary perspective provides a unifying theoretical perspective on language processing, language structure, and language change that is capable of bridging gaps between studies in acquisition, processing, evolution, description, and variation of language. Theory and practice should go hand in hand. What an evolutionary approach to language desperately needs is theories that link sociolinguistic processes to historical patterns (and vice versa). This modeling enterprise should be coupled with team projects that bring together modelers and sociolinguists, psycholinguists and historical linguists, as well as biologists, mathematicians, and many other specialists in their respective fields.

Conclusions

Linguistic systems offer a spectacular parallel to biological evolution, but in the cultural realm. They have extraordinary complexity, and because we cannot change them to suit ourselves they are largely beyond the ability of individuals to change consciously. Thus they offer us elaborate design without any designer, showing us the "blind watchmaker" of evolutionary processes hard at work. They have lineages of deep antiquity and wide diversity, rivaling some biological systems. Thanks to centuries of human thought about language, we have tools for describing the fundamental units and the processes that combine them. Many significant challenges remain even on the descriptive front, especially in how to find parameters of comparison across lineages. Nevertheless, by applying modern phylogenetic and bioinformatic techniques to current descriptive materials and databases, we are now able to extract deep phylogenies, quantify rates of change, measure degrees of reticulation or horizontal borrowing, or combine these with geographical databases to yield interesting inferences about the spread of languages (Levinson and Gray 2012). We can expect all these data and methods to improve dramatically over the next two decades, allowing many insights into the history and sociology of our species.

Religion

17

The Evolution of Prosocial Religions

Edward Slingerland, Joseph Henrich, and Ara Norenzayan

Abstract

Building on foundations from the cognitive science of religion, this chapter synthesizes theoretical insights and empirical evidence concerning the processes by which cultural evolutionary processes driven by intergroup competition may have shaped the package of beliefs, rituals, practices, and institutions that constitute modern world religions. Five different hypothesized mechanisms are presented through which cultural group selection may have operated to increase the scale of cooperation, expand the sphere of trustworthy interactions, galvanize group solidarity, and sustain group-beneficial beliefs and practices. The mechanisms discussed involve extravagant displays, supernatural monitoring and incentives, ritual practices, fictive kinship, and moral realism. Various lines of supporting evidence are reviewed and archaeological and historical evidence is summarized from early China (roughly 2000 BCE–220 BCE), where prosocial religion and rituals coevolved with societal complexity.

Introduction

In this chapter we summarize a growing body of work that jointly addresses two major evolutionary puzzles: the rise of large-scale human societies over the last 12 millennia and the origin of world religions. The origin of large-scale human societies that rely on substantial exchange and cooperation among ephemeral or anonymous interactants (Henrich et al. 2010b) stands as a major evolutionary puzzle (Jordan et al., this volume). While the standard evolutionary mechanisms associated with kinship, reciprocity, and reputation clearly influence human cooperation in important ways, they do not capture the full extent of our species' prosociality, and cannot explain the most important and peculiar aspects of human cooperation (see also Gintis and van

Schaik, this volume; Turchin, this volume). Kinship cannot explain cooperation among nonrelatives, and reciprocity and reputation do not suffice to explain cooperation beyond dense social networks, small villages, or tightly knit neighborhoods. Moreover, neither direct nor indirect reciprocity can explain cooperation in ephemeral interactions in large groups, where reputational information rapidly degrades, or in large-group interactions such as those associated with many kinds of public goods or commons dilemmas. Perhaps even more telling is that none of these mechanisms are able to explain the variation in cooperation among extant human societies, or the massive expansion of cooperation in some societies over the last twelve millennia (Chudek and Henrich 2011).

Religions are also puzzling: the existence of supernatural beliefs and ritualized behaviors is hard to explain from an evolutionary perspective. Since natural selection tends to filter out behaviors and beliefs which do not contribute to an organism's fitness, it is difficult to see how costly religious behaviors or counterintuitive supernatural beliefs (e.g., devoting time and resources toward elaborate rituals, building massive tombs, and observing debilitating taboos) could have originated, spread, and endured in so many societies.

We argue that these two puzzles are related: converging lines of field, experimental, and historical evidence indicate that particular religious beliefs, rituals, and practices have spread because groups possessing these cultural traits have expanded at the expense of groups possessing different traits or trait packages. Over time, a variety of cultural evolutionary processes, driven by intergroup competition, gradually assembled integrated packages of cultural elements (including beliefs, rituals, devotions, and social norms) to deepen group solidarity, sustain internal harmony, galvanize trust and cooperation on larger scales, and motivate their further spreading. Central to these packages are beliefs in supernatural agents or forces that (a) moralize human action in particular (and predictable) ways, (b) incentivize certain behaviors using supernatural rewards and punishments (see Norenzayan et al., this volume), and (c) manipulate our psychology in other ways that favor success in competition with other groups. These emerging cultural packages facilitated the origins of complex, large-scale societies and explain why religions with costly religious displays and moralistic high gods—which were likely rare over most of human history—have spread at the expense of other types of religious beliefs and practices. Our hypothesized link between religion, group identity, and morality may also explain the persistence of religious belief in the face of countervailing pressures; it provides a cultural evolutionary explanation for the emergence of the moral realism that now pervades both religious and secular discourses.

We supplement our review of the evolutionary and cognitive science literature with historical evidence from early China, chosen specifically because China is often held up—inaccurately—as an example of a complex society that emerged without high moralizing gods, dualistic thought, supernatural punishment, moral realism, or religious ritual strictures.

Points of Departure

Our approach here builds on and extends previous work within the cognitive science of religion. We take as our point of departure that mental representations related to religion are underpinned by the same reliably developing features of mind as nonreligious representations (Boyer 2001; Atran and Norenzayan 2004; Barrett 2004; Guthrie 1993), and that religious thinking is in many ways more intuitive than other kinds of thought (Shenhav et al. 2011; Gervais and Norenzayan 2012a), such as science (McCauley 2011). For example, core among the cognitive capacities that underpin religious representations is our ability to mentalize, or theory of mind, that allows people to think about goal-directed supernatural agents (Bering 2006a). Recent work shows that a decreased ability, or tendency, to use this mentalizing ability reduces belief in god (Norenzayan et al. 2012). Similarly, the ability to hold mental representations about souls or an afterlife may arise as a by-product of the separate evolutionary histories of our capacities for object tracking and mentalizing, leading to dualistic tendencies that permit us to readily conceptualize a separation of minds and bodies (Bloom 2004). Recent work has established this type of dualistic thinking in young children and adults from diverse societies (Chudek et al. under review; Cohen et al. 2011) and in ancient Chinese texts (Slingerland and Chudek 2011).

While foundational, this work leaves unexplained (a) the distribution of different kinds of supernatural beliefs, (b) the cultural evolution of religious representations over time, and (c) why people are emotionally committed to some supernatural beliefs or agents (gods) but not others (the so-called "Zeus Problem"; see Gervais and Henrich 2010). For example, although our cognitive capacities can readily entertain an immense range of god-beliefs, it is important to understand why beliefs in potent, morally concerned agents equipped with ample power to punish and reward became so common over the last 5,000 years. In small-scale societies, and likely in those of our Paleolithic ancestors, gods were quite different from those found in modern world religions, being relatively weak, whimsical, and morally ambiguous (Roes and Raymond 2003). Moreover, in the modern world, people often hold mental representations of many different gods (e.g., Zeus, Shiva, and Yahweh), but only believe in (i.e., are committed to and respond behaviorally to) one or a small subset of these. Therefore, when it comes to commitment or faith, factors are at play besides the content of the representations themselves (Gervais et al. 2011).

To address these issues, we incorporate basic insights from the cognitive science of religion into a cultural evolutionary (dual inheritance) framework. The central insight of this approach is that, unlike other animals, humans have evolved to rely heavily on acquiring behavior, beliefs, motivations, and strategies from other members of their group. The psychological processes that permit this cultural learning have been shaped by natural selection to focus our attention on those domains and those individuals most likely to possess fitness-enhancing information (Richerson and Boyd 2005). Human social learning generates vast bodies of know-how and complex practices that accumulate and improve over generations. Unlike other animals, human survival and reproduction, even in the smallest-scale societies, depends on acquiring cumulative bodies of cultural information related to hunting (animal behavior), edible plants (e.g., seasonality, toxicity), technical manufacture, and so on (Boyd et al. 2011). To exploit this accumulated body of adaptive information fully, learners often need to give priority to faith in their culturally acquired beliefs and practices over their own personal experience or basic intuitions. We have evolved to have faith in culture, with this faith being directed by certain salient cues (Atran and Henrich 2010).

What we describe below is, in part, a cultural evolutionary process in which cultural group selection "figures out" ways to exploit these learning abilities to spread effectively beliefs and practices that favor success in intergroup competition. Our ability to entertain supernatural and ritualized practices, as byproducts of our evolved cognitive capacities, provides the foundation for the rise of prosocial religions and of complex, cooperative societies.

The Cultural Group Selection of Beliefs, Extravagant Displays, and Rituals

A growing body of evidence suggests that religious beliefs, rituals, devotions, and social norms have coevolved in interlocking cultural complexes in a process driven by competition among alternative complexes. Cultural group selection can assemble those combinations of cultural traits that most effectively reinforce cooperative or other prosocial norms in a variety of interrelated ways. Here we focus on five. First, observation and participation in costly or extravagant rituals or devotions likely induces deeper emotional commitment to supernatural beliefs or agents, who can then be more effective monitors and punishers. Building on this transmission effect, extravagant displays can also evolve culturally to act as honest signals of group commitment or group membership, thereby favoring the associations that sustain cooperation. Second, supernatural policing and incentives (heaven vs. hell) can buttress more earthly norm-sustaining mechanisms, such as punishment, signaling, and reputation. Third, religions can extend the scope of cooperative tendencies by using collective rituals to forge unrelated individuals into emotionally connected, cooperative communities. Fourth, prosocial norms can be more readily transmitted by using fictive kinship, just as in many small-scale societies. Finally, the psychological force and endurance of prosocial norms can be increased by grounding them in the structure of the universe, either by directly attributing their creation to supernatural beings or portraying them as reflecting metaphysical truths. Our efforts here build on much prior work (Wilson 2002; Durkheim

1915/1965; Norenzayan and Shariff 2008; Atran and Henrich 2010; Wright 2009). Each of these proposed mechanisms will be explored in turn.

Costly Displays and Religious Faith

Once grounded in a cultural evolutionary framework, work on extravagant displays suggests three possible and interrelated explanations for the origin of religious asceticism, intense devotions, and some ritual practices. First, some rituals and devotional ascetic practices may have evolved culturally to deepen people's commitments to counterintuitive beliefs. Second, once such beliefs have been established, seemingly costly displays or signals evolve culturally to better demarcate group boundaries and discriminate those who share one's religious commitments and social norms (including cooperative norms) from those that do not. Third, this cultural evolutionary theory has the potential to explain the broad differences in rituals found in large- versus small-scale societies: specifically, the growing importance of low-arousal, high-frequency rituals (the "doctrinal" mode) and the relative de-emphasis of high-arousal, low-frequency rituals (the "imagistic" mode) with the expansion of societal complexity (Whitehouse 2004).

Evolutionary thinking suggests that humans possess a learning mechanism that gives weight to seemingly costly acts (credibility enhancing displays, CREDs) that are diagnostic of underlying beliefs and commitments (Henrich 2009a). Attention to these CREDs in acquiring one's degree of internal commitment to particular beliefs reduces one's chances of being manipulated by those seeking to transmit beliefs that one does not actually hold. Costly ritualized acts may have evolved as a means to convince learners effectively of the personal commitment of either the rest of the congregation (exploiting conformist biases in our learning) or of locally prestigious models. By exploiting our evolved reliance on CREDs, rituals and devotions can operate to deepen our commitments to counterintuitive beliefs. They also link performance of costly acts or extravagant displays to social success, thereby perpetuating the transmission of belief commitment across generations. Formal cultural evolutionary models show that costly displays can interlock with and sustain counterintuitive beliefs that would otherwise not be sustained by cultural evolution (Henrich 2009a).

This approach suggests that commitment to supernatural agents tends to spread in a population to the extent that it elicits, or is associated with, costly or extravagant displays. When community leaders and the congregation demonstrate commitment to supernatural beliefs by performing a costly ritual, observers who witness these commitments are more inclined to trust and learn from these actors, deepening their own belief commitments. If supernatural agents demand and incentivize certain behaviors, those with deeper commitment and beliefs in these agents are more likely to shift to behavior in compliance with these agents. This means that rituals with CREDs can influence costly prosocial behavior indirectly, by increasing belief commitments to agents who demand such behavior. This also explains why gods demand costly rituals. Supernatural demands for rituals, devotions, and sacrifices facilitate the intergenerational transmission of deep commitments, as children infer deep commitments from the costly or extravagant actions of adults (Alcorta and Sosis 2005).

Meanwhile, the psychological nature of commitments to culturally transmitted beliefs means that sacrifices and rituals need not seem (subjectively) costly for those who already deeply believe in the agent's incentives. Once culturally transmitted beliefs exist, and individuals are equipped psychologically to distinguish CREDs, cultural evolution may also harness diagnostic actions in ways that help believers identify each other or to exclude nonbelievers (and potential free-riders) from participation and the benefits of group members (Sosis and Alcorta 2003). By embedding ideas about signaling within this broader cultural evolutionary framework, we address a number of theoretical shortcomings without losing the core insights of work on signaling (Henrich 2009a).

Several lines of evidence support these hypotheses. The cultural evolution of the interrelationship between religious beliefs and costly rituals/devotions emerges from a study of 83 utopian communes in the nineteenth century (Sosis and Bressler 2003). Analyses show that religious groups with more costly rituals were more likely to survive over time than religious groups with fewer or less costly rituals. Differential group survival caused an increase in the mean number of costly rituals per group over time: cultural group selection in action, increasing the frequency of costly ritual, and devotional requirements over time via differential group extinction. Additional ethnohistorical evidence for the spread of rituals via cultural group selection can be found in Henrich (2009a). Similarly, among Israeli kibbutzim, individuals from religious kibbutzim cooperated more in a behavioral experiment than those from nonreligious kibbutzim, with the increased cooperativeness of religious members being accounted for by their ritual participation (Sosis and Ruffle 2003; Ruffle and Sosis 2006). Surveys and experiments in the West Bank and Gaza also show that a person's frequency of attendance at religious services predicts support for martyrdom missions. Convergent findings emerge for representative samples of Indian Hindus, Russian Orthodox, Mexican Catholics, British Protestants, and Indonesian Muslims. In these samples, greater ritual attendance predicts both declared willingness to die for one's god, or gods, and belief that other religions are responsible for problems in the world (Ginges et al. 2009). Moreover, a study of 60 small-scale societies reveals that males from groups in the most competitive socioecologies (with frequent warfare) endure the costliest rites (e.g., genital mutilation, scarification), which "signal commitment and promote solidarity among males who must organize for warfare" (Sosis et al. 2007:234). In such socioecologies, cultural group selection will shape religious rites and beliefs to manipulate our psychology to increase solidarity and commitment. A related analysis by Atkinson and Whitehouse (2011) documents the predicted relationship between ritual frequency and scale of society, with a pattern of more frequent, less dysphorically arousing rituals being associated with larger community size, more dependence on agriculture, and more influence by classical, "high god" religions.

The historical-archaeological record, combined with comparative ethnography, indicates that the costliness, size, specialization, and regularity of communal rituals increased with the scale and political complexity of societies. Archaeological research on the coevolution of ritual and society indicate that rituals became much more formal, elaborate, and costly as societies developed from foraging bands into chiefdoms and states (Marcus and Flannery 2004). In Mexico before 2000 BCE, for example, nomadic foraging bands relied on informal, unscheduled, and inclusive rituals. The same goes for contemporary foragers, such as the San of Africa's Kalahari desert, whose ad hoc rituals (e.g., trance dancing) include community members and are organized according to the contingencies of rainfall, hunting, and illnesses (Lee 1979). However, with the establishment of permanent villages and multivillage chiefdoms (2000–1000 BCE), rituals were managed by social achievers (prestigious "Big Men" and chiefs) and scheduled according to solar and astral events. This also appears to be the case for predynastic Egypt (4000–3000 BCE) and China (2500–1500 BCE), as well as for the chiefdoms of North America. After the state formed in Mexico (500 BCE), important rituals were performed by a class of full-time priests, subsidized by society, using religious calendars and occupying temples built at enormous costs in terms of labor and lives. This is also true for the earliest state-level societies of Mesopotamia (after 3500 BCE) and India (after 2500 BCE), which, as in Mesoamerica, practiced fearsome human sacrifice (Campbell 1974). Combining this with comparative ethnography suggests that high moralizing gods likely coevolved with costly regularized rituals, creating a mutually reenforcing cultural package capable of enhancing internal cooperation and harmony, while providing a justification to exploit out-groups.

Combining these observations with recent work in psychology may illuminate the linkage between monumental architecture and religion. The earliest civilizations are known for their stunning monumental architecture, usually in the form of temples, pyramids (tombs), and ziggurats (altars), all of which apparently served a religious function. The importance of such grandeur may serve at least two important psychological purposes. First, they may represent costly displays of commitment from the society's leaders, or of the society in general, to help instill in learners a deeper commitment to religious or group ideologies. Second, their visibility may act as an omnipresent "religious prime" that stimulates prosocial behavior (Norenzayan et al., this volume). A large temple in the market square may provide a salient cue that evokes, if only at the margins, more prosocial behavior in those interacting on the square.

Societies that exploit these aspects of human psychology will expand at the expense of those that do not.

In early China, the most prominent feature of the Chinese archaeological record is enormous tombs dedicated to deceased leaders. Though these were completely encased in earth, their elaborate structure and fabulous wealth which they contained were broadcast widely to the community in the course of public ceremonies dedicated to the entombment process, and the resulting mounds towered over the surrounding landscape. Our earliest written records of Chinese religious practices come from the Shang Dynasty, the first historically attested large-scale polity. In addition to constructing enormous tombs, it is clear that the life of at least the Shang elites was dominated by time-consuming and materially costly sacrifices to ancestral spirits and various deities, including the high god Shang Di ("Lord on High") (Eno 2009). By the Western Zhou period (1046–771 BCE), evidenced by longer and more discursive texts, codified ritual observances expanded to encompass every aspect of the elites' daily lives: their manner of dressing, eating, sleeping, and interaction with peers were all subject to a variety of taboos and injunctions that were viewed as being grounded in the basic structure of the cosmos. It is difficult to access the economic cost of mortuary and ritual practices in early China, but one estimate puts it at a full 10% of the society's gross domestic product (Sterckx 2009). Mandatory religious practices clearly occupied the majority of elites' waking lives.

Supernatural Policing

Our hypothesis suggests that, as we move from small-scale to large-scale societies, supernatural agents become increasingly morally concerned, more effective at monitoring norm violations (omniscience) and better equipped to provide punishment and rewards (heaven and hell) according to prescribed behavior. This view predicts many relationships, but among them is that belief in such gods should promote prosocial behavior toward co-religionists. There is now a substantial experimental and behavioral literature establishing this connection (for a detailed review, see Norenzayan et al., this volume). Here we confine ourselves to discussing the evidence for supernatural surveillance in early China.

Historically, although much evidence from Abrahamic religions is consistent with a supernatural surveillance view (Wright 2009), some researchers have suggested that a similar pattern did not emerge in early China. However, to the contrary, evidence from early China shows that supernatural monitoring played an important role in cobbling together large-scale cooperation. Even from the sparse records available from the Shang Dynasty, it is apparent that the uniquely broad power of the Lord on High to "order" a variety of events in the world led the Shang kings to feel a particular urgency about placating it with proper ritual offerings. As we move into the Western Zhou Dynasty, the

"Mandate of Heaven"—the idea that the right of the Zhou kings to rule was determined by the high god—makes obedience to the desires and standards of Heaven a central religious and political requirement. The term *tianming*, or "Mandate of Heaven," first appears in a bronze inscription from ca. 998 BCE and quickly became a central term of art in Zhou religious discourse. The idea that political power was the result of a supernatural mandate led to tremendous, and increasing, anxiety on part of the Zhou elite that the Heaven which gave them their power might, on the basis of its observations of their behavior, revoke this mandate.

Looking at the Eastern Zhou period (770–256 BCE), when the Zhou polity begins to fragment into a variety of independent, and often conflicting, states, supernatural surveillance and the threat of supernatural sanctions remain at the heart of interstate diplomacy and internal political, legal relations, and public morality. The fifth century BCE text *Mozi* argues that faith in ghosts and spirits must be encouraged among the people, because belief in, and fear of, supernatural agents is crucial to sustain moral behavior. For the majority of thinkers in early China, Heaven continued to function as a Boyerian "full access strategic agent" (Boyer 2001), aware of and prone to judge one's actions and inner thoughts.

Rituals of Collective Effervescence

The idea that religious practices may function to create larger, cooperative units out of collections of individuals is one that can, in the West, be traced back to the beginning of religious studies as an academic discipline. Émile Durkheim, for instance, famously argued that the apparent practical irrelevance of rituals is more than outweighed by the fact that they "put the group into action," serving to "bring individuals together, to multiply the relations between them and to make them more intimate with one another" (Durkheim 1915/1965:389) and create a state of "collective effervescence" (Durkheim 1915/1965:405). The theories of Durkheim and other pioneers have been revived by the cognitive science of religion, where evidence is accumulating that religious rituals appear to engage both emotions and motivations using music, rhythm, and synchrony to build group solidarity. For example, a growing body of evidence suggests that synchrony increases feelings of affiliation and may encourage acts of sacrifice for the group. Recent experimental studies have found that acting in synchrony—by marching, singing, or dancing in rhythm—increases feelings of affiliation, empathy, compassion, and connectedness (Valdesolo and DeSteno 2011; Valdesolo et al. 2010; Wiltermuth and Heath 2009), even among strangers. The joint experience of synchrony results in greater cooperation in subsequent group exercises, even in situations that require personal sacrifice. There is also some evidence that these effects emerge in childhood; for example, joint music making by preschoolers promotes prosocial behavior (Kirschner and Tomasello 2010). The ability of music, rhythm, and synchrony to instill commitment and trust is no doubt why militaries have employed for millennia drill routines to train soldiers and build armies (McNeill 1995). Such drill techniques appear to have spread by copying more successful groups, a form of cultural group selection.

In early China, most of the central religious practices are characterized by collective and coordinated physical movement, singing, dancing, and the intonation of sacred texts. A forerunner to Durkheim, Xunzi (third century BCE) was an early Confucian functionalist theorist of religion, who argued that the primary purpose of ritual activities was not—as most of his contemporaries believed—to serve the spirits, but rather to bind people together into effective cooperative wholes through synchronized group activities (Campany 1992).

Fictive Kin

Another strategy that religions appear to employ is to harness standard human familial emotions to foster cooperation within the larger religious "family" (Alexander 1987; Atran and Norenzayan 2004). Kinship terminology is common in religious groups. There are two different hypotheses about how these extensions of kinship may be used to influence our behavior. The strong version of this hypothesis is that by calling strangers "brothers," our kinship psychology is actually tricked into perceiving a genealogical relationship, and behavior is consequently adjusted in altruistic and sexually averse ways. An alternative view, the "extension hypothesis" is that using kinship labels facilitates the transmission process for social norms by helping people understand how they are *supposed* to feel and act toward others in that category. In this way, kin labels allow for the ready apprehension and transmission of social norms, without psychologically conflating kinship relationships with interactions among strangers.

The phenomenon of religiously grounded, metaphorically expanded kin group is clearly at work in early Chinese culture. In the Shang, it appears that the Lord on High was viewed with both trepidation and awe precisely because the Shang kings did not enjoy a special familial relationship with him. By the time we reach the Western Zhou, the Zhou rulers are attempting to cement their relationship with their similarly independent high god by creating metaphorical kinship ties with it. In the earliest Zhou texts, the relationship between the normative order of the cosmos and the political order of the Zhou is modeled on family relationships. Heaven—or the combination of "Heaven and Earth"—is often portrayed in these texts as the "father and mother" of the universe, and the Zhou king, in turn, as the metaphorical father and mother to the Chinese people. This metaphorical extension of the family is fundamentally linked to the supernatural normative order: it is the approval and support of the supreme god, Heaven, which makes the Zhou king the "Son of Heaven," and this status as Son of Heaven gives him his "Heavenly Mandate" to rule. It is perhaps no accident that this expansive conception of "all under Heaven," united in a great, metaphorical family—together with a quite sophisticated bureaucracy and other institutional innovations—allowed the Zhou to extend their sway over a remarkably large geographical area and to swallow up the other large and ancient cultures in neighboring regions (i.e., to expand via cultural group selection).

By the time of the Eastern Zhou period (770–221 BCE), and particularly the Warring States (479–221 BCE) period, metaphysically grounded family metaphors were foundational in ethical and political discourse (Schaberg 2001:137). Confucius, in what is arguably the foundational religious philosophical text of early China, declares that "all people within the Four Seas [i.e., the known world] are brothers." Indeed, one of the main tenets of early Confucianism is that public ethical behavior is a direct development of familial emotions, which are first to be perfected within the context of the biological family and then extended to the political realm (one's metaphorical family). Confucius's follower Mencius developed this idea into an explicit doctrine of "extension," whereby, for instance, innate feelings of compassion for one's genetic kin are to be gradually extended—through training in cultural learning and imaginative projection—to encompass strangers, and finally the entire world. Mencius can therefore be seen as anticipating the contemporary extension hypothesis.

Moral Realism

The fifth way of stabilizing norms is to postulate for them a supernatural origin, or otherwise provide them with some sort of supernatural authority. *Moral realism*, or the belief that one's moral intuitions are grounded in the metaphysical structure of the universe, both explains their psychological force and justifies their imposition on others (Haidt et al. 2008; Taylor 1989). Charles Taylor (1989) has argued for a basic distinction in human judgments between "weak" as opposed to "strong" evaluations. Weak evaluations, like one's preference for a particular flavor of ice cream, are subjective and arbitrary. Strong evaluations, on the other hand, derive their strength from being based on one or more explicit or implicit metaphysical claims, and are therefore perceived as having objective force rather than being a merely subjective whim. People are motivated to punish violations of strong evaluations and condemn such violations in metaphysical terms. For instance, a person might not particularly like chocolate ice cream and believe that the flavor of vanilla ice cream is superior. This individual does not, however, expect everyone to share this preference, and is certainly not moved to condemn others for preferring chocolate. People in modern Western societies are also generally not inclined to sexually abuse small children, but this is an entirely different sort of preference: abusing small children is felt to be *wrong*, and people condemn and are moved to punish anyone who acts in a manner that violates this feeling. Moreover, if pressed on the matter, this condemnation would be framed in metaphysical terms: defended on the basis of beliefs about the value of undamaged human personhood, or the need to safeguard innocence.

We hypothesize that moral realism emerged by first assembling the moral domain (category) using existing cognitive tools, such as essentialism, and then forging the link between this moral domain and supernatural agents by either (a) assigning the authorship of moral norms to universal gods or (b) tracing norms to the metaphysical structure of the universe. Moral realism sets up a bulwark against the spread of alternative views from powerful and self-interested coalitions. For example, the sacred quality of norms for monogamous marriage advanced by early Christianity—favored by cultural group selection in the environments of complex, trade-dependent, societies—allowed them to win over elite males who would otherwise be expected to resist the imposition of monogamy for their own fitness reasons (Henrich et al. 2012). In addition to creating a bulwark against interest groups, moral realism was also favored by cultural group selection because it motivates the assimilation of populations with alternative beliefs, and the active extermination of competing beliefs (think missionaries).

Much anthropological evidence indicates little or no connection between the moral and supernatural domains in small-scale human societies (Marshall 1962; Swanson 1960). On the basis of such evidence, many authors have argued that the connection between the moral and supernatural domains evolved over the course of human history (e.g., Swanson 1964). We agree with these observations, but also propose that cultural group selection drove these processes of change because moral realism influences the success of cultural complexes.

Recent experiments provide some support by showing that when norms are associated with the sacred (connected to the supernatural), they become emotionally charged and less subject to material calculations and practical tradeoffs (Tetlock 2003). In conflict situations, as in today's Middle East, material offers from one group to another to relax or abandon norms associated with sacred values generate moral outrage and increased readiness to support lethal violence (Ginges et al. 2007; Atran et al. 2007).

Turning to historical evidence from early China, our records of Shang religion are too sparse to tell us much in detail about the relationship of morality and meaning to the sacred, but the Shang supreme deity, Lord on High, was seen as the ultimate enforcer of at least ritual norms. By the time we reach the Western Zhou, Heaven and its Mandate are central to the moral order inhabited by the Zhou kings: the outlines of moral behavior have been dictated by Heaven and encoded in a set of cultural norms. A failure to adhere to these norms—either in outward behavior or one's inner life—was to invite instant supernatural punishment. As Eno (2009) observes, by the time of the Western Zhou, the idea of Heaven and the Heavenly Mandate had come to support a sophisticated and centrally important theodicy—a narrative detailing the religious and moral factors behind the Zhou's rise to power, and their continued hold on it—and becomes the basic organizing concept of Zhou religion. As he explains, by the Western Zhou, Heaven "has taken on the role of ethical guardian, rewarding and punishing rulers according to the quality of their stewardship of the state. The relationship of the ruler to the High Power has now added to worship the fulfillment of an imperative to govern according to moral standards" (Eno 2009:101) Eno quite plausibly sees the creation of this sort of ethical high god as an important contributing factor in the Western Zhou's unprecedented ability to expand militarily and politically, the clear theodicy and supernaturally mandated moral code both legitimizing the dynasty and providing a common sense of sacred history and destiny across the growing Zhou polity (Eno 2009).

By the time we reach the Warring States, we encounter a variety of views on the relationship of morality to supernatural authorities such as Heaven, reflecting the diversity that bloomed among the period's so-called "Hundred Schools" of thought. Confucius of the *Analects* believed himself to be on a mission from Heaven, charged with leading his contemporaries back to the practice of a set of traditional cultural norms revealed by Heaven to the ancient Zhou kings. One of Mozi's primary arguments in favor of his central doctrine of "impartial caring" was that it was modeled on the behavior of Heaven, who would actively punish those who went against its dictates and reward those who embraced them. Confucius's follower Mencius somewhat naturalized the Heavenly Mandate by turning it into an innate endowment embedded in each individual's nature. The primary warrant for valuing and developing this nature, however, was that it represented a gift from Heaven, and to neglect it would therefore be a direct affront to Heaven's will. In a similar vein, one of the recently discovered Confucian archaeological texts describes the cardinal human relationships and their attendant virtues as part of a "great constancy" (*dachang*) sent down by Heaven. *Tian* sometimes appears in a less anthropomorphic form in so-called "Daoist" texts, such as the *Daodejing* or *Zhuangzi*, but nonetheless continues to serve as the primary locus of normative value and meaning.

Religious Diversity and the Rise of Large-Scale Civilizations

An article in *The Economist* (2011), "Killings in Liberia: Nasty Business," documents the manner in which a recent spread of beliefs centered on witchcraft and sorcery-based killings have effectively paralyzed civil society throughout growing swathes of Liberia, creating an environment of such pervasive interpersonal suspicion and competition that not even the most basic forms of social cooperation can get off the ground. This case, which captures the antisocial effects of witchcraft in many societies (Knauft 1985),[1] illustrates that not all religious beliefs lead to prosocial behavior. While some, or all, of the features sketched out above are often taken to be typical of "religions" in general, there is reason to suspect that they actually represent relatively novel but successful products of a long cultural evolutionary process that has forged links between prosociality, morality, rituals, and deep commitments to supernatural agents or principles. Our central argument is that groups which succeeded in integrating the above features into packages of cultural elements (beliefs, rituals, and devotions) deepened group solidarity by incentivizing trust and cooperation with supernatural punishments and rewards, and were able to outcompete other groups. We believe that the gradual assembly of this cultural package was not only a key to the origin of large-scale societies, but also provides a convincing answer to the historical question of why religions with moralistic gods—rather rare among the panoply of human religious variety—have spread at the expense of other types of religion: cultural groups with religions that best promote within-group cooperation and harmony tend to outcompete other groups.

Significant advances in the study of religious cognition, the transmission of culture, and the evolution of cooperation are relatively recent. Bringing these new insights, in combination with older ideas, to bear on phenomena as complex as moralizing religions and large-scale societies is an ongoing challenge. The argument and evidence presented here provides a plausible scenario showing how synthetic progress is possible. More rigorous study is needed on the evolved psychology and cultural processes associated with the role of counterintuitive religious agents and costly rituals in up-scaling the scope of trust and exchange, of sacred values and taboos in sustaining large-scale cooperation against external threats, and also of maintaining social and political causes that defy self-interest. Empirical research that combines in-depth ethnography with both cognitive and behavior experiments among diverse societies, including those lacking a world religion, is crucial to understanding how religion influences our cognition, decision making, and judgments. The formal modeling of cultural evolutionary processes should be combined with historical and archaeological efforts to apply these emerging insights to broad patterns of history. Jointly, such efforts will further illuminate the origins of religions.

[1] For a discussion of how witchcraft may operate to enhance cooperation as societies expanded in scale, see Bulbulia et al. (this volume).

18

Rethinking Proximate Causation and Development in Religious Evolution

Harvey Whitehouse

Abstract

Efforts to understand cultural evolution, and its articulation with biological evolution, have tended to focus on problems of ultimate rather than proximate causation; that is on issues of function and selection rather than issues of mechanism and development. Although we now have sophisticated models of multilevel selection (Wilson 2002) and gene–culture coevolution (Boyd and Richerson 1985), we lack a similarly sophisticated account of the various levels at which proximate explanation needs to be understood. This chapter attempts to sketch out a more sophisticated framework for proximate explanation in religious evolution, inspired by C. H. Waddington's notion of the "epigenetic landscape." Building on this idea, three kinds of landscapes are disambiguated: epigenetic, cognitive-developmental, and social-historical. The discussion here focuses on religious phenotypes, but the general approach would be applicable to cultural practices more generally. The aim is to bring greater conceptual clarity and integration to a somewhat complex and messy cluster of research areas and, at the same time, open up new hypotheses ripe for investigation.

Introduction

The developmental pathways of biological organisms, minds, and social systems are intimately interconnected. This is not always obvious when conducting research at these different *explanatory* levels in light of discipline-specific questions, theories, and methods. Thus, most theories in the cognitive science of religion ignore efforts to establish the genetic and neurological foundations of religiosity. Social scientists are meanwhile notoriously skeptical of psychological and biological reductionism and seldom consider the shaping and constraining effects of cognitive and physiological processes. The resulting silo effect would not be a problem if processes unfolded at these different levels independently. But they do not. Efforts to show how they are related tend to approach the subject in a rather arbitrary and piecemeal fashion. What is needed is a more integrated conceptual scheme, one that generates systematic hypotheses and provides a more comprehensive and flexible understanding of proximate causation and development in religious evolution.

Waddingtonian Landscapes

A fruitful heuristic for thinking about proximate causation and development is provided by C. H. Waddington's famous notion of the "epigenetic landscape" (Waddington 1957). The basic idea is that the development of any phenotypic characteristic (whether morphological, physiological, or behavioral) is an outcome of both genetic and environmental factors in varying degrees. To represent this complex interaction, Waddington invited us to imagine a virtual landscape (Figure 18.1) in which the contours vary and to imagine developing traits (e.g., organs) as marbles rolling down through that landscape, their descent corresponding to a process of maturation over time. In this rather elegant metaphor, genes are represented as pegs and the effects of genes are represented as guy ropes. These guy ropes tug under the surface of the landscape so as to create furrows, canalizing development toward a steady end state (the mature phenotype). The idea is that where the tug of genes is weaker, the furrows in the landscape are shallower and therefore environmental influences can push the developing phenotype onto a new path, something that could not be accomplished by the effects of genes alone.

Waddington was admittedly proposing a mixed metaphor, combining the image of a tent (the canvas of which is held taught by pegs and guy ropes) and the image of a landscape (the contours of which are formed by quite different forces, such as erosion). Although mixed metaphors are considered a faux pas

Waddington	Epigenetic landscape
Pegs	Genes
Guy ropes	Biochemical or regulatory effects of genes
Landscape	Sum of the effects of genes and environment in producing a stable end state
Steepness	Genetic robustness Steep = genetic canalization Shallow = plasticity
End state (attractor)	Mature phenotype (morphology, physiology or behavior)

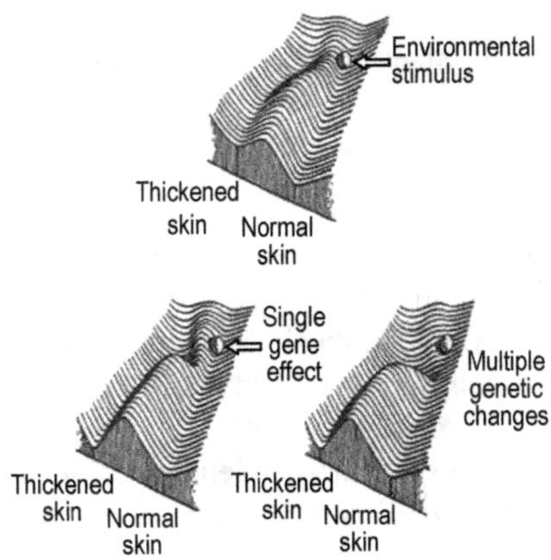

Figure 18.1 Waddington's epigenetic landscape.

in some literary circles, Waddington's works quite well because the surface of a hillside does in many ways resemble the wall of a tent. We could, however, dispense with the idea of a landscape altogether and simply think of the image of a badly pitched tent with furrowed walls and imagine developing phenotypic traits as raindrops sliding down the canvas. An added advantage of this modification is that it affords a source analog for the environment, in the form of a gusting wind that can tauten and relax the furrows of the fabric within the constraints imposed by genetic pegs and ropes. A similar modification is proposed below.

Others have also suggested thought-provoking revisions. Tavory et al. (2013) have recently extended Waddington's metaphor as a way of understanding the development of sociocultural systems. In their new version of the metaphor, pegs represent cultural traits of various kinds, and these can canalize the development of communities in much the same way as genes can canalize the development of an organ in the body. As in Waddington's original metaphor, the flatter parts of the landscape represent regions where the canalizing effects of pegs are less strongly exerted, allowing outside factors to push development in new directions. In the so-called "social-developmental landscape" these outside factors include the conscious strategies of agents in their efforts to accomplish various outcomes (Figure 18.2).

The general proposal advanced by Tavory et al. is original and thought provoking. Nevertheless it raises a host of unanswered questions. For example, what exactly do the pegs and guy ropes, etc., refer to in this social-developmental landscape? Do the pegs represent individual behaviors or recurrent *patterns* of behavior at a population level? Or do they represent *cultural maps* rather than behaviors? Is the mature phenotype a cultural system or a social group (or both or something else)?

Waddington's original analogy was not without its limitations. As noted above, the effects of the environment are not represented pictorially despite playing an important role in the story. In addition, simple one-to-one mappings of genes (pegs) and their expressions (guy ropes) do not capture well the kinds of processes described by animal geneticists nowadays. Such problems could be remedied, however, by extending Waddington's analogy such that *environmental* pegs and guy ropes *above* the landscape pull on its fabric so as to counter or exacerbate the effects of *genetic* pegs and guy ropes tugging from *below*. (These environmental dynamics need not be as Tavory et al. portray them, as we shall see.) To capture more of the complexity of the gene–phenotype relationship, instead of simple direct connections between pegs and their points of attachment to the landscape, one might imagine entangled guy ropes (somewhat more along the lines of Tavory et al.'s inverted landscape but without the inversion). Moreover, although Waddington envisaged the mature phenotype as a stable resting place in the epigenetic landscape, when applying this analogy to human development it would be more accurate to imagine the mature phenotype as a very gently descending valley floor where development

Jablonka	Social-developmental landscape
Pegs	Each peg is a distinctive cultural trait such as a norm, or skill, or story, or style of clothing, etc.
Guy ropes	Exert "pull" on the landscape via their entanglements
Landscape	Sum of the effects of cultural traits and agentive strategies in producing a stable end state.
Steepness	Steep = faithful inheritance of tradition? Shallow = innovation?
End state (attractor)	A more or less stable community or tradition?

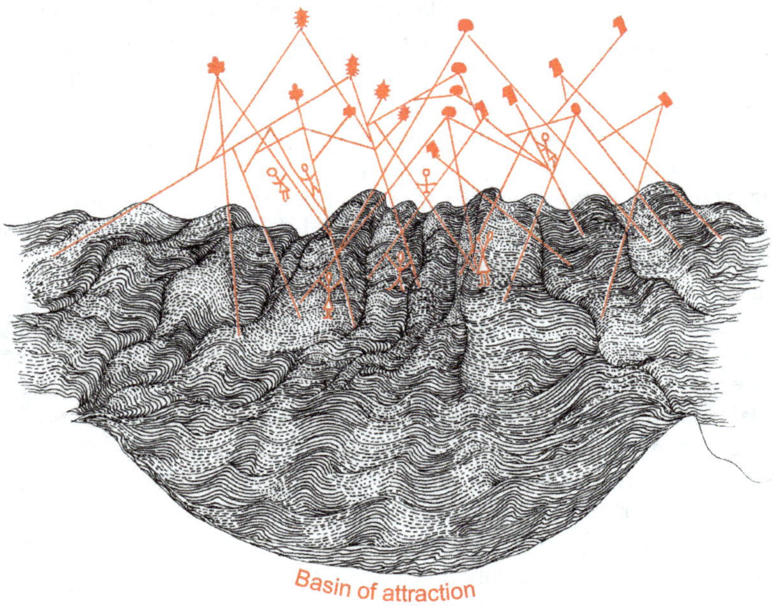

Figure 18.2 The social-developmental landscape (Tavory et al. 2013). Figure created by Anna Zeligowski and used with permission.

proceeds slowly but does not come to a halt (until death), in contrast with the much faster pace of development during immaturity.

Epigenetic Landscapes and Religious Bodies

Religion is intimately linked to physiological development in a wide range of ways through its influence, for instance, on diet, sexual behavior, drug use, and so on. Here we focus on some aspects of religiosity and brain physiology. The neurophysiology of religious thinking is little understood, but it is clearly the case that religious beliefs depend as much upon activity in the brain as would any other mental representations (Newberg and Waldman 2009). There has been some research linking religious experiences to temporal lobe epilepsy (Persinger 1983). Similar research by Urgesi et al. (2010), using brain-scanning techniques, suggests that feelings of transcendence induced by meditation are linked to decreased activity in the parietal lobe (a part of the brain involved in orienting the body to three-dimensional spaces). Other studies have sought to understand the *biochemistry* of certain aspects of religious experience, focusing on the role of neurotransmitters like dopamine in altered states of consciousness (Previc 2009). Whatever the long-term outcomes of these lines of research, they point to the presence of processes in the brain that are shaped by genes as well as culture and culturally constructed environments and practices. Religious traits are phenotypic characteristics, just like any other, and their development can be conceptualized in Waddingtonian terms in the same way as the development of organs.

To illustrate, let us consider one possible account, formulated by Boyer and Lienard (2006), of the neurological processes involved in the performance of religious rituals. Whether or not this account turns out to be correct in all (or indeed any) of its details, it can serve as an example of the epigenetic landscape of a religious phenotype. According to Boyer and Lienard, religious rituals (and cultural rituals of all kinds) activate a cluster of brain systems designed to respond to hazardous substances by triggering precautionary routines such as cleaning, separating, and straightening. They argue that these brain systems malfunction in patients suffering from obsessive-compulsive disorder (OCD), but where no such pathology is present, they serve a useful function biologically by causing people to handle potentially contaminating substances with special care. This biological mechanism, they suggest, is routinely hijacked by cultural systems that mimic the relevant input, in the form of religious rituals, for example, which similarly involve a concern with taboo or sacred materials and substances and stereotyped behavioral routines resembling those of OCD patients. Efforts to understand the neurological malfunction responsible for OCD have produced quite a detailed picture of the neural pathways involved. The most important network for understanding the hazard precaution system as described by Boyer and Lienard is a cortical-striato-pallidal-thalamic circuit (CSPT) that is connected to many other regions of the cortex via direct and indirect pathways. The basic idea is that this CSPT and its projections into the cortex, striatum, substantia nigra, and thalamus form the basic machinery for activating hazard precaution routines in normal individuals but that part of the system malfunctions in OCD patients, resulting in a felt need to repeat the routines over and over again.

Boyer and Lienard argue that the hazard precaution system has a distinctive developmental trajectory. They cite evidence that diagnostic features of the hazard precaution system, such as concern with "just right" object placement, cleaning, stereotypy, and repetition of routines, appear around two years of age and peak prior to puberty. This would seem to be consistent with the hypothesis that the evolved function of the hazard precaution system is to protect against infectious or contaminating materials during vulnerable phases of development, when exploratory play and learning is most intense and therefore the risk of exposure to hazards is most acute. Boyer and Lienard also present some evidence that OCD-like thinking peaks in women during pregnancy and in men after the birth of their first child.

To conceptualize the development of the hazard precaution system in terms of the epigenetic landscape metaphor, we might envisage a topography of relatively steep valleys during adolescence, indicating somewhat stronger genetic canalization during this phase of development (Figure 18.3). Here the "tug" of genes and their guy ropes underneath the landscape would be stronger than during most other life stages. (These effects would result from networks of genes rather than simple one-to-one mappings between genes and development.) There would then be further valleys during reproductive phases before the landscape flattens out and a stable or "mature" phenotype is achieved. I would propose that we extend Waddington's metaphor to represent the effects of the environment on this developmental process. Exposure to cues activating the hazard precaution system might be more or less frequent or intense for different individuals depending, for instance, on how ritualistic the environment is, how much concern there is with issues of hygiene and boundary marking, and how great the risk is of contamination and infection. A highly ritualistic religious system, such as Judaism, might serve to deepen the furrows in the landscape created by genetic canalization of the hazard precaution system and its neurological circuits. By contrast, a more iconoclastic religious system, such as a Protestant church eschewing ritual, might soften the contours of the landscape resulting in a different mature phenotype.

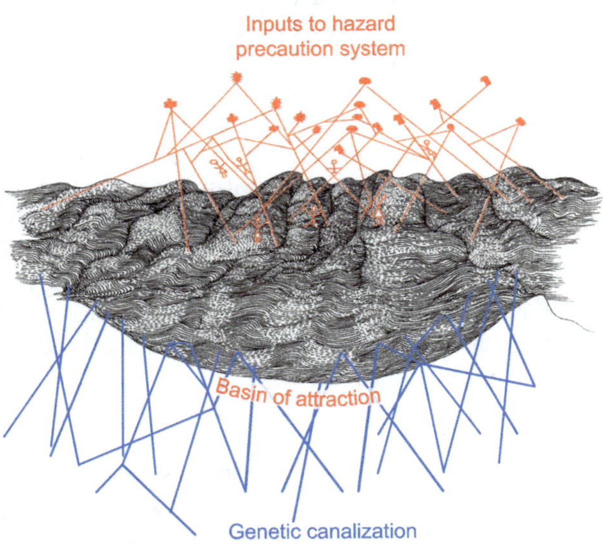

Figure 18.3 Picturing the modified epigenetic landscape.

Cognitive-Developmental Landscapes and Religious Thinking

Although cognitive processes are shaped and constrained by processes in the brain (and the genetic, cultural, and environmental influences shaping their expression and development), psychological systems are not wholly reducible to neurological ones. To understand the complexity of mentation, it is necessary to formulate theories of reasoning, memory, motivation, and emotional response at a level distinct from biological events. In short, we need to postulate a cognitive-developmental landscape as well as an epigenetic one. This cognitive-developmental landscape is closely analogous to Waddington's original scheme (Table 18.1). In this new scheme, however, pegs would represent species-specific, genetically canalized pathways in cognitive development and guy ropes would represent evolved cognitive constraints on learning. These constraints could be very strong (e.g., the impression that celestial objects move across the sky) but with a certain amount of learning and practice they can also be overcome to some extent (so we can appreciate that the earth is actually rotating).

The interaction of cognitive canalization and cultural learning in development results in more or less stable *semantic networks* in the minds of mature agents (Figure 18.4). Semantic networks are systems of representations, each representation being conceptualized as a node that is in turn linked to other nodes with varying frequency, credibility, and emotional salience. Not all nodes in a semantic network are equally easy to represent, believe, or remember. Some nodes or clusters of nodes are more intuitive than others, chiming more readily with maturationally natural and universal implicit beliefs

Table 18.1 Summary of the cognitive-developmental landscape.

	Epigenetic Landscape	Cognitive-Developmental Landscape
Pegs	Genes	Evolved cognitive capacities (does not require massive modularity)
Guy ropes	Biochemical or regulatory effects of genes	Cognitive constraints on learning
Landscape	Sum of the effects of genes and their effects plus environment on changing gravitational pull toward a stable end state (environment as geology?)	Sum of the effects of cognitive constraints plus cultural scaffolding on changing gravitational pull toward a more or less stable end state
Steepness	Steep = canalization Shallow = plasticity	Steep = cognitive canalization Shallow = cognitive plasticity
End state (attractor)	Mature phenotype (morphology, physiology, or behavior)	Mature phenotype (stable semantic networks)

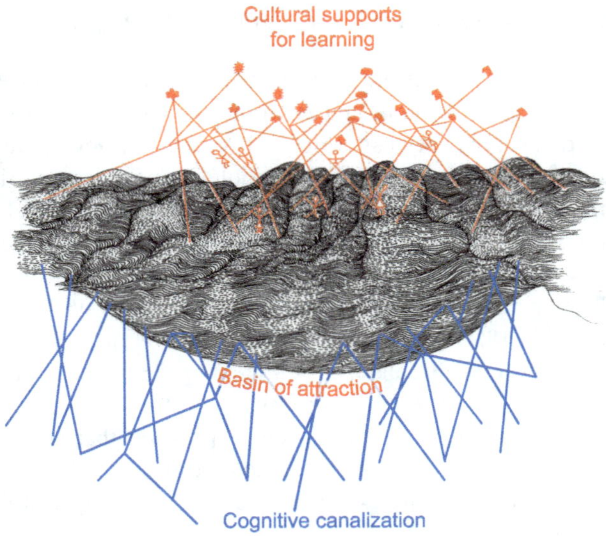

Figure 18.4 Picturing the cognitive-developmental landscape.

(McCauley 2011). By contrast, some nodes may be more counterintuitive or form part of more elaborated clusters, thus making them more difficult to acquire and maintain.

The theologically elaborated teachings of literate religions are often more counterintuitive than the "theologically incorrect" but nevertheless more widespread beliefs of rank and file laity. Consider some examples from the Roman Catholic cultural system. Many Catholics in Southern Europe cross themselves upon entering a church because they have an implicit belief that God is present in holy places even though he cannot be seen or heard. This belief forms part of a network of explicit representations supported by quite simple intuitive assumptions. Mind–body dualism, for instance, has a number of intuitive properties that would appear to recur in all human populations (Cohen et al. 2011). These include the expectation that the memories, beliefs, and desires of agents can occur outside bodies, can survive death, and can even move between bodies in the case of spirit possession or divine inspiration. Although disembodied agency is easy to represent and would therefore seem to be culturally universal, agency is intuitively tied to a location in space (Barrett and Keil 1996). Catholics do not appeal to relics several miles away but move within earshot of the relic in order to commune with it. By contrast, the Vatican teaches that God is omnipresent, a proposition that is much less intuitive and therefore difficult to implement as a guiding principle in worshipful practice. If the principle of omnipresence were really used as a guide to behavior, self-crossing would be no more necessary in churches than in other places, and it would be just as effective to address a holy relic from afar.

Similar disjunctions between intuitive lay beliefs and more counterintuitive theological systems have been observed with respect to many features of intuitive reasoning. Our predisposition to adopt teleological explanations for the functions of material objects makes it easier to represent features of the natural world as the purposeful creations of gods, ancestors, or other agents than as the products of erosion or descent with modification (Kelemen 2004). Likewise, our intuitions about immanent justice make it easier to imagine that wicked people will get their comeuppance than to absorb the intricate and often counterintuitive propositions of moral philosophy (Binmore 2005). To express this in Waddingtonian terms, some semantic networks are more strongly *canalized* than others in the course of development. In general, this means that intuitive nodes in semantic networks decay less rapidly than counterintuitive nodes.

Nevertheless, semantic networks can also exhibit remarkable plasticity. Rehearsal and review of a network of representations, even of a complex and counterintuitive network, can strengthen the links between nodes. Theologians overcome the limitations of intuitive reasoning producing religious phenotypes quite different from what would be expected based on processes of cognitive canalization alone (Slone 2004). There is some evidence that emotional arousal as well as rehearsal and repetition can aid the formation of elaborated counterintuitive religious systems. Costly signaling (Sosis and Alcorta 2003) and "credibility enhancing displays" (Henrich 2009a), such as self-flagellation or large charitable donations, can meanwhile increase the plausibility of semantic networks (see also Slingerland et al., this volume). Logic and narrative used in the teaching of religious doctrines can help to make a body of orthodox teachings more memorable as well as more coherent and believable (Whitehouse 2000). External mnemonics, such as sacred texts, can also help to preserve semantic networks over time.

Another approach that has proven useful in the study of religion's cognitive-developmental landscape is to construct agent-based models. This allows us to vary the effects of cognitive canalization, emotional salience, repetition, conformism bias, and prestige bias. Although computational simulations cannot tell us directly about the workings of the real world, they help us understand our own theories better, enabling us to generate more precise and testable hypotheses (Whitehouse et al. 2012).

Social-Historical Landscapes and Religious Traditions

Semantic networks can be communicated as public representations, for instance, by means of speech, text, or body decoration. Individuals sample public representations around them and update their semantic networks accordingly, with the result that meaning systems can be largely shared across entire populations. The sum of all people's semantic networks in a bounded population can be described as a "sociocultural system." As in Waddington's epigenetic landscape, the pegs in a sociocultural-historical landscape would represent information (Table 18.2), but rather than pegs encoding information in genes (or in minds in the case of the cognitive-developmental landscape) pegs would now represent a set of normative beliefs and behaviors in a population; in other words, its "social structure." Guy ropes would now represent the implementation of these rules in practice: sometimes people follow the rule book to the letter but at other times they innovate, as their individual strategies unfold on the ground (this is sometimes described as "social organization" as distinct from the normative rule book of social structure). The sum of the combined effects of social structure and social organization, on the one hand, and various environmental forces acting on a population (such as invasions and natural disasters), on the other, determine its historical trajectory. Inasmuch as some sociocultural systems eventually coalesce into relatively stable forms, they may be said to achieve their mature phenotype (change becoming very gradual like the aging process in the body). Like organisms, however, sociocultural systems have a finite life span (they "rise and fall" or evolve into something else). Some never accomplish stability or die young.

The formation of some religions is heavily canalized by social structure, for instance in the case of some of the stricter Protestant denominations of Christianity that maintain a rigid orthodoxy through the use of unrelenting repetition of the creed, its codification in text, and the supervisory prominence of doctrinal authorities. Other traditions, such as New Age cults and fashions,

Table 18.2 Summary of the social-historical landscape.

	Epigenetic Landscape	Social-Historical Landscape
Pegs	Genes	Shared semantic networks (aka "social structure": institutional instruction manual)
Guy ropes	Biochemical or regulatory effects of genes	Cumulative social-behavioral effects of the rules and models encoded in semantic networks (aka "social organization": agency in action)
Landscape	Sum of the effects of genes and their effects plus environment on changing gravitational pull toward a stable end state (environment as geology?)	Sum of the effects of social structure and organization plus environment (e.g., invasion, revolution, tsunami) on changing gravitational pull toward a more or less stable end state
Steepness	Steep = canalization Shallow = plasticity	Steep = social structural canalization Shallow = social structural plasticity
End state (attractor)	Mature phenotype (morphology, physiology, or behavior)	Mature phenotype (more or less stable cultural system, i.e., distributed and standardized semantic networks at the population level)

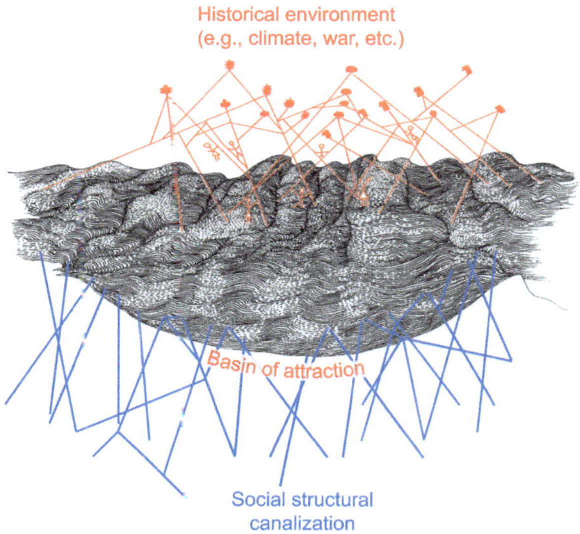

Figure 18.5 Picturing the social-historical landscape.

exhibit much greater plasticity, and the canalizing effects of social structure are weaker. Where the pull of social structure is weaker, the development of the tradition will be more easily influenced by changes in the historical environment. These environmental factors will exercise a correspondingly weaker impact in the more normatively canalized traditions (Figure 18.5).

Multilevel Landscapes in Religious Evolution

Epigenetic, cognitive-developmental, and social-historical landscapes shape and constrain each other. For example, genetically canalized neural systems constrain our psychological susceptibility to particular nodes in semantic networks. Practice and review of these networks can, however, extend our capacities for reasoning and memory (outcomes in the cognitive-developmental landscape), and thus the uniformity and stability of cultural representations across human populations (outcomes in the social-historical landscape). In other words, processes unfolding in any given landscape can have major consequences for processes in all the others.

The Evolution of Melanesian Cargo Cults

To illustrate, consider the many so-called "cargo cults"' of Melanesia that blossomed in the wake of colonization and missionization by Western powers, mainly during the twentieth century. The term "cargo cult" is typically used to

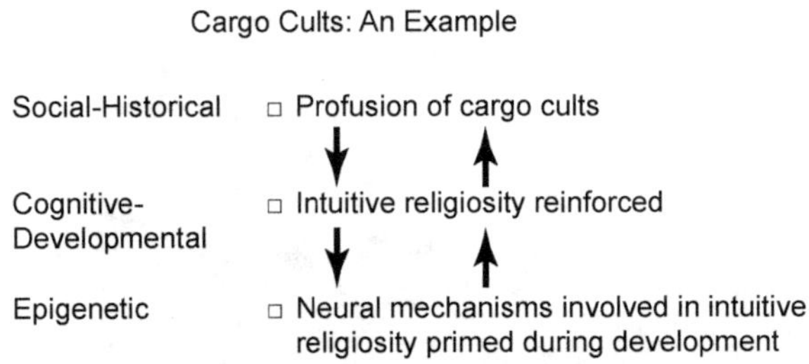

Figure 18.6 Some causal relations between epigenetic, cognitive-developmental, and social-historical landscapes in cargo cults.

refer to groups which believe that the wealth, technology, and high standard of living in postindustrial societies have a supernatural origin and can be obtained by appealing to spirits, ancestors, or gods. A variety of epigenetic, cognitive-developmental, and social-historical processes came into play in the evolution of the cargo cult phenomenon. Moral intuitions, for example, played an important role in the cognitive-developmental landscape. Cult leaders foretold that the great imbalances of wealth between native peoples and the colonists would be leveled or reversed and enemy tribes would be vanquished or punished. Prophets also appealed to intuitions about mind–body dualism (e.g., predicting that the spirits of the dead would be reincarnated), about the efficacy of ritual (e.g., by means of which the cargo would materialize), and about acts of creation (e.g., on the part of primordial ancestors). Little is currently known about the role of genetic processes in the development of the neural systems involved in intuitive reasoning about hazard precaution, immanent justice, mind–body dualism, and promiscuous teleology. Nevertheless, social-historical systems postulating supernatural agent concepts presumably fine-tune the maturation of intuitive reasoning in various ways, and this would naturally also impact the development of neural pathways. Thus, these three levels of analysis are interconnected (see Figure 18.6).

Interacting Landscapes in the Kivung

To explore how these interactions might unfold in practice, let us consider a concrete example: the Kivung movement of New Britain, Papua New Guinea (Whitehouse 1995). Like most cargo cults in Papua New Guinea, the Kivung was based on a set of highly intuitive propositions about human origins, the return of the ancestors, the arrival of cargo, and the righting of wrongs. However, whereas all other cargo cults in New Britain had flared up and then disappeared just as rapidly, the Kivung had unusual staying power and it spread to a much larger population than any of the earlier cults. Moreover, it offered a doctrinal system and orthopraxy more elaborate than any previous cult. Some of the movement's teachings went far beyond merely intuitive ideas, postulating a complex theology comparable to missionary Christianity, with all its attendant dogmas and a corpus of parables and stories linking them together. Kivung beliefs and practices formed an extensive religious system that became standardized in the 1960s and has remained much the same ever since. How are we to explain the success of this cultural system?

The Kivung was more persuasive, enduring, and widespread than other creeds in part because it exploited the cognitive-developmental landscape in a novel way: *It subjected its theories and stories to frequent repetition at public gatherings.* This seemingly minor alteration in the cognitive-developmental landscape allowed a single coherent body of teachings and practices to become standardized across the movement as a whole, changing the social-historical landscape (Whitehouse 1995, 2000, 2004). It meant that much more elaborate semantic networks than those featured in earlier cargo cults could now be sustained in memory. This would, of course, have had consequences for the development of neural processes involved in semantic memory, rote learning, narrative construction, oratorical expertise, and so on, but it would also facilitate the regional homogenization of the tradition and its orthodoxy at a population level (see Figure 18.7).

In the example just given, the ramifications of a change in one landscape had feedback effects in the others. That is, a change in cognitive development (doctrinal learning through rehearsal and review) occasioned changes at both epigenetic and social-historical levels, which fed back into cognitive development by strengthening certain forms of expertise and memory from below and stabilizing normative rules from above. Still, one can easily imagine other patterns, for instance where changes in the three landscapes occur in a cyclical fashion. Consider the following example.

When the Kivung was first established, its leaders declared that followers should no longer chew betel nut (a widespread and somewhat addictive practice in Papua New Guinea). The explicit rationale for tabooing betel nut was that the red substance produced (and spat around on the ground) was akin to menstrual blood, considered by some of the ethnic groups joining the Kivung to be polluting and dangerous. Linking betel nut to menstrual blood made the practice of chewing suddenly seem disgusting—a transformation in

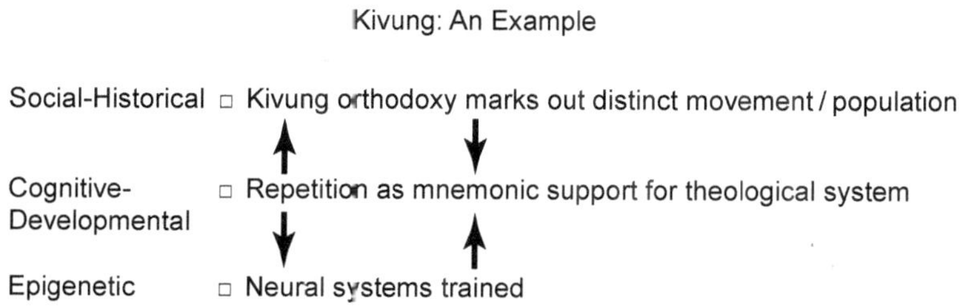

Figure 18.7 Some causal relations between epigenetic, cognitive-developmental, and social-historical landscapes in the Kivung.

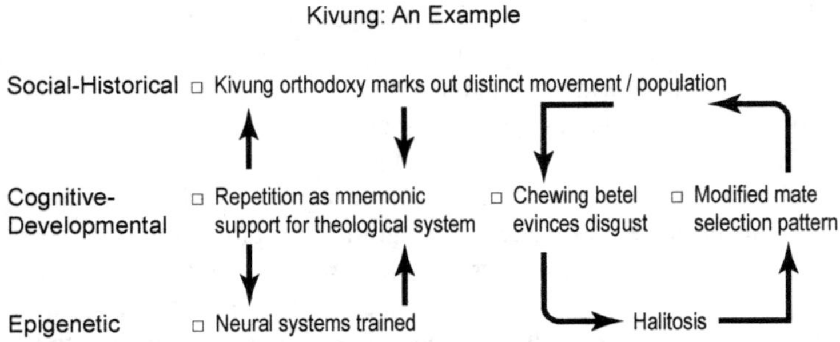

Figure 18.8 Further causal relations between epigenetic, cognitive-developmental, and social-historical landscapes in the Kivung.

the cognitive-developmental landscape of Kivung followers. An unintended consequence of this lifestyle change was the emergence of widespread dental problems, especially halitosis, which had previously been prevented by betel chewing. This change in the epigenetic landscape affected the perceived attractiveness of Kivung followers, at least to outsiders, and affected processes of mate selection in the cognitive-developmental landscape (see Figure 18.8). Resulting changes in patterns of marriage made the Kivung movement increasingly endogamous. This, in turn, sharpened the boundaries between in-group and out-group in the social-historical landscape. Sanctions were introduced to deter betel chewing among followers, reinforcing the normative social system but also impacting the psychological association of betel nut with sinfulness and pollution. In this way, a cyclical pattern of reinforcement for this novel aspect of life in the Kivung became established.

Of course there are many other patterns of this kind that could be hypothesized using the landscapes metaphor. The examples given here are merely to show how this can open up a new perspective on processes of religious evolution, and sociocultural evolution more generally.

Conclusions

The landscapes metaphor brings conceptual integration to a very complex set of multilevel relationships between proximate causation and development in religious evolution (and sociocultural evolution more generally). Magical thinking, mind–body dualism, creationism, and a host of other patterns of reasoning commonly associated with religion derive, at least partly, from evolved neural processes that are also influenced in development by social and cultural environments. This is hardly controversial and yet expressing it this way is vague and potentially confusing. What are the processes and what is their provenance? Efforts to answer that question have made progress but often only within narrow parameters, thus losing sight of the bigger picture. As a consequence, many researchers in the cognitive science of religion have argued that explaining religion is largely a matter of identifying universal cognitive predispositions and susceptibilities to believe certain things rather than others (e.g., Boyer 2001; Barrett 2004). Although this has produced much valuable research, it tends to overlook the fact that all religious representations are part of cultural systems with highly variable content and structure. These systems sometimes foreground intuitive ideas but can also exclude or obscure them with widely varying consequences for history, individual experience, and physiological processes—all of which often have important feedback effects. Scientific research on religion, therefore, suffers from a "silo effect" whereby neuroscientists look for religion "in the brain," cognitive scientists emphasize intuitive biases, and social scientists privilege systemic patterns at the level of populations. Hardly anyone seriously considers how these different aspects of religious evolution are interconnected.

As well as bringing greater clarity and integration, the landscapes metaphor opens up new hypotheses. If we look more closely at religion through the lens of the epigenetic landscape metaphor we begin to ask not only about processes in the brain but in the organism as a whole. The role of betel nut in dental hygiene is a fairly random example but consider how many religions taboo recreational drugs, most commonly alcohol. How has this affected or been affected by cognitive development for individuals growing up in culturally heterogeneous cities, where alcohol is freely available and widely used, and how has this impacted or been influenced by processes of sociocultural reproduction and change? Questions which traverse the boundaries of biology, individual psychology, and social systems are seldom asked or only posed in a way that narrowly focuses on the seemingly idiosyncratic interests of particular researchers, rather than being systematically derived from a single overarching conceptual framework. Although this chapter only sketches the contours of how such a framework might be developed, the aim is to generate discussion that would refine and improve the approach, so that it can be rendered more precisely and its implications for empirical research fleshed out more comprehensively.

Acknowledgments

This work was supported by an ESRC Large Grant (REF RES-060-25-0085) entitled "Ritual, Community, and Conflict." Some of the principal arguments were originally presented in the Chichele Series on the Evolution of Cognition at All Souls College, Oxford University (May 24, 2011), and I am grateful to its organizer, Celia Heyes, and to all the participants, especially Eva Jablonka and Kim Sterelny, for many insightful comments. I also benefited greatly from discussion of this material with Kevin Foster and Joanne P. Webster. Thanks are also due to Anna Zeligowski for permission to reproduce Figure 18.2.

19

Religious Prosociality

A Synthesis

Ara Norenzayan, Joseph Henrich,
and Edward Slingerland

Abstract

Religion is a ubiquitous aspect of human culture, yet until recently, relatively little was known about its natural origins and effects on human minds and societies. This is changing, as scientific interest in religion is on the rise. Debates about the evolutionary origins and functions of religion, including its origins in genetic and cultural evolution, hinge on a set of empirical claims about religious prosociality: whether, and through which particular pathways, certain religious beliefs and practices encourage prosocial behaviors. Here we synthesize and evaluate the scientific literature on religious prosociality, highlighting both gaps and open questions. Converging evidence from several fields suggests a nuanced pattern such that some religious beliefs and practices, under specific sociohistorical contexts, foster prosocial behaviors among strangers. This emerging picture is beginning to reveal the psychological mechanisms underlying religious prosociality. Further progress will depend on resolving outstanding puzzles, such as whether religious prosociality exists in small-scale societies, the extent to which it is constrained by in-group boundaries, and the psychology underlying various forms of disbelief.

Introduction

It has long been argued that religion facilitates acts which benefit others at a personal cost, a hypothesis that can be termed religious prosociality (Norenzayan and Shariff 2008). This idea has a long history in the social sciences (e.g., Durkheim 1915/1965; Wilson 2002; Darwin 1859) and has returned to center stage in recent debates about the evolutionary origins of religions (Bulbulia et al. 2008; Norenzayan and Shariff 2008; Atran and Henrich 2010). These debates revolve around (a) whether religion arose as a cognitive by-product of evolved cognitive biases (e.g., Boyer 2001, 2008; Barrett 2004; Lawson and McCauley

1990; Atran and Norenzayan 2004) or (b) whether religion (or some parts of it) is a genetic adaptation for cooperation either at the individual level (e.g., Bering 2006b, 2011; Johnson 2009; Sosis and Alcorta 2003; Schloss and Murray 2011) or through a process of multilevel selection (Wilson 2002). A third alternative synthesizes the cognitive by-product approach with cultural evolutionary theory (e.g., Richerson and Boyd 2005). In this view, a suite of cognitive biases lead to intuitions that support religious beliefs. Some cultural variants of these beliefs are then harnessed by cultural evolution and intergroup cultural competition to enable large-scale cooperation (Norenzayan 2013; Norenzayan and Shariff 2008; Norenzayan and Gervais 2012; Atran and Henrich 2010; Henrich 2009a; Slingerland et al., this volume; see also Wilson 2007a).

Progress on these debates critically depends on a number of empirical claims about whether, and through which specific pathways, religious beliefs and practices encourage prosociality. In surveying the evidence, we do not need, and do not offer, a strict definition of religion in terms of necessary and sufficient features, as religion is best seen as a family resemblance construct that consists of various converging elements (see, e.g., Atran and Norenzayan 2004; Boyer 2001). As in any other scientific enterprise, we think that an outline of the features of what is labeled "religion" cannot be decided a priori but emerges out of years of rigorous empirical and theoretical research. The theoretical claims and debates about the origins of religion are addressed elsewhere (Slingerland et al., this volume). In this chapter, we offer a nonexhaustive synthesis of the key aspects of the growing empirical literature for which competing (though not necessarily incompatible) evolutionary theories must account. We offer some conclusions, point to some apparent inconsistencies and possible resolutions, debate methodological challenges, and highlight outstanding questions for future research.

Surveys of Religiosity and Self-Reported Charitability

One of the earliest empirical works that links religion to prosocial behavior comes from sociology. A long line of survey findings conducted in the United States and elsewhere suggest that those who frequently pray and attend religious services (Christians and Jews of various denominations, as well as Muslims and Hindus) reliably report more prosocial behavior, such as more charitable donations and volunteerism (Brooks 2006). Brooks reports, for example, that 91% of religious people (defined as those who attend religious services weekly or more often) report donating money to charities, compared to only 66% secularists (defined as those who attend religious services a few times a year or less or those who declare no religious affiliation). The results for volunteering time are 67% (for religious people) versus 44% (for secularists). This "charity gap" is consistent across surveys and remains after statistically controlling for income disparities, political conservatism, marital status, education level, age, and gender. Some commentators cite these findings as evidence that religious people are more prosocial than nonreligious individuals (Brooks 2006).

There are, however, several limitations to these findings. One unresolved issue is whether this charity gap persists beyond the in-group boundaries of the religious groups (Monsma 2007). Another is the extent to which this finding generalizes to more secularized societies with stronger social safety nets, where governments have usurped the traditionally strong social functions of religious charities (Norris and Inglehart 2004). Third, a more serious limitation of these findings is that these surveys are based on self-reports of prosocial behavior, and are thus open to several alternative interpretations (for a critique, see Norenzayan and Shariff 2008). In psychology, a long line of work has shown that self-reports of socially desirable behaviors such as charitability or honesty are often exaggerated, and are strongly influenced by social desirability, impression management, or self-deception (Paulhus 1984). Therefore, the charity gap found in these surveys may be more reflective of "appearing good," rather than "doing good." This interpretation is plausible since religiosity is positively associated with socially desirable responding (e.g., Sedikides and Gebauer 2010). Finally, new experimental evidence suggests that this relationship is causal: religious reminders increase impression management concerns among believers (Gervais and Norenzayan 2012c). These findings raise serious questions about the validity of relying on self-reports to assess charitable behavior or generosity. To address these methodological limitations, experiments that assess prosocial behavior (not self-report of prosocial behavior) are necessary.

Correlating Religious Involvement and Prosocial Behavior

In social psychology, Batson and colleagues have systematically explored this question. In several behavioral studies under anonymous conditions, researchers failed to find any reliable association between religiosity and prosocial tendencies (Darley and Batson 1973). More recent studies have similarly found no strong evidence to associate religiosity with prosocial behavior in anonymous contexts in the United States (Paciotti et al. 2011). Subsequently, several laboratory studies with Christian university student participants in the United States have found that religious involvement does predict more prosocial behavior, but only when the prosocial act could promote a positive image for the participant, either in their own eyes or in the eyes of observers (Batson et al. 1993).

Other behavioral studies have also found reliable associations between various indicators of religiosity and prosociality, albeit under limited conditions. A study employing a common pool resource game allowed researchers to compare levels of cooperation and coordination between secular and religious kibbutzim in Israel (Sosis and Ruffle 2003). In this game, two members of the same kibbutz who remained anonymous to each other were given access to an envelope with a certain amount of money. Each participant simultaneously decided how much money to withdraw from the envelope and keep. Players only kept the money they requested if the sum of their requests did not exceed the total amount in the envelope. If it did, the players received nothing. Controlling for relevant variables, participants showed higher cooperation in the religious kibbutzim than in the secular ones; the effect was driven by highly religious men, who engaged in daily and communal prayer, and took the least amount of money from the common pool. A study conducted by Soler (2012), among members of an Afro-Brazilian religious group, showed similar results. In this public goods game, participants were divided into n-person anonymous groups. Each participant was given an equal monetary endowment, any portion of which they could keep for themselves or contribute to a common pool. Any contribution to the common pool would get doubled, then distributed equally back to the participants. Controlling for various sociodemographic variables, individuals who displayed higher levels of religious commitment behaved more generously and reported more instances of both giving and receiving within their religious community. Ahmed (2009) found similar results in a public goods game in a study conducted in rural India with a Muslim population. Devout Muslim students in a *madrassah* contributed more to a public good compared to a matched group of students in a secular school. The effect was sizable: whereas 15% of secular participants contributed nothing, only 2% did not contribute anything in the more religious group.

Prosocial religions, such as Christianity, Islam, and many variants of Hinduism, endorse a package of beliefs and practices that revolves around powerful, omniscient, and morally involved gods who demand credible displays of faith from their adherents. In an investigation spanning 15 societies of pastoralists and horticulturalists, Henrich et al. (2010a) measured the association between religious participation and prosocial behavior in three standard bargaining games. In the dictator game, two anonymous players are allotted a sum of real money (a day's wage) in a one-shot interaction. Player 1 must decide how to divide this sum between himself and Player 2. Player 2 then receives the allocation from Player 1. The ultimatum game is identical to the dictator game, except that Player 2 can accept or reject the offer. If Player 2 specifies that he would accept the amount of the actual offer, then he receives the amount of the offer and Player 1 receives the rest. If Player 2 specifies that he would reject the amount offered, both players receive zero. Unlike previous studies, this game specifically tested the idea that participation in prosocial religions engenders more prosocial behavior compared to participation in local religions that typically do not have a prosocial dimension. Henrich et al. found that, controlling for a host of demographic and economic variables, participation in a world religion (Christian or Muslim) increased offers in the dictator game by 6%, and in the ultimatum game by 10% (when the stake was standardized at 100).

Interestingly, however, world religion did not reliably predict offers in another economic game: the third-party punishment game, which allows people to punish others for not playing fairly. In this experiment, people in some societies also made lower (less equal) offers. Analyzing the data from all three experiments indicates that adding the third-party punishment drove out the component of prosociality created by religion. Combined with other recent findings which show that secular and divine sources of punishment are perceived to be interchangeable (Laurin et al. 2012), this suggests that adding a third-party punisher "replaces god" in a sense, leading to both lower offers and no impact of religion in this experiment.

There are several potential pathways through which religion might operate to increase prosociality. One possible pathway, which we explain further below, is the supernatural monitoring hypothesis: religious believers act prosocially to the extent that they experience being under supernatural surveillance by watchful, moralizing gods (Norenzayan and Shariff 2008; Bering 2011). Relatedly, another potential complementary pathway involves extravagant rituals or seemingly costly devotions (Slingerland et al., this volume; Xygalatas et al. 2013). Such practices can sustain greater prosociality and social solidarity because, as credible displays of deep faith, they lead to more successful cultural transmission of these belief-ritual complexes (Henrich 2009a). Alternatively, or in addition, ritual participation may, through various mechanisms, serve as a cooperative signal, encouraging greater prosocial behavior (Sosis and Alcorta 2003; Bulbulia 2004). Whitehouse (this volume) theorizes that infrequent, high-arousal rituals build solidarity at the level of relatively small social units, whereas the frequent, low-arousal rituals of larger-scale societies foster cultural cohesion on a broader social scale. There likely are other pathways as well (for discussion, see Slingerland et al., this volume).

The anthropological record is consistent with these ideas. In moving from the smallest-scale human societies to the largest and most complex, Big Gods (i.e., powerful, omniscient, interventionist supernatural watchers who demand extravagant displays of loyalty) go from relatively rare to increasingly common (Roes and Raymond 2003; Swanson 1966), and morality and religion move from largely disconnected to increasingly intertwined (Wright 2009). As societies get larger and more complex, ritual forms also change, becoming more frequent and dogmatic, increasingly used to transmit and reinforce religious orthodoxy (Atkinson and Whitehouse 2011). A recent cross-cultural study (Atkinson and Bourrat 2010) provides evidence that participation in prosocial religions goes hand in hand with a stricter moral enforcement of norms. In a large global sample of 87 countries from the World Value Survey, beliefs about two related sources of supernatural monitoring and punishment—God and the afterlife, as well as frequency of religious attendance—were found independently to predict harsher condemnation of a range of moral transgressions, such as cheating on taxes or fare-skipping on public transport. Importantly, belief in a personal God was more strongly related to these outcomes than belief in an abstract impersonal God.

Reconciling Inconsistent Findings on Religious Prosociality

In recent laboratory studies conducted in Western societies (mostly with university students), where prosocial behavior is measured in anonymity, individual differences in religious commitment typically fail to predict prosocial behavior reliably (for a discussion, see Norenzayan and Gervais 2012). This is similar to earlier findings which indicate that religious participants show greater prosocial tendencies when the prosocial act can enhance one's self-image, but that religiosity is a null predictor when no such reputational incentives are available (e.g., Batson et al. 1993). These findings deserve more scrutiny. Why does religious involvement predict prosocial behavior in some studies, but not others? Here we propose three explanations to resolve these inconsistencies.

One explanation is that, compared to a typical social psychology study with student samples, reminders of religion are likely to be more chronically present in religious kibbutz, *madrasahs*, and Candomblé communities, where religious prayer and attendance are a daily part of life. This is important because any behavior is more likely to occur to the extent that concepts associated with these behaviors are primed through situational cues (e.g., Bargh and Chartrand 1999).

A second explanation is that prosociality in these communities clearly benefits in-group members (despite being anonymous), whereas in psychological studies that are conducted in anonymous contexts, the victim or the recipient of generosity typically is a total stranger. In the classic "Good Samaritan" study (Darley and Batson 1973), for example, seminary students were led to walk past a stranger (actually, a confederate of the researcher) lying on the ground who appeared in need of help. Levels or types of religious involvement failed to predict helping rates.

A third important factor that helps reconcile these null findings with the literature reviewed above is cultural differences in the strength of secular institutions. Note that all of the studies which found weak or no reliable associations between religiosity and prosociality were conducted on Western, Educated, Industrialized, Rich, and Democratic (WEIRD) samples (Henrich et al. 2010b), whereas all of the studies which found reliable associations were typically conducted on non-WEIRD samples. In WEIRD societies, high trust levels toward secular institutions (e.g., the police, courts, governments) encourage high levels of prosocial behavior across the board (Hruschka and Henrich, submitted) and might crowd out the influence of religion on prosociality. Conversely, in societies with weak institutions, religion has no credible alternative and is the main driver of broad prosociality. Consistent with this idea, in societies with strong institutions such as Canada, experimentally induced subtle reminders of secular authority (e.g., concepts such as police, court, judge) reduce believers' reliance on religion as a source of morality (Gervais and Norenzayan 2012c). Furthermore, in a cross-national analysis that controlled for a number of relevant factors such as human development, general trust, and individualism, it was found that believers are more trusting of atheists in politics if they are culturally exposed to strong secular institutions as measured by the World Bank's index (Norenzayan and Gervais 2013b).

In summary, a growing number of behavioral studies have found associations between religious commitment and prosocial tendencies, especially when secular sources of prosocial behavior are unavailable (i.e., weak institutions), reputational cues are heightened (e.g., helping is not anonymous), and the targets of prosociality are in-group members (we will return to this latter point below). However, causal inference in these studies is limited by their reliance on correlational designs. If religious devotion is related to prosocial behavior in some contexts, it cannot be conclusively ruled out that having a prosocial disposition causes one to be religious or that a third variable, such as dispositional empathy or guilt proneness, causes both prosocial and religious tendencies. Recent controlled experiments have addressed this issue by experimentally inducing religious thinking and subsequently measuring prosocial behavior.

Experimental Evidence: Religious Priming

If religious belief has a causal effect on prosocial tendencies, then experimentally induced religious thoughts should increase prosocial behavior in controlled conditions. If so, subtle religious reminders may reduce cheating, curb selfish behavior, and increase generosity toward strangers. This hypothesis was tested and supported in two anonymous dictator game experiments: one used a sample of university students while the other used nonstudent adults in Canada (Shariff and Norenzayan 2007). In one experiment, adult nonstudent participants were randomly assigned to three groups:

1. Participants in the religious prime group unscrambled sentences that contained words such as *God*, *divine*, and *spirit*.
2. The secular prime group unscrambled sentences with words such as *civic*, *jury*, and *police*.
3. The control group unscrambled sentences with entirely neutral content.

Each participant subsequently played an anonymous one-shot dictator game (described above). Post-experimental interviews showed that participants were unaware of religious content and remained naïve concerning the hypothesis being tested. Compared to the control group, nearly twice as much money was offered by subjects in the religious prime group, who not only showed a quantitative increase in generosity but also a qualitative shift in social norms. In the control group, the modal response was selfish: most players pocketed the full ten dollar stake allotted to them. In the religious prime group, the mode shifted to equality: participants split the money evenly. Of particular interest, the secular prime group had as much effect as the religious prime group. This suggests that secular mechanisms, when available, can also encourage generosity.

These findings have been replicated with a Chilean Catholic sample and show similar religious priming effects on generosity in the dictator game and on cooperation levels in the prisoner's dilemma game. In the latter game, self-interest leads both parties to not cooperate, but cooperation leads to better reward for both (Ahmed 2011). Religious primes have also been shown to reduce cheating among student samples in North America (Randolph-Seng and Nielsen 2007), as well as in children (Piazza et al. 2011). McKay et al. (2011) found that subliminal religious priming increased third-party costly punishment of unfair behavior in a Swiss sample, but only for religious participants who had previously donated to a religious charity (for similar results regarding altruistic punishment, see Laurin et al. 2012). Taking a "situational priming" approach, Xygalatas (2013) randomly assigned Hindu participants in Mauritius to play a common pool resource game (described earlier), either in a religious setting (a temple) or in a comparable secular setting (a restaurant). He found that participants, regardless of their self-reported intensity of religiosity, withdrew less from the shared pool of money when they played the game in the temple compared to when playing in the restaurant.

There is some evidence that priming effects are to some extent parochial as well as prosocial, as prime-induced religious prosociality is sensitive to group boundaries. This question is open for detailed investigation. Currently we know of one preliminary study with Canadian Christians (Shariff and Norenzayan, unpublished) which suggests that, in a one-shot dictator game, religiously primed Christian givers were most generous toward a Christian receiver, less generous toward a stranger with unknown religious affiliation, and even less generous toward a Muslim receiver (playing with a Muslim receiver was the equivalent of not being primed with religious words). This is not surprising given that human prosocial behavior is shaped by parochial concerns (Koopmans and Rebers 2009).

In summary, a small but growing literature shows that the arrow of causality goes from religion to a variety of prosocial behaviors, including generosity, honesty, cooperation, and altruistic punishment. Next we examine the psychological mechanisms underlying these religious priming effects and explore evidence that these effects are due, at least in part, to perceptions of being under supernatural monitoring.

Why Do Religious Reminders Increase Prosociality?

What are the psychological processes that might explain the empirical link between religious primes and prosociality? Two distinct accounts suggest themselves (for the potential role of development, see Whitehouse, this volume). First, the *supernatural monitoring* account argues that heightened awareness of being under social surveillance increases prosociality. Thoughts of religions invariably activate reminders that God or gods—omniscient and morally concerned judges—are watching (Gervais and Norenzayan 2012b). Granted, as an ultrasocial species, humans can be prosocial even when no one is watching (Henrich and Henrich 2007; Barmettler et al. 2012). Nevertheless, being under social surveillance encourages prosociality. A large number of studies show that feelings of anonymity—even illusory anonymity, such as the act of wearing dark glasses or sitting in a dimly lit room—increase the likelihood of selfishness and cheating (Zhong et al. 2010; see also Hoffman et al. 1994). Conversely, social surveillance (e.g., being in front of cameras or audiences) has the opposite effect. Even incidental and subtle exposure to representations of eyes encourages good behavior toward strangers in the laboratory (Haley and Fessler 2005; Rigdon et al. 2009) as well as in real-world settings (Bateson et al. 2006; for a critique, see Fehr and Schneider 2010). As the saying goes, "watched people are nice people." It is no surprise, then, that the notion of supernatural watchers who observe, punish, and reward morally relevant behaviors has spread culturally in prosocial religions.

A second possibility is the *behavioral priming* or *ideomotor* account. The idea behind this hypothesis is that prosocial behavior is more likely if concepts related to benevolence or generosity are unconsciously activated (e.g., Bargh et al. 2001). If thoughts of God are associated with notions of benevolence and charity, then priming these thoughts may activate prosocial behavior, just as activating the social stereotype of the "elderly" increases behaviors consistent with it, such as slow walking speed (Bargh and Chartrand 1999; for this interpretation, see Pichon et al. 2007; Randolph-Seng and Nielsen 2007). To be clear, these two accounts are not mutually exclusive and in fact may operate together to produce prosocial effects of religion. The vital question is not whether ideomotor effects result from religious primes—they almost certainly do. Instead, it is important to ask whether supernatural monitoring effects *also* result from religious primes.

What evidence can distinguish the supernatural watcher account from behavioral priming processes? Norenzayan et al. (2010) discuss three empirical criteria. First, if the supernatural watcher account is in play, religious primes should arouse both feelings of external authorship for events and perceptions of being under social surveillance independent of any prosocial behavior. Second, if religious priming effects are weaker or nonexistent for nonbelievers, then the effect could not be solely due to ideomotor processes, which are argued to be impervious to prior explicit beliefs or attitudes associated with the behavior (e.g., see Bargh et al. 2001; Bargh and Chartrand 1999). This is because everyone, including nonbelievers, is aware of (although they do not necessarily endorse) the association between religious concepts and benevolence. Therefore, if ideomotor processes are solely responsible for these effects, awareness should be sufficient to trigger priming effects. Third, differing perceptions of supernatural agents can disentangle these two accounts. Specifically, the supernatural monitoring hypothesis predicts that the belief that God is punitive should encourage more prosociality, whereas the ideomotor account would lead to the contrary expectation; namely, that belief in a benevolent God is a stronger motivator for prosocial behavior.

Addressing the first question, several religious priming experiments clearly separate the felt presence of a supernatural agent from their prosocial outcomes. Dijksterhuis et al. (2008) found that after being subliminally primed with the word "God," believers (but not atheists) were more likely to ascribe an outcome to an external source of agency, rather than their own actions. In four studies, Gervais and Norenzayan (2012b) followed up on this line of reasoning and found that thinking of God does, indeed, influence variables that are sensitive to perceived social surveillance, independent of any ideomotor effects associated with benevolence or prosociality. The results suggest that religious primes trigger not only notions of benevolence, but also experiences associated with mind perception (i.e., feelings of being observed by an intentional agent) as the supernatural monitoring hypothesis predicts (for evidence that religious agents trigger mind perception, see also Norenzayan et al. 2012).

To address the second question, it is necessary to reexamine the priming literature in light of the second criterion: Do God primes influence behavior independent of prior belief, or are these effects confined to believers? Ideomotor processes typically do not interact with prior belief. A supernatural monitoring account, on the other hand, would suggest that people who believe in the actual existence of supernatural beings should be most susceptible to these primes, whereas nonbelievers should be less susceptible. The answer to this question is also crucially important for debates about evolutionary origins of religion. Genetic adaptationist accounts of religious prosociality (for a discussion, see Schloss and Murray 2011) would predict that everyone, even self-declared atheists, are responsive to supernatural monitoring effects (e.g., Bering 2011). Cultural evolutionary accounts of religious prosociality, on the other hand, are more compatible with the prediction that responsiveness to supernatural monitoring is culturally variable (e.g., Norenzayan and Shariff 2008; Henrich et al. 2010a). To be clear, socialization with culturally variable concepts of the divine could produce effects on prosociality that supplement or compete with universal religious tendencies to behave prosocially. Therefore, cultural variability is not incompatible with a genetic adaptationist account, provided there is no complete absence of an effect for nonbelievers. Moreover, the answer to this question reveals critical details about the psychology of atheism, a topic of great importance ripe for research, but unfortunately beyond the scope of this report (for further discussion, see Norenzayan and Gervais 2013a).

A review of the (admittedly limited) relevant evidence suggests that at least some nonbelievers are impervious to religious priming effects, a finding that is compatible with the idea that supernatural monitoring plays a part in religious priming effects. There is currently mixed evidence as to whether religious priming effects (typically bypassing conscious awareness) interact with explicit belief (for discussion, see Norenzayan et al. 2010). Some studies have found religious priming effects—irrespective of the explicit prior religious belief of participants—on honesty (Randolph-Seng and Nielsen 2007), generosity in the dictator game (Shariff and Norenzayan 2007, Study 1), public self-awareness (Gervais and Norenzayan 2012b, Study 3), and prosocial intentions (Pichon et al. 2007). Several other studies, however, found significant interaction with prior religious belief, reflecting null effects for nonbelievers (Dijksterhuis et al. 2008; Shariff and Norenzayan 2007, Study 2; McKay et al. 2011; Gervais and Norenzayan 2012b; Piazza et al. 2011; Laurin et al. 2012). In a recent meta-analysis of religious priming effects on prosocial behavior, Shariff, Willard, Andersen, and Norenzayan (unpublished) found a reliable and sizable effect for religious believers. However, on average, religious priming was unreliable and statistically nonsignificant for nonbelievers. Again, this suggests there is much variability in the extent to which nonbelievers are responsive to religious reminders. Laurin et al. (2012) found similarly that the effects of reminders of God were specific to believers only, and led to increased punishing behavior. Furthermore, believing that God is punishing caused *less* punishing behavior (presumably because participants could offload punishing duties to God). This last point is the opposite of what one would predict from the ideomotor account.

Further examination of the priming studies portrays a revealing pattern: all of the priming studies that have shown no interaction with prior belief have also recruited exclusively American university student samples. However, student atheists, particularly in religious America, might be "soft atheists." In one religious priming experiment that recruited a nonstudent adult sample in Vancouver, Canada (Shariff and Norenzayan 2007, Study 2), the effect of the prime emerged for believers, but disappeared entirely for "hard" atheists. Similarly, in the majority nonreligious Netherlands, Dutch student nonbelievers were not responsive to religious priming effects, even when they were presented subliminally (Dijksterhuis et al. 2008). Finally, in the more secular environment of Vancouver, no reliable priming effects were found on student nonbelievers across several studies (Gervais and Norenzayan 2012b).

Finally, consistent with the theoretical idea that punishment is superior to reward in sustaining prosocial behavior, there is a *negative* relationship between cheating behavior and the degree to which people endorse a vision of God as punitive and judging, whereas cheating rates *increase* with the belief that God is benevolent and forgiving (Shariff and Norenzayan 2011; Debono et al., unpublished). Consistent with these experimental findings, cross-national analyses (Shariff and Rhemtulla 2012) reveal that, controlling for a number of relevant socioeconomic and psychological variables such as gross domestic product, economic inequality, belief in God, and relevant personality dimensions, belief in hell is negatively related to crime rates, whereas belief in heaven has the opposite effect. As with the findings by Laurin et al. (2012), these results are difficult to reconcile with a purely ideomotor account, which presumably would lead to the opposite expectation (i.e., that a benevolent and kind God would more clearly fit the prosocial stereotype that causes greater prosocial behavior and less antisocial behavior, and that reminders of a benevolent God would reduce punishing behavior).

To summarize what we know about the psychological mechanisms underlying religious priming, several lines of evidence show that religious reminders increase the perception of external authorship of events and perceptions of social surveillance independent of any prosocial consequences. In addition, there is mounting evidence that the effects of religious primes are most effective among believers, and there is provocative (though preliminary) evidence that mature nonbelievers are less susceptible, and possibly immune, to these primes. A reasonable initial conclusion from the empirical evidence is that, at the very least, both accounts remain viable. Therefore, the supernatural monitoring hypothesis and the ideomotor hypothesis may reflect the operation of independent psychological mechanisms that link religion to prosocial tendencies. These mechanisms also have differing theoretical implications for the relationship between religion and prosociality. Whereas the ideomotor hypothesis posits that the link between religion and prosociality is the *consequence* of a cultural association reflected at the cognitive level, the supernatural monitoring hypothesis speaks to the more basic evolutionary question of *why* religion might *cause* large-scale anonymous prosociality in humans. If reminders of moralizing gods make people feel watched, then beliefs in moralizing gods, who can monitor social interactions even when no humans are watching, may have been instrumental in promoting large-scale human cooperation.

Ethnographic and Historical Evidence: How Supernatural Monitoring Contributed to Large-Scale Prosociality

Over time and as groups gain in size, morality and religion move from being disconnected to increasingly intertwined, and gods become more powerful, moralizing, and interventionist (Wright 2009). Ethnographic work shows that in foraging and hunting groups, such as the Hadza or the San, religion does not have a moral dimension and the gods are largely indifferent to human moral affairs (Boyer 2001; Swanson 1966). In an earlier assessment of the ethnographic record, Swanson (1966:153) concluded:

> The people of modern Western nations are so steeped in these beliefs which bind religion and morality that they find it hard to conceive of societies which separate the two. Yet most anthropologists see such a separation as prevailing in primitive societies.

Here we briefly highlight ethnographic and historical evidence that indicates that across groups and over time, supernatural monitoring coevolved with increasingly large, complex, cooperative societies.

In his review of 427 societies from the Ethnographic Atlas, Stark (2001) found that only 23.9% acknowledge a god who is active in human affairs and is specifically supportive of human morality. Religions with such gods are, in fact, peculiar. Yet, the vast majority of human beings today live in prosocial religious groups with big moralizing gods. Going further, in one notable analysis using the standard cross-cultural sample, Roes and Raymond (2003) showed that the variability in supernatural sanctioning found in the ethnographic record is correlated with group size: the larger the group size, the more likely the group has culturally sanctioned omniscient, all-powerful, morally concerned deities who directly observe, reward, and punish social behavior. This highlights one problem with much of the work in the psychology of religion, as Christianity is often used as a representative religion, when in fact it is a rather unusual religion.

These ethnographic findings converge with what can be gleaned from historical analyses. The archaeological record is, of course, limited, but available evidence hints at the possibility that the expansion of regular rituals and the construction of religiously significant monumental architecture co-emerged with increasing societal size, political complexity, and reliance on agriculture (Marcus and Flannery 2004). Evidence for this can be found in Çatalhöyük, a 9,500-year-old Neolithic site in southern Anatolia (for a discussion, see Whitehouse and Hodder 2010). The excavation of Göbekli Tepe, an 11,000-year-old complex of monumental architecture, suggests that it may have been one of the world's first temples, where hunter-gatherers possibly congregated and engaged in organized religious rituals (Schmidt 2010).

Once the written historical record begins, it becomes much easier to establish clear links between large-scale cooperation, ritual elaboration, and powerful gods who police human behavior. This historical work is ongoing, and many questions are being actively debated. However, some historical patterns have emerged. The best documented historical work looks at Abrahamic faiths. Wright (2009) provides a useful summary of textual evidence that reveals the gradual evolution of the Abrahamic God from a rather limited, whimsical, tribal war god—a subordinate in the Pantheon—to the unitary, supreme, moralizing deity of two of the world's largest religious communities. Evidence from early China also shows that supernatural monitoring played a key role in the emergence of the first large-scale societies in East Asia (see Slingerland et al., this volume). Turchin (2009) offers an account of how Axial Age religions fostered cohesion among agrarian societies. In an analysis that compares the longevity of religious and secular communes in nineteenth-century America, Sosis and Bressler (2003) found that religious communes outcompeted secular ones, and this survival advantage was statistically explained by the costly displays and restrictions on behaviors that religious communes imposed on their members (Henrich 2009a). (Presumably these behaviors increased in-group commitment and cooperation.) The ethnographic and historical record, taken together with the empirical evidence reviewed above, points to the idea that religious beliefs and practices played a key role in the spread of prosocial groups over the last 12,000 years.

Outstanding Questions

We conclude with some outstanding questions for further research which we believe has the potential to advance theoretical work on the origins of religious prosociality, and invite discussion about future directions:

- An important extension would be to conduct religious priming studies in smaller-scale societies, where reminders of morally indifferent gods could be compared to the Abrahamic God or the powerful, moralizing gods of other world religions. These comparisons would help researchers tease apart cultural evolutionary explanations from genetic adaptationist explanations of religious prosociality.
- A deeper understanding of the psychology underlying atheism may also shed light on competing explanations for the evolutionary origins of religion. For example, genetic adaptationist arguments for religion would presumably predict that even atheists are responsive to nonconscious religious priming. Cultural evolutionary explanations, in contrast, would predict that *at least some atheists* would be immune to religious priming. Studies could compare "atheist converts" with "lifetime atheists" to clarify the extent to which religious prosociality is culturally learned. These questions are ripe for empirical investigation.
- Historical and cross-cultural comparative work should be done to examine the extent to which secular alternatives to religious prosociality—institutions such as courts, contracts, and police—can culminate in the decline of religion in societies. This again could help us understand the extent to which religious prosociality is genetically fixed, culturally learned, or both.
- It is important to tease apart the relative effects of various elements that get labeled "religion" on prosociality. Future studies should assess in a more fine-grained fashion the extent to which religious prosociality is explained by belief in supernatural monitors and supernatural punishment mechanisms (such as belief in heaven vs. hell, karma, fate),

and by various forms and elements of ritual participation (such as synchrony, extravagance, and emotional intensity).

- Beyond anecdotal evidence, we know relatively little about the social boundaries of religious prosociality. Does it weaken, or break down, where the religious in-group ends and the out-group begins? Or is religious prosociality, in some respects, extended universally? Can religious prosociality be harnessed and co-opted to extend cooperation and solve collective action problems?

First column (top to bottom): Armin Geertz, Joe Henrich, Herb Gintis, Joe Henrich, Fiona Jordan, Harvey Whitehouse, Pieter François
Second column: Joseph Bulbulia, Ara Norenzayan, Quentin Atkinson, Armin Geertz, Joseph Bulbulia, Pete Richerson, Nick Evans
Third column: Emma Cohen, Peter Turchin, Harvey Whitehouse, Thomas Widlok, David Wilson, Ted Slingerland, Russell Gray

20

The Cultural Evolution of Religion

Joseph Bulbulia, Armin W. Geertz, Quentin D. Atkinson, Emma Cohen, Nicholas Evans, Pieter François, Herbert Gintis, Russell D. Gray, Joseph Henrich, Fiona M. Jordon, Ara Norenzayan, Peter J. Richerson, Edward Slingerland, Peter Turchin, Harvey Whitehouse, Thomas Widlok, and David S. Wilson

Abstract

Religion may be one factor that enabled large-scale complex human societies to evolve. Utilizing a cultural evolutionary approach, this chapter seeks explanations for patterns of complexity and variation in religion within and across groups, over time. Properties of religious systems (e.g., rituals, ritualized behaviors, overimitation, synchrony, sacred values) are examined at different social scales, from small-scale forager to large-scale urban societies. The role of religion in transitional societies is discussed, as well as the impact of witchcraft, superhuman policing, and the cultural evolution of moralizing gods. The shift from an imagistic to a doctrinal mode of religiosity is examined, as are the relationships between sacred values and secular worlds. Cultural evolutionary approaches to religion require evidence and methods from collaborative and multidisciplinary science. The chapter concludes with an overview of several projects that are working to provide conceptual, methodological, and empirical groundwork.

Why Take a Cultural Evolutionary Approach to Religion?

What Do We Mean by "Religion"?

Is a definition of "religion" essential to its study? Scholars of religion have long debated this question (Platvoet and Molendijk 1999). Some argue that we need a definition to distinguish religion from related domains of human behavior and concern (Clarke and Byrne 1993). Others suggest that between "religion" and other cultural and cognitive domains, only artificial lines can be drawn (James 1902). Some suggest that folk intuitions about religion mislead (Barrett and Keil 1996), whereas others contend that "religion" does not describe any natural kind (Boyer 1994; Saler 2010).

Discussions about definitions of religion have a role to play in the evolutionary study of religion, in the sense that researchers investigating religion must specify what they want to explain (Geertz 1999). A mature science, however, is one that gradually delineates its object of study, not one that fully describes that object in advance of its study. Specifying an explanatory target should not be confused with offering a once-and-for-all definition. In the natural and social sciences, researchers typically stipulate—or "operationalize"— meanings for their theoretical constructs. For example, in one of the foundational works in the cognitive science of religion, Lawson and McCauley (1990:5) began with a pragmatic clarification: "For the purposes of theorizing we construe a religious system as a symbolic-cultural system of ritual acts accompanied by an extensive and largely shared conceptual scheme that includes culturally postulated superhuman agents....for definitions of religion that emphasizes the role of culturally postulated superhuman beings, this book begins that exploration." Similarly, in his monograph on religious rituals, Whitehouse (2004:2) starts by pragmatically circumscribing his interest: "[f]or the present purposes, let us simply say that religion consists of any set of shared beliefs and actions appealing to supernatural agency."

The project of defining "religion" for "the purposes of theorizing" or for "present purposes" must be distinguished from the project of defining religion once-and-for-all. Most naturalists, including each author of this report, agree that religions are both complex and varying: no single study can be expected to capture all of this complexity and variation. Generally speaking, most evolutionary scholars of religion focus on symbolically and emotionally laden beliefs and practices regarding superhuman powers, and on the institutions that maintain and transmit such beliefs and practices. Unless otherwise stated, we use "religion" below to denote such beliefs, practices, and institutions.

The Scientific Interest of Religious Complexity and Variation

The very fact that religions are complex and varying may explain why they are the target of empirical interest. Religions vary, and the qualities of this variation change over time. Yet there appear to be predictable patterns for religious change and continuity. Put another way, religions are complex but they are not random.

Cultural evolutionary approaches to religion begin with the observation that symbolic-cultural systems of beliefs and rituals acts regarding the superhuman tend to come in "packages" (discussed at length by Slingerland et al., this volume). Items within such packages include beliefs in superhuman persons or powers, ritualized behaviors, devotions and pieties, mythologies, values, goals, and moral doctrines (Atran and Henrich 2010; Geertz 1999; Gervais et al. 2011). Notably, certain features, such as beliefs in superhuman beings and rites that respect superhuman beings, recur across religious groups (Paden 2013). Other features, such as specialized religious castes, appear to be restricted to specific groups. Patterns of continuity and variation, then, admit of historical and geographical regularities. Current horizons in the cultural evolutionary study of religion focus on the role that coordination and competition have played, and continue to play, in affecting historical and geopolitical patterns of religious variation and complexity. Cultural evolutionary approaches to religion seek explanations for such patterns of complexity and variation within and across groups, over time (Paden 2001). Devotion to interventionist, moralizing deities, for example, might have arisen only recently in human history, during the Holocene. Yet devotions are reflected in the religious doctrines of geographically dispersed communities (Norenzayan, this volume; Whitehouse 2004). As Slingerland et al. (this volume) state, cultural evolutionary models of religion hypothesize that patterns of complexity and variation in religious systems are the effects of the cultural evolutionary processes (see also Atran and Henrich 2010). The idea that religion promotes cooperation is hardly new. Anthropologists and historians have long hypothesized that religion fosters social cohesion and builds moral solidarity (Durkheim 1915/1965; Rappaport 1999). Cultural evolutionary approaches, however, break from the past in seeking appropriate evidence and methods of analysis by which to decide between hypotheses. Early results, reviewed below, show clear signs of progress.

In our discussions of strategies for refining and evaluating specific hypotheses for the role religion has played in the historical transition from small to large societies, we considered questions such as: How can we measure the complexity and diversity of religions? What methods are appropriate for identifying functions and functional change? How should we best organize collaborative databases to enable rigorous testing of cultural evolutionary hypotheses? Are there important evolutionary hypotheses that are being neglected? How might we better interest classically trained historians and specialist anthropologists to join our intensely collaborative teams, so that we can better address our questions, and theirs? There were, predictably, lively disagreements. We begin, however, by focusing on the framework of assumptions that is enabling progress.

Nine Points of Agreement

The interdisciplinary field of cultural evolution has developed rapidly over the past twenty years (Mesoudi 2011a), generating fascinating new insights into the mechanisms that enable human behavior and psychology, and about the evolutionary history of these mechanisms (Laland and Brown 2002; Richerson and Boyd 2005). The success of cultural evolutionary approaches has arisen from a combination of cogent theory, often grounded in mathematical models of both cultural and genetic evolution, and from a disciplinary inclusiveness that facilitates the integration of tools from diverse disciplines, including history, anthropology, psychology, archaeology, and economics (Geertz 2004; Henrich and McElreath 2007). As evidenced by the contributions in this volume, there has been impressive progress in the cultural evolutionary study of languages (Atkinson 2011; Gray et al. 2009), technology and science (McCauley 2011; Mesoudi 2011b; Mesoudi and O'Brien 2008a; Shennan 2002), and social complexity (Currie et al. 2010a; Jordan et al. 2009; Turchin 2011; Henrich and Boyd 2008). Initial forays inspire optimism for the cultural evolutionary study of religions (Atkinson and Bourrat 2011; Atkinson and Whitehouse 2011; Donald 1991b; Gervais et al. 2011; Henrich and Henrich 2010; Jensen 2002; Matthews 2012; Matthews et al. 2012b; Richerson and Newson 2008; Slingerland and Chudek 2011; Wilson 2005).

Although we do not wish to overstate agreement among all members of our discussion group, it is fair to say that cultural evolutionists have a rough working consensus about the following basic features of cultural evolutionary approaches:

1. Human minds exhibit reliably developing features of cognition and emotion, for example, that influence recurring patterns of behavior across diverse populations (Boyer 1990; Sperber 1990). Whereas certain human cognitive and behavioral traits are the products of natural selection, others arise from an interaction of genetic and cultural inheritance systems (Boyd and Richerson 1985; Cavalli-Sforza and Feldman 1981). Human populations are impressively adapted to their environments thanks to a large body of learned information that is transmitted across generations through cultural systems (Sterelny 2006).
2. Cultural systems accumulate design features from selection, biased adoption decisions, and nonrandom innovation (Chudek and Henrich 2011; Wilson and Wilson 2007). Such processes affect genetic evolution. The dynamics and effects of gene–culture coevolution are beginning to be studied scientifically (see Laland et al. 2010; Richerson et al. 2010).
3. Humans do not merely "acquire" cognitive and behavioral traits from cultural systems; the mechanisms of transmission themselves rely on both cultural and genetic adaptations (Chudek and Henrich 2011; Deacon 1997; Spuhler 1959). The human capacity for culture is itself a coevolutionary adaptation. Some such capacities, such as prestige and conformist biases, appear to be ancient and nearly universal (Raafat et al. 2009). Other skills, such as reading, mathematics, and clearing email, arrived more recently and are less diffuse (Donald 1991a). The underlying neural and psychological mechanisms of cultural transmission are only beginning to be studied scientifically, as discussed at length in the chapters by Haun and Over, Stout, Lieven,

and Whitehouse (this volume).[1] Initial studies reveal that we cannot understand aspects of human cognitive and behavioral evolution independently of understanding how culture affects what we learn and how we behave (Cavalli-Sforza and Feldman 1981; Geertz 2010). Nor can we understand gene–culture coevolution independently of understanding how culture supports cooperative teaching, interaction, and exchange (Sterelny 2011).

4. To underscore their inherently functional properties, cultural systems can be defined as "meaning systems." Meaning systems are designed to receive environmental information as input and to generate action as output.
5. We can better understand both human brains and human meanings systems by studying how they interact.
6. One of the most salient, general, and apparently ancient outcomes of the human cultural evolutionary process is also one of the most poorly understood, what might be called "religious meaning systems" or cultural systems and institutions that transact in symbolically and emotionally laden beliefs and practices respecting superhuman beings (Geertz 1966). Religious meaning systems appear to link environmental information, especially symbolic information, with behavioral outputs, especially social behaviors. A detailed understanding of how symbolically laden beliefs and practices that relate to superhuman powers variously affect social actions remains elusive.
7. The beliefs and practices that comprise religious meaning systems include pragmatic elements; that is, people doing things for utilitarian reasons that are explicitly understood. Recurring features across assemblies of religious traits include costly rituals and beliefs in superhuman entities. Despite their superficial lack of utility, religious meaning systems are eminently functional, exhibiting "practical realism" (Durkheim 1915/1965; McKay et al. 2007; Wilson 2002). One challenge for the study of meaning systems, in general, and religions, in particular, is to understand the functional elements of meaning systems where they exist (Henrich 2009a).
8. Centering the study of meaning systems on group-level functionality does not entail that every element is functional. Nor does it mean that individual- and group-level functions converge (Gervais et al. 2011). The mechanisms of cultural inheritance virtually guarantee that nonadaptive group-level traits will evolve along with adaptive traits (Richerson and Boyd 2005; Wilson 2008). Religious meaning systems

[1] See also the special issues on the neural and psychological mechanisms in *Philosophical Transactions of the Royal Society B* (2008, vol. 363; 2009, vol. 364; 2011, vol. 33), the *Social Cognitive and Affective Neuroscience* (2010, vol. 5), especially Chiao (2010), Roepstorff et al. (2010), and Vogeley and Roepstorff (2009).

must be comprehensively studied to understand how their components evolved, develop over the course of life histories, and function to produce effects that support their conservation and transmission, or not, as only attention to cases may decide (Wilson 2008).

9. Of course, religious meaning systems interact with political, technological, and linguistic meaning systems, as well as with exogenous environmental change. The study of religion or religious elements (e.g., belief in superhuman agents) per se must be situated within, and is relevant to, the study of meaning systems as a whole (Geertz 1966; Wilson 2002). The study of religion cannot be sharply separated from the study of other cultural domains (Turchin 2006). Some have suggested that a single meaning system can be crudely compared to a species occupying an ecological niche. Meaning systems interact with each other, similar to species in multispecies communities. The full range of ecological relationships among groups can potentially exist (competition, predation, parasitism, mutualism, communalism, and coexisting without interacting; see Wilson 2007b).

These nine points form the basis of widespread, but not complete, agreement among researchers who take a cultural evolutionary approach to religion. Next, we discuss hypotheses for the cultural evolution for properties of religious systems ("packages") in the context of evolutionary transitions from small-scale forager societies to large-scale urban societies.

Religious Elements at Different Social Scales

Religion in Small-Scale Societies: Rituals

When we talk about small-scale societies, we are not talking about groups of the same size and complexity typical of the other great apes. The smallest human societies are perhaps egalitarian foraging bands comprising fifty or so members, whose social ties and ritual obligations may almost always extend to much wider networks of crosscutting and overlapping bands (Boehm 1999). That is, even the least complex human societies are "tribal" in scale. Societies that are small, in this qualified sense, must overcome a wide range of collective action programs (see Jordan et al., this volume). These problems include the coordination of group members for big-game hunting, knowledge pooling against uncertainty through extended networks of unfamiliar conspecifics, coordinated defense against predators (including predatory human groups), alloparenting, warfare, the control of defectors, and the coordination of longer-term projects, including the intergenerational transmission of technological expertise. Sophisticated forms of cooperative sociality, then, appear even in the smallest human groups. In other species, complex cooperative societies exist only when their members are close relatives (Boyd and Richerson 2002b), an unparalleled evolutionary achievement for a vertebrate.

In recent millennia, the sphere of human cooperation has only expanded to include societies of millions with vast networks of global exchange. Something has happened, clearly, for such communities to exist. Current thought on social complexity looks to a unique combination of species-typical predispositions and cultural innovations. Features include a reverse dominance hierarchy and social-learning biases, a norm psychology that allowed for flexible rules and institutions to govern social interactions, symbolic behavior and language, and cumulative cultural transmission (for discussion, see Jordan et al., this volume). These requirements, in turn, depend on other more basic conditions: increasing returns to scale with group size and control of defectors. Only with such an alignment of conditions could gene–culture coevolutionary dynamics enable and sustain large-scale cooperative living among partners who are not closely related. this is what cultural evolutionists call "ultrasociality" (discussed in Turchin 2013; see also Turner and Maryanski 2008). We are interested in the role that religions have played, and continue to play, in enabling human ultrasociality at small and large social scales.

Ritualized Behaviors

A wealth of evidence suggests that religions forge solidarity and cooperation Cooperative effects have been observed from the level of small-scale foraging bands (Boehm 1993) to the level of complex nation-states (Bellah 1967). Evolutionary historians suggest the hypothesis that religions facilitate cooperation across multiethnic and linguistic divides (Bellah 2011; Turchin 2006). Yet how do religions variously promote high levels of solidarity, even among strangers? The anthropology of religion has documented a wide range of candidate mechanisms (Whitehouse 2008). Among these, religious rituals have been identified as ancient tackle in the human cooperative toolkit (Rappaport 1971).

Ritual performances emerge among even the most egalitarian foragers, and they appear to build solidarity and both broaden and tighten social ties (Katz 1984; Radcliffe-Brown 1922). For example, the San trance dance has been described as an arena for creating coherence and for mobilizing support for cooperative projects such as medical provision, entertainment, and expressive art (Widlok 1999, 2007). Researchers have also suggested that ceremonial and ritualized gift giving is a likely route for the creation of obligations and dependencies (Hayden 1987). Among Hadza foragers, for example, sacred meat rituals (and similar practices among Pygmy groups) through which senior men can claim privileged access to some of the meat have been described as possible entry points for inequality in a system governed by social leveling mechanisms (Woodburn 1970); still, such inequalities do not appear to be generally maintained through other inequality-deflation rituals, such as ritualized gift giving and healing dances (Widlok 2007). Collective and effervescent rituals, then, that seem "designed" to increase local or tribal solidarity are not limited to specific foraging communities: they are widespread (for studies on Australia and the Andaman Islands foragers, see Wade 2009). In healing and exchange rituals we find examples of systems that, though apparently lacking in factual utilities, underwrite the basic practical utilities of community making (Wilson 2002). What are the proximate mechanisms? Let us consider several plausible models.

Overimitation

Imitation is the process by which learners acquire the behaviors of teachers. In humans, learners not only copy behaviors but also copy the representations of intentions and goals (Tomasello 1999). We noted above that prestige and conformist biases equip humans for cultural learning. Recent research suggests that "overimitation" (i.e., the copying of causally opaque behaviors) may have been a crucial adaptation in the evolution both of language and of social norms (on the role of imitation for normative learning, see Donald 2001; Haun and Over, this volume). Developmental psychologists have tended to regard children as little scientists, exploring the affordances of their environments by informally (and often implicitly) testing hypotheses (Gopnik 2001). Overimitation has been interpreted within this general framework as a strategy for social learning that transmits technological knowledge through a copy-now-correct-later strategy, which assumes that there are sound instrumental reasons for modeled adult behaviors, even if superfluous information is later jettisoned (Lyons et al. 2007). Recently, investigators have shown that more rigid forms of overimitation support affiliation and the learning of norms (Legare and Whitehouse 2011). Results suggest that strict imitation may harbor social cognitive functions beyond the acquisition of technical skills (Chudek and Henrich 2011). Notably, where ritual actions are synchronized in groups, via collective dancing, singing, and marching, overimitation appears to stabilize norms, there being no better way of copying than the "proper" way modeled (see also Frith and Frith 2007).

Synchrony

We have observed that in their surface properties, ritual behaviors appear to lack "factual realism"; they appear purposeless, in the sense that the goals of ritual behaviors cannot be readily discerned from component behaviors (Sørensen 2006, 2007). Looking into the entrails of sheep and offering animals to statues would appear, on the face of it, to be an inefficient and ineffective means for planning action. Yet as we have seen, the cohesion of religious communities cross-culturally reveals tacit social bonding functions (Alcorta and Sosis 2005; Irons 1996; Rappaport 1979; Turner 1990). Some researchers have looked to synchronous group behaviors as a basic ingredient of ritual-induced cooperation, conjecturing that synchrony "coevolved biologically and culturally to serve as a technology of social bonding" (Freeman 2000:411). Informal accounts of the cooperative affects of synchronous rituals are abundant in both the ethnographic and historical records. For example, in recalling his World War II military cadet training, the historian William McNeill (1995:2) writes:

> Words are inadequate to describe the emotion aroused by prolonged movement in unison that drilling involved. A sense of pervasive well-being is what we recall; more specifically, a strange sense of personal enlargement; a sort of swelling out, becoming bigger than life, thanks to participation in collective ritual.... Obviously, something visceral was at work; something, we later concluded, far older than language and critically important to human history, because the emotion it arouses constitutes an indefinitely expansible basis for social cohesion among any and every group that keeps together in time, moving big muscles together and chanting, singing or shooting rhythmically.

Why should synchronous movement have the effect of building solidarity? Why is chanting, singing, shouting, and marching rhythmically effective at uniting a group? Why does social bonding benefit from being "muscular"? Indirect evidence suggests that synchronous rituals might activate pleasure centers in the brain by stimulating the opioidergic system (Cohen et al. 2010). Cohen and colleagues propose that synchronous rituals alter psychological states to promote a sense of trust and commitment toward others, which affects the development of social bonds. Consider recent evidence in favor of cooperation through synchrony.

Wiltermuth and Heath (2009) randomly assigned participants to one of four groups involving different levels of synchronous movement (passing cups) and vocalization (singing "O Canada"). To evaluate the prosocial consequences of synchrony, they asked participants questions about perceived unity and measured decisions in economic games. Results showed that participants in the synchronous singing and movement conditions sustained higher levels of cooperation over time than participants in the asynchronous and passive control conditions. Those in the synchronous singing and moving conditions also reported (a) enhanced feelings of being on the same team and (b) greater subjective perceptions of similarity to their counterparts—the sort of "personal enlargement" and "swelling out" that McNeill describes in his conjecture about muscular bonding. Synchronous participants also trusted each other more, and feelings of being on the same team were found to partially mediate the effect of synchrony on cooperation. These results offer some initial support for the theory that cooperation is evoked through synchronous performances (for further discussion, see Cohen et al. 2013; Kirschner and Tomasello 2010; Reddish et al. 2013; Slingerland et al., this volume).

Synchrony and Sacred Values

Synchronous movement affects cooperation, but how do muscular features interact with religious beliefs and values? A hundred years ago, Durkheim (1915/1965:19) conjectured:

> The [ritual] group regularly produces an intellectual and moral uniformity… [with which] everything is common to everyone. Movements are stereotyped; everyone executes the same ones in the same circumstances; and this conformity of conduct merely translates that of thought. Since all the consciousnesses are pulled along in the same current, the individual type virtually confounds itself with the generic type.

Durkheim surmises that it is both a physical and mental alignment during ritual performances that leads to a confounding of self and group; "swelling out" in McNeill's terms. According to Durkheim, such cooperative motivations are more strongly expressed when partners share conceptions of the "sacred," or "things set apart and forbidden" (Durkheim 1915/1965:44). On Durkheim's model, then, it is a combination of shared body movements and sacred values that intensifies solidarity during religious ritual performances. What is the evidence that explicit values interact with synchronous rituals to affect cooperation?

In a recent study, Fischer et al. (2013) investigated the prosocial effects of nine naturally occurring rituals in Wellington, New Zealand. The authors operationalized "prosociality" in two ways: (a) as attitudes about fellow ritual participants (stated prosociality) and (b) as donations to a common pool in a public goods game (revealed prosociality). The nine rituals varied in levels of synchrony and in levels of sacred attribution, ranging from poker games and running competitions, at the one extreme, to Christian choir singing and Kirtan chanting, at the other. The researchers found that rituals with synchronous body movements were more likely to increase prosocial attitudes. However, the team also found that rituals judged to be sacred were associated with the largest contributions in the public goods game. A path analysis using MPlus favored a model according to which sacred values mediated the effects of synchronous movements on prosocial behaviors. The analysis suggests that ritual synchrony inflates the perception of oneness with others, which in turn increases sacred values to amplify prosocial behaviors. It seems that when muscular synchrony is framed by a shared set of beliefs related to sacred themes—such as a religious narrative or theology (Geertz 2011b; Pyyssiäinen 2011)—there seems to be an intensification of social bonding and within-group cooperation. Such effects might contribute to the exceptional cooperation observed among religious communities at small and large social scales (Bulbulia 2012; Sosis and Bressler 2003).

Ritual Signaling

How do cooperators avoid defectors and assort? Ordinary language would appear to be an ineffective tool. Defectors might express cooperative intentions, only later to defect. Unreliable expressions cannot be used as the basis for cooperative assorting. Language, however, is not the only medium by which to communicate. Some expressions index cooperative commitments. Gazelle leap up and down (stotting) in the presence of predatory lions. Unfit gazelle are unable to produce convincing displays of health. Stotting appears to have evolved as a signaling device that indexes speed, enabling prey and predators to avoid costly chases (for even here, in a battle to the death, there is scope for cooperation). Biologists call such indices signals (Zahavi and Zahavi 1997). Irons (2001) and Cronk (1994) argue that rituals evolved as signaling devices because, according to hypothesis, rituals reliably discriminate between those who possess religious commitments and those who do not. (For evidence consistent with ritual signaling but also with credibility enhancing displays, discussed below, see Sosis 2000; Sosis and Bressler 2003.) Wherever religions are associated with cooperative sensibilities (e.g., from intrinsic and extrinsically motivating beliefs in superhuman agents and causation) and are difficult to perform without such commitments, there will be scope for rituals to identify cooperative commitments (Bulbulia 2004; Sosis 2003).[2] Atran and Henrich (2010) point out that wherever local social, economic, and ecological conditions can influence cultural evolution such that the functions of rituals may vary over time, those that do a better job of discriminating between the cooperators will gain an advantage: in short, costly signaling cultures evolve.

Are the functions of ritual signaling systems fixed? The evidence suggests that cultural evolutionary processes may lead to phase changes in religious functions (Bellah 2011). At certain historical stages, rituals might foster social integration and equality, whereas at other stages, the same rituals might lead to social differentiation and inequality. Indeed it has been argued that religious elites might manipulate cooperative signals to their own advantage (Cronk 1994). In thinking about the space of evolutionary possibilities, cultural evolution need not be unidirectional. Rituals need not always and everywhere perform identical functions. Consider the following example.

Bloch (1986) argues that the circumcision ritual of the Merina has undergone functional changes, from (a) an occasional familial ritual in 1780 to (b) a seven-year state ritual that culminates with royal circumcision during the following 100-year period to (c) an intermediate period with royal circumcision

---

2 Note: estimating the cost or value of a ritual brings all the problems of assessing a complex system. For this reason, so-called signaling theorists prefer "commitment" signaling or "honest" signaling because signals do not need to be costly to evolve (Bliege Bird and Smith 2005; Bulbulia 2008a; Matthews 2012; Widlok 2010).

that continues on a reduced scale around 1869, when Christianity becomes the Merina state religion. After this intermediary period, circumcision "evolved" into (d) a small-scale, familial and largely hidden ritual, at which point (e) circumcision again increased in public importance, taking on anti-Christian and antielite overtones by about 1960. Whereas at certain points in history circumcision rituals look like good candidates for hard-to-fake signals of group identity, at other times this interpretation looks implausible. It is possible that in most cases there are a lot of highly redundant markers of group membership. Where there are redundant markers, random drift-like effects or linage to other parts of the evolving symbolic system may be the main driver of evolution. When social divisions within a fairly homogeneous group arise, that might directly drive the evolution of new symbolic traits. For them to be effective, symbols often have to have a traditional or sacred justification. "Neotraditional" symbols may tap disused or formerly insignificant symbols that can be argued to be ancient. New sacred justifications can be had by conversion to a different religion. In line with the drift to functionality conjecture, Matthews (2012) finds evidence for rapid symbolic evolution at points of religious schism within Christianity, which is consistent with the prediction that symbolic differentiation and claims to sacred authority interact to define the boundaries of groups.

Of course, Bloch's study offers only one case, from which it is difficult to generalize (as is true for any $n = 1$ sample). However, even one case is sufficient to demonstrate that cultural evolution might harness existing cultural practices and patterns in surprising ways. Bloch's circumcision example also holds an important methodological lesson: the cultural evolutionary study of religion demands close attention to historical facts, ranging over long historical spans and across wide geographical domains. This highlights a need for a new collaborative science of history (Durkheim 1915/1965; Turchin 2009).

Dysphoric and Synchronous Arousal

Recent evidence suggests that rituals coordinate empathetic arousal among audiences and performers, at unfamiliar social scales, extending cooperative benefits beyond the circle of those who perform rituals (Bulbulia and Frean 2010). In a recent study of a Spanish fire-walking ritual, Konvalinka et al. (2011) quantified shared patterns in the heart rhythms between fire-walkers and spectators. They hypothesized that synchronous arousal keyed to focal ritual events—each of a series of fire-walks—would be detectable in the heart rhythms among both performers and spectators. Intriguingly, analysis revealed global similarities among fire-walkers and socially connected spectators, but not among unrelated spectators. Merely observing a ritual was insufficient to produce empathic responses. If an observer knew at least one firewalker, however, empathetic arousal extended to all fire-walkers. Prior personal investment in ritual was an essential condition for shared and expansive arousal (Xygalatas et al. 2011). How do people become invested in rituals before they partake of them? Does shared arousal translate to cooperative behaviors? How common are such effects across the great diversity of rituals? Presently, little is known (see Luhrmann 2012; Schjoedt et al. 2009).

Dysphoric (painful or frightening) rituals have been observed to bolster solidarity among initiates (Aronson and Mills 1959; Gerard et al. 1956, Xygalatas et al. 2013). Researchers have shown that the extreme dysphoria features prominently in a broader range of rituals, not merely in rituals that mark entry into the group. It has also been shown in a survey of 644 rituals that peak dysphoric arousal correlates negatively with agricultural intensity (Atkinson and Whitehouse 2011). This finding suggests that dysphoric rituals might be an adaptation to resource extraction problems, such as large game hunting and warfare, where cooperative problems are rampant (Whitehouse and Hodder 2010; Whitehouse et al. 2012). Recent studies point to enduring effects from dysphoric ritual, during late adolescence, which might explain the cross-cultural prevalence of painful initiatory ordeals during that phase of development (Alcorta 2008). Whitehouse's model of dysphoric ritual is important because it focuses attention on a large class of rituals that express exceptionally powerful forms of cohesion by recruiting pain, not pleasure. The class of dysphoric rituals is obscured when rituals are described purely as mechanisms for unleashing happiness and joy (for a joy-centered hypothesis, see Haidt et al. 2008).

Religion in Transitional Societies: Superhuman Policing

When most people think of "religion" they think of superhuman agents or powers: "gods," "witches," and "ancestors." Yet such superhuman agents have been conspicuously absent from our discussions. How might beliefs in gods have evolved, and where do they fit into the cooperative suite? Notably, the cognitive by-product model conjectures that superhuman beliefs, such as beliefs in witches, do not spread for specific functional purposes but rather because they are intuitively attractive (Boyer 1994). Synthetic culture–gene coevolutionary approaches, however, have explored how some of these cognitive by-products may have been favored by cultural evolution. Atran and Henrich (2010) argue that cultural evolutionary processes harness formerly functionless beliefs for cooperative effects. As societies expand in size, cultural evolutionary processes favor belief and ritual packages that more effectively galvanized compliance with prosocial or group-beneficial norms, for such systems are vital for the success of expanding social groups. Over time, cultural evolution, driven by intergroup competition, can aggregate and calibrate a system of interlocking beliefs, practices, and values that extend cooperation and enhance internal harmony.

Witchcraft

A recurrent feature in many societies is the belief in witches and in witchcraft, which in some case may have been favored by the kinds of processes described by Atran and Henrich. In complex agricultural populations, such beliefs appear to have exerted a significant influence on social life (Malotki and Gary 2001). Indeed, anthropologists have repeatedly documented the role that witchcraft accusations have in preventing the emergence of social and economic inequalities in small, egalitarian societies. Witchcraft beliefs also appear to facilitate conflict resolution in larger societies by mediating disputes between parties who are too deeply connected by social ties to resort to formal resolution in public courts, yet who are not closely enough related to resolve their differences through the distributed justice of kinship systems (Harris 1974). Might beliefs in witchcraft offer examples of cultural evolutionary processes operating on basic cognitive predispositions to believe in spiritual powers?

Consider the Hopi Indians of northeastern Arizona. Evidence suggests that the Hopis were once a small, egalitarian society. However by the year 1000 CE, the archaeology of Hopi villages points to population densities of as many as a thousand individuals (in the town of Oraibi). How did the Hopis make the transition from a small-scale foraging society to middle-sized semi-horticulturalists and agriculture society? Here we focus on the hypothesis that Hopi conceptions of witchcraft might have played a functional role in this transition. Notably, Hopi society consists of a large number of matrilineal clans organized in large phratries spread across a number of pueblo villages. The Hopis had well-developed priesthoods based in secret societies that cross-cut clan affinities, whose members performed time-consuming, frequent high-pageantry ceremonies supplemented by briefer but colorful masked dances. "To the Hopis, witches or evil-hearted persons deliberately try to destroy social harmony by sowing discontent, doubt, and criticism through evil gossip" (Geertz 2011a:379). Witches are also believed to excel at ritually combating medicine men, as well as the effects of the high-solidarity ceremonies of the ritual brotherhoods. Witches are assumed to prolong their own lives through the occult murder of family relatives (i.e., people from their own matrilineage or closest household members). Thus, what can be considered the most central social institution and the source of stability and nurture, can, at the same time, form an arena where this social cancer grows. Witchcraft, suspicion, and gossip often destabilize family and interpersonal relations. Witches are thought to be the source of unusual illnesses or deaths, strange natural phenomena, and unexpected negative turns of fortune. They are said to belong to a secret society that practices its rituals at night. Witches have two hearts: one human and the other animal. With their animal heart, they can transform themselves into their power animal and do superhuman things. Hopi oral traditions are filled with tales of evil witches and their exploits (Malotki and Gary 2001).

Hopi do not confront witches, nor do they conduct witch hunts. Such a direct confrontation, it is believed, tempts harm from witches on witch hunters and their families. For our purposes, it is notable that the best protection against witches is thought to be virtuous action. One can do no better to avoid witchcraft than by living up to the Hopi ideals, which include self-deprecation, low ambition, friendliness, and hospitable comport. Such prosocial behaviors, it is assumed, do not arouse witch jealousy, whereas bragging or showing off your wealth and fortune are invitations for trouble. Thus, witchcraft is not used merely to explain evil and misfortune, it is used to promote prosociality. Among the Hopi we find "an atmosphere permeated with witchcraft fears and fuelled by gossip, rumor, and slander…[involving] a mixture of confidential information, troubling yet unverified facts, misunderstandings, fantasy and irritating, egotistic neighbors" (Geertz 2011a:379). Thus gossip and witchcraft could be viewed as a kind of narrated ethics expressing more or less defined models of thought and behavior that stage social and personal identities in conversational narratives (Geertz 1974:213). "It defines and redefines these identities in terms of contemporary issues and helps people work their way through baffling problems, normative principles, and potential interpretations….[E]ven malicious gossip plays by the same rules—which is why people are so easily deceived by it. Gossip is a two-edged social instrument that ensures the on-going socialization of the individual. It is a powerful and merciless instrument" (Geertz 2011a:379).

Did beliefs in witches generally and invariably evolve to promote within-group prosociality? The best that can be said at present is that such beliefs sometimes lead to normative vigilance and sometimes lead to cascades of killings and violent retribution, both building and destabilizing normative orders (Knauft 1985). To repeat, historical dynamics can be cyclical. We need not expect a steady march from nonfunctional to functional moralizing witchcraft. Functions can, and do, oscillate over time (for a discussion on cyclical dynamics, see Turchin 2003). Given the instability of Hopi witchraft as an effective policing system, cultural evolutionary models would predict that Hopi witchraft would not survive intergroup competition against cultures with more effective systems. This possibility raises an important point: Although within-group dynamics can lead to cycles, between-group dynamics sometimes lead to longer-term evolutionary trends. Within groups, there are a variety of individual-level decisions that can slowly coordinate prosociality by favoring cultural elements that foster one's own interest or the interests of some subpopulation. Such tendencies will be ratified wherever the forces of between-group selection are strong. In the case of witchcraft, individuals sometimes use witchcraft allegations to seek vengeance and to settle old scores (Knauft 1985). Between-group processes, however, may favor those packages of beliefs and practices that galvanize and sustain group solidarity, leading to the decline of corrosive antisocial traits (Knauft 1985). Knowing nothing else, then, we would expect that the level of religious prosociality in a group will depend on the relative strength of competition between religious groups versus the competition of subgroups or individuals within religious groups. In this case, Hopi witchcraft does not look like it brings success to intergroup competition. Instead, it looks like in-group forces are winning. To evaluate such questions, a new collaborative evolutionary history is needed.

A fundamental challenge for testing hypotheses about the covariates of religious features across cultures and over time is that, like any aspect of transmitted culture, data points are not independent, invalidating standard statistical tests (Mace and Pagel 1994). This problem has come to be known as "Galton's problem," after a prominent objection by Francis Galton to an early statistical study of cultures by E. B. Tylor (Laland 1992). One approach to Galton's problem has been to sample cultural data sparsely, thus avoiding close relatives (Murdock 1966). However, this approach has the rather unfortunate consequence of dramatically reducing sample sizes, increasing statistical uncertainty. More problematic, there is no guarantee that sparse sampling avoids dependencies in the data (Dow and Eff 2008). Thus, Galton's problem remains.

Biologists use an alternative approach, called comparative phylogenetics, to solve the problem of nonindependence. Comparative phylogenetics involves explicitly modeling the process of trait evolution through time on known species phylogenies. Comparative phylogenetics allows statistical control for variation explained by shared ancestry (Felsenstein 1985), solving Galton's problem, and allowing the testing hypotheses about ancestral states, rates of change, sequences of change, and dependencies between traits.

Recently, phylogenetic approaches have been applied to the study of cultural evolution (Currie et al. 2010b). By mapping cultural features onto phylogenies representing the genealogical relationships between societies or cultural elements, it is possible to test hypotheses about ancestral cultural states (Fortunato 2011), sequences and rates of cultural change (Currie et al. 2010b), and dependencies between cultural traits (Holden and Mace 2009). Cultural phylogenetic methods[3] hold promise in addressing many of the core questions about the evolution of religious features raised above, such as: What features were present in the ancestral religion of a lineage? Do features evolve in a particular sequence through intermediate forms? Which features evolve most quickly? Which are more stable over time? Are changes in one feature predicted or conditioned by changes in another? Whereas formerly mathematical modeling was consigned to investigate "how possible" questions, computational phylogenetics is shedding new light on how the human past actually unfolded, and with respect to our interests, how features of religious cultures have affected what human populations have variously become.

[3] For recent articles which use cultural phylogenetics to investigate the evolution of religious groups, see Matthews et al. (2012); Matthews (2012).

Superhuman Policing and the Cultural Evolution of Moralizing Gods

In smaller-scale societies, groups are able to build local solidarity without appealing to moralizing heavenly agents (Gervais et al. 2011). In larger polities, however, moralizing gods appear to be effective in fostering cohesion across multitribal subunits (Swanson 1964). How did this transition occur? Some argue that as polities and societies have grown larger and more complex, the forces of cultural evolution have favored richer conceptions of superhuman reality, populated increasingly by potent moralizing gods or by a single moralizing god, equipped with ample powers to visit superhuman rewards and punishments on norm followers and infringers. If larger societies (by hypothesis) benefit from larger gods, the kind of intergroup competition that leads to large-scale civilizations should, all things equal, select for "moralizing gods" (Norenzayan 2013; Turner and Maryanski 2008). Wright (2009) is one who argues that in more complex societies, such as chiefdoms, ancestor gods appear to sanction various sorts of moral transgressions, including the failure to perform costly, faith-inducing rituals. Interestingly, while ancestor-god beliefs may provide some superhuman sanctioning, ancestor-god beliefs also appear to lack many of the features of the high, moralizing gods found in the most complex societies. For example, ancestor gods tend to be limited to specific places and serve only a narrowly defined group of people. Such gods lack omniscience, have limited powers, and are not universalizing, nor can they grant afterlife rewards (Wright 2009). Cultural evolutionary approaches to the evolution of moralizing god-beliefs are based on the idea that crucial elements of religion may be influenced by cultural selection and intergroup competition. If the features of religions that express solidarity vary, and if certain religious groups have a competitive edge over others, then group-level processes can, in principle, select for religions that more effectively support prosociality. Next we consider the logic of this evolutionary approach in more detail.

A central barrier to the evolution of large-scale societies is the risk of defection of anonymous partners. Norenzayan et al. (this volume) focus on the role of superhuman surveillance in addressing this problem. The evolution of beliefs in all-knowing moralizing gods serves to deter antisocial behavior beyond the reach of secular institutions (Atran and Henrich 2010; Boyer 2001). This adaptation emerged and spread not only among the Abrahamic religions but also among ancient Chinese religions (Slingerland et al., this volume). As Slingerland and colleagues point out, we should expect "packages" of religious traits that reliably express higher degrees of cooperation to be favored by cultural evolutionary processes wherever the resulting cooperation increases fertility or cultural transmission (Rowthorn 2011). In an important study, Johnson shows that omniscient moralizing gods with the power to mete out punishments and rewards are much more common in large-scale societies than in smaller ones (Johnson 2005). God-concepts of this kind might serve an important policing function in dense populations, where temptations to cheat, defect, and free-ride under the cloak of anonymity are particularly acute. Whereas classical cooperation models of religion focused on fixed effects, beliefs in gods may be associated with dynamic cultural evolutionary processes (Gervais and Henrich 2010). Studies of foragers underline the relative absence of moralizing gods, and the relative rarity of superhuman sanctions for antisocial behavior. Among the much studied hunter-gatherers in the Kalahari, Marshall (1962:245) wrote: "Man's wrong-doing against man is not left to ≠Gao!nas [the relevant god] punishment nor is it considered to be his concern. Man corrects or avenges such wrong-doings himself in his social context." Similarly, while Hadza foragers in Tanzania believe in a creator god (Haine), this deity cares little about human morality and does not intervene in human affairs. Far from being a reliably developing product of our evolved cognition, moralizing gods appear rather peculiar from a historical and anthropological perspective (Tylor 1873). Their popularity in the modern world is thus a puzzle.

Though some argue that the smallest-scale societies—especially foragers—do not reveal much explicit connection between superhuman beliefs and incentives regarding antisocial and prosocial behaviors, some have argued that the emergence of larger-scale chiefdoms, after the origins of sedentary food production, are associated with changes in religious beliefs, rituals, and institutions (Swanson 1964; Turner and Maryanski 2008). The ancestor gods of the simplest chiefdoms appear to be flawed: they occasionally punish errant individuals for violations such as theft, murder, and adultery using illness, accidents (e.g., shark attacks in Polynesia and Fiji), and bad luck (Handy 1927; Lowie 1948). Ancestor gods also punish people on a whim, demand payments in the form of sacrifice, and remain absent during critical times. Moreover, the religions of chiefdoms seem to favor political stability by endowing chiefs with divine wisdom and power (Nolan and Lenski 2004). Though not omnipotent, omniscient, and benevolent, these superhuman overseers may reveal the first footprints of the track created by the competition among religions. A similar pattern can be detected in the historical record. Once the written historical record begins, it becomes much easier to establish clear links between large-scale cooperation, ritual elaboration, moralizing gods, and morality. To date, most of the historical work related to this topic centers on the Abrahamic faiths. Wright (2009) provides a summary of what he takes to be textual evidence revealing the gradual evolution of the Abrahamic god from a rather limited, whimsical, tribal war god—a subordinate in the pantheon—to the unitary, supreme, moralizing deity of two of the world's largest religious communities. While an evolutionary cognitive study of Middle Eastern religions is still in its infancy, and there are many open questions, Wright's presentation is consistent with a cultural evolutionary hypothesis. Is Wright right? The case remains unresolved. We need to apply cultural evolutionary science to religious history.

Evidence for the Cultural Evolution of Moralizing Gods in China

Surveillance by morally concerned superhuman agents also appears as a prominent theme in early China. Even from the sparse records we have from the earliest recorded dynasty, the Shang, it is apparent that the uniquely broad power of the Lord on High to command a variety of events in the world led the Shang kings to feel a particular urgency about placating it with proper ritual offerings. As we move into the better-documented Eastern Zhou period (770–256 BCE), when the Chinese polity begins to fragment into a variety of independent, and often conflicting, states, superhuman surveillance and the threat of superhuman sanctions remain at the heart of interstate diplomacy and internal political and legal relations (Poo 2009). Finally, the written record reveals an increasingly clear connection in early China between morality and religious commitments. The outlines of moral behavior have been dictated by Heaven and encoded in a set of cultural norms, and a failure to adhere to these norms—either in outward behavior or one's inner life—was to invite instant superhuman punishment. Some scholars see the creation of ethical high gods as an important contributing factor in the Zhou's unprecedented ability to expand militarily and politically, the clear theodicy and superhumanly mandated moral code both legitimating the dynasty and providing a shared sense of sacred history and destiny across the growing Zhou polity (Eno 1990). Slingerland et al. (this volume) discusses this at greater length.

Future Research on the Cultural Evolution of Superhuman Policing

The studies reviewed above support a cultural evolutionary model for moralizing religions, yet as Norenzayan et al. (2013) point out, the evidence for the moralizing gods model is mixed. Sociological studies investigating the cooperative effects of religion have generally employed self-report measures rather than behavioral measures. Verbal reports, however, sometimes mislead. Participants are poor judges about how their minds work, and biases pervade in their reporting. In addition, religious people seem to be prone to social desirability biases, suggesting caution when interpreting how religious people report prosociality (Norenzayan et al., this volume). Database studies afford better evidence for religious prosociality (Atkinson and Bourrat 2011; Johnson 2005; Sosis et al. 2007; Swanson 1964). However, thus far database studies have yielded mainly correlational findings. Controlled behavioral studies, which better address causal questions, remain relatively scarce, and their results have been mixed. McKay, Efferson, Whitehouse, and Fehr (2011) showed that religious priming had effects only on people who had previously donated to religious charities. The result suggests that priming only affects a subset of the believers. Other studies of undergraduates suggest that religious priming expresses greater prosociality in both believers and disbelievers alike (Mazar et al. 2008; Paciotti et al. 2011; Shariff and Norenzayan 2007a).

Norenzayan et al. (this volume) offer the following suggestions for resolving such inconsistencies in the data. First, broader contextual features likely interact with religious cues to affect prosociality. This suggestion is plausible because contextual variables, such as being in a hurry, have long been known to modulate the sociocognitive effects on prosocial behavior stronger than dispositional variables such as type of religious orientation (Darley and Batson 1973). Note that in Darley and Batson's study, the situational variable (being in a hurry) swamps previous religious training as well as contextual religious cues (e.g., preparing a talk on a religious theme). There are many factors, then, which collectively conspire to affect behaviors. Religious cues need not dominate others. Second, disbelievers from countries with strong secular institutions exhibit a high degree of prosociality, which suppresses any contribution that individual differences in religiosity might bring to prosocial behaviors (Zuckerman 2008). This finding might explain why religious cues, more so than religious dispositions, affect prosociality in secular societies where the rule of law is strong. Third, there might be different psychological profiles of atheists, for it appears that not all atheists respond to religious reminders in the same way (Johnson 2012). The prospect for a diversity of atheisms suggests that developmental environments, in interaction with genetic polymorphisms perhaps, must enter into explanations for how religious situations affect prosocial responses (Geertz and Markusson 2010). The observation that situations can affect prosociality, when placed within an evolutionary framework, raises the fascinating question: Has cultural evolution coevolved human natural and social ecologies to afford cooperative norm compliance (Bulbulia 2008b, 2011; Norenzayan and Shariff 2008; Whitehouse and Hodder 2010)?

The Shift from an Imagistic to a Doctrinal Mode of Religiosity

One of the major challenges in understanding how and why religion changes as societies become larger and more complex relates to the changing structure and function of ritual. In small-scale societies, collective rituals tend to be less frequent and more emotionally intense, creating identity fusion in localized, face-to-face communities (Swann et al. 2012; Whitehouse 2000)—an adaptation to collective action problems entailing strong incentives for defection. Warfare and other forms of predation by out-groups present a salient set of problems of this kind but there are others, such as the coordination and cooperation problems posed by hunting large and dangerous animals with simple weapons (Whitehouse and Hodder 2010; Whitehouse et al. 2012). Whitehouse argues that with the evolution of social complexity, however, religious rituals become more routinized, dysphoric rituals become less widespread, doctrine and narrative becomes more standardized, beliefs become more universalistic, religion becomes more hierarchical, offices more professionalized, sacred texts help to codify and legitimate emergent orthodoxies, and religious guilds increasingly monopolize resources. Correlates of this "doctrinal" mode of religiosity (Whitehouse 2000, 2004) have recently been documented quantitatively using large samples of religious traditions from the ethnographic record. For example, Atkinson and Whitehouse have shown that as societies become larger and more hierarchical, rituals are more frequently performed (Atkinson and Whitehouse 2011) and low-frequency dysphoric rituals typical of small, cohesive social groups, such as warring tribes (Whitehouse 1996), come to be confined to specialized niches (e.g., hazing and initiation in military organizations). Whitehouse points out that small, tightly bonded groups with dysphoric rituals might be generally deleterious to cooperation in larger societies (creating opposing coalitions) which explains why they are "selected out" of the cultural repertoire, at least for the population at large, and relegated to confined organizations (e.g., militaries). Instead, the much more frequent rituals typical of regional and world religions sustain forms of group identification better suited to the kinds of collective action problems presented by interactions among strangers, or socially more distant individuals (Whitehouse 2004). Whitehouse (2000) argues that as rituals become more frequent, they also become less stimulating emotionally, and perhaps even more plain. According to Whitehouse's model, new rituals evolved to convey propositional information about superhuman beliefs through a combination of repetition and costly displays (such as animal sacrifices or monetary donations) that culturally transmit commitment to certain beliefs (Atran and Henrich 2010; Henrich 2009a).

Credibility Enhancing Displays

Henrich (2009a) offers a cultural evolutionary account of religious cooperation in large societies, based on teacher–learner models of cultural learning. The *credibility enhancing display* (CRED) *model* proposes that the transmission of otherwise difficult-to-accept beliefs (e.g., the existence of an invisible being in the sky who is worried about your sex life) is facilitated by the performance of seemingly costly actions by models or teachers; these actions are those that such a teacher would be unlikely to engage in unless the teacher were deeply committed (believed) in the aforementioned belief. This evolved bias, which allows learners to avoid manipulation by teachers, has been harnessed by cultural evolution in ways that enhance the transmission of the faith across the generations. CREDs in rituals, taboos, and devotions—such fire-walking, sacrifices, circumcision, and celibacy—deepen the faith of the learners who observe them. By incorporating elements that tap our CRED psychology, cultural evolution has equipped religions.

Sacred Values and Secular Worlds

We have noted that religious community making does not merely trade in rituals and beliefs, but also in sacred values, "things set apart and forbidden" (Durkheim 1915/1965:44). The idea that religion underpins sacred or inviolable values has a venerable history in the discipline of comparative religion (Paden 1994; Taves 2009). This idea has recently attracted the attention of moral psychologists (Graham et al. 2009; Haidt and Graham 2009), who operationalize "sacred values" as "those values that a moral community treats as possessing transcendental significance that precludes comparisons, trade-offs, or indeed any mingling with secular values" (Tetlock 2003:320).

A recent study conducted in the West Bank gives an intriguing insight into how sacred values function in political hotbeds (Atran et al. 2007). Notably, a substantial majority of the Jewish and Palestinian populations living on the West Bank value their land as sacred. These groups are in violent competition for the land. Such values cannot be bought. Indeed both groups react with outrage and disgust when cash is offered in exchange for sacred land and become more tolerant of violence to the other side. Importantly, Atran et al. find that sacred values need not result in violent attitudes to out-groups. When opposing groups sincerely acknowledge each other's sacred values, significant declines in tolerance for aggression were found. Those who hold ostensibly different sacred values will be motivated to act on their values, but they are not fated to decades of hatred and violence (Atran and Ginges 2012). How to foster mutual understanding in a context of reciprocal violence remains an important question on the horizon of policy research (see Matthews et al. 2012b; Rappaport 1971; Sosis and Alcorta 2003; Sosis 2011).

It seems that sacred values are not the exclusive possession of religious groups. Secular people, too, regard certain values to be sacred. Are secular judgments about moral rights and wrongs examples of superhuman thinking? Answers remain elusive. Some suggest that the distinction between conventional and moral judgments has been presented as a human cognitive universal (Turiel 2010). Others argue that the distinction between conventional and moral is absent in many small-scale societies, and that moral absolutes are part of the novel cultural package assembled with the rise of large-scale societies. The question whether moral judgment is an emergent property of an innate human psychology or a cultural evolutionary achievement has yet to be resolved.

Related puzzles arise for whether moral judgments are differentially linked to postulated metaphysical entities, either by being viewed as commanded by superhuman beings, or as embedded in some more impersonal, but nonetheless sacred, superhuman order. Put simply, do moral judgments require beliefs in metaphysical "stuff" to make them true? Charles Taylor has long argued that moral judgments are inextricably linked to metaphysical claims, and that even supposedly secular Enlightenment values can be seen as grounded in such "self-evident" objects of faith (human rights, human dignity, freedom)

(Taylor 1989; see also Anscombe 1958). Although Taylor's argument is, in the end, an a priori claim about transcendental personhood, Slingerland (2008) has proposed a naturalistic version that awaits empirical investigation. Do we find the "sacred" psychological profile (absolute commitment, resistance to trade-offs, strong emotion, punitive sentiments toward violators) even in self-professed atheists when it comes to their own moral values? Are these values tied to nebulous, perhaps not fully conscious metaphysical commitments, in the same way as more traditional religious-moral values? Are secular-moral values functionally and psychologically equivalent to more traditional religious values, or are there important differences? Such questions addressing the similarities and differences between traditional faiths and modern, secular societies (Huebner et al. 2010) could not be seriously raised even a decade ago. Cultural evolutionary researchers are spearheading a fiercely collaborative and multidisciplinary science which is laying the conceptual, methodological, and empirical groundwork on which progress toward answers depends.

Summary

In the domain of popular culture, the evolution of religion has been a hot topic, as evidenced by the recent spate of bestselling books and wide media coverage on the topic. A good deal of this recent literature has tended to view religion as a dispensable cognitive spandrel (Dennett 2006) or, worse, as a dangerous delusion (Dawkins 2006). Yet much research suggests that religion is not an aberrant disease or childish illusion, but rather that it may be one of the cornerstones of the evolution of large-scale complex human societies (Atran and Henrich 2010).

In this chapter, we have addressed questions which have troubled scholars in comparative religion, anthropology and philosophy for centuries (Preus 1987): How have religions changed? What is the extent of their complexity and variation? What have religions done for us? What are they doing for us now? Are religions dispensable?

Answers remain elusive. However, progress is possible when large questions are decomposed into smaller questions and addressed with appropriate methods.

In pursuit of these questions, we wish to highlight the following ongoing projects which were discussed at the Forum:

- The Binghamton Religion and Spirituality Project, which is using cultural evolution to understand how religion works in the context of an American city. For more information see: http://bnp.binghamton.edu/projects/brsp/
- Ritual, Community, and Conflict, Oxford University, which is assembling a large historical database from 5000 years BP that will enable the

testing of functional hypotheses of the kind described in this chapter. For further information see: http://www.icea.ox.ac.uk/large-grants/ritual/

- Cultural Phylogenies of Religion, Auckland University, New Zealand, which is applying cultural phylogenetics to the study of religious change in the Pacific and elsewhere. For more about this research see: http://www.psych.auckland.ac.nz/uoa/language-and-cultural-evolution-group
- Religion, Cognition and Culture Research, Aarhus University, Denmark, which is integrating the cultural evolutionary study of religion with experimental psychology and neuroscience to better understand the social and cognitive underpinnings of religion. For more information see: http://rcc.au.dk/
- A Global Consortium to Study the Evolution of Religion, University of British Columbia, Canada, which is uniting a transnational team of researchers to test cultural evolutionary hypotheses about the origins and maintenance of religion. For more information about the Cultural Evolution of Religion Research Consortium (CERC) go to: http://www.hecc.ubc.ca/cerc/project-summary/

Clearly, a greater union between natural scientists, social scientists, and humanities scholars is well under way. Further progress will require the fearless collaboration of experts across disciplinary boundaries (Slingerland and Collard 2012; Whitehouse 2011). We hope that readers will sense the excitement that cultural evolutionary approaches are bringing to the study of human society and religions, as understanding grows.

Acknowledgments

We are grateful to Dr. Andreas Strüngmann, Dr. Thomas Strüngmann, and the Ernst Strüngmann Foundation for providing the generous support which enabled this Forum. Ongoing research is supported by the Canadian Social Sciences and Humanities Research Council (SSHRC), "The Evolution of Religion and Morality"; the U.K.'s Economic and Social Research Council (ESRC), "Ritual, Community, and Conflict" (Large Grant RES-060-25-0085); the Danish Government UNIK "MINDLab" grant; the Aarhus University Interdisciplinary Centre grant "Interacting Minds"; The John F. Templeton Foundation, Testing the functional roles of religion in human society, ID 28745; The Royal Society of New Zealand, "The cultural evolution of religion," 11-UOA-239; and the Victoria URF Grant Award 8-3046-108855. We wish to thank Simon Greenhill for helpful comments to this chapter, and are grateful to Julia Lupp, Morten Christiansen, Pete Richerson, and the entire ES Forum team for their organizational and editorial efforts.

Appendix

Developmental Issues

Elena Lieven

with contributions from Morten H. Christiansen, Emma Flynn, Daniel Haun, Robert McCauley, Victoria Reyes-García, Pete Richerson, Claudio Tennie, Harvey Whitehouse, and Polly Wiessner

The basic argument put forth in this book is that many aspects of human endeavor can be better understood by adopting a cultural evolutionary perspective (see Richerson and Christiansen, this volume). The basic approach taken was to concentrate on four focal topics: social systems, technology, language, and religion. However, a number of issues cut across these areas; in particular, as relates to human development in infants, children, and adolescents. As an impetus for further discussion, this Appendix highlights some of these issues, starting with three general questions before moving to brief discussion of a number of more specific points:

1. Do the mechanisms of cultural evolution play similar roles in the acquisition of different cultural behaviors (e.g., language, tool use, social norms)?
2. How does the construction of the child's world by adults affect cultural evolution (Flynn et al. 2012)?
3. What can the different cultural practices used during childhood tell us about cultural evolution in different domains?

Biological Evolution

There is a huge literature on the biological changes that have occurred in human evolution and which underpin the extended learning period that is of such importance in cultural evolution (e.g., neotony, large brains, brain plasticity, and the length of the juvenile period) (for a review, see Locke and Bogin 2006). We will not reiterate these here. However, an area that warrants further investigation is whether there are specifically human characteristics of puberty which might have important implications for cultural evolution.

For example, are there particular ways in which adolescents confront or change norms and/or innovate which should be considered by explanations of cultural evolution? A second example is that the age of puberty in technological cultures has fallen dramatically (for girls, at least) over the last hundred years. Does this have implications for cultural evolution, and, if so, what are they? More generally, what role does the timetable of biological development play in understanding cultural evolution? For instance, if the prefrontal cortex is not fully mature until a person is in their mid-twenties, this covers the same time period as the transition to full adulthood. It would be interesting to work out the implications of this. Is this a particularly experimental/innovative stage of human development across cultures?

How Do We Measure Childhood across Cultures?

Different cultures have very different definitions of childhood and different expectations about when a person leaves childhood to become a full member of society, with rights and obligations) (e.g., Lamb and Hewlett 2005; Konner 2010). Variations in definitions of childhood as well as in the ages at which the following occur are important as are any universal regularities. Given this, is it methodologically possible to attempt a universal definition of childhood? Some ideas for further consideration:

- The end of childhood could be defined as the point in time when an individual produces a net gain of calories; that is, when energy production is higher than energy consumption (cf. Kaplan 1997; Kramer 2011).
- Many cultures have rites of passage into adulthood, particularly for boys. Are there commonalities in the age at which these occur and what other factors might modulate this?
- Is the average age of marriage correlated with net energy production and/or changes in the relations between the parental generation and children in terms of labor contribution?
- Are changes in rights and obligations between parental and child generations, which may signal the end of cultural childhood, related to the age at which the first child is born?

Adaptation of Human Systems to be Learned by Human Cognitive Apparatus

In theories of language evolution, a well-developed position suggests that language has evolved to be learnable by the human cognitive apparatus; that is, language has been shaped by just those cognitive biases that children bring to learning it (for a review, see Christiansen and Chater 2008). This then places constraints on the kinds of language systems that can emerge. Can these ideas be extended to other aspects of cultural evolution?

It is important, however, to point out that just because something is easy to learn does not mean that it will be propagated. This will depend on the broader landscape of constraints and biases, and much work remains to be done on how this might play out in terms of the interaction between learning and cultural evolution. In addition, although, other things being equal, we will be biased to learn what is easy to learn in a domain, there will be exceptions when some form of social selection favors learning hard things as a social signal. For instance, if learning to speak "well" is used as a social sign of intelligence (as it certainly used to be in the British educational context), then competition to speak "better" than others could lead to the evolution of aspects of language that are hard to learn. One theory we could have about why English spelling is much harder than it needs to be is that being able to spell accurately is a sign of educational accomplishment. Perhaps every language has some hard elements just to serve this function. Similarly, this may also apply for other domains of culture (e.g., medicine, the legal world, engineering).

Is There a Universal Basis for Cultural Learning?

Despite considerable differences in the ways in which infants are cared for during the first year of life, current evidence suggests a near universal timetable for the emergence of basic sociocognitive skills involving the communication and reading of intentions: joint attention, pointing, imitation, collaboration (Lieven, this volume; Callaghan et al. 2011). These appear to be the foundational skills for many aspects of cultural evolution and to be uniquely characteristic of human development, in that they are found for all typically developing infants, and show a reliable timetable of emergence. Arguably, they underpin the capacity for acquiring nongenetically coded information, which must be at the root of human cultural evolution (Konner 2010, especially Part IV on Enculturation). However, more research is needed on within- and across-cultural differences in learning strategies for acquiring information. What may start out as universals of development (or small differences) could be amplified by the environment (e.g., demonstrated differences in early language attainment as a function of social economic status).

What Role Do Children Play in Promoting/Restricting Cultural Variation?

Haun and Over (this volume) suggest that children restrict in-group variation but contribute to cross-cultural variation. Norms, the enforcement of norms, ostracism aversion, and the identification of social in-groups and out-groups are argued to underpin cooperation in small-scale societies and the development of institutions in the move to large-scale societies. Thus, Haun and Over (this volume) argue that from "early in development, children prefer to interact with, and learn from, individuals who are similar to themselves"; they are "ostracism averse" and they "show a tendency to match their behavior to that of the majority." The suggestion is that this maintains culture-specific characteristics and thus promotes variation between cultures upon which selection can operate (Whiten and Flynn 2010). On the other hand, a number of sociolinguists have suggested that adolescents are at the forefront of propagating language changes which are already in progress (see, e.g., Labov 1982). In contrast, Christiansen and Chater (2008) argue that the human cognitive apparatus restricts what can be learned and thus shapes the universal features of language in a probabilistic manner. Thus general learning biases may provide a regularizing pressure on cultural evolution, not only in the case of children (Chater and Christiansen 2010) but also through adult second language learners (Lupyan and Dale 2010). Clearly, there are important issues to be addressed here in terms of the ways in which human development may restrict or promote innovation and variation, how this may change across the lifespan, the role of prior states of the system (in terms of knowledge, skills, etc.), and how this impacts on different aspects of cultural evolution.

Childhood: The Period during which Children Become or Develop "Embodied Capital"

Is the length of childhood related to the overall complexity of society? The suggestion is that one of the landmark accomplishments in human cultural evolution was the creation of cultural means for prolonging childhood, and that this goes hand-in-hand with the development of complex technology. Hugely significant in this respect was the invention of writing systems, the development of literacy and, much later, the invention of the printing press with the consequent proliferation of books and the laborious process of teaching and learning to read. In its turn, this may have given rise to new forms of human conceptual and cognitive processing. New cognitive skills such as reading and mathematics are unlikely to be the result of biological adaption since the time span is too short. Instead they most likely involve the formation of novel networks via recruitment during development of prior brain areas (e.g., as suggested by Dehaene and Cohen 2007).

Relating the length of childhood to the level of complexity in a society can, in principle, be tested. Kaplan and Lancaster (2003) have looked at the caloric productivity of people in small-scale simple societies as a function of age. They found that children do not become net producers until about age 18 and do not peak until age 30 or so. However, when production is evaluated in terms of time children spend in economic activities, children become net producers much earlier (Kramer 2011). This may be a more inclusive measure of children's roles since in most smaller-scale societies they spend more time in food-processing activities than food procurement.

The main differences in complex societies seem to be in formal teaching and written storage. Arguably, children in formal education learn more per unit time. Written records reduce the burden on memory, which is presumably greater for those who cannot read, but there is also the issue of whether there are differences between simple and complex cultures in how much needs to be remembered.

Bibliography

Abelson, R. P. 1981. Psychological Status of the Script Concept. *Am. Psychol.* **36**:715. [9]

Abreu, D. 1988. On the Theory of Infinitely Repeated Games with Discounting. *Econometrica* **56** 383–396. [3]

Adamson, R. E. 1952. Functional Fixedness as Related to Problem Solving: A Repetition of Three Experiments. *J. Exp. Psychol.* **44**:288–291. [11]

Adolph, K. E., B. Vereijken, and P. E. Shrout. 2003. What Changes in Infant Walking and Why. *Child. Devel.* **74**:475–497. [9]

Agapow, P.-M., and A. Purvis. 2002. Power of Eight Tree Shape Statistics to Detect Nonrandom Diversification: A Comparison by Simulation of Two Models of Cladogenesis. *Syst. Biol.* **51**:866–872. [15]

Agnetta, B., and P. Rochat. 2004. Imitative Games by 9-, 14-, and 18-Month-Old Infants. *Infancy* **6**:1–36. [5]

Ahmed, A. M. 2009. Are Religious People More Prosocial? A Quasi-Experimental Study with Madrasah Pupils in a Rural Community in India. *J. Sci. Study Rel.* **48**:368–374. [19]

———. 2011. Implicit Influences of Christian Religious Representations on Dictator and Prisoner's Dilemma Game Decisions. *J. Socioecon.* **40**:242–246. [19]

Alchian, A. A. 1950. Uncertainty, Evolution, and Economic Theory. *J. Political Econ.* **58**:211–221. [1]

Alcorta, C. S. 2008 iPods, Gods, and the Adolescent Brain. In: The Evolution of Religion: Studies, Theories, Critiques, ed. J. Bulbulia et al., pp. 263–270. Santa Margarita, CA: Collins Foundation Press. [20]

Alcorta, C. S., and R. Sosis. 2005. Ritual, Emotion, and Sacred Symbols: The Evolution of Religion As an Adaptive Complex. *Hum. Nature* **16**:323–359. [17, 20]

Alexander, R. D. 1974. The Evolution of Social Behavior. *Annu. Rev. Ecol. System.* **5**:325–383. [1, 2]

———. 1979. Darwinism and Human Affairs. The Jessie and John Danz Lectures. Seattle: Univ. of Washington Press. [1]

———. 1987. The Biology of Moral Systems. New York: Aldine De Gruyter. [17]

Al-Khalili, J. 2011. The House of Wisdom: How Arabic Science Saved Ancient Knowledge and Gave Us the Renaissance. New York: Penguin. [10]

Allen, R. C. 1983. Collective Invention. *J. Econ. Behav. Org.* **4**:1–24. [7]

Allen-Arave, W., M. Gurven, and K. Hill. 2008. Reciprocal Altruism, Rather Than Kin Selection, Maintains Nepotistic Food Transfers on an Aché Reservation. *Evol. Hum. Behav.* **29**:305–318. [3]

Alperson-Afil, N. 2008. Continual Fire-Making by Hominins at Gesher Benot Ya'aqov, Israel. *Quat. Sci. Rev.* **27**:1733–1739. [2]

Alpher, B., and D. Nash. 1999. Lexical Replacement and Cognate Equilibrium in Australia. *Aust. J. Linguis.* **19**:5–56. [13]

Ambridge, B., and E. V. M. Lieven. 2011. Child Language Acquisition: Contrasting Theoretical Approaches. Cambridge: Cambridge Univ. Press. [14]

Anderson, M. L. 2010. Neural Reuse: A Fundamental Organizational Principle of the Brain. *Behav. Brain Sci.* **33**:245–266. [9, 16]

Anderson, P. W. 1972. More Is Different. *Science* **177**:393–396. [16]

André, J.-B., and O. Morin. 2011. Questioning the Cultural Evolution of Altruism. *J. Evol. Biol.* **24**:2531–2542. [6]

Andreski, S. 1971. Military Organization and Society. Berkeley: Univ. of California Press. [4]

Andrews, L. 1836. Remarks on the Hawaiian Dialect of the Polynesian Language. *Chinese Repository* **5**:12–21. [15]

Andronov, M. S. 1962. Razgovornyj Tamil'skij Jazyk I Ego Dialekty. Moscow: Institute of Peoples of Asia. [13]

Anscombe, G. E. 1958. Modern Moral Philosophy. In: The Collected Philosophical Papers of G. E. M. Anscombe, vol. 3, pp. 26–42. Oxford: Basil Blackwell. [20]

Anthony, D. W. 2007. The Horse, the Wheel, and Language: How Bronze-Age Riders from the Eurasian Steppes Shaped the Modern World. Princeton: Princeton Univ. Press. [15]

Aoki, K. 2010. Evolution of the Social-Learner-Explorer Strategy in an Environmentally Heterogeneous Two-Island Model. *Evolution* **64**:2575–2586. [7]

Aoki, K., and Y. Kobayashi. 2012. Innovativeness, Population Size, and Cumulative Cultural Evolution. *Theor. Pop. Biol.* **8**:38–47. [7]

Aoki, K., L. Lehmann, and M. W. Feldman. 2011. Rates of Cultural Change and Patterns of Cultural Accumulation in Stochastic Models of Social Transmission. *Theor. Pop. Biol.* **79**:192–202. [11]

Apicella, C. L., F. W. Marlowe, J. H. Fowler, and N. A. Christakis. 2012. Social Networks and Cooperation in Hunter-Gatherers. *Nature* **481**:497–501. [5, 6]

Appiah, K. A. 2010. The Honor Code: How Moral Revolutions Happen. New York: W. W. Norton. [3]

Arbib, M., J. Bonaiuto, S. Jacobs, and S. Frey. 2009. Tool Use and the Distalization of the End-Effector. *Psychol. Res.* **73**:441–462. [9]

Ardrey, R. 1966/1997. The Territorial Imperative: A Personal Inquiry into the Animals Origins of Property and Nations. New York: Kodansha America. [2]

Aris-Brosou, S. 2005. Determinants of Adaptive Evolution at the Molecular Level: The Extended Complexity Hypothesis. *Mol. Biol. Evol.* **22**:200–209. [16]

Aristotle. 350 BC/2002. Nichomachean Ethics. Newburyport, MA: Focus Publishing. [2]

Armstrong, J. S. 1997. Peer Review for Journals: Evidence on Quality Control, Fairness, and Innovation. *Sci. Eng. Ethics* **3**:63–84. [10]

Arnold, M. L. 2008. Reticulate Evolution and Humans: Origins and Ecology. Oxford: Oxford Univ. Press. [16]

Arnos, K. S., K. O. Welch, M. Tekin, et al. 2008. A Comparative Analysis of the Genetic Epidemiology of Deafness in the United States in Two Sets of Pedigrees Collected More Than a Century Apart. *Am. J. Hum. Genet.* **83**:200–207. [12]

Aronson, E., and J. Mills. 1959. The Effect of Severity of Initiation on Liking for a Group. *J. Abnorm. Soc. Psychol.* **59**:177–181. [20]

Arrow, K. 1962. The Economic Implications of Learning by Doing. *Rev. Econ. Stud.* **29**:155–173. [7]

Arthur, W. B. 2009. The Nature of Technology. New York: Free Press. [8, 11]

Asendorpf, J. B., V. Warkentin, and P.-M. Baudonnière. 1996. Self-Awareness and Other-Awareness II: Mirror Self-Recognition, Social Contingency Awareness, and Synchronic Imitation. *Devel. Psychol.* **32**:313–321. [5]

Aslin, R. N., and E. L. Newport. 2008. What Statistical Learning Can and Can't Tell Us About Language Acquisition. In: Infant Pathways to Language: Methods, Models, and Research Directions, ed. J. Colombo et al. Mahwah, NJ: Lawrence Erlbaum. [14]

Atkinson, Q. D. 2011. Phonemic Diversity Supports a Serial Founder Effect Model of Language Expansion from Africa. *Science* **332**:346–349. [15, 20]

Atkinson, Q. D., and P. Bourrat. 2010. Beliefs About God, the Afterlife and Morality Support the Role of Supernatural Policing in Human Cooperation. *Evol. Hum. Behav.* **32**:41–49. [19]

———. 2011. Beliefs About God, the Afterlife and Morality Support the Role of Supernatural Policing in Human Cooperation. *Evol. Hum. Behav.* **32**:41–49. [20]

Atkinson, Q. D., and R. D. Gray. 2005. Curious Parallels and Curious Connections: Phylogenetic Thinking in Biology and Historical Linguistics. *Syst. Biol.* **54**:513–526. [13, 15]

Atkinson, Q. D., A. M. Meade, C. Venditti, S. J. Greenhill, and M. Pagel. 2008. Languages Evolve in Punctuational Bursts. *Science* **319**:588. [15, 16]

Atkinson, Q. D., and H. Whitehouse. 2011. The Cultural Morphospace of Ritual Form Examining Modes of Religiosity Cross-Culturally. *Evol. Hum. Behav.* **32**:50–62. [17, 19, 20]

Atran, S., R. Axelrod, and R. Davis. 2007. Sacred Barriers to Conflict Resolution. *Science* **317**:1039–1040. [17, 20]

Atran, S., and J. Ginges. 2012. Religious and Sacred Imperatives in Human Conflict. *Science* **336**:855–857. [20]

Atran, S., and J. Henrich. 2010. The Evolution of Religion: How Cognitive By-Products, Adaptive Learning Heuristics, Ritual Displays, and Group Competition Generate Deep Commitments to Prosocial Religions. *Biol. Theory* **5**:18–30. [17, 19, 20]

Atran, S., and A. Norenzayan. 2004. Religion's Evolutionary Landscape: Counterintuition, Commitment, Compassion, Communion. *Behav. Brain Sci.* **27**:713–770. [17, 19]

Auguste, A. J., P. Lemey, O. G. Pybus, et al. 2010. Yellow Fever Virus Maintenance in Trinidad and Its Dispersal Throughout the Americas. *J. Virol.* **84**:9967–9977. [15]

Aunger, R. 1994. Are Food Avoidances Maladaptive in the Ituri Forest of Zaire. *J. Anthropol. Res.* **50**:277–310. [1]

———, ed. 2000. Darwinizing Culture: The Status of Memetics as a Science. Oxford: Oxford Univ. Press. [11]

———. 2002. The Electric Meme: A New Theory of How We Think. New York: Free Press. [1]

Axelrod, R., and W. D. Hamilton. 1981. The Evolution of Cooperation. *Science* **211**:1390–1396. [3]

Bacharach, M. 2006. Beyond Individual Choice: Teams and Games in Game Theory, N. Gold and R. Sugden, series ed. Princeton: Princeton Univ. Press. [2]

Badre, D., and M. D'Esposito. 2009. Is the Rostro-Caudal Axis of the Frontal Lobe Hierarchical? *Nat. Rev. Neurosci.* **10**:659–669. [9]

Baker, M. C. 1996. The Polysynthesis Parameter. New York: Oxford Univ. Press. [13]

Baker, T. 1992. Bow Design and Performance. In: The Traditional Bowyer's Bible, ed. J. Hamm, vol. 1. Guilford: Lyons Press. [7]

Bakker, P. 1997. A Language of Our Own: The Genesis of Michif, the Mixed Cree-French Language of the Canadian Métis. New York: Oxford Univ. Press. [13]

Bakker, P., A. Daval-Markussen, M. Parkvall, and I. Plag. 2011. Creoles Are Typologically Distinct from Non-Creoles. *J. Pidgen and Creole Languages* **26**:5–42. [15]

Balikci, A. 1989. The Netsilik Eskimo. Long Grove: Waveland Press. [7]

Bamforth, D. B. 1994. Indigenous People, Indigenous Violence: Precontact Warfare on the North American Great Plains. *Man* **29**:95–115. [3]

Bandura, A., and R. H. Walters. 1963. Social Learning and Personality Development. New York: Holt Rinehart and Winston. [1]

Bane, M. 2008. Quantifying and Measuring Morphological Complexity. In: Proc. 26th West Coast Conf. on Formal Linguistics, ed. C. B. Chang and H. J. Haynie, pp. 69–76. Somerville, MA: Cascadilla Proceedings Project. [16]

Bannister, R. C. 1979. Social Darwinism: Science and Myth in Anglo-American Social Thought. American Civilization. Philadelphia: Temple Univ. Press. [1]

Barbey, A. K., F. Krueger, and J. Grafman. 2009. Structured Event Complexes in the Medial Prefrontal Cortex Support Counterfactual Representations for Future Planning. *Phil. Trans. R. Soc. B* **364**:1291–1300. [9]

Barbujani, G., and V. Colonna. 2010. Human Genome Diversity: Frequently Asked Questions. *Trends Genet.* **26**:285–295. [12, 16]

Bargh, J. A., and T. L. Chartrand. 1999. The Unbearable Automaticity of Being. *Am. Psychol.* **54**:462–479. [19]

Bargh, J. A., P. M. Gollwitzer, A. Lee-Chai, K. Barndollar, and R. Troetschel. 2001. Automating the Will: Nonconscious Activation and Pursuit of Behavioral Goals. *J. Pers. Soc. Psychol.* **81**:1014–1027. [19]

Barkow, J. H., L. Cosmides, and J. Tooby. 1992. The Adapted Mind: Evolutionary Psychology and the Generation of Culture. New York: Oxford Univ. Press. [2]

Barmettler, F., E. Fehr, and C. Zehnder. 2012. Big Experimenter Is Watching You! Anonymity and Prosocial Behavior in the Laboratory. *Games Econ. Behav.* **75**:17–34. [19]

Baronchelli, A., N. Chater, R. Pastor-Satorras, and M. H. Christiansen. 2012. The Biological Origin of Linguistic Diversity. *PLoS One* **7**:e48029. [16]

Baronchelli, A., M. Felici, V. Loreto, E. Caglioti, and L. Steels. 2006. Sharp Transition Towards Shared Vocabularies in Multi-Agent Systems. *J. Statis. Mechan.* **6**:P06014. [16]

Baronchelli, A., T. Gong, A. Puglisi, and V. Loreto. 2010. Modeling the Emergence of Universality in Color Naming Patterns. *PNAS* **107**:2403. [16]

Barrett, H. C. 2013. The Shape of Thought: How Mental Adaptations Evolve. New York: Oxford Univ. Press, in press. [7]

Barrett, H. C., L. Cosmides, and J. Tooby. 2007a. The Hominid Entry into the Cognitive Niche. In: Evolution of Mind, Fundamental Questions and Controversies, ed. S. Gangestad and J. Simpson, pp. 241–248. New York: Guilford Press. [7]

Barrett, J. L. 2004. Why Would Anyone Believe in God? Cognitive Science of Religion Series. Walnut Creek: AltaMira Press. [17–19]

Barrett, J. L., and F. C. Keil. 1996. Conceptualizing a Nonnatural Entity: Anthropomorphism in God Concepts. *Cogn. Psychol.* **31**:219–247. [18, 20]

Barrett, T. M., E. F. Davis, and A. Needham. 2007b. Learning About Tools in Infancy. *Devel. Psychol.* **43**:352–368. [11]

Barrickman, N. L., M. L. Bastian, K. Isler, and C. P. van Schaik. 2008. Life History Costs and Benefits of Encephalization: A Comparative Test Using Data from Long-Term Studies of Primates in the Wild. *J. Hum. Evol.* **54**:568–590. [2]

Bartley, W. W., III. 1984. The Retreat to Commitment (2nd edition). London: Open Court Publishing Co. [1]

Basalla, G. 1988. The Evolution of Technology. Cambridge: Cambridge Univ. Press. [7, 11]

Basu, S., M. Kirk, and G. Waymire. 2009. Memory, Transaction Records, and the Wealth of Nations. *Acct. Org. Soc.* **34**:895–917. [6]

Bates, T. C., M. Luciano, S. E. Medland, et al. 2011. Genetic Variance in a Component of the Language Acquisition Device: ROBO1 Polymorphisms Associated with Phonological Buffer Deficits. *Behav. Genet.* **41**:50–57. [12, 16]

Bateson, G. 1935. Culture Contact and Schismogenesis. *Man* **35**:178–183. [15]

Bateson, M., D. Nettle, and G. Roberts. 2006. Cues of Being Watched Enhance Cooperation in a Real-World Setting. *Biol. Lett.* **2**:412–414. [19]

Batson, C. D., P. Schoenrade, and W. L. Ventis. 1993. Religion and the Individual: A Social-Psychological Perspective. New York: Oxford Univ. Press. [19]

Battye, A., and I. Roberts. 1995. Introduction. In: Clause Structure and Language Change, ed. A. Battye and I. Roberts, pp. 3–28. Oxford: Oxford Univ. Press. [16]

Baxter, G. J., R. A. Blythe, W. Croft, and A. J. McKane. 2006. Utterance Selection Model of Language Change. *Phys. Rev. E* **73**:046118. [16]

———. 2009. Modeling Language Change: An Evaluation of Trudgill's Theory of the Emergence of New Zealand English. *Lang. Var. Change* **21**:157–196. [16]

Baxter, P. T. W. 1978. Borana Age-Sets and Generation-Sets: Gada, a Maze? In: Age, Generation and Feature of East African Age Organizations, ed. P. T. W. Baxter and U. Almagor. London: C. Hurst. [6]

Becker, G. S. 1978. The Economic Approach to Human Behavior. Chicago: Univ. of Chicago Press. [4]

Beckner, C., R. Blythe, J. Bybee, et al. 2009. Language Is a Complex Adaptive System: Position Paper. *Lang. Learn.* **59**:1–27. [13]

Beheim, B. A., and R. Baldini. 2012. Evolutionary Decomposition and the Mechanisms of Cultural Change. *Cliodynamics* **3**:217–233. [11]

Bekkering, H., and W. Prinz. 2002. Goal Representations in Imitative Actions. In: Imitation in Animals and Artifacts, ed. K. Dautenhahn and L. Nehaniv, pp. 555–572. Cambridge, MA: MIT Press. [9]

Bell, A. V. 2013. The Dynamics of Culture Lost and Conserved: Demic Migration as a Force in New Diaspora Communities. *Evol. Hum. Behav.* **34**:23–28. [1]

Bell, A. V., P. J. Richerson, and R. McElreath. 2009. Culture Rather Than Genes Provides Greater Scope for the Evolution of Large-Scale Human Prosociality. *PNAS* **106**:17,671–17,674. [6]

Bellah, R. N. 1967. Civil Religion in America. *Daedalus* **96**: [20]

———. 2011. Religion in Human Evolution: From the Paleolithic to the Axial Age. Cambridge, MA: Harvard Univ. Press. [4, 20]

Bellwood, P. S., and C. Renfrew, eds. 2002. Examining the Farming/Language Dispersal Hypothesis. Cambridge: McDonald Institute. [13]

Bendor, J., and P. Swistak. 1995. Types of Evolutionary Stability and the Problem of Cooperation. *PNAS* **92**:3596–3600. [3]

Bentz, C., and M. H. Christiansen. 2010. Linguistic Adaptation at Work? The Change of Word Order and Case System from Latin to the Romance Languages. In: Proc. 8th Intl. Conf. on the Evolution of Language, ed. A. Smith et al., pp. 26–33. London: World Scientific Publishing. [13]

Benveniste, É. 1966. Problèmes De Linguistique Générale. Paris: Gallimard. [13]

Berenda, R. W. 1950. The Influence of the Group on the Judgments of Children: An Experimental Investigation. New York: King's Crown Press. [5]

Berg, M. 2002. From Imitation to Invention: Creating Commodities in Eighteenth Century Britain. *Econ. Hist. Rev.* **55**:1–30. [8]

Bergsland, K., and H. Vogt. 1962. On the Validity of Glottochronology. *Curr. Anthropol.* **3**:115–153. [15]

Bergstrom, C., and M. Lachmann. 1997. Signaling among Relatives I: Is Signaling Too Costly? *Proc. R. Soc. B* **352**:609–617. [3]

———. 1998. Signaling among Relatives III: Talk Is Cheap. *PNAS* **95**:5100–5105. [3]

Bering, J. M. 2006a. The Cognitive Science of Souls: Clarifications and Extensions of the Evolutionary Model. *Behav. Brain Sci.* **29**:486–498. [17]

———. 2006b. The Folk Psychology of Souls. *Behav. Brain Sci.* **29**:453–498. [19]

———. 2011. The Belief Instinct: The Psychology of Souls, Destiny, and the Meaning of Life. New York: W. W. Norton. [19]

Berlin, B., and P. Kay. 1969. Basic Color Terms: Their Universality and Evolution. Berkeley: Univ. of California Press. [13]

Berna, F., P. Goldberg, L. K. Horwitz, et al. 2012. Microstratigraphic Evidence of In Situ Fire in the Acheulean Strata of Wonderwerk Cave, Northern Cape Province, South Africa. *PNAS* **109**:215–220. [6]

Bernstein, N. A. 1996. On Dexterity and Its Development. In: Dexterity and Its Development, ed. M. L. Latash and M. T. Turvey. Mahwah, NJ: Lawrence Erlbaum. [9]

Bersaglieri, T., P. C. Sabeti, N. Patterson, et al. 2004. Genetic Signatures of Strong Recent Positive Selection at the Lactase Gene. *Am. J. Hum. Genet.* **74**:1111–1120. [8]

Berti, A., and F. Frassinetti. 2000. When Far Becomes Near: Remapping of Space by Tool Use. *J. Cogn. Neurosci.* **12**:415–420. [9]

Berwick, R. C., A. D. Friederici, N. Chomsky, and J. J. Bolhuis. 2013. Evolution, Brain, and the Nature of Language. *Trends Cogn. Sci.* **17**:91–100. [16]

Bethell, T. 1976. Darwin's Mistake. *Harper's* **252**:1509. [7]

Bettinger, R. L., and J. W. Eerkens. 1997. Evolutionary Implications of Metrical Variation in Great Basin Projectile Points. In: Rediscovering Darwin: Evolutionary Theory and Archaeological Explanation, ed. C. M. Barton and G. A. Clark, vol. Archaeological Papers 7, pp. 177–191. Arlington: American Anthropological Association. [1]

Bingham, P. M. 1999. Human Uniqueness: A General Theory. *Q. Rev. Biol.* **74**:133–169. [2]

Binmore, K. G. 1994. Game Theory and the Social Contract: Playing Fair, vol. 1. Cambridge, MA: MIT Press. [3]

———. 1998. Game Theory and the Social Contract: Just Playing, vol. 2. Cambridge, MA: MIT Press. [3]

———. 2005. Natural Justice. New York: Oxford Univ. Press. [18]

Birch, S. A. J., N. Akmal, and K. L. Frampton. 2010. Two-Year-Olds Are Vigilant of Others' Non-Verbal Cues to Credibility. *Devel. Sci.* **13**:363–369. [7]

Birch, S. A. J., S. A. Vauthier, and P. Bloom. 2008. Three- and Four-Year-Olds Spontaneously Use Others' Past Performance to Guide Their Learning. *Cognition* **107**:1018–1034. [7]

Bishop, D. V. M. 2009. Genes, Cognition, and Communication: Insights from Neurodevelopmental Disorders. *Ann. NY Acad. Sci.* **1156**:1–18. [16]

Blackmore, S. J. 1999. The Meme Machine. Oxford: Oxford Univ. Press. [16]

Blakemore, S. J., D. M. Wolpert, and C. D. Frith. 1998. Central Cancellation of Self-Produced Tickle Sensation. *Nat. Neurosci.* **1**:635–640. [9]

Blevins, J. 2004. Evolutionary Phonology: The Emergence of Sound Patterns. Cambridge: Cambridge Univ. Press. [13, 16]

Bliege Bird, R. L., and E. A. Smith. 2005. Signaling Theory, Strategic Interaction, and Symbolic Capital. *Curr. Anthropol.* **46**:221–248. [20]

Bloch, M. 1986. From Blessing to Violence: History and Ideology in the Circumcision Ritual of the Merina. Cambridge Studies in Social and Cultural Anthropology. Cambridge: Cambridge Univ. Press. [20]

Blome, M. W., A. S. Cohen, C. A. Tryon, A. S. Brooks, and J. Russell. 2012. The Environmental Context for the Origins of Modern Human Diversity: A Synthesis of Regional Variability in African Climate 150,000–30,000 Years Ago. *J. Hum. Evol.* **62**:563–592. [8]

Bloom, P. 2000. How Children Learn the Meanings of Words. Cambridge, MA: MIT Press. [1]

———. 2004. Descartes' Baby. How the Science of Child Development Explains What Makes Us Human. New York: Basic Books. [17]

Bloor, D. 1971. Essay Review: Two Paradigms for Scientific Knowledge? *Soc. Stud. Sci.* **1**:101–115. [1]

———. 1991. Knowledge and Social Imagery (2nd edition). Chicago: Univ. of Chicago Press. [10]

Blumenschine, R. J., J. A. Cavallo, and S. D. Capaldo. 1994. Competition for Carcasses and Early Hominid Behavioral Ecology: A Case Study and Conceptual Framework. *J. Hum. Evol.* **27**:197–213. [2]

Blust, R. 2000. Why Lexicostatistics Doesn't Work: The "Universal Constant" Hypothesis and the Austronesian Languages. In: Time Depth in Historical Linguistics, ed. C. Renfrew et al., pp. 311–332. Cambridge: McDonald Institute. [15]

Blyth, J. 2013. Preference Organization Driving Structuration: Evidence from Australian Aboriginal Interaction for Pragmatically Motivated Grammaticalization. *Language*, in press. [13]

Blythe, R. A., and W. Croft. 2012. S-Curves and the Mechanisms of Propagation in Language Change. *Language* **88**:269–304. [16]

Boatwright, M. T. 2000. Hadrian and the Cities of the Roman Empire. Princeton: Princeton Univ. Press. [6]

Bocquet-Appel, J. P. 2002. Paleoanthropological Traces of a Neolithic Demographic Transition. *Curr. Anthropol.* **43**:637–647. [8]

Boehm, C. 1993. Egalitarian Behavior and Reverse Dominance Hierarchy. *Curr. Anthropol.* **34**:227–254. [6, 20]

———. 1999. Hierarchy in the Forest: The Evolution of Egalitarian Behavior. Cambridge, MA: Harvard Univ. Press. [2, 4, 20]

———. 2011. Moral Origins: The Evolution of Virute, Altruism, and Shame. New York: Basic Books. [2, 4, 6]

Boehm, C., and J. C. Flack. 2010. The Emergence of Simple and Complex Power Structures through Social Niche Construction. In: Social Psychology of Power, ed. A. Guinote and T. K. Vescio. New York: Guilford Press. [2]

Boesch, C., G. Hohmann, and L. Marchant. 1998. Behavioural Diversity in Chimpanzees and Bonobos. Cambridge: Cambridge Univ. Press. [2]

Boesch, C., G. Kohou, H. Nànà, and L. Vigilant. 2006. Male Competition and Paternity in Wild Chimpanzees of the Tai Forest. *Am. J. Phys. Anthropol.* **130**:103–115. [2]

Boltz, W. G., J. Renn, and M. Schemmel. 2003. Mechanics in the Mohist Canon and Its European Counterpart: Texts and Contexts. http://www.mpiwg-berlin.mpg.de/Preprints/P241.PDF. (accessed May 13, 2013). [10]

Borenstein, E., M. W. Feldman, and K. Aoki. 2008. Evolution of Learning in Fluctuating Environments: When Selection Favors Both Social and Exploratory Individual Learning. *Evolution* **62**:586–602. [7]

Borgerhoff Mulder, M., S. Bowles, T. Hertz, and A. V. Bell. 2009. Intergenerational Wealth Transmission and the Dynamics of Inequality in Small-Scale Societies. *Science* **326**:682–688. [2]

Bouchard-Côté, A., D. Hall, T. L. Griffiths, and D. Klein. 2013. Automated Reconstruction of Ancient Languages Using Probabilistic Models of Sound Change. *PNAS* **110**:4224–4229. [15, 16]

Bouckaert, R., P. Lemey, M. Dunn, et al. 2012. Mapping the Origins and Expansion of the Indo-European Language Family. *Science* **337**:957–960. [1, 11, 15, 16]

Bowerman, M., and S. Choi. 2001. Shaping Meanings for Language: Universal and Language-Specific in the Acquisition of Spatial Semantic Categories. In: Language Acquisition and Conceptual Development, ed. M. Bowerman and S. C. Levinson, pp. 475–511. Cambridge: Cambridge Univ. Press. [14]

Bowern, C., and Q. Atkinson. 2012. Computational Phylogenetics and the Internal Structure of Pama-Nyungan. *Language* **88**:817–845. [15]

Bowler, P. J. 1983. The Eclipse of Darwinism: Anti-Darwinian Evolution Theories in the Decades around 1900. Baltimore: Johns Hopkins Univ. Press. [1]

Bowles, S. 2006. Group Competition, Reproductive Leveling and the Evolution of Human Altruism. *Science* **314**:1569–1572. [2, 4]

———. 2009. Did Warfare among Ancestral Hunter-Gatherers Affect the Evolution of Human Social Behaviors? *Science* **324**:1293–1298. [4]

———. 2011. Cultivation of Cereals by the First Farmers Was Not More Productive Than Foraging. *PNAS* **108**:4760–4765. [8]

Bowles, S., and H. Gintis. 1986. Democracy and Capitalism: Property, Community, and the Contradictions of Modern Social Thought. New York: Basic Books. [2]

———. 2011. A Cooperative Species: Human Reciprocity and Its Evolution. Princeton: Princeton Univ. Press. [1, 2, 6]

Boyd, R. 1982. Density-Dependent Mortality and the Evolution of Social Interactions. *Anim. Behav.* **30**:972–982. [3]

Boyd, R., H. Gintis, and S. Bowles. 2010. Coordinated Punishment of Defectors Sustains Cooperation and Can Proliferate When Rare. *Science* **328**:617–620. [2, 3]

Boyd, R., H. Gintis, S. Bowles, and P. J. Richerson. 2003. The Evolution of Altruistic Punishment. *PNAS* **100**:3531–3535. [3, 6]

Boyd, R., and P. J. Richerson. 1985. Culture and the Evolutionary Process. Chicago: Univ. of Chicago Press. [1–4, 6–8, 11–13, 18, 20]

———. 1987a. The Evolution of Ethnic Markers. *Cult. Anthropol.* **2**:65–79. [16]

———. 1987b. An Evolutionary Model of Social Learning: The Effects of Spatial and Temporal Variation. In: Social Learning: Psychological and Biological Perspectives, ed. T. R. Zentall and B. G. Galef, pp. 29–48. Hillsdale: Lawrence Erlbaum. [5, 7]

———. 1990. Group Selection among Alternative Evolutionarily Stable Strategies. *J. Theor. Biol.* **145**:331–342. [3]

———. 1992. How Microevolutionary Processes Give Rise to History. In: History and Evolution, ed. M. Nitecki and D. V. Nitecki, pp. 178–209. Albany: SUNY Press. [1, 11]

———. 1995. Why Does Culture Increase Human Adaptability? *Ethol. Sociobiol.* **16**:125–143. [7]

———. 1996. Why Culture Is Common, but Cultural Evolution Is Rare. *Proc. Br. Acad.* **88**:77–93. [11]

———. 2002a. Group Beneficial Norms Can Spread Rapidly in a Structured Population. *J. Theor. Biol.* **215**:287–296. [3]

———. 2002b. Solving the Puzzle of Human Cooperation. In: Evolution and Culture, ed. S. Levinson. Cambridge, MA: MIT Press. [20]

———. 2005. The Origin and Evolution of Cultures. New York: Oxford Univ. Press. [5, 8, 11]

———. 2009. Culture and the Evolution of Human Cooperation. *Phil. Trans. R. Soc. B* **364**:3281–3288. [5]

Boyd, R., P. J. Richerson, and J. Henrich. 2011. The Cultural Niche: Why Social Learning Is Essential for Human Adaptation. *PNAS* **108**:10,918–10,925. [9, 17]

Boyd, R., and J. B. Silk. 2002. How Humans Evolved (3rd edition). New York: W. W. Norton. [2]

Boyer, P. 1990. Tradition as Truth and Communication: A Cognitive Description of Traditional Discourse. Cambridge: Cambridge Univ. Press. [20]

———. 1994. The Naturalness of Religious Ideas. Berkeley: Univ. of California Press. [20]

———. 1998. Cognitive Tracks of Cultural Inheritance: How Evolved Intuitive Ontology Governs Cultural Transmission. *Am. Anthropol.* **100**:876–889. [7]

———. 2001. Religion Explained: The Evolutionary Origins of Religious Thought. New York: Basic Books. [17–20]

———. 2008. Religion: Bound to Believe? *Nature* **455**:1038–1039. [19]

Boyer, P., and P. Lienard. 2006. Why Ritualized Behavior? Precaution Systems and Action Parsing in Developmental, Pathological and Cultural Rituals. *Behav. Brain Sci.* **29**:1–56. [18]

Boysson-Bardies, B. 1999. How Language Comes to Children: From Birth to Two Years. Cambridge, MA: MIT Press. [1]

Boysson-Bardies, B., and M. Vihman. 1991. Adaptation to Language: Evidence from Babbling and First Words in Four Languages. *Language* **67**:297–319. [14]

Brandt, H., C. Hauert, and K. Sigmund. 2006. Punishing and Abstaining for Public Goods. *PNAS* **103**:495–497. [3]

Brandt, S., A. Verhagen, E. Lieven, and M. Tomasello. 2011. German Children's Productivity with Simple Transitive and Complement-Clause Constructions: Testing the Effects of Frequency and Variability. *Cogn. Ling.* **22**:325–357. [14]

Brass, M., and C. Heyes. 2005. Imitation: Is Cognitive Neuroscience Solving the Correspondence Problem? *Trends Cogn. Sci.* **9**:489–495. [9]

Bratman, M. E. 1993. Shared Intention. *Ethics* **104**:97–113. [2]

Bremer, S. A., and M. Mihalka. 1977. Machiavelli in Machina: Or Politics among Hexagons. In: Problems of the World Modeling: Political and Social Implications, ed. K. W. Deutsch et al., pp. 303–337. Cambridge: Ballinger Publishing. [4]

Bril, B., R. Rein, T. Nonaka, F. Wenban-Smith, and G. Dietrich. 2010. The Role of Expertise in Tool Use: Skill Differences in Functional Action Adaptations to Task Constraints. *J. Exp. Psychol. Hum. Percep. Perform.* **36**:825–839. [9]

Bril, B., V. Roux, and G. Dietrich. 2000. Habilites Impliquees dans la Taille des Perles en Roches Dure: Characteristiques Motrices et Cognitives d'une Action Situee Complexe. In: Les Perles de Cambay: Des Practiques Techniques aux Technosystemes de l'Orient Ancien, ed. V. Roux, pp. 211–329. Paris: Editions de la MSH. [9]

Brooks, A. C. 2006. Who Really Cares: The Surprising Truth About Compassionate Conservatism. New York: Basic Books. [19]

Brown, D. E. 1988. Hierarchy, History, and Human Nature: The Social Origins of Historical Consciousness. Tucson: Univ. of Arizona Press. [1]

———. 1991. Human Universals. New York: McGraw-Hill. [2]

Brown, P. 2001. Learning to Talk About Motion Up and Down in Tzeltal: Is There a Language-Specific Bias for Verb Learning? In: Language Acquisition and Conceptual Development, ed. M. Bowerman and S. C. Levinson, pp. 512–543. Cambridge: Cambridge Univ. Press. [14]

———. 2011. The Cultural Organization of Attention. In: The Handbook of Language Socialization, ed. A. Duranti et al., pp. 29–55. Oxford: Wiley-Blackwell. [14]

Bruner, J. 1964. The Course of Cognitive Growth. *Am. Psychol.* **19**:1–15. [11]

———. 1975. From Communication to Language. *Cognition* **3**:255–287. [14]

———. 1990. Acts of Meaning. Cambridge, MA: Harvard Univ. Press. [9]

———. 1993. Commentary on Tomasello et al. "Cultural Learning." *Behav. Brain Sci.* **16**:515–516. [5]

Bshary, R., and R. Bergmüller. 2007. Distinguishing Four Fundamental Approaches to the Evolution of Helping. *J. Evol. Biol.* **21**:405–420. [6]

Buccino, G., S. Vogt, A. Ritzl, et al. 2004. Neural Circuits Underlying Imitation Learning of Hand Actions: An Event-Related fMRI Study. *Neuron* **42**:323–334. [9]

Buchsbaum, D., A. Gopnik, T. L. Griffiths, and P. Shafto. 2011. Children's Imitation of Causal Action Sequences Is Influenced by Statistical and Pedagogical Evidence. *Cognition* **120**:331–340. [9]

Bulbulia, J. 2004. Religious Costs as Adaptations That Signal Altruistic Intention. *Evol. Cogn.* **10**:19–38. [19, 20]

———. 2008a. Free Love: Religious Solidarity on the Cheap. In: The Evolution of Religion: Studies, Theories, and Critiques, ed. J. Bulbulia et al., pp. 153–160. Santa Margarita, CA: Collins Foundation Press. [20]

———. 2008b. Meme Infection or Religious Niche Construction? An Adaptationist Alternative to the Cultural Maladaptationist Hypothesis. *Meth. Theory Stud. Rel.* **20**:67–107. [20]

———. 2011. The Hypnotic Stag Hunt. *J. Cogn. Culture* **11**:353–365. [20]

———. 2012. Spreading Order: Religion, Cooperative Niche Construction, and Risky Coordination Problems. *Biol. Philos.* **27**:1–27. [20]

Bulbulia, J., and M. Frean. 2010. The Evolution of Charismatic Cultures. *Meth. Theory Stud. Rel.* **22**:254–271. [20]

Bulbulia, J., R. Sosis, C. Genet, et al., eds. 2008. The Evolution of Religion: Studies, Theories, and Critiques. Santa Margarita, CA: Collins Foundation Press. [19]

Burkart, J. M., S. B. Hrdy, and C. P. van Schaik. 2009. Cooperative Breeding and Human Cognitive Evolution. *Evol. Anthropol.* **18**:175–186. [2, 6]

Burkart, J. M., and C. P. van Schaik. 2010. Cognitive Consequences of Cooperative Breeding in Primates. *Anim. Cogn.* **13**:1–19. [2]

Burnham, K. P., and D. Anderson. 2002. Model Selection and Multi-Model Inference. Berlin: Springer-Verlag. [1]

Butcher, A. R. 2006. Australian Aboriginal Languages: Consonant-Salient Phonologies and the "Place-of-Articulation Imperative." In: Speech Production: Models, Phonetic Processes, and Techniques, pp. 187–210. New York: Psychology Press. [13, 16]

———. 2010. The Phonetics of Australian Indigenous Languages. HCSNet Winterfest. [13]

Buttelmann, D., N. Zmyj, M. M. Daum, and M. Carpenter. 2012. Selective Imitation of In-Group over Out-Group Members in 14-Month-Old Infants. *Child. Devel.* **84**:1467–8624. [5]

Buxbaum, L. J., K. M. Kyle, and R. Menon. 2005. On Beyond Mirror Neurons: Internal Representations Subserving Imitation and Recognition of Skilled Object-Related Actions in Humans. *Cogn. Brain Res.* **25**:226–239. [9]

Bybee, J. 2007. Frequency of Use and the Organization of Language. Oxford: Oxford Univ. Press. [13]

Bybee, J., and J. Scheibman. 1999. The Effect of Usage on Degrees of Constituency: The Reduction of Don't in English. *Linguistics* **37**:575–596. [13]

Byers, S. N., S. E. Churchill, and B. Curran. 1997. Identification of Euro-Americans, Afro-Americans, and Amerindians from Palatal Dimensions. *J. Forensic Sci.* **42**:3–9. [12]

Byrne, R. 1999. Imitation without Intentionality. Using String Parsing to Copy the Organization of Behaviour. *Anim. Cogn.* **2**:63–72. [9]

Byrne, R., and J. M. E. Byrne. 1993. The Complex Leaf-Gathering Skills of Mountain Gorillas (*Gorilla g. beringei*): Variability and Standardization. *Am. J. Primatol.* **31**:241–261. [9]

Byrne, R., and A. E. Russon. 1998. Learning by Imitation: A Hierarchical Approach. *Behav. Brain Sci.* **21**:667–721. [9]

Byrne, R., and A. Whiten. 1988. Machiavellian Intelligence: Social Expertise and the Evolution of Intellect in Monkeys, Apes, and Humans. Oxford: Clarendon. [2, 4]

Caldarelli, G. 2007. Scale-Free Networks: Complex Webs in Nature and Technology. Oxford: Oxford Univ. Press. [16]

Caldwell, C. A., and A. E. Millen. 2008. Studying Cumulative Cultural Evolution in the Laboratory. *Phil. Trans. R. Soc. B* **363**:3529–3539. [1]

———. 2009. Social Learning Mechanisms and Cumulative Cultural Evolution: Is Imitation Necessary? *Psychol. Sci.* **20**:1478–1483. [9, 11]

Call, J., M. Carpenter, and M. Tomasello. 2005. Focusing on Outcomes and Focusing on Actions in the Process of Social Learning: Chimpanzees and Human Children. *Anim. Cogn.* **8**:151–163. [5]

Callaghan, T., H. Moll, H. Rakoczy, et al. 2011. Early Social Cognition in Three Cultural Contexts. In: Monographs of the Society for Research in Child Development, vol. 76, W. A. Collins, series ed. Oxford: Wiley-Blackwell. [14, Appendix]

Calvin, W. H. 1983. A Stone's Throw and Its Launch Window: Timing Precision and Its Implications for Language and Hominid Brains. *J. Theor. Biol.* **104**:121–135. [2]

Camerer, C. F., and R. M. Hogarth. 1999. The Effects of Financial Incentives in Experiments: A Review and Capital-Labor Production Framework. *J. Risk Uncert.* **19**:7–42. [10]

Campany, R. 1992. Xunzi and Durkheim as Theorists of Ritual Practice. In: Discourse and Practice, ed. F. Reynolds and D. Tracy. Albany: SUNY Press. [17]

Campbell, D. T. 1960. Blind Variation and Selective Retention in Creative Thought as in Other Knowledge Processes. *Psychol. Rev.* **67**:380–400. [1]

———. 1965. Variation and Selective Retention in Socio-Cultural Evolution. In: Social Change in Developing Areas: A Reinterpretation of Evolutionary Theory, ed. H. R. Barringer et al., pp. 19–49. Cambridge, MA: Schenkman Publishing Company. [1]

———. 1975. On the Conflicts between Biological and Social Evolution and between Psychology and Moral Tradition. *Am. Psychol.* **30**:1103–1126. [1]

———. 1979. A Tribal Model of the Social System Vehicle Carrying Scientific Knowledge. *Knowledge: Creation, Diffusion, Utilization* **1**:181–201. [1]

Campbell, D. T. 1983. The Two Distinct Routes Beyond Kin Selection to Ultrasociality: Implications for the Humanities and Social Sciences. In: The Nature of Prosocial Development: Theories and Strategies, ed. D. Bridgeman, pp. 11–39. New York: Academic Press. [4]

Campbell, J. 1974. The Mythic Image. Princeton: Princeton Univ. Press. [17]

Capps, L., and E. Ochs. 1995. Constructing Panic: The Discourse of Agarophobia. Cambridge MA: Harvard Univ. Press. [14]

Carey, S. 2009. The Origin of Concepts. New York: Oxford Univ. Press. [1, 7]

Carey, S., and E. Spelke. 1994. Domain-Specific Knowledge and Conceptual Change. In: Mapping the Mind: Domain Specificity in Cognition and Culture, ed. L. A. Hirschfeld and S. A. Gelman, pp. 169–200. Cambridge: Cambridge Univ. Press. [1]

Carneiro, R. L. 1967. Editor's Introduction. In: The Evolution of Society: Selections from Herbert Spencer's Principles of Sociology, pp. i–vii. Chicago: Univ. of Chicago Press. [1]

———. 1998. What Happened at the Flashpoint? Conjectures on Chiefdom Formation at the Very Moment of Conception. In: Chiefdoms and Chieftaincy in the Americas, ed. E. M. Redmond, pp. 18–42. Gainsville: Univ. of Florida Press. [4]

Carpenter, M. 2006. Instrumental, Social, and Shared Goals and Intentions in Imitation. Imitation and the Development of the Social Mind: Lessons from Typical Development and Autism. In: Imitation and the Social Mind, ed. S. J. Rogers and J. H. G. Williams, pp. 48–70. New York: Guilford Press. [5]

Carpenter, M., and J. Call. 2009. Comparing the Imitative Skills of Children and Nonhuman Apes. *Rev. Primatol.* **1**: [5]

Cartwright, D., and F. Harary. 1956. Structural Balance: A Generalization of Heider's Theory. *Psychol. Rev.* **63**:277–293. [6]

Catani, M., P. G. A. Allin, M. Husain, et al. 2007. Symmetries in Human Brain Pathways Predict Verbal Recall. *PNAS* **104**:17,163–17,168. [12]

Cavalli-Sforza, L. L., and M. W. Feldman. 1973. Models for Cultural Inheritance: Within Group Variation. *Theor. Pop. Biol.* **4**:42–55. [1, 2]

———. 1981. Cultural Transmission and Evolution: A Quantitative Approach. Monographs in Population Biology. Princeton: Princeton Univ. Press. [1, 2, 6, 8, 11, 20]

Cavalli-Sforza, L. L., P. Menozzi, and A. Piazza. 1994. The History and Geography of Human Genes. Princeton: Princeton Univ. Press. [12]

Cederman, L. E. 1997. Emergent Actors in World Politics: How States and Nations Develop and Dissolve. Princeton: Princeton Univ. Press. [4]

Cederman, L. E., and L. Girardin. 2010. Growing Sovereignty: Modeling the Shift from Indirect to Direct Rule. *Intl. Stud. Q.* **54**:27–48. [4]

Chagnon, N. A., and W. Irons. 1979. Evolutionary Biology and Human Social Behavior: An Anthropological Perspective. North Scituate, MA: Duxbury Press. [1]

Chaminade, T., A. N. Meltzoff, and J. Decety. 2002. Does the End Justify the Means? A Pet Exploration of the Mechanisms Involved in Human Imitation. *NeuroImage* **15**:318–328. [9]

———. 2005. An fMRI Study of Imitation: Action Representation and Body Schema. *Neuropsychologia* **43**:115–127. [9]

Chapais, B. 2008. Primeval Kinship: How Pair-Bonding Gave Birth to Human Society. Cambridge, MA: Harvard Univ. Press. [3, 5]

Charlton, M. F. 2009. Identifying Iron Production Lineages: A Case-Study in North-West Wales. In: Pattern and Process in Cultural Evolution, ed. S. J. Shennan, pp. 133–144. Berkeley: Univ. of California Press. [8]

Charlton, M. F., P. Crew, T. Rehren, and S. J. Shennan. 2010. Explaining the Evolution of Ironmaking Recipes: An Example from Northwest Wales. *J. Anthropol. Archaeol.* **29**:352–367. [8]

Charney, E. 2012. Behavior Genetics and Postgenomics. *Behav. Brain Sci.* **35**:331–358. [12, 16]

Chartrand, T. L., and J. A. Bargh. 1999. The Chameleon Effect: The Perception–Behaviour Link and Social Interaction. *J. Pers. Soc. Psychol.* **76**:893–910. [5]

Chase-Dunn, C. K., and T. D. Hall. 1997. Rise and Demise: Comparing World-Systems. New Perspectives in Sociology. Boulder: Westview Press. [4]

Chater, N., and M. H. Christiansen. 2010. Language Acquisition Meets Language Evolution. *Cognitive Science* **34**:1131–1157. [13, 16, Appendix]

Chater, N., F. Reali, and M. H. Christiansen. 2009. Restrictions on Biological Adaptation in Language Evolution. *PNAS* **106**:1015–1020. [1]

Chen, C. 1999. Population Migration and the Variation of Dopamine D4 Receptor (DRD4) Allele Frequencies around the Globe. *Evol. Hum. Behav.* **20**:309–324. [6]

Chiao, J. Y. 2010. At the Frontier of Cultural Neuroscience: Introduction to the Special Issue. *Soc. Cogn. Aff. Neurosci.* **5**:109–110. [20]

Chick, G. 1997. Cultural Complexity: The Concept and Its Measurement. *Cross-Cultural Res.* **31**:275–307. [6]

Choi, J.-K., and S. Bowles. 2007. The Coevolution of Parochial Altruism and War. *Science* **318**:636–640. [3, 4]

Chomsky, N. 1981. Lectures on Government and Binding. Dordrecht: Foris. [13]

———. 2010. Some Simple Evo Devo Theses: How True Might They Be for Language? In: The Evolution of Human Language, ed. R. K. Larson et al., pp. 45–62. Cambridge: Cambridge Univ. Press. [16]

Chouniard, M. M., and E. V. Clark. 2003. Adult Reformulations of Child Errors as Negative Evidence. *J. Child Lang.* **30**:637–669. [14]

Christiansen, M. H., and N. Chater. 2008. Language as Shaped by the Brain. *Behav. Brain Sci.* **31**:489–558. [13, 16, Appendix]

Christiansen, M. H., and S. Kirby. 2003. Language Evolution, vol. 3. New York: Oxford Univ. Press. [1]

Chudek, M., S. Heller, S. Birch, and J. Henrich. 2012. Prestige-Biased Cultural Learning: Bystander's Differential Attention to Potential Models Influences Children's Learning. *Evol. Hum. Behav.* **33**:46–56. [7]

Chudek, M., and J. Henrich. 2011. Culture-Gene Coevolution, Norm-Psychology, and the Emergence of Human Prosociality. *Trends Cogn. Sci.* **15**:218–226. [1, 5, 6, 17, 20]

Cioffi-Revilla, C. 2005. A Canonical Theory of Origins and Development of Social Complexity. *J. Math. Soc.* **29**:133–153. [4]

Claidière, N., and D. Sperber. 2007. The Role of Attractors in Cultural Evolution. *J Cogn. Culture* **7**:89–111. [7]

Clancy, P. M. 1985. The Acquisition of Japanese. In: The Crosslinguistic Study of Language Acquisition, ed. D. I. Slobin, vol. 1, pp. 373–524. Mahweh, NJ: Lawrence Erlbaum. [14]

Clancy, P. M., M. Akatsuka, and S. Strauss. 1997. Deontic Modality and Conditionality in Discourse: A Cross-Linguistic Study of Adult Speech to Young Children. In Directions in Functional Linguistics, ed. A. Kamio, pp. 19–57. Amsterdam: John Benjamins. [14]

Clarke, P., and P. Byrne. 1993. Religion Defined and Explained. London: Macmillian. [20]

Cloak, F. T., Jr. 1975. Is a Cultural Ethology Possible? *Hum. Ecol.* **3**:161–182. [1]

Clutton-Brock, T. H. 1974. Primate Social Organization and Ecology. *Nature* **250**:539–542. [2]

———. 2006. Cooperative Breeding in Mammals. In: Cooperation in Primates and Humans: Mechanisms and Evolution, ed. P. M. Kappeler and C. P. van Schaik, pp. 173–190. Berlin: Springer-Verlag. [3]

———. 2009. Cooperation between Non-Kin in Animal Societies. *Nature* **462**:51–57. [2, 3]

Clutton-Brock, T. H., P. N. M. Brotherton, A. F. Russell, et al. 2001. Cooperation, Control, and Concession in Meerkat Groups. *Science* **291**:478–481. [3]

Cohen, E. 2012. The Evolution of Tag-Based Cooperation in Humans: The Case for Accent. *Curr. Anthropol.* **53**:588–616. [5, 16]

Cohen, E., E. Burdett, N. Knight, and J. Barrett. 2011. Cross-Cultural Similarities and Differences in Person-Body Reasoning: Experimental Evidence from the United Kingdom and Brazilian Amazon. *Cogn. Sci.* **35**:1282–1304. [17, 18]

Cohen, E., R. Ejsmond-Frey, N. Knight, and R. I. M. Dunbar. 2010. Rowers' High: Behavioral Synchrony Is Correlated with Elevated Pain Thresholds. *Biol. Lett.* **6**:106–108. [20]

Cohen, E., and D. B. M. Haun. 2013. The Development of Tag-Based Cooperation via a Socially Acquired Trait. *Evol. Hum. Behav.* **34**:230–235. [5]

Cohen, E., R. Mundry, and S. Kirschner. 2013. Religion, Synchrony, and Cooperation. *Relig. Brain Behav.* http://www.tandfonline.com/doi/abs/10.1080/2153599X.2012.741075#.UaIDZsqjhoo. (accessed May 26, 2013). [20]

Cole, P., G. Hermon, Y. Tjung, C.-Y. Sim, and C. Kim. 2008. A Binding Theory Exempt Anaphor. In: Reciprocals and Reflexives. Theoretical and Typological Explorations, ed. E. König and V. Gast, pp. 577–590. Berlin: Mouton de Gruyter. [13]

Coleman, J. 1990. Foundations of Social Theory. Cambridge, MA: Harvard Univ. Press. [6]

Collard, M., K. Edinborough, S. Shennan, and M. G. Thomas. 2010. Radiocarbon Evidence Indicates That Migrants Introduced Farming to Britain. *J. Archaeol. Sci.* **37**:866–870. [1]

Collard, M., M. Kemery, and S. Banks. 2005. Causes of Toolkit Variation among Hunter-Gatherers: A Test of Four Competing Hypotheses. *J. Can. Archaeol.* **29**:1–19. [7]

Collias, E., and N. Collias. 1964. The Development of Nest-Building Behavior in a Weaverbird. *Auk* **81**:42–52. [7]

Coop, G., K. Bullaughey, F. Luca, and M. Przeworski. 2008. The Timing of Selection at the Human *FOXP2* Gene. *Mol. Biol. Evol.* **25**:1257–1259. [1]

Corbett, G. 2012. Features. Cambridge: Cambridge Univ. Press. [13]

Cordes, C. 2005. Veblen's "Instinct of Workmanship," Its Cognitive Foundations, and Some Implications for Economic Theory. *J. Econ. Issues* **39**:1–20. [1]

Corriveau, K. H., and P. L. Harris. 2010. Preschoolers (Sometimes) Defer to the Majority in Making Simple Perceptual Judgments. *Devel. Psychol.* **46**:437–445. [5]

Cosmides, L., and J. Tooby. 2005. Neurocognitive Adaptations Designed for Social Exchange. In: The Handbook of Evolutionary Psychology, ed. D. M. Buss, pp. 584–627. Hoboken: Wiley. [10]

Couzin, I. D., and N. R. Franks. 2003. Self-Organized Lane Formation and Optimized Traffic Flow in Army Ants. *Proc. R. Soc. B* **270**:139–146. [3]

Crain, S., and P. Pietroski. 2002. Why Language Acquisition Is a Snap. *Ling. Rev.* **19**:163–183. [16]

Crespi, B. J. 2001. The Evolution of Social Behavior in Microorganisms. *Trends Ecol. Evol.* **16**:178–183. [3]

Croft, W. 2000. Explaining Language Change: An Evolutionary Approach. Harlow: Longman. [16]

Croft, W., T. Bhattacharya, D. Kleinschmidt, D. E. Smith, and T. F. Jaeger. 2011. Greenbergian Universals, Diachrony and Statistical Analyses: Commentary on Dunn et al., "Evolved Structure of Language Shows Lineage-Specific Trends in Word Order Universals." *Ling. Typol.* **15**:433–453. [16]

Cronk, L. 1994. Evolutionary Theories of Morality and the Manipulative Use of Signals. *Zygon* **29**:81–101. [20]

Crowley, T. 1982. The Paamese Language of Vanuatu. Canberra: Pacific Linguistics. [13]

Csibra, G., and G. Gergely. 2009. Natural Pedagogy. *Trends Cogn. Sci.* **13**:148–153. [11]

———. 2011. Natural Pedagogy as an Evolutionary Adaptation. *Phil. Trans. R. Soc. B* **366**:1149–1157. [1]

Currie, T. E., S. J. Greenhill, R. D. Gray, T. Hasegawa, and R. Mace. 2010a. Rise and Fall of Political Complexity in Island South-East Asia and the Pacific. *Nature* **467**:801–804. [1, 6, 20]

Currie, T. E., S. J. Greenhill, and R. Mace. 2010b. Is Horizontal Transmission Really a Problem for Phylogenetic Comparative Methods? A Simulation Study Using Continuous Cultural Traits. *Phil. Trans. R. Soc. B* **365**:3903–3912. [16, 20]

Currie, T. E., and R. Mace. 2009. Political Complexity Predicts the Spread of Ethnolinguistic Groups. *PNAS* **106**:7339–7344. [15]

Cusack, T. R., and R. J. Stoll. 1990. Exploring Realpolitik: Probing International Relations Theory with Computer Simulations. Boulder: Lynne Rienner Publishers. [4]

Custance, D. M., A. Whiten, and K. A. Bard. 1995. Can Young Chimpanzees (*Pan troglodytes*) Imitate Arbitrary Actions? Hayes & Hayes (1952) Revisited. *Behaviour* **132**:837–859. [9]

Dąbrowska, E. 1997. The Lad Goes to School: A Cautionary Tale for Nativists. *Linguistics* **35**:735–766. [12]

Dacey, M. F. 1968. A Probability Model for the Rise and Decline of States. *Peace Res. Soc.* **14**:147–153. [4]

Dagan, T., and W. Martin. 2006. The Tree of One Percent. *Genome Biol.* **7**:118. [16]

Dall'Asta, L., A. Baronchelli, A. Barrat, and V. Loreto. 2006. Nonequilibrium Dynamics of Language Games on Complex Networks. *Phys. Rev. E* **74**:036105. [16]

Damerow, P., G. Freudenthal, P. McClaughlin, and J. Renn. 2004. Exploring the Limits of Preclassical Mechanics. New York: Springer-Verlag. [11]

Danchin, É., L.-A. Giraldeau, T. J. Valone, and R. H. Wagner. 2004. Public Information: From Nosy Neighbors to Cultural Evolution. *Science* **305**:487–490. [1]

Darley, J. M., and C. D. Batson. 1973. From Jerusalem to Jericho: A Study of Situational and Dispositional Variables in Helping Behavior. *J. Pers. Soc. Psychol.* **27**:100–108. [19, 20]

Darlington, P. J. 1975. Group Selection, Altruism, Reinforcement and Throwing in Human Evolution. *PNAS* **72**:3748–3752. [2]

Dart, R. A. 1925. *Australopithecus Africanus*: The Man-Ape of South Africa. *Nature* **115**:195–199. [2]

Darwin, C. 1859. The Origin of Species by Means of Natural Selection (6th edition). London: John Murray. [2, 16, 19]

———. 1869. The Variation of Animals and Plants under Domestication, vol. 2. London: Murray [1]

———. 1871. The Descent of Man, and Selection in Relation to Sex (2nd edition). London: John Murray. [1, 2, 6, 12, 15]

———. 1877. A Biographical Sketch of an Infant. *Mind* **2**:285–294. [1]

Dawkins, R. 1976. The Selfish Gene. New York: Oxford Univ. Press. [4, 8, 16]

———. 1982. The Extended Phenotype. Oxford: Oxford Univ. Press. [16]

———. 2006. The God Delusion. Boston: Houghton Mifflin. [20]

Daxinger, L., and E. Whitelaw. 2012. Understanding Transgenerational Epigenetic Inheritance via the Gametes in Mammals. *Nat. Rev. Genet.* **13**:153–162. [1]

Deacon, T. W. 1997. The Symbolic Species: The Co-Evolution of Language and the Human Brain. New York: W. W. Norton. [11, 20]

———. 2012. Incomplete Nature: How Mind Emerged from Matter. New York: W. W. Norton. [9]

Dean, L. G., R. L. Kendal, S. J. Schapiro, B. Thierry, and K. N. Laland. 2012. Identification of the Social and Cognitive Processes Underlying Human Cumulative Culture. *Science* **335**:1114–1118. [1, 8, 11]

De Boer, B. 2001. The Origins of Vowel Systems. Oxford: Oxford Univ. Press. [15]

Dediu, D. 2008. The Role of Genetic Biases in Shaping Language-Genes Correlations. *J. Theor. Biol.* **254**:400–407. [12]

———. 2011a. A Bayesian Phylogenetic Approach to Estimating the Stability of Linguistic Features and the Genetic Biasing of Tone. *Proc. R. Soc. B* **278**:474–479. [16]

———. 2011b. Are Languages Really Independent from Genes? If Not, What Would a Genetic Bias Affecting Language Diversity Look Like? *Hum. Biol.* **83**:279–296. [12]

Dediu, D., and D. R. Ladd. 2007. Linguistic Tone Is Related to the Population Frequency of the Adaptive Haplogroups of Two Brain Size Genes, ASPM and Microcephalin. *PNAS* **104**:10,944–10,949. [6, 12, 13]

Dediu, D., and S. C. Levinson. 2012. Abstract Profiles of Structural Stability Point to Universal Tendencies, Family-Specific Factors, and Ancient Connections between Languages. *PLoS One* **7**:e45198. [16]

———. 2013. The Time Frame of the Emergence of Modern Language and Its Implications. In: The Social Origins of Language: Early Society, Communication and Polymodality, ed. D. Dor et al. Oxford: Oxford Univ. Press, in press. [16]

Dehaene, S., and L. Cohen. 2007. Cultural Recycling of Cortical Maps. *Neuron* **56**:384–398. [16, Appendix]

Dehaene-Lambertz, G., S. Dehaene, and L. Hertz-Pannier. 2002. Functional Neuroimaging of Speech Perception in Infants. *Science* **298**:2013–2015. [14]

Dellavia, C., C. Sforza, F. Orlando, et al. 2007. Three-Dimensional Hard Tissue Palatal Size and Shape in Down Syndrome Subjects. *Eur. J. Orthod.* **29**:417–422. [12]

deMenocal, P. 2011. Climate and Human Evolution. *Science* **331**:540–542. [2]

Denison, S., C. Reed, and F. Xu. 2013. The Emergence of Probabilistic Reasoning in Very Young Infants: Evidence from 4.5- and 6-Month-Olds. *Devel. Psychol.* **49**:243–249. [10]

Denison, S., and F. Xu. 2009. Twelve- to 14-Month-Old Infants Can Predict Single-Event Probability with Large Set Sizes. *Devel. Sci.* **13**:798–803. [10]

Dennett, D. 2006. Breaking the Spell: Religion as a Natural Phenomenon. New York: Viking Adult. [20]

Derbyshire, D. 1977. Word Order Universals and the Existence of OVS Languages. *Ling. Inquiry* **8**:590–599. [13]

de Vignemont, F., and P. Haggard. 2008. Action Observation and Execution: What Is Shared? *Soc. Neurosci.* **3**:421–433. [9]

de Villiers, J., and P. de Villiers. 2000. Linguistic Determinism and the Understanding of False Belief. In: Children's Reasoning and the Mind, ed. P. Mitchell and K. J. Riggs, pp. 191–228. Hove, UK: Psychology Press. [14]

De Vos, C. 2012. Sign-Spatiality in Kata Kolok: How Village Sign Language of Bali Inscribes Its Signing Space. PhD dissertation, Max Planck Institute for Psycholinguistics, Radboud University, Nijmegen. [12]

de Waal, F. B. M. 1997a. Bonobo: The Forgotten Ape. Berkeley: Univ. of California Press. [2]

———. 1997b. Good Natured: The Origins of Right and Wrong in Humans and Other Animals. Cambridge, MA: Harvard Univ. Press. [2]

———. 1998. Chimpanzee Politics: Sex and Power among the Apes. Baltimore: Johns Hopkins Univ. Press. [2]

Diamond, J. M. 1978. Tasmanians: The Longest Isolation, the Simplest Technology. *Nature* **273**:185. [7]

Diamond, J. M., and P. Bellwood. 2003. Farmers and Their Languages: The First Expansions. *Science* **300**:597–603. [11, 15]

Di Cosmo, N. 2002. Ancient China and Its Enemies: The Rise of Nomadic Power in East Asian History. Cambridge: Cambridge Univ. Press. [11]

Diessel, H., and M. Tomasello. 2001. The Acquisition of Finite Complement Clauses in English: A Corpus-Based Analysis. *Cogn. Ling.* **12**:97–141. [14]

Dijksterhuis, A., J. Preston, D. M. Wegner, and H. Aarts. 2008. Effects of Subliminal Priming of Self and God on Self-Attribution of Authorship for Events. *J. Exp. Social Psychol.* **44**:2–9. [19]

Ding, Y. C., H. C. Chi, D. L. Grady, et al. 2002. Evidence of Positive Selection Acting at the Human Dopamine Receptor D4 Gene Locus. *PNAS* **99**:309–314. [6]

Dominguez-Rodrigoa, M., and R. Barba. 2006. New Estimates of Tooth Mark and Percussion Mark Frequencies at the FLK Zinj Site: The Carnivore-Hominid-Carnivore Hypothesis Falsified. *J. Hum. Evol.* **50**:170–194. [2]

Donald, M. 1991a. Origins of the Modern Mind: Three Stages in the Evolution of Culture and Cognition. Cambridge, MA: Harvard Univ Press. [20]

———. 2001. A Mind So Rare: The Evolution of Human Consciousness. New York: W. W. Norton. [20]

Donald, T. C. 1991b. A Naturalistic Theory of Archaic Moral Order. *Zygon* **26**:91–114. [20]

Doolittle, W. F., and E. Bapteste. 2007. Pattern Pluralism and the Tree of Life Hypothesis. *PNAS* **104**:2043–2049. [16]

Dow, M. M., and E. A. Eff. 2008. Global, Regional, and Local Network Auto-Correlation in the Standard Cross-Cultural Sample. *Cross-Cultural Res.* **42**:148–171. [20]

Downs, A. 1957. An Economic Theory of Democracy. Boston: Harper Row. [2]

Drummond, A. J., and A. Rambaut. 2007. BEAST: Bayesian Evolutionary Analysis by Sampling Trees. *BMC Evol. Biol.* **7**:214. [15]

Drummond, A. J., and M. A. Suchard. 2010. Bayesian Random Local Clocks, or One Rate to Rule Them All. *BMC Biol.* **8**:114. [15]

Dryer, M. S. 1992. The Greenbergian Word Order Correlations. *Language* **68**:81–138. [15]

———. 1998. Why Statistical Universals Are Better Than Absolute Universals. *CLS The Panels* **33**:123–145. [13]

———. 2011. Order of Subject Object and Verb. ed M. S. Dryer and M. Haspelmath. Munich: Max Planck Digital Library. http://wals.info/feature/81A (accessed May 15, 2013). [15, 16]

Dryer, M. S., and Haspelmath, M., eds. 2011. The World Atlas of Language Structures Online. Munich: Max Planck Digital Library. http://wals.info/ (accessed June 11, 2013). [15]

DuBois, J. 1987. The Discourse Basis of Ergativity. *Language* **55**:59–138. [13]

Dugatkin, L. A. 2003. Principles of Animal Behavior. New York: W. W. Norton. [6]

Dunbar, R. I. M. 1992. Neocortex Size as a Constraint on Group Size in Primates. *J. Hum. Evol.* **22**:469–493. [4]

———. 1993. Coevolution of Neocortical Size, Group Size and Language in Humans. *Behav. Brain Sci.* **16**:681–735. [16]

———. 1995. The Trouble with Science. London: Faber and Faber. [10]

Dunbar, R. I. M., and S. Shultz . 2007. Evolution in the Social Brain. *Science* **317**:1344–1347. [4]

Dunham, Y., A. S. Baron, and S. Carey. 2011. Consequences of "Minimal" Group Affiliations in Children. *Child. Devel.* **82**:793–811. [5]

Dunn, J., and C. Kendrick. 1982. Siblings: Love, Envy and Understanding. London: Grant McIntyre. [14]

Dunn, M., N. Burenhult, N. Kruspe, S. Tufvesson, and N. Becker. 2011a. Aslian Linguistic Prehistory: A Case Study in Computational Phylogenetics. *Diachronica* **28**:291–323. [15]

Dunn, M., S. J. Greenhill, S. C. Levinson, and R. D. Gray. 2011b. Evolved Structure of Language Shows Lineage-Specific Trends in Word-Order Universals. *Nature* **473**:79–82. [13, 15, 16]

Durham, W. 1982. The Relationship of Genetic and Cultural Evolution: Models and Examples. *Hum. Ecol.* **10**:289–323. [1]

———. 1991. Coevolution. Stanford: Stanford Univ. Press. [13]

Durkheim, E. 1915/1965. Elementary Forms of Religious Life. New York: George Allen & Unwin Ltd. [17, 19, 20]

———. 1933/1902. The Division of Labor in Society. New York: Free Press. [2]

Dyson, G. 1991. Form and Function of the Baidarka: The Framework of Design. Bellingham: Baidarka Historical Society. [7]

Earle, T., ed. 1991. Chiefdoms: Power, Economy, and Ideology. Cambridge: Cambridge Univ. Press. [6]

Edelman, G. M. 1987. Neural Darwinism: The Theory of Neuronal Group Selection. New York: Basic Books. [1]

Edelman, G. M., and J. A. Gally. 2001. Degeneracy and Complexity in Biological Systems. *PNAS* **98**:13,763–13,768. [9]

Eerkens, J. W., and C. P. Lipo. 2005. Cultural Transmission, Copying Errors, and the Generation of Variation in Material Culture and the Archaeological Record. *J. Anthropol. Arch.* **24**:316–334. [11]

Efferson, C., R. Lalive, and E. Fehr. 2008a. The Coevolution of Cultural Groups and In-Group Favoritism. *Science* **321**:1844–1849. [1, 6]

Efferson, C., R. Lalive, P. J. Richerson, R. McElreath, and M. Lubell. 2008b. Conformists and Mavericks in the Lab: The Structure of Frequency Dependent Learning. *Evol. Hum. Behav.* **29**:56–64. [1]

Ehn, M., and K. Laland. 2012. Adaptive Strategies for Cumulative Cultural Learning. *J. Theor. Biol.* **301**:103–111. [11]

Eibl-Eibesfeldt, I. 1989. Human Ethology. New York: Aldine de Gruyter. [2]

Eisenberg, J. F., N. A. Muckenhirn, and R. Rudran. 1972. The Relation between Ecology and Social Structure in Primates. *Science* **176**:863–874. [2]

Ellickson, R. C. 1991. Order without Law: How Neighbors Settle Disputes. Cambridge, MA: Harvard Univ. Press. [6]

Endler, J. A. 1986. Natural Selection in the Wild. In: Monographs in Population Biology, vol. 21, pp. xiii, 336. Princeton: Princeton Univ. Press. [1]

Enfield, N. J., ed. 2002. Ethnosyntax: Explorations in Grammar and Culture. Oxford: Oxford Univ. Press. [13]

Eno, R. 1990. The Confucian Creation of Heaven: Philosophy and the Defense of Ritual Mastery. SUNY Series in Chinese Philosophy and Culture. Albany: SUNY Press. [20]

———. 2009. Shang State Religion and the Pantheon of the Oracle Texts. In: Early Chinese Religion: Part One: Shang through Han (1250 BC–22 AD), ed. J. Lagerwey and M. Kalinowski. Leiden Brill. [17]

Enquist, M., K. Eriksson, and S. Ghirlanda. 2007. Critical Social Learning: A Solution to Rogers's Paradox of Nonadaptive Culture. *Am. Anthropol.* [6, 7]

Enquist, M., S. Ghirlanda, and K. Eriksson. 2011. Modelling the Evolution and Diversity of Cumulative Culture. *Phil. Trans. R. Soc. B* **366**:412–423. [11]

Enquist, M., S. Ghirlanda, A. Jarrick, and C.-A. Wachtmeister. 2008. Why Does Human Culture Increase Exponentially? *Theor. Pop. Biol.* **74**:46–55. [7, 11]

Ericsson, K. A., R. T. Krampe, and C. Tesch-Romer. 1993. The Role of Deliberate Practice in the Acquisition of Expert Performance. *Psychol. Rev.* **100**:363–406. [9]

Errington, J. 1988. Structure and Style in Javanese: A Semiotic View of Linguistic Etiquette. Philadelphia: Univ. of Pennsylvania Press. [13]

Eshel, I., and L. L. Cavalli-Sforza. 1982. Assortment of Encounters and Evolution of Cooperativeness. *PNAS* **79**:1331–1335. [6]

Etienne, R. S., B. Haegeman, T. Stadler, et al. 2012. Diversity-Dependence Brings Molecular Phylogenies Closer to Agreement with the Fossil Record. *Proc. R. Soc. B* **279**:1300–1309. [15]

Evans, N. 1995a. A Grammar of Kayardild. Berlin: Mouton de Gruyter. [13]

———. 1995b. Multiple Case in Kayardild: Anti-Iconicity and the Diachronic Filter. In: Double Case. Agreement by Suffixaufnahme, ed. F. Plank, pp. 396–428. Oxford: Oxford Univ. Press. [13]

———. 1998. Iwaidja Mutation and Its Origins. In: Case, Typology and Grammar: In Honour of Barry J. Blake, ed. A. Siewierska and J. J. Song, pp. 115–149. Amsterdam: John Benjamins. [13]

———. 2003a. Bininj Gun-Wok: A Pan-Dialectal Grammar of Mayali, Kunwinjku and Kune. Canberra: Pacific Linguistics. [13, 16]

———. 2003b. Context, Culture and Structuration in the Languages of Australia. *Annu. Rev. Anthropol.* **32**:13–40. [13]

———. 2003c. Introduction: Comparative Non-Pama-Nyungan and Australian Historical Linguistic. In: The Non-Pama-Nyungan Languages of Northern Australia: Comparative Studies of the Continent's Most Linguistically Complex Region, ed. N. Evans, pp. 3–25. Canberra: Pacific Linguistics. [13]

———. 2005. Australian Languages Reconsidered: A Review of Dixon (2002). *Oceanic Ling.* **44**:216–260. [16]

———. 2007. Standing up Your Mind: Remembering in Dalabon. In: The Language of Memory in a Crosslinguistic Perspective, ed. M. Amberber, pp. 67–95. Amsterdam: John Benjamins. [13]

———. 2008. Reciprocal Constructions: Towards a Structural Typology. In: Reciprocals and Reflexives: Cross-Linguistic and Theoretical Explorations, ed. E. König and V. Gast, pp. 33–104. Berlin: Mouton de Gruyter. [13]

———. 2010. Complex Events, Propositional Overlay and the Special Status of Reciprocal Clauses. In: Empirical and Experimental Methods in Cognitive/Functional Research, ed. S. Rice and J. Newman, pp. 1–40. Stanford: CSLI. [13]

Evans, N., A. Gaby, S. C. D. Levinson, and A. Majid, eds. 2011. Reciprocals and Semantic Typology. Amsterdam: John Benjamins. [13]

Evans, N., A. Gaby, and R. Nordlinger. 2007. Valency Mismatches and the Coding of Reciprocity in Australian Languages. *Ling. Typol.* **11**:543–599. [13]

Evans, N., and S. C. Levinson. 2009a. The Myth of Language Universals: Language Diversity and Its Importance for Cognitive Science. *Behav. Brain Sci.* **32**:429–448. [12, 13, 16]

———. 2009b. With Diversity in Mind: Freeing the Language Sciences from Universal Grammar. *Behav. Brain Sci.* **32**:472–492. [13]

Evans, N., and D. Wilkins. 2000. In the Mind's Ear: The Semantic Extensions of Perception Verbs in Australian Languages. *Language* **76**:546–592. [13]

Everett, D. L. 2012. Language: The Cultural Tool. New York: Pantheon. [11]

Fagg, A., and M. A. Arbib. 1998. Modeling Parietal-Premotor Interaction in Primate Control of Grasping. *Neural Netw.* **11**:1277–1303. [9]

Faisal, A., D. Stout, J. Apel, and B. Bradley. 2010. The Manipulative Complexity of Lower Paleolithic Stone Toolmaking. *PLoS One* **5**:e13718. [9]

Farmer, T. A., J. B. Misyak, and M. H. Christiansen. 2012. Individual Differences in Sentence Processing. In: Cambridge Handbook of Psycholinguistics, ed. M. J. Spivey et al., pp. 353–364. Cambridge: Cambridge Univ. Press. [12, 16]

Fawcett, C. A., and L. Markson. 2010. Similarity Predicts Liking in 3-Year-Old Children. *J. Exp. Child. Psychol.* **105**:345–358. [5]

Fay, N., S. Garrod, and J. Carletta. 2000. Group Discussion as Interactive Dialogue or as Serial Monologue: The Influence of Group Size. *Psychol. Sci.* **11**:487–492. [16]

Fehr, E., and S. Gächter. 2000. Cooperation and Punishment. *Am. Econ. Rev.* **90**:980–994. [2]

———. 2002. Altruistic Punishment in Humans. *Nature* **415**:137–140. [2]

Fehr, E., and F. Schneider. 2010. Eyes Are on Us, but Nobody Cares: Are Eye Cues Relevant for Strong Reciprocity? *Proc. R. Soc. B* **277**:1315–1323. [19]

Feldman, H. M., B. MacWhinney, and K. Sacco. 2002. Sentence Processing in Children with Early Unilateral Brain Injury. *Brain Lang.* **83**:335–352. [14]

Feldman, M. W., and K. Aoki. 1992. Assortative Mating and Persistence of a Sign Language. *Theor. Pop. Biol.* **42**:107–116. [12]

Felsenstein, J. 1985. Phylogenies and the Comparative Method. *Am. Natural.* **125**:1–15. [20]

Feyerabend, P. K. 1975. Against Method. London: Verso. [10]

Fifer, F. C. 1987. The Adoption of Bipedalism by the Hominids: A New Hypothesis. *Hum. Evol.* **2**:135–147. [2]

Finlay, B. L., and R. Darlington. 1995. Linked Regularities in the Development and Evolution of Mammalian Brains. *Science* **268**:1578–1584. [16]

Finlay, B. L., R. Darlington, and N. Nicastro. 2001. Developmental Structure of Brain Evolution. *Behav. Brain Sci.* **24**:263–308. [16]

Fischer, R., R. Callander, P. Reddish, and J. Bulbulia. 2013. How Do Rituals Affect Cooperation? An Experimental Field Study Comparing Nine Ritual Types. *Hum. Nature* **24**:115–125. [20]

Fisher, R. A. 1918. The Correlation between Relatives on the Supposition of Mendelian Inheritance. *Trans. R. Soc. Edinb.* **52**:399–433. [1]

———. 1930. The Genetical Theory of Natural Selection. Oxford: Oxford Univ. Press. [2, 7]

Fisher, S. E. 2006. Tangled Webs: Tracing the Connections between Genes and Cognition. *Cognition* **101**:270–297. [12]

Fisher, S. E., and C. Scharff. 2009. FOXP2 as a Molecular Window into Speech and Language. *Trends Genet.* **25**:66–177. [1, 12]

Fitch, W. T. 2011. Unity and Diversity in Human Language. *Phil. Trans. R. Soc. B* **366**:376–388. [16]

Fitzhugh, B. 2001. Risk and Invention in Human Technological Evolution. *J. Anthropol. Arch.* **20**:125–167. [8, 11]

FitzJohn, R. G. 2010. Quantitative Traits and Diversification. *Syst. Biol.* **59**:619–663. [15]

Flannery, K. V. 1972. The Cultural Evolution of Civilizations. *Annu. Rev. Ecol. Evol. System.* **3**:399–426. [6]

———. 1999. Process and Agency in Early State Formation. *Cambridge Arch. J.* **9**:3–21. [6]

Floss, H. 2003. Did They Meet or Not? Observations on Châtelperronian and Aurignacian Settlement Patterns in Eastern France. The Chronology of the Aurignacian and of the Transitional Technocomplexes. Dating, Stratigraphies, Cultural Implications. *Trabalhos de Arqueologia* **33**:273–287. [16]

Flynn, E., K. N. Laland, R. L. Kendal, and J. R. Kendal. 2012. Developmental Niche Construction. *Devel. Sci.* **16**:296–313. [Appendix]

Flynn, E., and K. Smith. 2012. Investigating the Mechanisms of Cultural Acquisition: How Pervasive Is Adults' Overimitation? *Soc. Psychol.* **43**:185–195. [11]

Flynn, E., and A. Whiten. 2008a. Cultural Transmission of Tool Use in Young Children: A Diffusion Chain Study. *Soc. Develop.* **17**:699–718. [11]

———. 2008b. Imitation of Hierarchical Structure Versus Component Details of Complex Actions by 3- and 5-Year-Olds. *J. Exp. Child. Psychol.* **101**:228–240. [9]

———. 2012. Experimental "Microcultures" in Young Children: Identifying Biographic, Cognitive, and Social Predictors of Information Transmission. *Child. Devel.* **83**:911–925. [11]

Fogassi, L., P. F. Ferrari, B. Gesierich, et al. 2005. Parietal Lobe: From Action Organization to Intention Understanding. *Science* **308**:662–667. [9]

Fortunato, L. 2011. Reconstructing the History of Marriage and Residence Strategies in Indo-European-Speaking Societies. *Hum. Biol.* **83**:129–135. [20]

Fortunato, L., C. Holden, and R. Mace. 2006. From Bridewealth to Dowry? A Bayesian Estimation of Ancestral States of Marriage Transfers in Indo-European Groups. *Hum. Nature* **17**:355–376. [6]

Fortunato, L., and F. M. Jordan. 2011. Your Place or Mine? A Phylogenetic Comparative Analysis of Marital Residence in Indo-European and Austronesian Societies. *Phil. Trans. R. Soc. B* **365**:3913–3922. [6]

Fortunato, L., and R. Mace. 2009. Testing Functional Hypotheses About Cross-Cultural Variation: A Maximum-Likelihood Comparative Analysis of Indo-European Marriage Practices. In: Pattern and Process in Cultural Evolution, ed. S. Shennan, pp. 235–250. Berkeley: Univ. of California Press. [4]

Fracchia, J., and R. C. Lewontin. 2005. The Price of Metaphor. *Hist. Theory* **44**:14–29. [1]

Frank, R. E., and J. B. Silk. 2009. Impatient Traders or Contingent Reciprocators? Evidence for the Extended Time-Course of Grooming Exchanges in Baboons. *Behaviour* **146**:1123–1135. [3]

Frank, S. A. 1998. Foundations of Social Evolution. Princeton: Princeton Univ. Press. [6]

Franks, N. R. 1986. Teams in Social Insects: Group Retrieval of Prey by Army Ants (*Eciton burchelli*, Hymenoptera: Formicidae). *Behav. Ecol. Sociobiol.* **18**:425–429. [3]

Franz, M., and C. L. Nunn. 2009a. Network-Based Diffusion Analysis: A New Method for Detecting Social Learning. *Proc. R. Soc. B* **276**:1829–1836. [11]

———. 2009b. Rapid Evolution of Social Learning. *J. Evol. Biol.* **22**:1914–1922. [7]

Freeman, D., C. J. Bajema, J. Blacking, et al. 1974. The Evolutionary Theories of Charles Darwin and Herbert Spencer [and Comments and Replies]. *Curr. Anthropol.* **15**:211–237. [1]

Freeman, W. J. 2000. A Neurobiological Role of Music in Social Bonding. In: The Origins of Music, ed. N. Wallin et al., pp. 411–424. Cambridge, MA: MIT Press. [20]

Frey, S. H. 2007. What Puts the How in Where? Tool Use and the Divided Visual Streams Hypothesis. *Cortex* **43**:368–375. [9]

Frey, S. H., and V. Gerry. 2006. Modulation of Neural Activity During Observational Learning of Action and Their Sequential Orders. *J. Neurosci.* **26**:13194–13201. [9]

Frey, S. H., D. Vinton, R. Norlund, and S. T. Grafton. 2005. Cortical Topography of Human Anterior Intraparietal Cortex Active During Visually Guided Grasping. *Cogn. Brain Res.* **23**:397–405. [9]

Fried, M. H. 1967. The Evolution of Political Society: An Essay in Political Anthropology. New York: Random House. [2]

Friederici, A. D. 2009. Neurocognition of Language Development. In: The Cambridge Handbook of Child Language, ed. E. Bavin, pp. 51–67. Cambridge: Cambridge Univ. Press. [14]

Frith, C. D., and U. Frith. 2007. Social Cognition in Humans. *Curr. Biol.* **17**:R724–R732. [20]

Fudenberg, D., and E. Maskin. 1986. The Folk Theorem in Repeated Games with Discounting or with Incomplete Information. *Econometrica* **54**:533–554. [3]

Fuentes, A. 2009. A New Synthesis: Resituating Approaches to the Evolution of Human Behaviour. *Anthrop. Today* **25**:12–17. [6]

Fujita, M., A. Venables, and P. R. Krugman. 1999. The Spatial Economy: Cities, Regions and International Trade. Cambridge, MA: MIT Press. [6]

Furuichi, T. 1987. Sexual Swelling, Receptivity, and Grouping of Wild Pygmy Chimpanzee Females at Wamba, Zaire. *Primates* **28**:309–318. [2]

———. 1989. Social Interactions and the Life History of Female (*Pan paniscus*) at Wamba, Republic of Zaire. *Intl. J. Primatol.* **10**:173–198. [2]

———. 1997. Agonistic Interactions and Matrifocal Dominance Rank of Wild Bonobos (*Pan paniscus*) at Wamba. *Intl. J. Primatol.* **18**:855–875. [2]

Fusco, G., and Q. C. B. Cronk. 1995. A New Method for Evaluating the Shape of Large Phylogenies. *J. Theor. Biol.* **175**: 235–243. [15]

Galantucci, B., S. Garrod, and G. Roberts. 2012. Experimental Semiotics. *Lang. Ling. Compass* **6**:477–493. [16]

Gallese, V., M. Rochat, G. Cossu, and C. Sinigaglia. 2009. Motor Cognition and Its Role in the Phylogeny and Ontogeny of Action Understanding. *Devel. Psychol.* **45**:103–113. [9]

Gamble, C. 1999. The Paleolithic Societies of Europe. Cambridge: Cambridge Univ. Press. [6]

Gardner, A., S. A. West, and G. Wild. 2011. The Genetical Theory of Kin Selection. *J. Evol. Biol.* **24**:1020–1043. [6]

Garrod, S., and A. Anderson. 1987. Saying What You Mean in Dialogue: A Study in Conceptual and Semantic Co-Ordination. *Cognition* **27**:181–218. [16]

Garrod, S., and G. Doherty. 1994. Conversation, Co-Ordination and Convention: An Empirical Investigation of How Groups Establish Linguistic Conventions. *Cognition* **53**:181–215. [16]

Gaskins, S. 2006. Cultural Perspectives on Infant-Caregiver Interaction. In: Roots of Human Sociality: Culture, Cognition and Interaction, ed. N. J. Enfield and S. C. Levinson, pp. 279–298. Oxford: Berg. [14]

Gat, A. 2006. War in Human Civilization. New York: Oxford Univ. Press. [3]

Gavin, M. C., and N. Sibanda. 2012. The Island Biogeography of Languages. *Glob. Ecol. Biogeography* **21**:958–967. [15]

Gavrilets, S., D. Anderson, and P. Turchin. 2010. Cycling in the Complexity of Early Societies. *Cliodynamics* **1**:58–80. [4]

Geertz, A. W. 1999. Definition as Analytical Strategy in the Study of Religion. *Hist. Reflect.* **25**:445–475. [20]

———. 2004. Cognitive Approaches to the Study of Religion. In: New Approaches to the Study of Religion, ed. P. Antes et al., vol. 2. Berlin: Walter de Gruyter. [20]

———. 2010. Brain, Body and Culture: A Biocultural Theory of Religion. *Meth. Theory Stud. Rel.* **22**:304–321. [20]

———. 2011a. Hopi Indian Witchcraft and Healing: On Good, Evil, and Gossip. *Am. Indian Q.* **35**:372–393. [20]

———. 2011b. Religious Narrative, Cognition and Culture: Approaches and Definitions. In: Religious Narrative, Cognition and Culture: Image and Word in the Mind of Narrative, ed. A. W. Geertz and J. S. Jensen, pp. 9–29. London: Equinox. [20]

Geertz, A. W., and G. I. Markusson. 2010. Religion Is Natural, Atheism Is Not: On Why Everybody Is Both Right and Wrong. *Religion* **40**:152–165. [20]

Geertz, C. 1966. Religion as a Cultural System. In: Anthropological Approaches to the Study of Religion, ed. M. Banton, pp. 1–46. London: Tavistock Publications, Ltd. [20]

———. 1974. From the Native's Point of View: On the Nature of Anthropological Understanding. *Bull. Am. Acad. Arts Sci.* **28**:26–45. [20]

Gentner, D., F. K. Anggoro, and R. S. Klibanoff. 2011. Structure-Mapping and Relational Language Support Children's Learning of Relational Categories. *Child. Devel.* **82**:1173–1188. [14]

Gentner, D., and S. Goldwin-Meadow, eds. 2003. Language in Mind: Advances in the Study of Language and Thought. Cambridge, MA: MIT Press. [13]

Gerard, R. W., C. Kluckhohn, and A. Rapoport. 1956. Biological and Cultural Evolution Some Analogies and Explorations. *Behav. Sci.* **1**:6–34. [1, 20]

Gergely, G., and G. Csibra. 2006. Sylvia's Recipe: The Role of Imitation and Pedagogy in the Transmission of Cultural Knowledge. In: Roots of Human Sociality: Culture, Cognition, and Human Interaction, ed. N. J. Enfield and S. C. Levinson, pp. 229–255. Oxford: Berg. [5]

Gerken, L. A. 2006. Decisions, Decisions: Infant Language Learning When Multiple Generalizations Are Possible. *Cognition* **98**:B67–B74. [14]

German, T., and C. Barrett. 2005. Functional Fixedness in a Technologically Sparse Culture. *Psychol. Sci.* **16**:1–5. [7]

Gervais, W. M., and J. Henrich. 2010. The Zeus Problem. *J. Cogn. Culture* **10**:383–389. [17, 20]

Gervais, W. M., and A. Norenzayan. 2012a. Analytic Thinking Promotes Religious Disbelief. *Science* **336**:493–496. [17]

———. 2012b. Like a Camera in the Sky? Thinking About God Increases Public Self-Awareness and Socially Desirable Responding. *J. Exp. Social Psychol.* **48**:298–302. [19]

———. 2012c. Reminders of Secular Authority Reduce Believers' Distrust of Atheists. *Psychol. Sci.* **23**:483–491. [19]

Gervais, W. M., A. Willard, A. Norenzayan, and J. Henrich. 2011. The Cultural Transmission of Faith: Why Innate Intuitions Are Necessary, but Insufficient to Explain Religious Belief. *Religion* **41**:389–410. [17, 20]

Gibbon, E. 1782. History of the Decline and Fall of Rome. http://www.gutenberg.org/files/25717/25717-h/25717-h.htm. (accessed April 28, 2013). [1]

Gieryn, T. F. 1982. Relativist/Constructivist Programmes in the Sociology of Science: Redundance and Retreat. *Soc. Stud. Sci.* **12**:279–297. [1]

Gignoux, C. R., B. M. Henn, and J. L. Mountain. 2011. Rapid, Global Demographic Expansions after the Origins of Agriculture. *PNAS* **108**:6044–6049. [8]

Gilbert, M. 1987. Modeling Collective Belief. *Synthese* **73**:185–204. [2]

Gilbert, S. F., and D. Epel. 2008. Ecological Developmental Biology: Integrating Epigenetics, Medicine and Evolution. Sunderland, MA: Sinauer. [16]

Gilligan, I. 2010. The Prehistoric Development of Clothing: Archaeological Implications of a Thermal Model. *J. Arch. Method Theory* **17**:15–80. [7]

Ginges, J., S. Atran, D. L. Medin, and K. Shikaki. 2007. Sacred Bounds on Rational Resolution of Violent Political Conflict. *PNAS* **104**:7357–7360. [17]

Ginges, J., I. Hansen, and A. Norenzayan. 2009. Religious and Popular Support for Suicide Attacks. *Psychol. Sci.* **20**:224–230. [17]

Gingrich, P. D. 1983. Rates of Evolution: Effects of Time and Temporal Scaling. *Science* **222**:159–161. [7]

Gintis, H. 2000. Game Theory Evolving. Princeton: Princeton Univ. Press. [11]

———. 2007. A Framework for the Unification of the Behavioral Sciences. *Behav. Brain Sci.* **30**:1–61. [1]

———. 2009. Bounds of Reason: Game Theory and the Unification of the Behavioral Sciences. Princeton: Princeton Univ. Press. [2]

Gintis, H., S. Bowles, R. Boyd, and E. Fehr. 2005. Moral Sentiments and Material Interests: On the Foundations of Cooperation in Economic Life. Cambridge, MA: MIT Press. [2]

Gintis, H., E. A. Smith, and S. Bowles. 2001. Costly Signaling and Cooperation. *J. Theor. Biol.* **213**:103–119. [3]

Giugni, M. G., D. McAdam, and C. Tilly. 1998. From Contention to Democracy. London: Rowman & Littlefield. [2]

Göksun, T., K. Hirsh-Pasek, and R. Golinkoff. 2010. Trading Spaces: Carving up Events for Learning Language. *Persp. Psychol. Sci.* **5**:33–42. [14]

Gómez, J. C. 2004. Apes, Monkeys, Children and the Growth of Mind. Cambridge, MA: Harvard Univ. Press. [14]

Goodall, J. 1964. Tool-Using and Aimed Throwing in a Community of Free-Living Chimpanzees. *Nature* **201**:1264–1266. [2]

———. 1986. The Chimpanzees of Gombe: Patterns of Behavior. Cambridge, MA: Belknap Press. [2, 5]

Goodreau, S., J. A. Kitts, and M. Morris. 2009. Birds of a Feather or Friend of a Friend? Using Exponential Random Graph Models to Investigate Adolescent Friendship Networks. *Demography* **46**:103–125. [6]

Goody, J. 1973. Evolution and Communication: The Domestication of the Savage Mind. *Br. J. Sociol.* **24**:1–12. [11]

———. 1987. The Interface between the Written and the Oral. Cambridge: Cambridge Univ. Press. [14]

Gopnik, A. 2001. The Scientist in the Crib: What Early Learning Tells Us About the Mind. London: HarperCollins. [20]

———. 2010. How Babies Think. *Sci. Am.* **7**:76–81. [10]

Gopnik, A., A. Meltzoff, and P. Kuhl. 1999. The Scientist in the Crib: Minds, Brains, and How Children Learn. New York: W. Morrow. [10]

Gopnik, A., and L. Schulz. 2004. Mechanisms of Theory Formation in Young Children. *Trends Cogn. Sci.* **8**:371–377. [7]

Gosden, C. 2005. What Do Objects Want? *J. Archaeol. Meth. Theory* **12**:193–211. [9]

Gould, J. L., and C. Gould. 2007. Animal Architects: Building and the Evolution of Intelligence. New York: Basic Books. [7]

Gould, S. J. 1989. Wonderful Life: Burgess Shale and the Nature of History. New York: W. W. Norton. [15]

Grafen, A. 1990a. Biological Signals as Handicaps. *J. Theor. Biol.* **144**:517–546. [3]

———. 1990b. Sexual Selection Unhandicapped by the Fisher Process. *J. Theor. Biol.* **144**:473–516. [3]

Grafton, S. T. 2009. Embodied Cognition and the Simulation of Action to Understand Others. *Ann. NY Acad. Sci.* **1156**:97–117. [9]

Grafton, S. T., and A. F. Hamilton. 2007. Evidence for a Distributed Hierarchy of Action Representation in the Brain. *Hum. Mov. Sci.* **26**:590–616. [9]

Graham, J., J. Haidt, and B. A. Nosek. 2009. Liberals and Conservatives Rely on Different Sets of Moral Foundations. *J. Pers. Soc. Psychol.* **96**:1029–1046. [20]

Graham, S. A., and S. E. Fisher. 2013. Decoding the Genetics of Speech and Language. *Curr. Opin. Neurobiol.* **23**:43–51. [16]

Grant, E. 1996. The Foundations of Modern Science in the Middle Ages: Their Religious, Institutional, and Intellectual Contexts. Cambridge: Cambridge Univ. Press. [10]

Grant, P., and R. Grant. 1986. Ecology and Evolution of Darwin's Finches. Princeton: Princeton Univ. Press. [7]

Gratzer, W. 2000. The Undergrowth of Science: Delusion, Self-Deception and Human Frailty. Oxford: Oxford Univ. Press. [10]

Gray, R. D. 2001. Selfish Genes or Developmental Systems? In: Thinking About Evolution: Historical, Philosophical and Political Perspectives, ed. R. S. Singh et al., pp. 184–207. Cambridge: Cambridge Univ. Press. [6]

Gray, R. D., and Q. Atkinson. 2003. Language-Tree Divergence Times Support the Anatolian Theory of Indo-European Origins. *Nature* **426**:435–439. [15]

Gray, R. D., Q. D. Atkinson, and S. J. Greenhill. 2011. Language Evolution and Human History: What a Difference a Date Makes. *Phil. Trans. R. Soc. B* **355**:1090–1100. [15]

Gray, R. D., A. J. Drummond, and S. J. Greenhill. 2009. Language Phylogenies Reveal Expansion Pulses and Pauses in Pacific Settlement. *Science* **323**:479–483. [15, 20]

Gray, R. D., S. J. Greenhill, and D. Bryant. 2010. On the Shape and Fabric of Human History. *Phil. Trans. R. Soc B* **365**:3923–3933. [15, 16]

Gray, R. D., S. J. Greenhill, and R. M. Ross. 2007. The Pleasures and Perils of Darwinizing Culture (with Phylogenies). *Biol. Theory* **2**:360–375. [15]

Green, R. 2003. Gurr-Goni: A Minority Language in a Multilingual Community: Surviving into the 21st Century. In: Endangered Language and Identity, ed. J. Blythe and R. M. Brown, pp. 127–134. Bath: Foundation for Endangered Languages. [16]

Green, R. E., J. Krause, A. W. Briggs, et al. 2010. A Draft Sequence of the Neandertal Genome. *Science* **328**:710–722. [12, 16]

Greenberg, J. H. 1963. Some Universals of Grammar with Particular Reference to the Order of Meaningful Elements. In: Universals of Language, ed. J. H. Greenberg, pp. 58–90. Cambridge, MA: MIT Press. [13]

———. 1966. Universals of Language (2nd edition). Cambridge, MA: MIT Press. [16]

———. 1990. Universals of Kinship Terminology: Their Nature and the Problem of Their Explanation. In: On Language: Selected Writings of Joseph H. Greenberg, ed. K. Denning and S. Kemmer. Stanford: Stanford Univ. Press. [13]

Greenfield, P. 2003. Historical Change, Cultural Learning, and Cognitive Representation in Zinacantec Maya Children. *Cogn. Devel.* **18**:455. [9]

Greenhill, S. J., Q. Atkinson, A. M. Meade, and R. D. Gray. 2010. The Shape and Tempo of Language Evolution. *Proc. R. Soc. B* **277**:2443–2450. [15]

Greenhill, S. J., R. Blust, and R. D. Gray. 2008. The Austronesian Basic Vocabulary Database: From Bioinformatics to Lexomics. *Evol. Bioinform.* **4**:271–283. [15]

Greenhill, S. J., and R. D. Gray. 2009. Austronesian Language Phylogenies: Myths and Misconceptions About Bayesian Computational Methods. In: Austronesian Historical Linguistics and Culture History: A Festschrift for Robert Blust, ed. A. Adelaar and A. Pawley, pp. 375–397. Canberra: Pacific Linguistics. [15]

———. 2010. How Accurate and Robust Are the Phylogenetic Estimates of Austronesian Language Relationships? *PLoS One* **5**:e9573. [15]

Greif, A. 1993. Contract Enforceability and Economic Institutions in Early Trade: The Maghribi Traders' Coalition. *Am. Econ. Rev.* **83**:525–548. [11]

Gribbin, J. 2003. Science: A History 1543–2001. London: Penguin. [10]

Griffiths, P. E., and K. Stotz. 2006. Genes in the Postgenomic Era? *Theor. Med. Bioeth.* **27**:499–521. [16]

Griffiths, T. L., and F. Reali. 2011. Modelling Minds as Well as Populations. *Proc. R. Soc.* **278**:1773–1776. [7]

Griliches, Z. 1957. Hybrid Corn: An Exploration of the Economics of Technological Change. *Econometrica* **48**:501–522. [1]

Grinin, L. E., and A. V. Korotayev. 2009. The Epoch of the Initial Politogenesis. *Soc. Evol. History* **8**:52–91. [4]

Grinnell, J., C. Packer, and A. E. Pusey. 1995. Cooperation in Male Lions: Kinship, Reciprocity or Mutualism? *Anim. Behav.* **49**:95–105. [3]

Gross, P. R., and N. Levitt. 1997. Higher Superstition: The Academic Left and Its Quarrels with Science. Baltimore: Johns Hopkins Univ. Press. [1]

Grossniklaus, U., B. Kelly, A. C. Ferguson-Smith, M. Pembrey, and S. Lindquist. 2013. Transgenerational Epigenetic Inheritance: How Important Is It? *Nat. Rev. Genet.* **14**:228–235. [1]

Gruenfeld, D. H., and L. Z. Tiedens. 2010. Organizational Preferences and Their Consequences. In: The Handbook of Social Psychology, ed. S. T. Fiske et al. New York: Wiley. [5]

Gurven, M. 2004. To Give and to Give Not: The Behavioral Ecology of Human Food Transfers. *Behav. Brain Sci.* **27**:543–583. [3]

Gurven, M., K. Hill, H. Kaplan, A. Hurtado, and R. Lyles. 2000. Food Transfers among Hiwi Foragers of Venezuela: Tests of Reciprocity. *Hum. Ecol.* **28**:171–218. [3]

Güth, W., R. Schmittberger, and B. Schwarze. 1982. An Experimental Analysis of Ultimatum Bargaining. *J. Econ. Behav. Org.* **3**:367–388. [2]

Guthrie, R. D. 2005. The Nature of Paleolithic Art. Chicago: Univ. of Chicago Press. [1]

Guthrie, S. 1993. Faces in the Clouds: A New Theory of Religion. New York: Oxford Univ. Press. [17]

Haidt, J., and J. Graham. 2009. Planet of the Durkheimians, Where Community, Authority, and Sacredness Are Foundations of Morality. New York: Oxford Univ. Press. [20]

Haidt, J., J. P. Seder, and S. Kesebir. 2008. Hive Psychology, Happiness, and Public Policy. *J. Legal Stud.* **37**:S133–S156. [17, 20]

Haiman, J. 1980. The Iconicity of Grammar: Isomorphism and Motivation. *Language* **56**:515–540. [13]

———. 1983. Iconic and Economic Motivation. *Language* **59**:781–819. [13]

Haley, K. J., and D. M. T. Fessler. 2005. Nobody's Watching? Subtle Cues Affect Generosity in an Anonymous Economic Game. *Evol. Hum. Behav.* **26**:245–256. [19]

Hamilton, A. F., and S. T. Grafton. 2008. Action Outcomes Are Represented in Human Inferior Frontoparietal Cortex. *Cereb. Cortex* **18**:1160–1168. [9]

Hamilton, M. J., and B. Buchanan. 2009. The Accumulation of Stochastic Copying Errors Causes Drift in Culturally Transmitted Technologies: Quantifying Clovis Evolutionary Dynamics. *J. Anthropol. Arch.* **28**:55–69. [11]

Hamilton, W. D. 1964. The Genetical Evolution of Social Behaviour. *J. Theor. Biol.* **7**:17–52. [3, 6]

———. 1971. Selection of Selfish and Altruistic Behaviour in Some Extreme Models. In: Man and Beast: Comparative Social Behavior, ed. J. F. Eisenberg and W. S. Dillon, pp. 59–91. Washington, DC: Smithsonian Institutions Press. [6]

Hammerstein, P. 2003. Why Is Reciprocity So Rare in Social Animals? A Protestant Appeal. In: Genetic and Cultural Evolution of Cooperation, pp. 83–93. Dahlem Workshop Reports, J. Lupp, series editor. Cambridge, MA: MIT Press. [3]

Handy, E. S. C. 1927. Polynesian Religion. Honolulu: The Museum. [20]

Hansell, M. 2005. Animal Architecture. New York: Oxford Univ. Press. [7]

Hare, B., A. P. Melis, V. Woods, S. Hastings, and R. Wrangham. 2007. Tolerance Allows Bonobos to Outperform Chimpanzees on a Cooperative Task. *Curr. Biol.* **17**:619–623. [2]

Harris, A. 2008. On the Explanation of Typologically Unusual Structures. In: Language Universals and Language Change, ed. J. Good, pp. 54–76. Oxford: Oxford Univ. Press. [13]

Harris, M. 1974. Cows, Pigs, Wars, and Witches: The Riddles of Culture. Toronto: Random House. [20]

Harris, P. L. 2012. Trusting What You're Told: How Children Learn from Others. Cambridge, MA: Belknap Press. [5]

Harris, P. L., and M. A. Koenig. 2006. Trust in Testimony: How Children Learn About Science and Religion. *Child Devel.* **77**:505–524. [1]

Harrison, D., and M. A. Brockhurst. 2012. Plasmid-Mediated Horizontal Gene Transfer Is a Coevolutionary Process. *Trends Microbiol.* **20**:262–267. [16]

Hart, B., and T. R. Risley. 1995. Meaningful Differences in the Everyday Experience of Young American Children. Baltimore: Paul Brookes Publishing. [14]

Hartmann, K., G. Goldenberg, M. Daumuller, and J. Hermsdorfer. 2005. It Takes the Whole Brain to Make a Cup of Coffee: The Neuropsychology of Naturalistic Actions Involving Technical Devices. *Neuropsychologia* **43**:625–637. [9]

Haspelmath, M. 1997. Indefinite Pronouns. Oxford: Oxford Univ. Press. [13]

Hauert, C., A. Traulsen, H. Brandt, M. A. Nowak, and K. Sigmund. 2007. Via Freedom to Coercion:The Emergence of Costly Punishment. *Science* **316**:1905–1907. [3]

Haun, D. B. M., C. J. Rapold, J. Call, G. Janzen, and S. C. Levinson. 2006. Cognitive Cladistics and Cultural Override in Hominid Spatial Cognition. *PNAS* **103**:17,568–17,573. [5]

Haun, D. B. M., Y. Rekers, and M. Tomasello. 2012. Majority-Biased Transmission in Chimpanzees and Human Children, but Not Orangutans. *Curr. Biol.* **22**:727–731. [5, 6]

Haun, D. B. M., and M. Tomasello. 2011. Conformity to Peer Pressure in Preschool Children. *Child. Devel.* **82**:1759–1767. [5]

Haun, D. B. M., E. J. C. Van Leeuwen, and M. G. Edelson. 2013. Majority Influence in Children and Other Animals. *Devel. Cogn. Neurosci.* **3**:61–71. [5]

Hauser, M. D. 2009. On the Possibility of Impossible Cultures. *Nature* **460**:190–196. [15]

Hauser, M. D., N. Chomsky, and W. T. Fitch. 2002. The Faculty of Language: What Is It, Who Has It, and How Did It Evolve? *Science* **298**:1569–1579. [16]

Hawkes, K., and R. L. Bliege Bird. 2002. Showing Off, Handicap Signaling, and the Evolution of Men's Work. *Evol. Anthropol.* **11**:58–67. [3]

Hawkes, K., J. F. O'Connell, and N. G. B. Jones. 2001. Hunting and Nuclear Families: Some Lessons from the Hadza About Men's Work. *Curr. Anthropol.* **42**:681–709. [3]

Hawkins, J. A. 1994. A Performance Theory of Order and Constituency. Cambridge: Cambridge Univ. Press. [16]

Hawks, J., E. T. Wang, G. M. Cochran, H. C. Harpending, and R. K. Moyzis. 2007. Recent Acceleration of Human Adaptive Evolution. *PNAS* **104**:20,753–20,758. [1, 4, 12]

Hayden, B. 1987. Alliances and Ritual Ecstasy: Human Responses to Resource Stress. *J. Sci. Study Rel.* **26**:81–91. [20]

Headrick, D. R. 1981. The Tools of Empire: Technology and European Imperialism in the Nineteenth Century. New York: Oxford Univ. Press. [6]

Heath, C., C. Bell, and E. Sternberg. 2001. Emotional Selection in Memes: The Case of Urban Legends. *J. Pers. Soc. Psychol.* **81**:1028–1041. [11]

Heath, C., and D. Heath. 2007. Made to Stick. London: Random House. [11]

Heath, J. 1982. Where Is That [Knee]? Basic and Supplementary Terms in Dhuwal (Yuulngu/Murngin). In: Languages of Kinship in Aboriginal Australia, ed. J. Heath et al., pp. 40–63. Sydney: Oceania Linguistic Monographs. [13]

Heath, S. B. 1983. Ways with Words. New York: Cambridge Univ. Press. [14]

Hecht, E. E., D. A. Gutman, T. M. Preuss, et al. 2012. Process Versus Product in Social Learning: Comparative Diffusion Tensor Imaging of Neural Systems for Action Execution–Observation Matching in Macaques, Chimpanzees, and Humans. *Cereb. Cortex* **23**:1014–1024. [9]

Hechter, M. 1987. Principles of Group Solidarity. Berkeley: Univ. of California Press. [6]

Hechter, M., and K. D. Opp. 2001. Social Norms. Berkeley: Univ. of California Press. [6]

Heggarty, P., and C. Renfrew. 2013. Introduction: Languages. In: The Cambridge World Prehistory, ed. C. Renfrew and P. Bahn. Cambridge: Cambridge Univ. Press, in press. [15]

Heimbauer, L. A., C. M. Conway, M. H. Christiansen, M. J. Beran, and M. J. Owren. 2010. Grammar Rule-Based Sequence Learning by Rhesus Macaques (*Macaca Mulatta*). *Am. J. Primatol.* **72**:65. [14]

Heinz, H. J. 1979. The Nexus Complex among The !Xo Bushmen of Botswana. *Anthropos* **74**:465–480. [6]

Heled, J., and A. J. Drummond. 2010. Bayesian Inference of Species Trees from Multilocus Data. *Mol. Biol. Evol.* **27**:570–580. [16]

Hennig, W. 1950. Grundzüge Einer Theorie Der Phylogenetischen Systematik. Berlin: Deutscher Zentralverlag. [13]

Henrich, J. 2001. Cultural Transmission and the Diffusion of Innovations. *Am. Anthropol.* **103**:992–1013. [11]

———. 2004a. Cultural Group Selection, Coevolutionary Processes and Large-Scale Cooperation. *J. Econ. Behav. Org.* **53**:3–35. [3, 6]

———. 2004b. Demography and Cultural Evolution: How Adaptive Cultural Processes Can Produce Maladaptive Losses: The Tasmanian Case. *Am. Antiq.* **69**:197–214. [6–8, 11, 15]

———. 2006. Understanding Cultural Evolutionary Models: A Reply to Read's Critique. *Am. Antiq.* **71**:771–782. [7]

———. 2008. A Cultural Species. In: Explaining Culture Scientifically, ed. M. Brown, pp. 184–210. Seattle: Univ. of Washington Press. [7]

———. 2009a. The Evolution of Costly Displays, Cooperation, and Religion: Credibility Enhancing Displays and Their Implications for Cultural Evolution. *Evol. Hum. Behav.* **30**:244–260. [17–20]

———. 2009b. The Evolution of Innovation-Enhancing Institutions. In: Innovation in Cultural Systems: Contributions in Evolution Anthropology, ed. S. J. Shennan and M. J. O'Brien, pp. 99–120. Cambridge, MA: MIT Press. [7, 8]

Henrich, J., and R. Boyd. 1998. The Evolution of Conformist Transmission and the Emergence of Between-Group Differences. *Evol. Hum. Behav.* **19**:215–241. [5]

———. 2002a. Culture and Cognition: Why Cultural Evolution Does Not Require Replication of Representations. *J. Culture Cogn.* **2**:87–112. [7]

———. 2002b. On Modelling Cognition and Culture: Why Cultural Evolution Does Not Require Replication of Representations. *J. Cogn. Culture* **2**:87–122. [8]

———. 2008. Division of Labor, Economic Specialization and the Evolution of Social Stratification. *Curr. Anthropol.* **49**:715–724. [1, 20]

Henrich, J., R. Boyd, S. Bowles, et al. 2005. Economic Man in Cross-Cultural Perspective: Behavioral Experiments in 15 Small-Scale Societies. *Behav. Brain Sci.* **28**:795–815. [2]

Henrich, J., R. Boyd, and P. J. Richerson. 2012. The Puzzle of Monogamous Marriage. *Phil. Trans. R. Soc. B* **367**:657–669. [17]

Henrich, J., J. Ensminger, R. McElreath, et al. 2010a. Market, Religion, Community Size and the Evolution of Fairness and Punishment. *Science* **327**:1480–1484. [19]

Henrich, J., and F. Gil-White. 2001. The Evolution of Prestige: Freely Conferred Deference as a Mechanism for Enhancing the Benefits of Cultural Transmission. *Evol. Hum. Behav.* **22**:165–196. [7]

Henrich, J., S. J. Heine, and A. Norenzayan. 2010b. The Weirdest People in the World. *Behav. Brain Sci.* **33**:61–83. [1, 12, 17, 19]

Henrich, J., and N. Henrich. 2010. The Evolution of Cultural Adaptations: Fijian Food Taboos Protect against Dangerous Marine Toxins. *Proc. R. Soc. B* **277**:3715–3724. [1, 7, 11, 20]

Henrich, J., and R. McElreath. 2003. The Evolution of Cultural Evolution. *Evol. Anthropol.* **12**:123–135. [1]

———. 2007. Dual Inheritance Theory: The Evolution of Human Cultural Capacities and Cultural Evolution. In: Oxford Handbook of Evolutionary Psychology, ed. R. Dunbar and L. Barrett, pp. 555–570. Oxford: Oxford Univ. Press. [20]

Henrich, J., R. McElreath, A. Barr, et al. 2006. Costly Punishment across Human Societies. *Science* **312**:1767–1770. [1, 5]

Henrich, N., and J. Henrich. 2007. Why Humans Cooperate: A Cultural and Evolutionary Explanation. Oxford: Oxford Univ. Press. [1, 19]

Herrmann, B., C. Thöni, and S. Gächter. 2008. Antisocial Punishment across Societies. *Science* **319**:1362–1367. [1]

Herrmann, E., J. Call, M. V. Hernandez-Lloreda, B. Hare, and M. Tomasello. 2007. Humans Have Evolved Specialized Skills of Social Cognition: The Cultural Intelligence Hypothesis. *Science* **317**:1360–1366. [11, 14]

Hihara, S., T. Notoya, M. Tanaka, et al. 2006. Extension of Corticocortical Afferents into the Anterior Bank of the Intraparietal Sulcus by Tool-Use Training in Adult Monkeys. *Neuropsychologia* **44**:2636–2646. [9]

Hill, K. 2002. Altruistic Cooperation During Foraging by the Ache, and the Evolved Human Predisposition to Cooperate. *Hum. Nature* **13**:105–128. [3]

Hill, K., M. Barton, and A. M. Hurtado. 2009. The Emergence of Human Uniqueness: Characters Underlying Behavioral Modernity. *Evol. Anthropol.* **18**:187–200. [6]

Hill, K., and M. Hurtado. 2009. Cooperative Breeding in South American Hunter-Gatherers. *Proc. R. Soc. B* **276**:3863–3870. [3]

Hill, K. R., R. S. Walker, M. Božičević, et al. 2011. Co-Residence Patterns in Hunter-Gatherer Societies Show Unique Human Social Structure. *Science* **331**:1286–1289. [2, 5, 6]

Hirshleifer, J., and M. J. Martinez Coll. 1988. What Strategies Can Support the Evolutionary Emergence of Cooperation? *J. Conflict Res.* **32**:367–398. [3]

Hock, H. H., and B. D. Joseph. 2009. Language History, Language Change, and Language Relationship: An Introduction to Historical and Comparative Linguistics. Berlin: Mouton de Gruyter. [1]

Hockett, C. F. 1960. The Origin of Speech. *Sci. Am.* **203**:89–96. [13]

Hodgson, G. M. 1993. Economics and Evolution: Bringing Life Back into Economics. Economics, Cognition, and Society. Ann Arbor: Univ. of Michigan Press. [1]

———. 2004. The Evolution of Institutional Economics: Agency, Structure and Darwinism in American Institutionalism. London: Routledge. [1]

Hoff-Ginsberg, E. 1991. Mother-Child Conversation in Different Social Classes and Communicative Settings. *Child. Devel.* **62**:782–796. [14]

Hoffman, E., K. McCabe, K. Shachat, and V. Smith. 1994. Preferences, Property Rights and Anonymity in Bargaining Games. *Games Econ. Behav.* **7**:346–380. [19]

Hofstadter, R. 1945. Social Darwinism in American Thought, 1860–1915. Philadelphia: Univ. of Pennsylvania Press. [1]

Holden, C. J. 2002. Bantu Language Trees Reflect the Spread of Farming across Sub-Saharan Africa: A Maximum-Parsimony Analysis. *Proc. R. Soc. B* **269**:793–799. [15]

Holden, C. J., and R. D. Gray. 2006 Rapid Radiation, Borrowing and Dialect Continua in the Bantu Languages. In: Phylogenetic Methods and the Prehistory of Languages, ed. P. Forster and C. Renfrew, pp. 19–31. Cambridge, MA: McDonald Institute. [15]

Holden, C. J., and R. Mace. 2003. Spread of Cattle Led to the Loss of Matrilineal Descent in Africa: A Coevolutionary Analysis. *Proc. R. Soc. B* **270**:2425–2433. [6]

———. 2009. Phylogenetic Analysis of the Evolution of Lactose Digestion in Adults. *Hum. Biol.* **81**:597–619. [20]

Holland, D., and N. Quinn. 1987. Cultural Models in Language and Thought. Cambridge: Cambridge Univ. Press. [9]

Holland, P., and S. Leinhardt. 1970. Detecting Structure in Sociometric Data. *Am. J. Sociol.* **76**:492–513. [6]

Holman, E. W. 2010. Do Languages Originate and Become Extinct at Constant Rates? *Diachronica* **27**:214–225. [15]

Hopper, L. M., S. J. Schapiro, S. P. Lambeth, and S. F. Brosnan. 2011. Chimpanzees' Socially Maintained Food Preferences Indicate Both Conservatism and Conformity. *Anim. Behav.* **81**:1195–1202. [5]

Hopper, P. J., and E. C. Traugott. 2003. Grammaticalization. Cambridge: Cambridge Univ. Press. [1]

Hoppitt, W., N. J. Boogert, and K. N. Laland. 2010. Detecting Social Transmission in Networks. *J. Theor. Biol.* **263**:544–555. [11]

Hoppitt, W., and K. N. Laland. 2008. Social Processes Influencing Learning in Animals: A Review of the Evidence. *Adv. Stud. Behav.* **38**:105–165. [11]

Horner, V., and A. Whiten. 2005. Causal Knowledge and Imitation/Emulation Switching in Chimpanzees *(Pan troglodytes)* and Children (*Homo sapiens*). *Anim. Cogn.* **8**:164–181. [5, 11]

Horton, R. 1993. Patterns of Thought in Africa and the West: Essays on Magic, Religion, and Science. Cambridge: Cambridge Univ. Press. [10]

Hrdy, S. B. 2000. Mother Nature: Maternal Instincts and How They Shape the Human Species. New York: Ballantine [2]

———. 2009. Mothers and Others: The Evolutionary Origins of Mutual Understanding. Cambridge, MA: Harvard Univ. Press. [2, 3, 6]

———. 2010. Estimating the Prevalence of Shared Care and Cooperative Breeding in the Order Primates. http://www.citrona.com/hrdy/documents/AppendixI.pdf. (accessed March 5, 2013). [2]

Hruschka, D. J. 2010. Friendship: Development, Ecology, and Evolution of a Relationship. Berkeley: Univ. of California Press. [6]

Hruschka, D. J., M. H. Christiansen, R. A. Blythe, et al. 2009. Building Social Cognitive Models of Language Change. *Trends Cogn. Sci.* **13**:464–469. [13, 16]

Hublin, J. J. 2009. Out of Africa: Modern Human Origins Special Feature: The Origin of Neandertals. *PNAS* **106**:16,022–16,027. [16]

Huebner, B., J. J. Lee, and M. D. Hauser. 2010. The Moral-Conventional Distinction in Mature Moral Competence. *J. Cogn. Culture* **10**:1–26. [20]

Huff, T. 1993. The Rise of Early Modern Science: Islam, China, and the West Cambridge: Cambridge Univ. Press. [10]

Hughes, W. O. H., B. P. Oldroyd, M. Beekman, and F. L. W. Ratnieks. 2008. Ancestral Monogamy Shows Kin Selection Is Key to the Evolution of Eusociality. *Science* **320**:1213–1216. [3]

Hull, D. L. 1988. Science as a Process: An Evolutionary Account of the Social and Conceptual Development of Science. Chicago: Chicago Univ. Press. [11, 16]

———. 2001. Science and Selection: Essays on Biological Evolution and the Philosophy of Science. Cambridge: Cambridge Univ. Press. [16]

Humphrey, N. 1976. The Social Function of Intellect. In: Growing Points in Ethology, ed. P. P. G. Bateson and R. A. Hinde, pp. 303–317. Cambridge: Cambridge Univ. Press. [2]

Hurford, J. R. 1989. Biological Evolution of the Saussurean Sign as a Component of the Language Acquisition Device. *Lingua* **77**:187–222. [1]

———. 2011. The Origins of Grammar: Language in the Light of Evolution II. Oxford: Oxford Univ. Press. [1]

Hurtado, N., V. A. Marchman, and A. Fernald. 2008. Does Input Influence Uptake? Links between Maternal Talk, Processing Speed and Vocabulary Size in Spanish-Learning Children. *Devel. Sci.* **11**:F31–F39. [14]

Huttenlocher, J., M. Vasilyeva, E. Cymerman, and S. Levine. 2002. Language Input and Child Syntax. *Cogn. Psychol.* **45**:337–374. [14]

Hyman, M. D., and J. Renn. 2012. Survey: From Technology Transfer to the Origins of Science. Max Planck Research Library for the History and Development of Knowledge, Studies 1. http://www.edition-open-access.de/studies/1/chapter_6.html. (accessed April 6, 2013). [11]

Ingram, C. J., C. A. Mulcare, Y. Itan, M. G. Thomas, and D. M. Swallow. 2009. Lactose Digestion and the Evolutionary Genetics of Lactase Persistence. *Hum. Genet.* **124**:579–591. [7]

Insko, C. A., R. Gilmore, S. Drenan, et al. 1983. Trade Versus Expropriation in Open Groups: A Comparison of Two Types of Social Power. *J. Person. Soc. Psychol.* **44**:977–999. [1]

Iriki, A., M. Tanaka, and Y. Iwamura. 1996. Coding of Modified Body Schema During Tool Use by Macaque Postcentral Neurones. *Neuroreport* **7**:2325–2330. [9]

Irons, W. 1996. Morality, Religion, and Evolution. In: Religion and Science: History, Method, and Dialogue, ed. W. M. Richardson, pp. 375–399. New York: Routledge. [20]

———. 2001. Religion as Hard-to-Fake Sign of Commitment. In: Evolution and the Capacity for Commitment, ed. R. Nesse, pp. 292–309. New York: Russell Sage Foundation. [20]

Isaac, B. 1987. Throwing and Human Evolution. *African Arch. Rev.* **5**:3–17. [2]

Isaac, G. L. 1978a. The Food-Sharing Behavior of Protohuman Hominids. *Sci. Am.* **238**:90–108. [2]

———. 1978b. Food Sharing and Human Evolution: Archaeological Evidence from the Plio-Pleistocene of East Africa. *J. Anthropol. Res.* **34**:311–325. [2]

Issenman, B. K. 1997. The Sinews of Survival. Vancouver: UBC Press. [7]

Itan, Y., A. Powell, M. A. Beaumont, J. Burger, and M. G. Thomas. 2009. The Origins of Lactase Persistence in Europe. *PLoS Comp. Biol.* **5**:e1000491. [1]

Jablonka, E. 2013. Epigenetic Inheritance and Plasticity: The Responsive Germline. *Prog. Biophys. Molec. Biol.* **111**:99–107. [1]

Jablonka, E., and M. J. Lamb. 2005. Evolution in Four Dimensions: Epigenetic, Behavioral and Symbolic Variation in the History of Life. Cambridge, MA: MIT Press. [16]

Jablonka, E., and G. Raz. 2009. Transgenerational Epigenetic Inheritance: Prevalence, Mechanisms, and Implications for the Study of Heredity and Evolution. *Q. Rev. Biol.* **84**:131–176. [16]

Jablonski, N. G., and G. Chaplin. 2010. Human Skin Pigmentation as an Adaptation to UV Radiation. *PNAS* **107**:8962–8968. [12]

Jackson, C. M., and J. W. Smedley. 2008. Medieval and Post-Medieval Glass Technology: The Chemical Composition of Bracken from Different Habitats through a Growing Season. *Eur. J. Glass Sci. Technol.* **49**:240–245. [8]

Jacobs, R. C., and D. T. Campbell. 1961. The Perpetuation of an Arbitrary Tradition through Several Generations of Laboratory Microculture. *J. Abnorm. Soc. Psychol.* **62**:649–568. [1]

Jaeggi, A. V., J. M. G. Stevens, and C. P. van Schaik. 2010. Tolerant Food Sharing and Reciprocity Is Precluded by Despotism among Bonobos but Not Chimpanzees. *Am. J. Phys. Anthropol.* **143**:41–51. [2]

Jäger, H., A. Baronchelli, E. Briscoe, et al. 2009. What Can Mathematical, Computational and Robotic Models Tell Us About the Origins of Syntax? In: Biological Foundations and Origin of Syntax, ed. D. Bickerton and E. Szathmáry, pp. 385–410, Strüngmann Forum Reports, vol. 3, J. Lupp, series ed. Cambridge, MA: MIT Press. [1, 16]

Jain, R., M. C. Rivera, and J. A. Lake. 1999. Horizontal Gene Transfer among Genomes: The Complexity Hypothesis. *PNAS* **96**:3801–3806. [16]

Jakobson, R. 1960. Closing Statements: Linguistics and Poetics, Style in Language. New York: T. A. Sebeok. [16]

James, W. 1902. The Varieties of Religious Experience: A Study in Human Nature. Seattle: Seven Treasures Publications. [20]

Jardine, L. 2000. Ingenious Pursuits: Building the Scientific Revolution. London: Abacus. [10]

Jelinek, E. 1995. Quantification in Straits Salish. In: Quantification in Natural Languages, ed. E. Bach et al., pp. 487–540. Dordrecht: Kluwer. [13]

Jennifer, U., M. Jarvis, J. O'Riain, N. C. Bennett, and P. W. Sherman. 1994. Mammalian Eusociality: A Family Affair. *Trends Ecol. Evol.* **9**:47–51. [3]

Jensen, J. S. 2002. Complex Worlds of Religion: Connecting Cultural and Cognitive Analysis. In: Current Approaches in the Cognitive Science of Religion, ed. I. Pyysiäinen and V. Anttonen, pp. 203–228. London: Continuum. [20]

Jerison, H. J. 1973. Evolution of Brain and Intelligence. New York: Academic Press. [2]

Jobling, M. A., M. Hurles, and C. Tyler-Smith. 2004. Human Evolutionary Genetics: Origins, Peoples and Disease. New York: Garland Science. [12]

Johnson, A. W., and T. Earle. 2000. The Evolution of Human Societies: From Foraging Group to Agrarian State. Stanford: Stanford Univ. Press. [6]

Johnson, D. D. P. 2005. God's Punishment and Public Goods: A Test of the Supernatural Punishment Hypothesis in 186 World Cultures. *Hum. Nature* **16**:410–446. [20]

———. 2009. The Error of God: Error Management Theory, Religion, and the Evolution of Cooperation. In: Games, Groups, and the Global Good, ed. S. A. Levin, pp. 169–180. Berlin: Springer-Verlag. [19]

———. 2012. Atheists: Accidents of Nature? *Relig. Brain Behav.* **2**:91–99. [20]

Johnson-Frey, S. H. 2004. The Neural Bases of Complex Tool Use in Humans. *Trends Cogn. Sci.* **8**:71–78. [9]

Johnstone, R. A. 1999. Signaling of Need, Signaling Competition, and the Cost of Honesty. *PNAS* **96**:12,644–12,649. [3]

Jolly, A. 1972. The Evolution of Primate Behavior. New York: MacMillan. [2]

Jones, B. F. 2010. Age and Great Invention. *Rev. Econ. Statis.* **92**:1–14. [11]

Jones, J. T., B. W. Pelham, M. Carvallo, and M. C. Mirenberg. 2004. How Do I Love Thee? Let Me Count the Js: Implicit Egotism and Interpersonal Attraction. *J. Pers. Soc. Psychol.* **87** 665–683. [5]

Jones, S. W. 1786/2013. Third Anniversary Discourse: On the Hindus. http://www.utexas.edu/cola/centers/lrc/books/read01.html. (accessed April 27, 2013). [1, 15]

Jordan, F. M. 2013. Comparative Phylogenetic Methods and the Study of Pattern and Process in Kinship. In: Kinship Systems: Change and Reconstruction, ed. P. McConvell et al., pp. 43–53. Salt Lake City: Univ. of Utah Press. [13]

Jordan, F. M., R. D. Gray, S. J. Greenhill, and R. Mace. 2009. Matrilocal Residence is Ancestral in Austronesian Societies. *Proc. R. Soc. B* **276**:1957–1964. [6, 20]

Jordan, P. 2009. Linking Pattern to Process in Cultural Evolution: Explaining Material Culture Diversity among the Northern Khanty of Northwest Siberia. In: Pattern and Process in Cultural Evolution, ed. S. Shennan, pp. 61–84. Berkeley: Univ. of California Press. [7]

Juleff, G. 2009. Technology and Evolution: A Root and Branch View of Asian Iron from First Millennium BC Sri Lanka to Japanese Steel. *Word Archaeol.* **41**:557–577. [8]

Kahlenberg, S. M., M. E. Thompson, M. N. Muller, and R. W. Wrangham. 2008. Immigration Costs for Female Chimpanzees and Male Protection as an Immigrant Counterstrategy to Intrasexual Aggression. *Anim. Behav.* **76**:1497–1509. [5]

Kahneman, D. 2011. Thinking Fast and Slow. New York: Farrar, Straus, & Giroux. [10]

Kalia, V. C., S. Lal, and S. Cheema. 2007. Insight in to the Phylogeny of Polyhydroxyalkanoate Biosynthesis: Horizontal Gene Transfer. *Gene* **389**:19–26. [15]

Kallgren, C. A., R. R. Reno, and R. B. Cialdini. 2000. A Focus Theory of Normative Conduct: When Norms Do and Do Not Affect Behaviour. *Pers. Social Psychol. Bull.* **26**:1002–1012. [5]

Kandel, D. B. 1978. Homophily, Selection, and Socialization in Adolescent Friendships. *Am. J. Sociol.* **84**:427–436. [6]

Kaplan, H., and K. Hill. 1985a. Food Sharing among Ache Foragers: Tests of Explanatory Hypotheses. *Curr. Anthropol.* **26**:223–246. [2, 3]

Kaplan, H., and K. Hill. 1985b. Hunting Ability and Reproductive Success among Male Aché Foragers: Preliminary Results. *Curr. Anthropol.* **26**:131–133. [2]

Kaplan, H., K. Hill, J. Lancaster, and A. M. Hurtado. 2000. A Theory of Human Life History Evolution: Diet, Intelligence, and Longevity. *Evol. Anthropol.* **9**:156–185. [2, 3]

Kaplan, H. S., S. W. Gangestad, M. Gurven, et al. 2007. The Evolution of Diet, Brain and Life History among Primates and Humans. In: Guts and Brains: An Interative Approach to the Hominin Record, ed. W. Roebroeks, pp. 47–81. Leiden, NL: Leiden Univ. Press. [2]

Karlsson, F. 2007. Constraints on Multiple Center-Embedding of Clauses. *J. Linguistics* **43**:365–392. [16]

Karmiloff-Smith, A. 1994. Precis of Beyond Modularity: A Developmental Perspective on Cognitive Science (with Peer Commentary). *Behav. Brain Sci.* **17** 693–706. [14]

Katz, R. 1984. Boiling Energy: Community Healing among the Kalahari Kung. Cambridge, MA: Harvard Univ. Press. [20]

Kauffman, H. J. 2007. American Axes. Morgantown: Masthof Press. [7]

Kauffman, S. A. 1993. The Origins of Order: Self-Organization and Selection in Evolution. New York: Oxford Univ. Press. [8, 11]

Kauffman, S. A., J. Lobo, and W. G. Macready. 2000. Optimal Search on a Technology Landscape. *J. Econ. Behav. Org.* **43**:141–166. [8]

Kavanagh, T. W. 1996. The Comanches: A History 1706–1875. Lincoln: Univ. of Nebraska Press. [6]

Keech McIntosh, S., ed. 2005. Beyond Chiefdoms: Pathways to Complexity in Africa. Cambridge: Cambridge Univ. Press. [6]

Keeley, L. 1996. War before Civilization: The Myth of the Peaceful Savage. New York: Oxford Univ. Press. [3]

Keeling, P. J., and J. D. Palmer. 2008. Horizontal Gene Transfer in Eukaryotic Evolution. *Nat. Rev. Genet.* **9**:605–618. [16]

Kelemen, D. 2004. Are Children "Intuitive Theists"? Reasoning About Purpose and Design in Nature. *Psychol. Sci.* **15**:295–301. [18]

Keller, C. M., and J. D. Keller. 1996. Cognition and Tool Use: The Blacksmith at Work. Cambridge: Cambridge Univ. Press. [9]

Keller, H. 2007. Cultures of Infancy. Mahwah, NJ: Lawrence Erlbaum. [14]

Keller, R. 1994. On Language Change: The Invisible Hand in Language. London: Routledge. [13, 16]

———. 1998. A Theory of Linguistic Signs. Oxford: Oxford Univ. Press. [13]

Kelly, R. C. 1985. The Nuer Conquest: The Structure and Development of an Expansionist System. Ann Arbor: Univ. of Michigan Press. [6]

Kelly, R. L. 1995. The Foraging Spectrum: Diversity in Hunter-Gatherer Lifeways. Washington, DC: Smithsonian Institution Press. [2]

Kemp, C., and T. Regier. 2012. Kinship Categories across Language Reflect Communicative Principles. *Science* **336**:1049–1054. [15]

Kempe, M., S. J. Lycett, and A. Mesoudi. 2012. An Experimental Test of the Accumulated Copying Error Model of Cultural Mutation for Acheulean Handaxe Size. *PLoS One* **7**:e48333. [11]

Kendon, A. 1988. Sign Languages of Aboriginal Australia: Cultural, Semiotic and Communicative Perspectives. Cambridge: Cambridge Univ. Press. [13]

Kenward, B. 2012. Over-Imitating Preschoolers Believe Unnecessary Actions Are Normative and Enforce Their Performance by a Third Party. *J. Exp. Child. Psychol.* **112**:195–207. [11]

Kenward, B., A. A. S. Weir, C. Rutz, and A. Kacelnik. 2005. Tool Use by Naive Juvenile Crows. *Nature* **433**:121. [10]

Kim, E., and S. Pak. 2002. Students Do Not Overcome Conceptual Difficulties after Solving 1000 Traditional Problems. *Am. J. Physics* **79**:759–765. [10]

Kinzler, K. D., K. H. Corriveau, and P. L. Harris. 2011. Children's Selective Trust in Native-Accented Speakers. *Devel. Sci.* **14**:106–111. [1, 5]

Kinzler, K. D., E. Dupoux, and E. S. Spelke. 2007. The Native Language of Social Cognition. *PNAS* **104**:12,577–12,580. [5]

Kinzler, K. D., K. Shutts, J. Dejesus, and E. S. Spelke. 2009. Accent Trumps Race in Guiding Children's Social Preferences. *Soc. Cogn.* **27**:623–634. [5]

Kinzler, K. D., K. Shutts, and E. S. Spelke. 2012. Language-Based Social Preferences among Children in South Africa. *Lang. Learn. Devel.* **8**:215–232. [5]

Kirby, S. 2002. Learning, Bottlenecks and the Evolution of Recursive Syntax. In: Linguistic Evolution through Language Acquisition: Formal and Computational Models, ed. T. Briscoe, pp. 173–204. Cambridge: Cambridge Univ. Press. [13]

Kirby, S., H. Cornish, and K. Smith. 2008. Cumulative Cultural Evolution in the Laboratory: An Experimental Approach to the Origins of Structure in Human Language. *PNAS* **105**:10,681–10,686. [12]

Kirby, S., M. Dowman, and T. L. Griffiths. 2007. Innateness and Culture in the Evolution of Language. *PNAS* **104**:5241–5245. [12]

Kirch, P. V. 1984. The Evolution of the Polynesian Chiefdoms. Cambridge: Cambridge Univ. Press. [6]

Kirschner, S., and M. Tomasello. 2010. Joint Music Making Promotes Prosocial Behavior in 4-Year-Old Children. *Evol. Hum. Behav.* **31**:354–364. [17, 20]

Kitchen, A., C. Ehret, S. Assefa, and C. J. Mulligan. 2009. Bayesian Phylogenetic Analysis of Semitic Languages Identifies an Early Bronze Age Origin of Semitic in the near East. *Proc. R. Soc. B* **276**:2703–2710. [15]

Kitts, J. A. 2006. Collective Action, Rival Incentives, and the Emergence of Antisocial Norms. *Am. Sociol. Rev.* **71**:235–259. [6]

Klein, R. G. 2009. The Human Career: Human Biological and Cultural Origins (3rd edition). Chicago: Univ. of Chicago Press. [1, 16]

Kline, M. A., and R. Boyd. 2010. Population Size Predicts Technological Complexity in Oceania. *Proc. R. Soc. B* **277**:2559–2564. [6, 7, 16]

Knauft, B. M. 1985. Good Company and Violence: Sorcery and Social Action in a Lowland New Guinea Society. *Am. Anthropol.* **89**:173–174. [17, 20]

———. 1991. Violence and Sociality in Human Evolution. *Curr. Anthropol.* **32**:391–428. [2]

Kobayashi, Y., and J. Y. Wakano. 2012. Evolution of Social Versus Individual Learning in an Infinite Island Model. *Evolution* **66**:1624–1635. [7]

Koivulehto, J. 2001. The Earliest Contacts between Indo-European and Uralic Speakers in the Light of Lexical Loans. In: Early Contacts between Uralic and Indo-European: Linguistic and Archaeological Considerations, ed. C. Carpelan et al., vol. 242, pp. 235–263. Helsinki: Suomalais-Ugrilaisen Seuran toimituksia. [15]

König, E., and S. Kokutani. 2006. Towards a Typology of Reciprocal Constructions: Focus on German and Japanese. *Linguistics* **44**:271–302. [13]

Konvalinka, I., D. Xygalatas, J. Bulbulia, et al. 2011. Synchronized Arousal between Performers and Related Spectators in a Fire-Walking Ritual. *PNAS* **108**:8514–8519. [20]

Koonin, E. V. 2009. Darwinian Evolution in the Light of Genomics. *Nucleic Acids Res.* **37**:1011–1034. [16]

Koonin, E. V. 2012. The Logic of Chance: The Nature and Origin of Biological Evolution. Upper Saddle River, NJ: Pearson Education. [16]

Koopmans, R., and S. Rebers. 2009. Collective Action in Culturally Similar and Dissimilar Groups: An Experiment on Parochialism, Conditional Cooperation, and Their Linkages. *Evol. Hum. Behav.* **30**:201–211. [19]

Krause, J., C. Lalueza-Fox, L. Orlando, et al. 2007. The Derived FOXP2 Variant of Modern Humans Was Shared with Neandertals. *Curr. Biol.* **17**:1908–1912. [12, 16]

Kreps, D. M., and J. Sobel. 1994. Signaling. In: Handbook of Game Theory with Economic Applications, ed. R. Aumann and S. Hart, vol. 2, pp. 849–867. Amsterdam: North Holland. [3]

Kristiansen, K., and T. B. Larsson. 2005. The Rise of Bronze Age Society: Travels, Transmissions and Transformations. Cambridge: Cambridge Univ. Press. [6]

Kroeber, A. L. 1939. Cultural and Natural Areas of Native North America. Berkeley: Univ. of California Press. [7]

Krugman, P. 1991. Increasing Returns and Economic Geography. *J. Polit. Econ.* **99**:483–499. [6]

Kuhl, P. 1991. Human Adults and Human Infants Show a "Perceptual Magnet Effect" for the Prototypes of Speech Categories, Monkeys Do Not. *Percept. Psychophys.* **50**:93–107. [14]

Kuhn, T. 1970. The Structure of Scientific Revolutions (2nd edition). Chicago: Univ. of Chicago Press. [10]

Kulick, D., and B. B. Schieffelin. 2004. Language Socialization. In: A Companion to Linguistic Anthropology, ed. A. Duranti, pp. 349–368. Malden, MA: Blackwell. [14]

Kushnir, T., and A. Gopnik. 2007. Conditional Probability Versus Spatial Contiguity in Causal Learning: Preschoolers Use New Contingency Evidence to Overcome Prior Spatial Assumptions. *Devel. Psychol.* **43**:186–196. [10]

Kylafis, G., and M. Loreau. 2008. Ecological and Evolutionary Consequences of Niche Construction for Its Agent. *Ecol. Lett.* **11**:1072–1081. [11]

Labov, W. 1963. The Social Motivation of a Sound Change. *Word* **19**:1966. [1, 15]

———. 1966. The Social Stratification of English in New York City. Washington, DC: Center for Applied Linguistics. [13]

———. 1969. The Logic of Non-Standard English. Georgetown Monographs on Language and Linguistics, vol. 22. Washington, DC: Georgetown Univ. Press. [14]

———. 1972. Sociolinguistic Patterns. Philadelphia: Univ. of Pennsylvania Press. [13]

———. 1982. Building on Empirical Foundations. In: Perspectives on Historical Linguistics, ed. W. P. Lehmann and Y. Malkiel, pp. 17–92. Amsterdam: John Benjamins. [20, Appendix]

———. 1994. Principles of Linguistic Change: Internal Factors. Cambridge, MA: Blackwell. [1, 13]

———. 2001. Principles of Linguistic Change: Social Factors, vol. 2, Language in Society, P. Trudgill, series ed. Malden, MA: Blackwell. [1, 13, 16]

———. 2010. Principles of Linguistic Change, vol. 3, Cognitive and Cultural Factors. Oxford: Wiley-Blackwell. [13]

Lachmann, M., and C. Bergstrom. 1997. Signaling among Relatives II: Beyond the Tower of Babel. *Theor. Pop. Biol.* **54**:146–160. [3]

Ladd, D. R. 2011. Phonetics in Phonology. In: The Handbook of Phonological Theory (2nd edition), ed. J. Goldsmith et al. Oxford: Wiley-Blackwell. [13]

Ladd, D. R., D. Dediu, and A. R. Kinsella. 2008. Languages and Genes: Reflections on Biolinguistics and the Nature-Nurture Question. *Biolinguistics* **2**:114–126. [12, 16]

Ladefoged, P. 1984. Out of Chaos Comes Order: Physical, Biological, and Structural Patterns in Phonetics. In: Proc. 10th Intl. Congress of Phonetic Sciences, ed. A. S. Cohen and M. P. R. van den Broecke, pp. 83–95. Dordrecht: Foris. [12, 16]

Lai, C. S., S. E. Fisher, J. A. Hurst, F. Vargha-Khadem, and A. P. Monaco. 2001. A Forkhead-Domain Gene Is Mutated in a Severe Speech and Language Disorder. *Nature* **413**:519–523. [12]

Lake, M. W. 1998. Digging for Memes: The Role of Material Objects in Cultural Evolution. In: Cognition and Material Culture: The Archaeology of Symbolic Storage, ed. C. Renfrew and C. Scarre. Cambridge: McDonald Institute. [8]

Lake, M. W., and J. Venti. 2009. Quantitative Analysis of Macroevolutionary Patterning in Technological Evolution: Bicycle Design from 1800 to 2000. In: Pattern and Process in Cultural Evolution, ed. S. J. Shennan, vol. 2, pp. 147–162. Berkeley: Univ. of California Press. [8, 11]

Laland, K. N. 1992. A Theoretical Investigation of the Role of Social Transmission in Evolution. *Ethol. Sociobiol.* **13**:87–113. [20]

———. 2004. Social Learning Strategies. *Learn. Behav.* **32**:4–14. [5]

Laland, K. N., and P. Bateson. 2001. The Mechanisms of Imitation. *Cybern. Syst.* **32**:195–224. [9]

Laland, K. N., and G. R. Brown. 2002. Sense and Nonsense: Evolutionary Perspectives on Human Behavior. New York: Oxford Univ Press. [20]

Laland, K. N., F. J. Odling-Smee, and M. W. Feldman. 1996. The Evolutionary Consequences of Niche Construction: A Theoretical Investigation Using Two-Locus Theory. *J. Evol. Biol.* **9**:293–316. [11]

———. 1999. Evolutionary Consequences of Niche Construction and Their Implications for Ecology. *PNAS* **96**:10,242–10,247. [11]

———. 2000. Niche Construction, Biological Evolution, and Cultural Change. *Behav. Brain Sci.* **23**:131–146. [6]

Laland, K. N., F. J. Odling-Smee, and S. Myles. 2010. How Culture Shaped the Human Genome: Bringing Genetics and the Human Sciences Together. *Nat. Rev. Genet.* **11**:137–148. [12, 16, 20]

Laland, K. N., K. Sterelny, F. J. Odling-Smee, W. Hoppitt, and T. Uller. 2011. Cause and Effect in Biology Revisited Is Mayr's Proximate-Ultimate Dichotomy Still Useful? *Science* **334**:1512–1516. [1, 6, 11]

Lambert, A. 2011. The Gates of Hell: Sir John Franklin's Tragic Quest for the Northwest Passage. New Haven: Yale Univ. Press. [7]

Lambert, P. 2002. The Archaeology of War: A North American Perspective. *J. Archaeol. Res.* **10**:207–241. [3]

Langergraber, K. E., J. C. Mitani, and L. Vigilant. 2007. The Limited Impact of Kinship on Cooperation in Wild Chimpanzees. *PNAS* **104**:7786–7790. [3]

Lashley, K. 1951. The Problem of Serial Order in Behavior. In: Cerebral Mechanisms in Behavior, ed. L. A. Jeffress, pp. 112–136. New York: John Wiley. [9]

Latour, B. 1993. We Have Never Been Modern (Translated by C. Porter). Cambridge, MA: Harvard Univ. Press. [10]

Laurin, K., A. F. Shariff, J. Henrich, and A. C. Kay. 2012. Outsourcing Punishment to God: Beliefs in Divine Control Reduce Earthly Punishment. *Proc. R. Soc. B* **279**:3272–3281. [1, 19]

Lave, J., and E. Wenger. 1991. Situated Learning: Legitimate Peripheral Participation. Cambridge: Cambridge Univ. Press. [9]

Lawson, E. T., and R. N. McCauley. 1990. Rethinking Religion: Connecting Cognition and Culture. Cambridge: Cambridge Univ. Press. [19, 20]

Lee, R. B. 1979. The !Kung San: Men, Women and Work in a Foraging Society. New York: Cambridge Univ. Press. [17]

Lee, R. B., and I. DeVore, eds. 1968. Man the Hunter. Chicago: Aldine. [2, 6]

Lee, S., and T. Hasegawa. 2011. Bayesian Phylogenetic Analysis Supports an Agricultural Origin of Japonic Languages. *Proc. R. Soc. B* **278**:3662–3669. [15]

Leeson, P. T. 2005. Endogenizing Fractionalization. *J. Inst. Econ.*75–98. [6]

Legare, C. H. 2012. Exploring Explanation: Explaining Inconsistent Evidence Informs Exploratory, Hypothesis-Testing Behavior in Young Children. *Child. Devel.* **83**:173–185. [10]

Legare, C. H., S. A. Gelman, and H. M. Wellman. 2010. Inconsistency with Prior Knowledge Triggers Children's Causal Explanatory Reasoning. *Child. Devel.* **81**:929–944. [10]

Legare, C. H., and H. Whitehouse. 2011. The Cognitive Underpinnings of Ritual. In: Biennial Meeting of the Society for Research in Child Development. Montreal. [20]

Lehmann, L. 2007. The Evolution of Trans-Generational Altruism: Kin Selection Meets Niche Construction. *J. Evol. Biol.* **20**:181–189. [11]

———. 2008. The Adaptive Dynamics of Niche Constructing Traits in Spatially Subdivided Populations: Evolving Posthumous Extended Phenotypes. *Evolution* **62**:549–566. [11]

Lehmann, L., and M. Feldman. 2008. War and the Evolution of Belligerence and Bravery. *Proc. R. Soc. B* **275**:2877–2885. [4]

Lehmann, L., M. W. Feldman, and R. Kaeuffer. 2010. Cumulative Cultural Dynamics and the Coevolution of Cultural Innovation and Transmission: An ESS Model for Panmictic and Structured Populations. *J. Evol. Biol.* **23**:2356–2369. [7]

Leijonhufvud, A. 1997. Models and Theories. *J. Econ. Method.* **4**:193–198. [1]

Lemey, P., A. Rambaut, J. J. Welch, and M. A. Suchard. 2010. Phylogeography Takes a Relaxed Random Walk in Continuous Space and Time. *Mol. Biol. Evol.* **27**:1877–1885. [15]

Lemey, P., M. A. Suchard, and A. Rambaut. 2009. Reconstructing the Initial Global Spread of a Human Influenza Pandemic: A Bayesian Spatial-Temporal Model for the Global Spread of H1N1pdm. *PLoS Curr.* RRN1031. [15]

Lenski, G. E., and J. Lenski. 1982. Human Societies: An Introduction to Macrosociology (4th edition). New York: McGraw-Hill. [1]

Leonardi, M., P. Gerbault, M. G. Thomas, and J. Burger. 2012. The Evolution of Lactase Persistence in Europe: A Synthesis of Archaeological and Genetic Evidence. *Intl. Dairy J.* **22**:88–97. [1]

Leonti, M. 2011. The Future Is Written: Impact of Scripts on the Cognition, Selection, Knowledge and Transmission of Medicinal Plant Use and Its Implications for Ethnobotany and Ethnopharmacology. *J. Ethnopharmacol.* **134**:542–555. [11]

Levins, R. 1966. The Strategy of Model Building in Population Biology. *Am. Sci.* **54**:421–431. [11]

Levinson, S. C. 2006. On the Human Interactional Engine. Roots of Human Sociality. In: Culture, Cognition and Human Interaction, ed. N. J. Enfield and S. C. Levinson, pp. 39–69. Oxford: Berg. [12]

———. 2012a. The Original Sin of Cognitive Science. *Top. Cogn. Sci.* **4**:396–403. [12]

———. 2012b. Psychology. Kinship and Human Thought. *Science* **336**:988–989. [15]

Levinson, S. C., and N. Evans. 2010. Time for a Sea-Change in Linguistics. Response to Comments on the "Myth of Language Universals." *Lingua* **120**:2733–2758. [13, 16]

Levinson, S. C., and R. D. Gray. 2012. Tools from Evolutionary Biology Shed New Light on the Diversification of Languages. *Trends Cogn. Sci.* **16**:167–173. [15, 16]

Levinson, S. C., S. J. Greenhill, R. D. Gray, and M. Dunn. 2011. Universal Typological Dependencies Should Be Detectable in the History of Language Families. *Ling. Typol.* **15**:509–534. [15]

Lewis, H. M., and K. N. Laland. 2012. Transmission Fidelity Is the Key to the Build-up of Cumulative Culture. *Phil. Trans. R. Soc. B* **367**:2171–2180. [8, 11]

Lewis, J. W. 2006. Cortical Networks Related to Human Use of Tools. *Neuroscientist* **12**:211–231. [9]

Lewis, P. M. 2009. Ethnologue: Languages of the World (16th edition). Dallas: SIL Intl. [15]

Liang, Y., A. Wang, F. J. Probst, et al. 1998. Genetic Mapping Refines DFNB3 to 17p11.2, Suggests Multiple Alleles of DFNB3, and Supports Homology to the Mouse Model Shaker-2. *Am. J. Hum. Genet.* **62**:904–915. [12]

Liebenberg, L. 2006. Persistence Hunting by Modern Hunter-Gatherers. *Curr. Anthropol.* **47**:1017–1026. [2]

Lieberman, P. 2007. The Evolution of Human Speech: Its Anatomical and Neural Bases. *Curr. Anthropol.* **48**:39–66. [16]

Lieberman, V. 2003. Strange Parallels: Southeast Asia in Global Context, c. 800–1830, vol. 1: Integration on the Mainland. Cambridge: Cambridge Univ. Press. [4]

———. 2010. Strange Parallels: Southeast Asia in Global Context, c. 800–1830, vol. 2: Mainland Mirrors, Europe, China, South Asia, and the Islands. Cambridge: Cambridge Univ. Press. [4]

Lieven, E. 1994. Crosslinguistic and Crosscultural Aspects of Language Addressed to Children. In: Input and Interaction in Language Acquisition, ed. C. Gallaway and B. J. Richards, pp. 56–73. Cambridge: Cambridge Univ. Press. [14]

———. 2010. Input and First Language Acquisition: Evaluating the Role of Frequency. *Lingua* **120**:2546–2556. [14]

Lieven, E., and S. Stoll. 2013. Early Communicative Development in Two Cultures. *Hum. Devel.*, in press. [14]

Lightfoot, D. 1999. The Development of Language: Acquisition, Change, and Evolution. Chichester: Wiley-Blackwell. [16]

Lipo, C. P., M. J. O'Brien, M. Collard, and S. Shennan, eds. 2006. Mapping Our Ancestors: Phylogenetic Approaches in Anthropology and Prehistory. New York: Aldine. [6, 11]

Liszkowski, U., P. Brown, T. Callaghan, A. Takada, and C. de Vos. 2012. A Prelinguistic Gestural Universal of Human Communication. Cognitive. *Science* **36**:698–713. [14]

Liu, X., and D. MacIsaac. 2005. An Investigation of Factors Affecting the Degree of Naïve Impetus Theory Application. *J. Sci. Ed. Technol.* **14**:101–116. [10]

Lockman, J. 2006. Tool Use from a Perception-Action Perspective: Developmental and Evolutionary Considerations. In: Stone Knapping: The Necessary Conditions for a Uniquely Hominin Behavior, ed. V. Roux and B. Bril, pp. 319–330. Cambridge: McDonald Institute. [9]

Logan, M. H., and D. A. Schmittou. 1998. The Uniqueness of Crow Art: A Glimpse into the History of an Embattled People. *Montana* **48**:58–71. [5]

Lohmann, H., and M. Tomasello. 2003. The Role of Language in the Development of False Belief Understanding: A Training Study. *Child. Devel.* **74**:1130–1144. [14]

Lombard, M. 2012. Thinking through the Middle Stone Age of Sub-Saharan Africa. *Quat. Intl.* **270**:140–155. [8]

Lombard, M., and M. N. Haidle. 2012. Thinking a Bow-and-Arrow Set: Cognitive Implications of Middle Stone Age Bow and Stone-Tipped Arrow Technology. *Cambridge Arch. J.* **22**:237–264. [8]

Lombard, M., and L. Phillipson. 2010. Indications of Bow and Stone-Tipped Arrow Use 64,000 Years Ago in Kwazulu-Natal, South Africa. *Antiquity* **84**:635–648. [8]

Lorenz, K. 1963. On Aggression. New York: Harcourt, Brace and World. [2]

Loreto, V., A. Mukherjee, and A. Tria. 2012. On the Origin of the Hierarchy of Color Names. *PNAS* **109**: 6819–6824. [13]

Lowie, R. H. 1948. Primitive Religion. New York: Liveright Publ. Corp. [20]

Lucy, J., and S. Gaskins. 2001. Grammatical Categories and the Development of Classification Preferences: A Comparative Approach. In: Language Acquisition and Conceptual Development, ed. M. Bowerman and S. C. Levinson, pp. 257–283. Cambridge: Cambridge Univ. Press. [14]

Luhrmann, T. 2012. When God Talks Back: Understanding the American Evangelical Relationship with God. New York: Knopf. [20]

Lumsden, C. J., and E. O. Wilson. 1981. Genes, Mind, and Culture: The Coevolutionary Process. Cambridge, MA: Harvard Univ. Press. [1, 2]

Lupo, K. D., and J. F. O'Connell. 2002. Cut and Tooth Mark Distributions on Large Animal Bones: Ethnoarchaeological Data from the Hadza and Their Implications for Current Ideas About Early Human Carnivory. *J. Archaeol. Sci.* **29**:85–109. [2]

Lupyan, G., and R. Dale. 2010. Language Structure Is Partly Determined by Social Structure. *PLoS One* **5**:e8559. [13, 16, Appendix]

Luttwak, E. 1976. The Grand Strategy of the Roman Empire: From the First Century A.D. To the Third. London: Weidenweld & Nicholson. [6]

Lyle, H. F., III, and E. A. Smith. 2012. How Conservative Are Evolutionary Anthropologists? *Hum. Nature* **23**:306–322. [1]

Lyman, R. L., T. VanPool, and M. J. O'Brien. 2009. The Diversity of North American Projectile-Point Types, before and after the Bow and Arrow. *J. Anthropol. Arch.* **28**:1–13. [8]

Lyons, D. E. 2009. The Rational Continuum of Human Imitation. In: Handbook of Environmental Engineering: Mirror Neuron Systems, ed. J. A. Pineda, pp. 77–103. New York: Humana Press. [5]

Lyons, D. E., D. H. Damrosch, J. K. Lin, D. M. Macris, and F. C. Keil. 2011. The Scope and Limits of Overimitation in the Transmission of Artifact Culture. *Phil. Trans. R. Soc. B* **366**:1158–1167. [5]

Lyons, D. E., A. G. Young, and F. C. Keil. 2007. The Hidden Structure of Overimitation. *PNAS* **104**:19,751–19,756. [5, 6, 11, 14, 20]

Macdonald, D. W., and C. Sillero-Zubiri. 2004. Biology and Conservation of Wild Canids. New York: Oxford Univ. Press. [2]

Mace, R., and C. J. Holden. 2005. A Phylogenetic Approach to Cultural Evolution. *Trends Ecol. Evol.* **20**:116–121. [4]

Mace, R., and M. Pagel. 1994. The Comparative Method in Anthropology. *Curr. Anthropol.* **35**:549–564. [20]

Mackay, A., and B. Marwick. 2011. Costs and Benefits in Technological Decision Making under Variable Conditions: Examples from the Late Pleistocene in Southern Africa. In: Keeping Your Edge: Recent Approaches to the Organisation of Stone Artefact Technology, ed. B. Marwick and A. Mackay, pp. 119–134. Oxford: Oxbow Books. [8]

MacLarnon, A. M., and G. P. Hewitt. 1999. The Evolution of Human Speech: The Role of Enhanced Breathing Control. *Am. J. Phys. Anthropol.* **109**:341–363. [16]

MacLaurin, J., and K. Sterelny. 2008. What Is Biodiversity? Chicago: Univ. of Chicago Press. [15]

MacNeilage, P. 2008. The Origin of Speech. Oxford: Oxford Univ. Press. [16]

Maddison, W. P., P. E. Midford, and S. P. Otto. 2007. Estimating a Binary Character's Effect on Speciation and Extinction. *Syst. Biol.* **56**:701–710. [15]

Maestripieri, D. 2007. Machiavellian Intelligence: How Rhesus Macaques and Humans Have Conquered the World. Chicago: Univ. of Chicago Press. [2]

Majolo, B., J. Lehmann, A. Bortoli de Vizioli, and G. Schino. 2012. Fitness-Related Benefits of Dominance in Primates. *Am. J. Phys. Anthropol.* **147**:652–660. [2]

Malaurie, J. 1985. The Last Kings of Thule: With the Polar Eskimos, as They Face Destiny. Chicago: Univ. of Chicago Press. [7]

Malkin, I. 2011. A Small Greek World: Networks in the Ancient Mediterranean. New York: Oxford Univ. Press. [6]

Malotki, E., and K. Gary. 2001. Hopi Stories of Witchcraft, Shamanism, and Magic. Lincoln: Univ. of Nebraska Press. [20]

Malvern, D., B. J. Richards, N. Chipere, and P. Durán. 2009. Lexical Diversity and Language Development: Quantification and Assessment. Basingstoke: Palgrave Macmillan. [14]

Mandler, J. M. 2000. Perceptual and Conceptual Processes in Infancy. *J. Cogn. Develop.* **1**:3–36. [14]

Mann, M. 1986. The Sources of Social Power. I. A History of Power from the Beginning to A.D. 1760. Cambridge: Cambridge Univ. Press. [4]

Maravita, A., and A. Iriki. 2004. Tools for the Body (Schema). *Trends Cogn. Sci.* **8**:79–86. [9]

Marchman, V. A., and A. Fernald. 2008. Speed of Word Recognition and Vocabulary Knowledge in Infancy Predict Cognitive and Language Outcomes in Later Childhood. *Devel. Sci.* **11**:F9–F16. [14]

Marcus, G. F., S. Vijayan, S. Bandi Rao, and P. M. Vishton. 1999. Rule Learning by Seven-Month-Old Infants. *Science* **283**:77–80. [14]

Marcus, J., and K. V. Flannery. 2004. The Coevolution of Ritual and Society: New C-14 Dates from Ancient Mexico. *PNAS* **101**:18,257–18,261. [17, 19]

Marean, C. W., M. Bar-Matthews, J. Bernatchez, et al. 2007. Early Human Use of Marine Resources and Pigment in South Africa in the Middle Pleistocene. *Nature* **449**:905–908. [4]

Marlowe, F. W. 2004. What Explains Hadza Food Sharing? *Res. Econ. Anthropol.* **23**:69–88. [3]

Marshall, L. 1962. !Kung Bushman Religious Beliefs. *J. Intl. African Inst.* **32**:221–252. [17, 20]

Martin, G. 2000. Stasis in Complex Artefacts. In: Technological Innovation as an Evolutionary Process, ed. J. Ziman, pp. 90–100. Cambridge: Cambridge Univ. Press. [8, 9, 11]

Martin, T. R. 2009. Herodotus and Sima Qian: The First Great Historians of Greece and China. Boston: Bedford/St. Martin's. [1]

Martínez, I., J. L. Arsuaga, R. Quam, et al. 2008a. Human Hyoid Bones from the Middle Pleistocene Site of the Sima De Los Huesos (Sierra De Atapuerca, Spain). *J. Hum. Evol.* **54**:118–124. [2]

Martínez, I., R. M. Quam, M. Rosa, et al. 2008b. Auditory Capacities of Human Fossils: A New Approach to the Origin of Speech. *J. Acoust. Soc. Am.* **123**:3606–3606. [16]

Martinez, I., M. Rosa, J.-L. Arsuaga, et al. 2004. Auditory Capacities in Middle Pleistocene Humans from the Sierra De Atapuerca in Spain. *PNAS* **101**:9997–9981. [2, 16]

Marwell, G., and P. Oliver. 1993. The Critical Mass in Collective Action: A Micro-Social Theory. Cambridge: Cambridge Univ. Press. [6]

Mary-Rousselière, G. 1996. Qitdlarssuaq: The Story of a Polar Migration. Winnepeg: Wuerz Publishing Ltd. [7]

Masters, R. D. 1998. On the Evolution of Political Communities: The Paradox of Eastern and Western Europe in the 1980s. In: Ethnic Conflict and Indoctrination, ed. I. Eibl-Eibesfeldt and F. K. Salter. Oxford: Berghahn Books. [4]

Mathew, S., and R. Boyd. 2011. Punishment Sustains Large-Scale Cooperation in Prestate Warfare. *PNAS* **108**:11,375–11,380. [1, 3, 6]

Matthews, D., T. Behne, E. Lieven, and M. Tomasello. 2012a. What Are the Origins of the Human Pointing Gesture? A Training Study with 9- to 11-Month Olds. *Devel. Sci.* **15**:817–829. [14]

Matthews, L. J. 2012. The Recognition Signal Hypothesis for the Adaptive Evolution of Religion. *Hum. Nature* **23**:218–249. [20]

Matthews, L. J., J. Edmonds, W. J. Wildman, and C. L. Nunn. 2012b. Cultural Inheritance or Cultural Diffusion of Religious Violence? A Quantitative Case Study of the Radical Reformation. *Relig. Brain Behav.* **3**:3–15. [20]

Mauss, M. 1973. Techniques of the Body. *Econ. Soc.* **2**:70–88. [12]

Mavisakalyan, A. 2011. Gender in Language and Gender in Employment. *ANU Working Papers Econ. Economet.* No. 563, http://ideas.repec.org/p/acb/cbeeco/2011-563.html. (accessed May 12, 2013). [13]

May, K. O. 1966. Quantitative Growth of the Mathematical Literature. *Science* **154**:1672. [11]

Maynard Smith, J., and E. Szathmáry. 1995. The Major Transitions in Evolution. New York: Wiley. [4, 6]

Mayr, E. 1970. Populations, Species, and Evolution. Cambridge, MA: Belknap Press. [6]

———. 1994. Recapitulation Reinterpreted: The Somatic Program. *Q. Rev. Biol.* **69**:223–232. [9]

Mazar, N., O. Amir, and D. Ariely. 2008. The Dishonesty of Honest People: A Theory of Self-Concept Maintenance. *J. Mark. Res.* **45**:633–644. [20]

McCauley, R. N. 2001. Explanatory Pluralism and the Coevolution of Theories in Science. In: Philosophy and the Neurosciences, ed. W. Bechtel et al., pp. 431–456. Oxford: Blackwell. [11]

———. 2011. Why Religion Is Natural and Science Is Not. New York: Oxford Univ. Press. [10, 17, 18, 20]

McCloskey, M. 1983. Intuitive Physics. *Sci. Am.* **248**:122–130. [10]

McDaniel, L. D., E. Young, J. Delaney, et al. 2010. High Frequency of Horizontal Gene Transfer in the Oceans. *Science* **330**:59. [16]

McElreath, R., A. V. Bell, C. Efferson, et al. 2008. Beyond Existence and Aiming Outside the Laboratory: Estimating Frequency-Dependent and Pay-Off-Biased Social Learning Strategies. *Phil. Trans. R. Soc. B* **363**:3515–3528. [1, 7]

McElreath, R., R. Boyd, and P. J. Richerson. 2003. Shared Norms and the Evolution of Ethnic Markers. *Curr. Anthropol.* **44**:122–130. [5, 13]

McGrew, W. C. 2004. The Cultured Chimpanzee: Reflections on Cultural Primatology. Cambridge: Cambridge Univ. Press. [2]

McGuigan, N., and A. Whiten. 2009. Emulation and "Overemulation" in the Social Learning of Causally Opaque Versus Causally Transparent Tool Use by 23- and 30-Month-Olds. *J. Exp. Child. Psychol.* **104**:367–381. [5]

McGuigan, N., A. Whiten, E. Flynn, and V. Horner. 2007. Imitation of Causally Opaque Versus Causally Transparent Tool Use by 3-and 5-Year-Old Children. *Cogn. Devel.* **22**:353–364. [5, 11]

McKay, R., C. Efferson, H. Whitehouse, and E. Fehr. 2011. Wrath of God: Religious Primes and Punishment. *Proc. R. Soc. B* **278**:1858–1863. [19, 20]

McKay, R., R. Langdon, and M. Coltheart. 2007. Models of Misbelief: Integrating Motivational and Deficit Theories of Delusions. *Conscious. Cogn.* **16**:932–941. [20]

McNeill, W. H. 1995. Keeping Together in Time: Dance and Drill in Human History. Cambridge, MA: Harvard Univ. Press. [17, 20]

McPherson, M., L. Smith-Lovin, and J. M. Cook. 2001. Birds of a Feather: Homophily in Social Networks. *Annu. Rev. Sociol.* **27**:415–444. [6]

McWhorter, J. 2011. Linguistic Simplicity and Complexity: Why Do Languages Undress? Berlin: Mouton de Gruyter. [16]

Mead, M. 1963. Sex and Temperament in Three Primitive Societies. New York: Morrow. [2]

Meillet, A. 1906. Comme les Mots Changent de Sens. *L'année Sociol.* **10**:1–38. [13]

Meir, I., W. Sandler, C. Padden, and M. Aronoff. 2010. Emerging Sign Languages. In: Oxford Handbook of Deaf Studies, Language, and Education, ed. M. Marschark and P. Spencer, pp. 267–280. Oxford: Oxford Univ. Press. [12]

Meltzoff, A. N. 1990. Foundations for Developing a Concept of Self: The Role of Imitation in Relating Self to Other and the Value of Social Mirroring, Social Modeling, and Self Practice in Infancy. In: The Self in Transition: Infancy to Childhood, ed. D. Cicchetti and M. Beeghly, pp. 139–164. Chicago: Univ. of Chicago Press. [5]

Menz, M., A. McNamara, J. Klemen, and F. Binkofski. 2009. Dissociating Networks of Imitation. *Hum. Brain Map.* **30**:3339–3350. [9]

Menzel, R., and J. Fischer, eds. 2011. Animal Thinking: Contemporary Issues in Comparative Cognition. Strüngmann Forum Reports, vol. 8, J. Lupp, series ed. Cambridge, MA: MIT Press. [1]

Merker, B. 2009. Returning Language to Culture by Way of Biology. *Behav. Brain Sci.* **32**:460–461. [13]

Mesoudi, A. 2008. Foresight in Cultural Evolution. *Biol. Philos.* **23**:243–255. [11]

———. 2009. How Cultural Evolutionary Theory Can Inform Social Psychology and Vice Versa. *Psychol. Rev.* **116**:929–952. [11]

———. 2011a. Cultural Evolution: How Darwinian Theory Can Explain Human Culture and Synthesize the Social Sciences. Chicago: Univ. of Chicago Press. [6, 11, 20]

———. 2011b. An Experimental Comparison of Human Social Learning Strategies: Payoff-Biased Social Learning Is Adaptive but Underused. *Evol. Hum. Behav.* **32**:334–342. [1, 7, 20]

———. 2011c. Variable Cultural Acquisition Costs Constrain Cumulative Cultural Evolution. *PLoS One* **6**:e18239. [7, 11]

Mesoudi, A., and S. J. Lycett. 2009. Random Copying, Frequency-Dependent Copying and Culture Change. *Evol. Hum. Behav.* **30**:41–48. [11]

Mesoudi, A., and M. J. O'Brien. 2008a. The Cultural Transmission of Great Basin Projectile Point Technology I: An Experimental Simulation. *Am. Antiq.* **73**:3–28. [11, 20]

———. 2008b. The Cultural Transmission of Great Basin Projectile Point Technology II: An Agent-Based Computer Simulation. *Am. Antiq.* **73**:627–644. [11]

———. 2008c. The Learning and Transmission of Hierarchical Cultural Recipies. *Biol. Theory* **3**:63–72. [9]

Mesoudi, A., and A. Whiten. 2004. The Hierarchical Transformation of Event Knowledge in Human Cultural Transmission. *J. Cogn. Culture* **4**:1–24. [9]

Mesoudi, A., A. Whiten, and K. N. Laland. 2006. Towards a Unified Science of Cultural Evolution. *Behav. Brain Sci.* **29**:329–347. [9]

Mézard, M., G. Parisi, and M. Virasoro. 1987. Spin Glass Theory and Beyond. World Scientific Lecture Notes in Physics. New York: World Scientific. [16]

Michels, R. 1915. Political Parties: A Sociological Study of the Oligarchical Tendencies of Modern Democracy. New York: Hearst's International Library. [4]

Miller, G. 2001. The Mating Mind: How Sexual Choice Shaped the Evolution of Human Nature. New York: Anchor. [2]

Miller, J. E., and S. Weinert. 1998. Spontaneous Spoken Language: Syntax and Discourse. Oxford: Clarendon Press. [14]

Milner, A. D., and M. A. Goodale. 2008. Two Visual Systems Re-Viewed. *Neuropsychologia* **46**:774–785. [9]

Milroy, J., and L. Milroy. 1985. Linguistic Change, Social Network and Speaker Innovation. *J. Linguistics* **21**:339–384. [16]

Milton, K. 1984. Habitat, Diet, and Activity Patterns of Free-Ranging Woolly Spider Monkeys. *Intl. J. Primatol.* **5**:491–514. [2]

Mokyr, J. 2000. Evolutionary Phenomena in Technological Change. In: Technological Innovation as an Evolutionary Process, ed. J. Ziman, pp. 52–65. Cambridge: Cambridge Univ. Press. [8]

Molenberghs, P., R. Cunnington, and J. B. Mattingley. 2009. Is the Mirror Neuron System Involved in Imitation? A Short Review and Meta-Analysis. *Neurosci. Biobehav. Rev.* **33**:975–980. [9]

———. 2012. Brain Regions with Mirror Properties: A Meta-Analysis of 125 Human fMRI Studies. *Neurosci. Biobehav. Rev.* **36**:341–349. [1]

Monaghan, P., M. H. Christiansen, and N. Chater. 2007. The Phonological-Distributional Coherence Hypothesis: Cross-Linguistic Evidence in Language Acquisition. *Cogn. Psychol.* **55**:259–305. [13]

Monaghan, P., M. H. Christiansen, and S. A. Fitneva. 2011. The Arbitrariness of the Sign: Learning Advantages from the Structure of the Vocabulary. *J. Exp. Psychol. Gen.* **140**:325–347. [13]

Monsma, S. V. 2007. Religion and Philanthropic Giving and Volunteering: Building Blocks for Civic Responsibility. *Interdis. J. Res. Relig.* **3**:1–27. [19]

Morris, D. 1999/1967. The Naked Ape: A Zoologist's Study of the Human Animal. New York: Delta. [2]

Mosca, G. 1939. The Ruling Class (*Elementi Di Scienza Politica*). New York: McGraw-Hill. [4]

Mott, L. V. 1997. The Development of the Rudder: A Technological Tale. College Station: Texas A&M Univ. Press. [7]

Mukherjee, A., F. Tria, A. Baronchelli, A. Puglisi, and V. Loreto. 2011. Aging in Language Dynamics. *PLoS One* **6**:e16677. [16]

Mukherjee, S., N. Sarkar-Roy, D. K. Wagener, and P. P. Majumder. 2009. Signatures of Natural Selection Are Not Uniform across Genes of Innate Immune System, but Purifying Selection Is the Dominant Signature. *PNAS* **106**:7073–7078. [12]

Mulcahy, N., and J. Call. 2006. How Great Apes Perform on a Modified Trap-Tube Task. *Anim. Cogn.* **9**:193–199. [9]

Müller, M. 1862/2010. Lectures on the Science of Language. New York: Scribner/Project Gutenberg. http://www.gutenberg.org/ebooks/32856 (accessed April 27, 2013). [1]

Murdock, G. P. 1966. Cross-Cultural Sampling. *Ethnology* **5**:97–114. [20]

———. 1967. Ethnographic Atlas. Pittsburgh: Univ. of Pittsburgh Press. [4]

Murdock, G. P., and C. Provost. 1973. Factors in the Division of Labor by Sex: A Cross-Cultural Analysis. *Ethnology* **12**:203–225. [6]

Murdock, G. P., and D. R. White. 1969. Standard Cross-Cultural Sample. *Ethnology* **8**:329–369. [4, 6]

Mynatt, C. R., M. E. Doherty, and R. Tweney. 1981. A Simulated Research Environment. In: On Scientific Thinking, ed. R. Tweney et al., pp. 145–157. New York: Columbia Univ. Press. [10]

Nadel, J. 2002. Imitation and Imitation Recognition: Functional Use in Preverbal Infants and Nonverbal Children with Autism. In: The Imitative Mind: Development, Evolution, and Brain Bases, ed. A. N. Meltzoff and W. Prinz, pp. 42–62. Cambridge: Cambridge Univ. Press. [5]

Nagell, K., R. S. Olguin, and M. Tomasello. 1993. Processes of Social Learning in the Tool Use of Chimpanzees (*Pan troglodytes*) and Human Children (*Homo sapiens*). *J. Comp. Psychol.* **107**:174–186. [5]

Nance, W. E., and M. J. Kearsey. 2004. Relevance of Connexin Deafness (DFNB1) to Human Evolution. *Am. J. Hum. Genet.* **74**:1081–1087. [12]

Narasimhan, B. 2007. Cutting and Breaking Verbs in Hindi and Tamil. *Cogn. Ling.* **18**:195–205. [14]

Narasimhan, B., and P. Brown. 2009. Getting the Inside Story: Learning to Talk About Containment in Tzeltal and Hindi. In: Routes to Language: Studies in Honour of Melissa Bowerman, ed. V. C. Mueller Gathercole, pp. 97–132. Mahwah, NJ: Lawrence Erlbaum. [14]

Naroll, R. 1956. A Preliminary Index of Social Development. *Am. Anthropol.* **58**:687–715. [6]

Navarrete, A., and C. P. van Schaik. 2011. Energetics and the Evolution of Human Brain Size. *Nature* **480**:91–93. [2]

Nee, S. 2006. Birth-Death Models in Macroevolution. *Annu. Rev. Ecol. Evol. System.* **37**:1–17. [15]

Neiman, F. D. 1995. Stylistic Variation in Evolutionary Perspective: Inferences from Decorative Diversity and Interassemblage Distance in Illinois Woodland Ceramic Assemblages. *Am. Antiq.* **60**:7–36. [7, 11]

Nelissen, K., G. Luppino, W. Vanduffel, G. Rizzolatti, and G. A. Orban. 2005. Observing Others: Multiple Action Representation in the Frontal Lobe. *Science* **310**:332–336. [9]

Nelson, R. R., and S. G. Winter. 1982. An Evolutionary Theory of Economic Change. Cambridge, MA: Belknap Press. [1]

Nelson-Sathi, S., J.-M. List, H. Geisler, et al. 2011. Networks Uncover Hidden Lexical Borrowing in Indo-European Language Evolution. *Proc. R. Soc. B* **278**:1794–1803. [15, 16]

Nettle, D. 1999. Linguistic Diversity. Oxford: Oxford Univ. Press. [13, 15, 16]

Nevalainen, T., H. Raumolin-Brunberg, and H. Mannila. 2011. The Diffusion of Language Change in Real Time: Progressive and Conservative Individuals and the Time Depth of Change. *Lang. Var. Change* **23**:1–43. [16]

Newberg, A. B., and M. R. Waldman. 2009. How God Changes Your Brain□: Breakthrough Findings from a Leading Neuroscientist. New York: Ballantine. [18]

Newport, E. L., M. D. Hauser, S. G., and R. N. Aslin. 2004. Learning at a Distance: II. Statistical Learning of Non-Adjacent Dependencies in a Non-Human Primate. *Cogn. Psychol.* **49**:85–117. [14]

Nichols, J. 1992. Linguistic Diversity in Space and Time. Chicago: Chicago Univ. Press. [13]

Nichols, J. 1997. Modeling Ancient Population Structures and Movement in Linguistics. *Annu. Rev. Anthropol.* **26**:359–384. [15]

Nielsen, M. 2006. Copying Actions and Copying Outcomes: Social Learning through the Second Year. *Devel. Psychol.* **42**:555–565. [5]

———. 2009. The Imitative Behaviour of Children and Chimpanzees: A Window on the Transmission of Cultural Traditions. *Rev. Primatol.* http://primatologie.revues.org/254 (accessed June 11, 2016). [5]

Nielsen, M., and C. Blank. 2011. Imitation in Young Children: When Who Gets Copied Is More Important Than What Gets Copied. *Devel. Psychol.* **47**:1050–1053. [5]

Nielsen, M., G. Simcock, and L. Jenkins. 2008. The Effect of Social Engagement on 24-Month-Olds' Imitation from Live and Televised Models. *Devel. Sci.* **11**:722–731. [5]

Niklas, K. J. 1994. Morphological Evolution through Complex Domains of Fitness. *PNAS* **91**:6772–6779. [13]

———. 2004. Computer Models of Early Land Plant Evolution. *Annu. Rev. Earth Planet Sci.* **32**:47–66. [13]

Nishida, T., and K. Hosaka. 1996. Coalition Strategies among Adult Male Chimpanzees of the Mahale Mountains, Tanzania. In: Great Ape Societies, ed. W. C. McGrew et al., pp. 114–134. Cambridge: Cambridge Univ. Press. [2]

Nitecki, M. H. 1988. Evolutionary Progress. Chicago: Univ. of Chicago Press. [1]

Nitecki, M. H., and D. V. Nitecki, eds. 1992. History and Evolution. Albany: SUNY Press. [1]

Nolan, P., and G. Lenski. 2004. Human Societies: An Introduction to Macrosociology (9th edition). Boulder: Paradigm Publishers. [20]

Norenzayan, A. 2013. Big Gods: How Religion Transformed Cooperation and Conflict. Princeton: Princeton Univ. Press. [19, 20, Appendix]

Norenzayan, A., and W. M. Gervais. 2012. The Cultural Evolution of Religion. In: Creating Consilience: Integrating Science and the Humanities, ed. E. Slingerland and M. Collard, pp. 243–265. Oxford: Oxford Univ. Press. [19]

———. 2013a. The Origins of Religious Disbelief. *Trends Cogn. Sci.* **17**:20–25. [19]

———. 2013b. Secular Rule of Law Erodes Believers' Political Intolerance of Atheists. *Relig. Brain Behav.*, in press. [19]

Norenzayan, A., W. M. Gervais, and K. H. Trzesniekwski. 2012. Mentalizing Deficits Constrain Belief in a Personal God. *PLoS One* **7**:e36880. [17, 19]

Norenzayan, A., and A. F. Shariff. 2008. The Origin and Evolution of Religious Prosociality. *Science* **322**:58–62. [17, 19, 20]

Norenzayan, A., A. F. Shariff, and W. M. Gervais. 2010. The Evolution of Religious Misbelief. *Behav. Brain Sci.* **32**:531–532. [19]

Norris, P., and R. Inglehart. 2004. Sacred and Secular: Religion and Politics Worldwide. Cambridge: Cambridge Univ. Press. [19]

Novembre, J., T. Johnson, K. Bryc, et al. 2008. Genes Mirror Geography within Europe. *Nature* **456**:274. [12, 16]

Nowak, M. A., N. L. Komarova, and P. Niyogi. 2001. Evolution of Universal Grammar. *Science* **291**:114–118. [16]

Nowak, M. A., and K. Sigmund. 1993. A Strategy of Win-Stay, Lose-Shift That Outperforms Tit-for-Tat in Prisoner's Dilemma. *Nature* **364**:56–58. [3]

———. 1998. Evolution of Indirect Reciprocity by Image Scoring. *Nature* **393**:673–676. [3]

O'Brien, M. J., and R. A. Bentley. 2011. Stimulated Variation and Cascades: Two Processes in the Evolution of Complex Technological Systems. *J. Archaeol. Meth. Theory* **18**:309–335. [8, 11]

O'Brien, M. J., J. Darwent, and R. L. Lyman. 2001. Cladistics Is Useful for Reconstructing Archaeological Phylogenies: Palaeoindian Points from the Southeastern United States. *J. Archaeol. Sci.* **28**:1115–1136. [9]

O'Brien, M. J., and R. L. Lyman. 2003a. Cladistics and Archaeology. Salt Lake City: Univ. of Utah Press. [11]

———. 2003b. Style, Function, Transmission. Foundations of Archaeological Inquiry, J. M. Skibo, series ed. Salt Lake City: Univ. of Utah Press. [11]

O'Brien, M. J., R. L. Lyman, A. Mesoudi, and T. L. VanPool. 2010. Cultural Traits as Units of Analysis. *Phil. Trans. R. Soc. B* **365**:3797–3806. [8]

O'Brien, M. J., and S. J. Shennan. 2010. Innovation in Cultural Systems: Contributions from Evolutionary Anthropology. Cambridge, MA: MIT Press. [11]

O'Connell, J. F., K. Hawkes, K. D. Lupo, and N. B. Jones. 2002. Male Strategies and Plio-Pleistocene Archaeology. *J. Hum. Evol.* **43**:831–872. [2, 6]

O'Neil, D. 2012. Analysis of Early Hominins. http://anthro.palomar.edu/hominid/australo_2.htm. (accessed March 17, 2013). [2]

O'Hara, R. J. 1997. Population Thinking and Tree Thinking in Systematics. *Zool. Scr.* **26**:323–329. [15]

Ochs, E. 1982. Talking to Children in Western Samoa. *Lang. Soc.* **11**:77–104. [14]

Ochs, E., C. Pontecorvo, and A. Fasulo. 1996. Socializing Taste. *Ethnos* **61**:5–42. [14]

Ochs, E., and B. B. Schieffelin. 1983. Acquiring Conversational Competence. London: Routledge & Kegan Paul. [14]

Odling-Smee, F. J., K. Laland, and M. Feldman. 1996. Niche Construction. *Am. Natural.* **147**:641–648. [9]

———. 2003. Niche Construction: The Neglected Process in Evolution. Princeton: Princeton Univ. Press. [6, 11, 12]

Okasha, S. 2007. Evolution and the Levels of Selection New York: Oxford Univ. Press. [4]

Okasha, S., and K. Binmore. 2012. Evolution and Rationality: Decisions, Co-Operation, and Strategic Behavior. Cambridge: Cambridge Univ. Press. [2]

Oller, D., L. Wiemar, W. Doyle, and C. Ross. 1975. Infant Babbling and Speech. *J. Child Lang.* **3**:1–11. [14]

Olson, M. 1965. The Logic of Collective Action: Public Goods and the Theory of Groups. Cambridge, MA: Harvard Univ. Press. [2]

Orban, G. A., K. Claeys, K. Nelissen, et al. 2006. Mapping the Parietal Cortex of Human and Non-Human Primates. *Neuropsychologia* **44**: 2647–2667. [9]

Ostler, N. 2005. Empires of the Word. A Language History of the World. London: Harper Perennial. [16]

Ostrom, E. 1990. Governing the Commons: The Evolution of Institutions for Collective Action. Cambridge: Cambridge Univ. Press. [6]

———. 2005. Understanding Institutional Diversity. Princeton: Princeton Univ. Press. [6]

Otterbein, K. F. 2004. How War Began. College Station: Texas A&M Univ. Press. [2]

Over, H., and M. Carpenter. 2009. Priming Third-Party Ostracism Increases Affiliative Imitation in Children. *Devel. Sci.* **12**:F1–F8. [5]

———. 2012. Putting the Social into Social Learning: Explaining Both Selectivity and Fidelity in Children's Copying Behaviour. *J. Comp. Psychol.* **126**:182–192. [5]

———. 2013. The Social Side of Imitation. *Child Devel. Perspect.* **7**:6–11. [5]

Over, H., M. Carpenter, R. Spears, and M. Gattis. 2013. Children Selectively Trust Individuals Who Have Imitated Them. *Soc. Develop.*, in press. [5]

Oyama, S., P. E. Griffiths, and R. D. Gray, eds. 2001. Cycles of Contingency: Developmental Systems and Evolution. Cambridge, MA: MIT Press. [6]

Paciotti, B., P. J. Richerson, B. Baum, et al. 2011. Are Religious Individuals More Generous, Trusting, and Cooperative? An Experimental Test of the Effect of Religion on Prosociality. In: The Economics of Religion: Anthropological Approaches, ed. L. Obedia and D. C. Wood, pp. 267–305. Bingley, UK: Emerald. [19, 20]

Paden, W. E. 1994. Religious Worlds: The Comparative Study of Religion. Boston: Beacon Press. [20]

———. 2001. Universals Revisited: Human Behaviors and Cultural Variations. *Numen* **48**:276–289. [20]

———. 2013. The Prestige of the Gods: Evolutionary Continuities in the Formation of Sacred Objects. In: Origins of Religion, Cognition and Culture, ed. A. Geertz. London: Equinox, in press. [20]

Pagel, M. 2009. Human Language as a Culturally Transmitted Replicator. *Nat. Rev. Genet.* **10**:405–415. [16]

———. 2012. Wired for Culture. New York: W. W. Norton. [2]

Pagel, M., Q. D. Atkinson, and A. Meade. 2007. Frequency of Word-Use Predicts Rates of Lexical Evolution Throughout Indo-European History. *Nature* **449**:717–720. [15]

Pan, B., M. Rowe, J. Singer, and C. Snow. 2005. Maternal Correlates of Growth in Toddler Vocabulary Production in Low-Income Families. *Child. Devel.* **76**:763–782. [14]

Panchanathan, K., and R. Boyd. 2004. Indirect Reciprocity Can Stabilize Cooperation without the Second-Order Free Rider Problem. *Nature* **432**:499–502. [3]

Panda, S. 2013. Alipur Sign Language: A Sociolinguistic and Cultural Profile. In: Sign Languages in Village Communities, ed. U. Zeshan and C. De Vos, pp. 353–360. Berlin: Walter de Gruyter. [13]

Pandit, S., and C. P. van Schaik. 2003. A Model for Leveling Coalitions among Primate Males: Toward a Theory of Egalitarianism. *Behav. Ecol. Sociobiol.* **55**:161–168. [2]

Parsons, T. 1967. Sociological Theory and Modern Society. New York: Free Press. [2]

Paschou, P., J. Lewis, A. Javed, and P. Drineas. 2010. Ancestry Informative Markers for Fine-Scale Individual Assignment to Worldwide Populations. *J. Med. Genet.* **47**:835–847. [12]

Patterson, K., P. J. Nestor, and T. T. Rogers. 2007. Where Do You Know What You Know? The Representation of Semantic Knowledge in the Human Brain. *Nat. Rev. Neurosci.* **8**:976–987. [9]

Paukner, A., S. J. Suomi, E. Visalberghi, and P. F. Ferrari. 2009. Capuchin Monkeys Display Affiliation toward Humans Who Imitate Them. *Science* **325**:880–883. [5]

Paul, A., S. Preuschoft, and C. P. van Schaik. 2000. The Other Side of the Coin: Infanticide and the Evolution of Affiliative Male-Infant Interactions in Old World Primates. In: Infanticide by Males and Its Implications, ed. C. P. van Schaik and C. H. Janson, pp. 269–292. Cambridge: Cambridge Univ. Press. [2]

Paulesu, E., J. F. Démonet, F. Fazio, et al. 2001. Dyslexia: Cultural Diversity and Biological Unity. *Science* **291**:2165–2167. [11]

Paulhus, D. L. 1984. Two-Component Models of Socially Desirable Responding. *J. Pers. Soc. Psychol.* **46**:598–609. [19]

Pawley, A. 2006. On the Size of the Lexicon in Preliterate Language Communities: Comparing Dictionaries of Australian, Austronesian and Papuan Languages. In: Favete Linguis: Studies in Honour of Viktor Krupa, ed. J. Genzor and M. Buckov, pp. 171–191. Bratislava: Institute of Oriental Studies. [16]

Pawley, A., and M. Pawley. 1994. Early Austronesian Terms for Canoe Parts and Seafaring. In: Austronesian Terminologies: Continuity and Change, ed. A. Pawley and M. D. Ross, pp. 329–361. Canberra: Pacific Linguistics. [15]

Peeters, R., L. Simone, K. Nelissen, et al. 2009. The Representation of Tool Use in Humans and Monkeys: Common and Uniquely Human Features. *J. Neurosci.* **29**:11,523–11,539. [9]

Pelegrin, J. 1990. Prehistoric Lithic Technology: Some Aspects of Research. *Arch. Rev. Cambridge* **9**:116–125. [9]

Perfors, A. F., and J. B. Tenenbaum. 2009. Learning to Learn Categories. Proc. 42nd Conf. of the Cognitive Science Society. http://csjarchive.cogsci.rpi.edu/Proceedings/2009/papers/26/index.html. (accessed May 27, 2013). [7]

Perkins, D. 2000. The Evolution of Adaptive Form. In: Technological Innovation as an Evolutionary Process, ed. J. Ziman, pp. 159–173. Cambridge: Cambridge Univ. Press. [8]

Perkins, R. 1995. Deixis, Grammar and Culture. Amsterdam: John Benjamins. [13, 16]

Perreault, C. 2012. The Pace of Cultural Evolution. *PLoS One* **7**:e45150. [7, 11]

Perreault, C., C. Moya, and R. Boyd. 2012. A Bayesian Approach to the Evolution of Social Learning. *Evol. Hum. Behav.* **33**:449–459. [7]

Persinger, M. A. 1983. Religious and Mystical Experiences as Artefacts of Temporal Lobe Function: A General Hypothesis. *Percept. Motor Skills* **57**:1255–1262. [18]

Peterson, N., ed. 1976. Tribes and Boundaries in Australia. Social Anthropology Series, vol. 19 Canberra: Australian Institute of Aboriginal Studies. [6]

Petroski, H. 1985. To Engineer Is Human: The Role of Failure in Successful Design. New York: St. Martin's Press. [7]

———. 1994. The Evolution of Useful Things. New York: Vintage. [7, 11]

———. 2006. Success through Failure: The Paradox of Design. Princeton: Princeton Univ. Press. [7]

Phillimore, A. B., and T. D. Price. 2008. Density-Dependent Cladogenesis in Birds. *PLoS Biol.* **6**:483–489. [15]

Piazza, J., J. M. Bering, and G. Ingram. 2011. Princess Alice Is Watching You: Children's Belief in an Invisible Person Inhibits Cheating. *J. Exp. Child. Psychol.* **109**:311–320. [19]

Pichon, I., G. Boccato, and V. Saroglou. 2007. Nonconscious Influences of Religion on Prosociality: A Priming Study. *Eur. J. Social Psychol.* **37**:1032–1045. [19]

Pinel, P., F. Fauchereau, A. Moreno, et al. 2012. Genetic Variants of FOXP2 and KIAA0319/TTRAP/THEM2 Locus Are Associated with Altered Brain Activation in Distinct Language-Related Regions. *J. Neurosci.* **32**:817–825. [12]

Pinhasi, R., M. G. Thomas, M. Hofreiter, M. Currat, and J. Burger. 2012. The Genetic History of Europeans. *Trends Genet.* **28**:496–505. [1]

Pinker, S. 1994. The Language Instinct. New York: W. Morrow. [1, 16]

———. 1997. How the Mind Works. New York: Norton. [16]

———. 2010. The Cognitive Niche: Coevolution of Intelligence, Sociality, and Language. *PNAS* **107**:8993–8999. [2, 3, 7, 11]

Pinker, S., and P. Bloom. 1990. Natural Language and Natural Selection. *Behav. Brain Sci.* **13**:707–784. [1]

Platvoet, J., and A. L. Molendijk. 1999. The Pragmatics of Defining Religion: Contexts, Concepts, and Contests. Leiden: Brill. [20]

Plooij, F. X. 1978. Tool-Using During Chimpanzees' Bushpig Hunt. *Carnivore* **1**:103–106. [2]

Plourde, A. 2009. Human Power and Prestige Systems. In: Mind the Gap: Tracing the Origins of Human Universals, ed. P. M. Kappeler and J. B. Silk. Berlin: Springer-Verlag. [2]

Poo, M.-C. 2009. Rethinking Ghosts in World Religions. Numen Book Series. Leiden: Brill. [20]

Popper, K. R. 1947. The Open Society and Its Enemies, vol. 2. London: Routledge. [1]

Potts, R. 1996. Humanity's Descent: The Consequences of Ecological Instability. New York: Aldine de Gruyter. [2]

———. 1998. Environmental Hypotheses of Hominin Evolution. *Yearb. Phys. Anthropol.* **41**:93–138. [2]

Povinelli, D. J., J. E. Reaux, and S. H. Frey. 2010. Chimpanzees' Context-Dependent Tool Use Provides Evidence for Separable Representations of Hand and Tool Even During Active Use within Peripersonal Space. *Neuropsychologia* **48**:243–247. [9]

Powell, A., S. Shennan, and M. G. Thomas. 2009. Late Pleistocene Demography and the Appearance of Modern Human Behavior. *Science* **324**:1298–1301. [6–8, 11]

Pradhan, G. R., C. Tennie, and C. P. van Schaik. 2012. Social Organization and the Evolution of Cumulative Technology in Apes and Hominins. *J. Hum. Evol.* **63**:180–190. [11]

Preus, J. S. 1987. Explaining Religion: Criticism and Theory from Bodin to Freud. New Haven: Yale Univ. Press. [20]

Preuschoft, S., and C. P. van Schaik. 2000. Dominance, Social Relationships and Conflict Management. In: Conflict Management, ed. F. Aureli and F. B. M. de Waal, pp. 77–105. Berkely: California Univ. Press. [6]

Previc, F. 2009. The Dopaminergic Mind in Human Evolution and History. Cambridge: Cambridge Univ. Press. [18]

Price, C. J. 2010. The Anatomy of Language: A Review of 100 fMRI Studies Published in 2009. *Ann. NY Acad. Sci.* **1191**:62–88. [9]

Price, C. J., and K. J. Friston. 2002. Degeneracy and Cognitive Anatomy. *Trends Cogn. Sci.* **6**:416–421. [9]

Price, D. J. S. 1963. Little Science, Big Science. New York: Columbia Univ. Press. [11]

Price, G. R. 1970. Selection and Covariance. *Nature* **227**:520–521. [11]

———. 1972. Extension of Covariance Selection Mathematics. *Ann. Hum. Genet.* **35**:485–490. [4]

Proffitt, D. R., and D. L. Gilden. 1989. Understanding Natural Dynamics. *J. Exp. Psychol. Hum. Percep. Perform.* **15**:384–393. [10]

Provençal, N., M. J. Suderman, C. Guillemin, et al. 2012. The Signature of Maternal Rearing in the Methylome in Rhesus Macaque Prefrontal Cortex and T Cells. *J. Neurosci.* **32**:15,626–15,642. [1]

Provine, W. B. 1971. The Origins of Theoretical Population Genetics. Chicago: Univ. of Chicago Press. [1]

Pruetz, J. D., and P. Bertolani. 2007. Savanna Chimpanzees, *Pan troglodytes Verus*, Hunt with Tools. *Curr. Biol.* **17**:412–417. [2]

Pugach, I., F. Delfin, E. Gunnarsdóttir, M. Kayser, and M. Stoneking. 2013. Genome-Wide Data Substantiate Holocene Gene Flow from India to Australia. *PNAS* **110**:1803–1808. [16]

Puglisi, A., A. Baronchelli, and V. Loreto. 2008. Cultural Route to the Emergence of Linguistic Categories. *PNAS* **105**:7936–7940. [16]

Puigbò, P., Y. I. Wolf, and E. V. Koonin. 2009. Search for a "Tree of Life" in the Thicket of the Phylogenetic Forest. *J. Biol.* **8**:59. [16]

Pulliam, H. R., and C. Dunford. 1980. Programmed to Learn: An Essay on the Evolution of Culture. New York: Columbia Univ. Press. [1]

Putnam, R. D. 2000. Bowling Alone: The Collapse and Revival of American Community. New York: Simon and Schuster. [4]

Pyyssiäinen, I. 2011. Believing and Doing: How Ritual Action Enhances Religious Belief. In: Religious Narrative, Cognition and Culture: Image and Word in the Mind of Narrative, ed. A. W. Geertz and J. S. Jensen, pp. 147–162. London: Equinox. [20]

Raafat, R. M., N. Chater, and C. Frith. 2009. Herding in Humans. *Trends Cogn. Sci.* **13**:420–428. [20]

Radcliffe-Brown, A. R. 1922. The Andaman Islanders. New York: Free Press. [20]

———. 1952. Structure and Function in Primitive Society. London: Cohen & West Ltd. [6]

Rainey, P. B., and K. Rainey. 2003. Evolution of Cooperation and Conflict in Experimental Bacterial Populations. *Nature* **425**:72–74. [3]

Rakison, D. H., and Y. Yermolayeva. 2010. Infant Categorization. *Wiley Interdis. Rev. Cogn. Sci.* **1**:894–905. [14]

Rakoczy, H., F. Warneken, and M. Tomasello. 2007. "This Way!" "No! That Way!" 3-Year Olds Know That Two People Can Have Mutually Incompatible Desires. *Cogn. Devel.* **22**:47–68. [6]

———. 2008. The Sources of Normativity: Young Children's Awareness of the Normative Structure of Games. *Devel. Psychol.* **44**:875–881. [5]

Randolph-Seng, B., and M. E. Nielsen. 2007. Honesty: One Effect of Primed Religious Representations. *Intl. J. Psychol. Relig.* **17**:303–315. [19]

Rapoport, A. 1957. A Contribution to the Theory of Random and Biased Nets. *Bull. Math. Biophys.* **19**:257–271. [6]

Rappaport, R. A. 1971. The Sacred in Human Evolution. *Annu. Rev. Ecol. System.* **2**:23–44. [20]

———. 1979. The Obvious Aspects of Ritual. In: Ecology, Meaning and Religion, ed. R. A. Rappaport, pp. 173–1221. Richmond, CA: North Atlantic Books. [20]

———. 1999. Ritual and Religion in the Making of Humanity. Cambridge: Cambridge Univ. Press. [20]

Ratner, N. B., and C. Pye. 1984. Higher Pitch in BT Is Not Universal: Acoustic Evidence from Quiche Mayan. *J. Child Lang.* **11**:515–522. [14]

Raup, D. M. 1967. Geometric Analysis of Shell Coiling: Coiling in Ammonoids. *J. Palaeontol.* **41**:43–65. [15]

Read, D. 2006. Tasmanian Knowledge and Skill: Maladaptive Imitation or Adequate Technology? *Am. Antiq.* **71**:164–184. [7]

Reali, F., and M. H. Christiansen. 2009. Sequential Learning and the Interaction between Biological and Linguistic Adaptation in Language Evolution. *Interaction Stud.* **10**:5–30. [13]

Reddish, P., J. Bulbulia, and R. Fischer. 2013. Does Synchrony Promote Generalized Prosociality? *Relig. Brain Behav.* doi: 10.1080/2153599X.2152013.2764545. [20]

Reesink, G., R. Singer, and M. Dunn. 2009. Explaining the Linguistic Diversity of Sahul Using Population Models. *PLoS Biol.* **7**:e1000241. [13]

Reich, D., N. Patterson, M. Kircher, et al. 2011. Denisova Admixture and the First Modern Human Dispersals into Southeast Asia and Oceania. *Am. J. Hum. Genet.* **89**:516–528. [12]

Rendell, L., L. Fogarty, W. J. E. Hoppitt, et al. 2011. Cognitive Culture: Theoretical and Empirical Insights into Social Learning Strategies. *Trends Cogn. Sci.* **15**:68–76. [11]

Rendell, L., L. Fogarty, and K. N. Laland. 2010. Rogers' Paradox Recast and Resolved: Population Structure and the Evolution of Social Learning Strategies. *Evolution* **64**:534–548. [7]

Renfrew, C., A. McMahon, and L. Trask, eds. 2000. Time Depth in Historical Linguistics. Cambridge: McDonald Institute. [16]

Renfrew, C., and D. Nettle, eds. 1999. Nostratic: Examining a Linguistic Macrofamily. Cambridge: McDonald Institute. [16]

Renn, J. 1995. Historical Epistemology and Interdisciplinarity. In: Physics, Philosophy and the Scientific Community: Essays in the Philosophy and History of the Natural Sciences and Mathematics in Honor of Robert S. Cohen, ed. K. Gavroglu et al., pp. 241–251. Dordrecht: Kluwer. [11]

———, ed. 2012. The Globalization of Knowledge in History. Max Planck Research Library for the History and Development of Knowledge Studies 1. Berlin: Edition Open Access. http://www.edition-open-access.de/studies/1/index.html (accessed April 6, 2013). [11]

Reuland, E. 2008. Anaphoric Dependencies: How Are They Encoded? Towards a Derivation-Based Typology. In: Reciprocals and Reflexives. Theoretical and Typological Explorations, ed. E. König and V. Gast, pp. 499–556. Berlin: Mouton de Gruyter. [13]

Reyes-García, V., J. Broesch, L. Calvet-Mir, et al. 2009. Cultural Transmission of Ethnobotanical Knowledge and Skills: An Empirical Analysis from an Amerindian Society. *Evol. Hum. Behav.* **30**:274–285. [11]

Reyes-García, V., J. L. Molina, J. Broesch, et al. 2008. Do the Aged and Knowledgeable Men Enjoy More Prestige? A Test of Predictions from the Prestige-Bias Model of Cultural Transmission. *Evol. Hum. Behav.* **29**:275–281. [11]

Reyes-García, V., V. Vadez, T. Huanca, W. Leonard, and D. Wilkie. 2005. Knowledge and Consumption of Wild Plants: A Comparative Study in Two Tsimane' Villages in the Bolivian Amazon. *Ethnobot. Res. Appl.* **3**:201–207. [11]

Richards, R. J. 1987. Darwin and the Emergence of Evolutionary Theories of Mind and Behavior. Chicago: Univ. of Chicago Press. [1]

Richerson, P. J. 1977. Ecology and Human Ecology: A Comparison of Theories in the Biological and Social Sciences. *Am. Ethol.* **4**:1–26. [1]

Richerson, P. J., and R. Boyd. 1976. A Simple Dual Inheritance Model of the Conflict between Social and Biological Evolution. *Zygon* **11**:254–262. [1]

———. 1998. The Evolution of Human Ultrasociality. In: Ethnic Conflict and Indoctrination, ed. I. Eibl-Eibesfeldt and F. K. Salter, pp. 71–95. Oxford: Berghahn Books. [4]

———. 1999. Complex Societies: The Evolutionary Origins of a Crude Superorganism. *Hum. Nature* **10**:253–289. [6]

———. 2001. Built for Speed, Not for Comfort: Darwinian Theory and Human Culture. *Hist. Philos. Life Sci.* **23**:423–463. [1]

———. 2005. Not by Genes Alone: How Culture Transformed Human Evolution. Chicago: Univ. of Chicago Press. [1, 2, 4, 6, 7, 11, 17, 19, 20]

———. 2010. Why Possibly Language Evolved. *Biolinguistics* **4**:289–306. [13]

Richerson, P. J., R. Boyd, and R. L. Bettinger. 2001. Was Agriculture Impossible During the Pleistocene but Mandatory During the Holocene? A Climate Change Hypothesis. *Am. Antiq.* **66**:387–411. [2, 11]

———. 2009. Cultural Innovations and Demographic Change. *Hum. Biol.* **81**:211–235. [6]

Richerson, P. J., R. Boyd, and J. Henrich. 2010. Gene-Culture Coevolution in the Age of Genomics. *PNAS* **107**:8985–8992. [1, 20]

Richerson, P. J., and J. Henrich. 2012. Tribal Social Instincts and the Cultural Evolution of Institutions to Solve Collective Action Problems. *Cliodynamics* **3**:38–80. [4]

Richerson, P. J., and L. Newson. 2008. Is Religion Adaptive? Yes, No, Neutral, but Mostly, We Don't Know. In: The Evolution of Religion: Studies, Theories, and Critiques, ed. J. Bulbulia et al., pp. 73–78. Santa Margarita, CA: Collins Foundation Press. [20]

Ridley, M. 2009. Modern Darwins. *Natl. Geographic* **215**:56–73. [1]

Riede, F. 2011. Adaptation and Niche Construction in Human Prehistory: A Case Study from the Southern Scandinavian Late Glacial. *Phil. Trans. R. Soc. B* **366**:793–808. [11]

Riedl, K., K. Jensen, J. Call, and M. Tomasello. 2012. No Third-Party Punishment in Chimpanzees. *PNAS* **109**:14,824–14,829. [5]

Riel-Salvatore, J. 2010. A Niche Construction Perspective on the Middle-Upper Paleolithic Transition in Italy. *J. Archaeol. Meth. Theory* **17**:323–355. [11]

Rigdon, M. L., K. Ishii, M. Watabe, and S. Kitayama. 2009. Minimal Social Cues in the Dictator Game. *J. Econ. Psychol.* **30**:358–367. [19]

Ritt, N. 2004. Selfish Sounds and Linguistic Evolution: A Darwinian Approach to Language Change. Cambridge: Cambridge Univ. Press. [16]

Rivers, W. H. R. 1926. Psychology and Ethnology. New York: Harcourt, Brace & Co, Inc. [7, 8]

Rizzolatti, G. 2005. The Mirror Neuron System and Imitation. In: Perspectives on Imitation: From Neuroscience to Social Science, ed. S. Hurley and N. Chater, vol. 1, pp. 55–76. Cambridge, MA: MIT Press. [1]

Rizzolatti, G., and L. Craighero. 2004. The Mirror-Neuron System. *Annu. Rev. Neurosci.* **27**:169–192. [9]

Roche, A. F. 1964. Aural Exostoses in Australian Aboriginal Skulls. *Ann. Otol. Rhinol. Laryngol.* **73**:82–91. [13]

Rodseth, L., R. W. Wrangham, A. M. Harrigan, and B. B. Smuts. 1991. The Human Community as a Primate Society. *Curr. Anthropol.* **32**:221–254. [6]

Roebroeks, W., and A. Verpoorte. 2009. A "Language-Free" Explanation for Differences between the European Middle and Upper Paleolithic Record. In: The Cradle of Language, ed. R. Botha and C. Knight, pp. 150–166. Oxford: Oxford Univ. Press. [16]

Roebroeks, W., and P. Villa. 2011. On the Earliest Evidence for Habitual Use of Fire in Europe. *PNAS* **108**:5209–5214. [2, 6]

Roepstorff, A., J. Niewöhner, and S. Beck. 2010. Enculturing Brains through Patterned Practices. *Neural Netw.* **23**:1051–1059. [20]

Roes, F. L., and M. Raymond. 2003. Belief in Moralizing Gods. *Evol. Hum. Behav.* **24**:126–135. [17, 19]

Rogers, A. R. 1988. Does Biology Constrain Culture. *Am. Anthropol.* **90**:819–831. [6, 7]

Rogers, D. S., and P. R. Ehrlich. 2008. Natural Selection and Cultural Rates of Change. *PNAS* **105**:3416–3420. [8, 11]

Rogers, E. M. 1983. Diffusion of Innovations (3rd edition). New York: Free Press. [8]

———. 1995. The Diffusion of Innovations. New York: Free Press. [11, 16]

Rogers, E. M., and F. F. Shoemaker. 1971. Communication of Innovations: A Cross-Cultural Approach (2nd edition). New York: Free Press. [1, 7]

Romer, P. 1993. Endogenous Technological Change. *J. Polit. Econ.* **98**:71–102. [7]

Roth, A. E., V. Prasnikar, M. Okuno-Fujiwara, and S. Zamir. 1991. Bargaining and Market Behavior in Jerusalem, Ljubljana, Pittsburgh, and Tokyo: An Experimental Study. *Am. Econ. Rev.* **81**:1068–1095. [2]

Roux, V. 1990. The Psychological Analysis of Technical Activities: A Contribution to the Study of Craft Specialization. *Arch. Rev. Cambridge* **9**:142–153. [9]

———. 2010. Technological Innovations and Developmental Trajectories: Social Factors as Evolutionary Forces. In: Innovations in Cultural Systems: Contributions from Evolutionary Anthropology, ed. M. J. O'Brien and S. J. Shennan, pp. 217–234. Cambridge, MA: MIT Press. [8]

Rowley-Conwy, P., and R. Layton. 2011. Foraging and Farming as Niche Construction: Stable and Unstable Adaptations. *Phil. Trans. R. Soc. B* **366**:849–862. [11]

Rowthorn, R. 2011. Religion, Fertility and Genes: A Dual Inheritance Model. *Proc. R. Soc. B* **278**:2519–2527. [20]

Rudolf von Rohr, C., S. E. Koski, J. M. Burkart, et al. 2012. Impartial Third-Party Interventions in Captive Chimpanzees: A Reflection of Community Concern. *PLoS One* **7**:e32494. [2]

Ruffle, B. J., and R. Sosis. 2006. Cooperation and the Ingroup–Outgroup Bias: A Field Test on Israeli Kibbutz Members and City Residents. *J. Econ. Behav. Org.* **60**:147–163. [17]

Ruyle, E. E. 1973. Genetic and Cultural Pools: Some Suggestion for a Unified Theory of Biocultural Evolution. *Hum. Ecol.* **1**:201–215. [1]

Sabeti, P. C., D. E. Reich, J. M. Higgins, et al. 2002. Detecting Recent Positive Selection in the Human Genome from Haplotype Structure. *Nature* **419**:832–837. [1]

Sacks, O. 1998. Island of the Colorblind. New York: A. Knopf Inc. [7]

Saler, B. 2010. Theory and Criticism: The Cognitive Science of Religion. *Meth. Theory Stud. Rel.* **22**:330–339. [20]

Sampson, G., D. Gill, and P. Trudgill, eds. 2009. Language Complexity as an Evolving Variable. Oxford: Oxford Univ. Press. [16]

Sanderson, S. K. 1999. Social Transformations: A General Theory for Historical Development. Lanham, MD: Rowman & Littlefield. [4]

Sandgathe, D. M., H. L. Dibble, P. Goldberg, et al. 2011. On the Role of Fire in Neandertal Adaptations in Western Europe: Evidence from Pech De L'azé IV and Roc De Marsal, France. *Paleoanthropology* **2011**:216–242. [2]

Sandler, W., M. Aronoff, I. Meir, and C. Padden. 2011. The Gradual Emergence of Phonological Form in a New Language. *Nat. Lang. Ling. Theory* **29**:503–543. [16]

Sandler, W., I. Meier, C. Padden, and M. Aronoff. 2005. The Emergence of Grammar: Systematic Structure in a New Language. *PNAS* **102**:2661–2665. [12]

Sankoff, G., and H. Blondeau. 2007. Language Change across the Lifespan: /R/ in Montreal French. *Language* **83**: 566–588. [16]

Sapir, E. 1916/1949. Time Perspective in Aboriginal American Culture: A Study in Method. In: Selected Writings of Edward Sapir, ed. D. G. Mandelbaum, pp. 389–467. Berkeley: Univ. of California Press. [15]

Scerri, T. S., and G. Schulte-Körne. 2010. Genetics of Developmental Dyslexia. *Eur. Child Adolesc. Psychiatry* **19**:179–197. [12]

Schaberg, D. 2001. A Patterned Past: Form and Thought in Early Chinese Historiography. Cambridge, MA: Harvard Univ. Press. [17]

Schalley, A., and D. Zaefferer, eds. 2007. Ontolinguistics. How Ontological Status Shapes the Linguistic Coding of Concepts. Berlin: Mouton de Gruyter. [13]

Schemmel, M. 2008. The English Galileo: Thomas Harriot's Work on Motion as an Example of Preclassical Mechanics. New York: Springer-Verlag. [11]

Schieffelin, B. B. 1985. The Acquisition of Kaluli. In: The Crosslinguistic Study of Language Acquisition, ed. D. I. Slobin, vol. 1, pp. 525–593. Hillsdale: Lawrence Erlbaum. [14]

Schjoedt, U., H. Stødkilde-Jørgensen, A. W. Geertz, and A. Roepstorff. 2009. Highly Religious Participants Recruit Areas of Social Cognition in Personal Prayer. *Soc. Cogn. Aff. Neurosci.* **4**:199–207. [20]

Schloss, J. P., and M. J. Murray. 2011. Evolutionary Accounts of Belief in Supernatural Punishment: A Critical Review. *Relig. Brain Behav.* **1**:46–99. [19]

Schmidt, K. 2010. Göbekli Tepe: The Stone Age Sanctuaries. New Results of Ongoing Excavations with a Special Focus on Sculptures and High Reliefs. *Documenta Praehistorica* **37**:239–256. [19]

Schmidt, M. F., H. Rakoczy, and M. Tomasello. 2011. Young Children Attribute Normativity to Novel Actions without Pedagogy or Normative Language. *Devel. Sci.* **14**:530–539. [5]

Schütze, C. T. 1996. The Empirical Base of Linguistics: Grammaticality Judgments and Linguistic Methodology. Chicago: Univ. of Chicago Press. [12]

Scott, D. A., R. Carmi, K. Elbedour, et al. 1995. Nonsyndromic Autosomal Recessive Deafness Is Linked to the DFNB1 Locus in a Large Inbred Bedouin Family from Israel. *Am. J. Hum. Genet.* **57**:965–968. [12]

Scott-Phillips, T. C., and S. Kirby. 2010. Language Evolution in the Laboratory. *Trends Cogn. Sci.* **14**:411–417. [1]

Sedikides, C., and J. E. Gebauer. 2010. Religiosity as Self-Enhancement: A Meta-Analysis of the Relation between Socially Desirable Responding and Religiosity. *Pers. Social Psychol. Rev.* **14**:17–36. [19]

Segerstråle, U. 2000. Defenders of the Truth: The Battle for Science in the Sociobiology Debate and Beyond. New York: Oxford Univ. Press. [1]

Seiffert, E. 2007. A New Estimate of Afrotherian Phylogeny Based on Simultaneous Analysis of Genomic, Morphological, and Fossil Evidence. *BMC Evol. Biol.* **7**:224. [15]

Senghas, A. 2005. Language Emergence: Clues from a New Bedouin Sign. *Curr. Biol.* **15**:R463–R465. [12]

Senghas, R. J., A. Senghas, and J. E. Pyers. 2005. The Emergence of Nicaraguan Sign Language: Questions of Development, Acquisition, and Evolution. In: Biology and Knowledge Revisited: From Neurogenesis to Psychogenesis, ed. J. Langer et al. Mahwah, NJ: Lawrence Erlbaum. [12]

Service, E. R. 1975. Origin of the State and Civilization: The Process of Cultural Evolution. New York: Norton. [2]

Seyfarth, R. M., and D. L. Cheney. 2012. The Evolutionary Origins of Friendship. *Annu. Rev. Psychol.* **63**:153–177. [6]

Shariff, A. F., and A. Norenzayan. 2007. God Is Watching You: Priming God Concepts Increases Prosocial Behavior in an Anonymous Economic Game. *Psychol. Sci.* **18**:803–809. [19, 20]

———. 2011. Mean Gods Make Good People: Different Views of God Predict Cheating Behavior. *Intl. J. Psychol. Relig.* **21**:85–96. [19]

Shariff, A. F., A. Norenzayan, and J. Henrich. 2010. The Birth of High Gods. In: Evolution, Culture, and the Human Mind, ed. M. Schaller et al. New York: Psychology Press. [19]

Shariff, A. F., and M. Rhemtulla. 2012. Divergent Effects of Beliefs in Heaven and Hell on National Crime Rates. *PLoS One* **7**:e39048. [19]

Shaul, D. 1986. Linguistic Adaptation and the Great Basin. *Am. Antiq.* **51**:415–416. [13]

Shaw, R. P., and Y. Wong. 1989. Genetic Seeds of Warfare. Boston: Unwin and Hyman. [4]

Shenhav, A., D. G. Rand, and J. D. Greene. 2011. Divine Intuition: Cognitive Styles Influence Belief in God. *J. Exp. Psychol.* **141**:423–428.[17]

Shennan, S. J. 2001. Demography and Cultural Innovation: A Model and Its Implications for the Emergence of Modern Human Culture. *Cambridge Arch. J.* **11**:5–16. [6–8]

———. 2002. Genes, Memes, and Human History: Darwinian Archaeology and Cultural Evolution. London: Thames and Hudson. [20]

———. 2011. Descent with Modification and the Archaeological Record. *Phil. Trans. R. Soc. B* **366**:1070–1079. [9]

Shennan, S. J., and J. R. Wilkinson. 2001. Ceramic Style Change and Neutral Evolution: A Case Study from Neolithic Europe. *Am. Antiq.* **66**:577–593. [11]

Shultz, S., C. Opie, and Q. D. Atkinson. 2011. Stepwise Evolution of Stable Sociality in Primates. *Nature* **479**:219. [6]

Silk, J. B. 2006. Practicing Hamilton's Rule: Kin Selection in Primate Groups. In: Cooperation in Primates and Humans: Mechanisms and Evolution, ed. P. M. Kappeler and C. P. van Schaik, pp. 25–46. Berlin: Springer-Verlag. [3]

Silk, J. B., S. C. Alberts, and J. Altmann. 2004 Patterns of Coalition Formation by Adult Female Baboons in Amboseli, Kenya. *Anim. Behav.* **67**:573–582. [3]

Silver, M., and E. D. Paolo. 2006. Spatial Effects Favour the Evolution of Niche Construction. *Theor. Pop. Biol.* **70**:387–400. [11]

Simon, H. A. 1962. The Architecture of Complexity. *Proc. Am. Phil. Soc.* **106**:467–482. [9]

Simpson, J. 2002. From Common Ground to Syntactic Construction: Associated Path in Warlpiri. In: Ethnosyntax: Explorations in Grammar and Culture, ed. N. J. Enfield, pp. 287–307. Oxford: Oxford Univ. Press. [13]

Sims-Williams, P. 1998. Genetics, Linguistics, and Prehistory: Thinking Big and Thinking Straight. *Antiquity* **72**:505–527. [12]

Slingerland, E. 2008. What Science Offers the Humanities: Integrating Body and Culture. Cambridge: Cambridge Univ. Press. [1, 20]

Slingerland, E., and M. Chudek. 2011. The Prevalence of Mind-Body Dualism in Early China. *Cogn. Sci.* **35**:997–1007. [17, 20]

Slingerland, E., and M. Collard. 2012. Creating Consilience: Toward a Second Wave. New York: Oxford Univ. Press. [20]

Slobin, D. 2003. Language and Thought Online. In: Language in Mind. Advances in the Study of Language and Thought, ed. D. Gentner and S. Goldin-Meadow, pp. 157–191. Cambridge, MA: MIT Press. [13]

Slone, J. 2004. Theological Incorrectness: Why Religious People Believe Things They Shouldn't. New York: Oxford Univ. Press. [18]

Smith, A. 1776. An Inquiry into the Nature and Causes of the Wealth of Nations. Indianapolis: Liberty Fund. [3]

Smith, B. D. 2007. Niche Construction and the Behavioral Context of Plant and Animal Domestication. *Evol. Anthropol.* **16**:188–199. [11]

Smith, B. H. 2009. Natural Reflections: Human Cognition at the Nexus of Science and Religion. New Haven: Yale Univ. Press. [10]

Smith, E. A. 2010. Communication and Collective Action: Language and the Evolution of Human Cooperation. *Evol. Hum. Behav.* **31**:231–245. [3]

Smith, E. A., and R. L. Bliege Bird. 2000. Turtle Hunting and Tombstone Opening: Public Generosity as Costly Signaling. *Evol. Hum. Behav.* **21**:245–261. [3]

Smith, K., and E. Wonnacott. 2010. Eliminating Unpredictable Variation through Iterated Learning. *Cognition* **116**:444–449. [12]

Snow, C. E., S. Burns, and P. Griffin, eds. 1998. Preventing Reading Difficulties in Young Children. Washington, DC: National Academy Press. [14]

Sober, E., and D. S. Wilson. 1991. Unto Others: The Evolution and Psychology of Unselfish Behavior. Cambridge, MA: Harvard Univ. Press. [4]

———. 1994. Reintroducing Group Selection to the Human Behavioral Sciences. *Behav. Brain Sci.* **17**:585–654. [3]

Soler, M. 2012. Costly Signaling, Ritual and Cooperation: Evidence from Candomblé, an Afro-Brazilian Religion. *Evol. Hum. Behav.* **33**:346–356. [19]

Solomon, M. 2001. Social Empiricism. Cambridge, MA: MIT Press. [10]

Solomon, N. G., and J. A. French, eds. 1997. Cooperative Breeding in Mammals. Cambridge: Cambridge Univ. Press. [3]

Soltis, J., R. Boyd, and P. J. Richerson. 1995. Can Group-Functional Behaviors Evolve by Cultural-Group Selection: An Empirical Test. *Curr. Anthropol.* **36**:473–494. [1, 6]

Sorensen, A. P., Jr. 1967. Multilingualism in the Northwest Amazon. *Am. Anthropol.* **69**:70–84. [16]

Sørensen, J. 2006. A Cognitive Theory of Magic. Lanham, MD: AltaMira Press. [20]

———. 2007. Acts That Work: A Cognitive Approach to Ritual Agency. *Meth. Theory Stud. Rel.* **19**:281–300. [20]

Sorrenson, R. 1996. The Ship as a Scientific Instrument in the Eighteenth Century. *Osiris* **11**:221–235. [1]

Sosis, R. 2000. Religion and Intragroup Cooperation: Preliminary Results of a Comparative Analysis of Utopian Communities. *Cross-Cultural Res.* **34**:70–87. [20]

———. 2003. Why Aren't We All Hutterities? Costly Signaling Theory and Religious Behavior. *Hum. Nature* **14**:91–127. [20]

———. 2011. Why Sacred Lands Are Not Indivisible: The Cognitive Foundations of Sacralising Land. *J. Terrorism Res.* **2**:17–44. [20]

Sosis, R., and C. Alcorta. 2003. Signalling, Solidarity, and the Sacred: The Evolution of Religious Behavior. *Evol. Anthropol.* **12**:264–274. [17–20]

Sosis, R., and E. R. Bressler. 2003. Cooperation and Commune Longevity: A Test of the Costly Signaling Theory of Religion. *Cross-Cultural Res.* **37**:211–239. [1, 17, 19, 20]

Sosis, R., H. Kress, and J. Boster. 2007. Scars for War: Evaluating Signaling Explanations for Cross-Cultural Variance in Ritual Costs. *Evol. Hum. Behav.* **28**:234–247. [17, 20]

Sosis, R., and B. J. Ruffle. 2003. Religious Ritual and Cooperation: Testing for a Relationship on Israeli Religious and Secular Kibbutzim. *Curr. Anthropol.* **44**:713–722. [17, 19]

Spencer, C. S. 1998. A Mathematical Model of Primary State Formation. *Cult. Dynam.* **10**:5–20. [4]

Spencer, H. 1862. First Principles. New York: A. L. Burt. [1]

Spencer, J. P., M. S. Blumberg, B. McMurray, et al. 2009. Short Arms and Talking Eggs: Why We Should No Longer Abide the Nativist–Empiricist Debate. *Child Devel. Perspect.* **3**:79–87. [1]

Sperber, D. 1984. Anthropology and Psychology: Towards an Epidemiology of Representations. *Man* **20**:73–89. [1]

———. 1990. The Epidemiology of Beliefs. In: The Social Psychological Study of Widespread Beliefs, ed. C. Fraser and G. Gaskell. New York: Oxford Univ. Press. [20]

Spuhler, J. N. 1959. The Evolution of Man's Capacity for Culture. Detroit: Duke Univ. Press. [20]

Stanford, J. N. 2008a. Child Dialect Acquisition: New Perspectives on Parent/Peer Influence. *J. Socioling.* **12**:567–596. [13]

Stanford, J. N. 2008b. A Sociotonetic Analysis of Sui Dialect Contact. *Lang. Var. Change* **20**:409–450. [13]

Stark, R. 2001. Gods, Ritual, and the Moral Order. *J. Sci. Study Rel.* **40**:619–636. [19]

Steele, J., and S. Shennan. 2009. Special Theme Issue: Demography and Cultural Macroevolution. *Hum. Biol.* 81, http://digitalcommons.wayne.edu/cgi/viewcontent.cgi?article=1048&context=humbiol. (accessed April 27, 2013). [1]

Steels, L. 1995. A Self-Organizing Spatial Vocabulary. *Artif. Life* **2**:319–332. [16]

———. 1997. The Synthetic Modeling of Language Origins. *Evol. Comm.* **1**:1–34. [1]

Stenberg, G. 2009. Selectivity in Infant Social Referencing. *Infancy* **14**:457–473. [7]

Sterck, E. H. M., D. P. Watts, and C. P. van Schaik. 1997. The Evolution of Female Social Relationships in Nonhuman Primates. *Behav. Ecol. Sociobiol.* **41**:291–309. [2]

Sterckx, R. 2009. The Economics of Religion in Warring States and Early Imperial China. In: Early Chinese Religion: Part One: Shang through Han (1250 BC–22 AD), ed. J. Lagerwey and M. Kalinowski. Leiden: Brill. [17]

Sterelny, K. 2006. The Evolution and Evolvability of Culture. *Mind Lang.* **21**:137–165. [20]

———. 2011. The Evolved Apprentice. Cambridge, MA: MIT Press. [20]

———. 2012. The Evolved Apprentice: How Evolution Made Humans Unique. Cambridge, MA: MIT Press. [1, 6]

Stevens, J. M. G., H. Vervaecke, H. de Vries, and L. van Elsacker. 2007. Sex Differences in the Steepness of Dominance Hierarchies in Captive Bonobo Groups. *Intl. J. Primatol.* **28**:1417–1430. [2]

Steward, J. H. 1955. Theory of Culture Change: The Methodology of Multilinear Evolution. Urbana: Univ. of Illinois Press. [1]

Stiner, M. C. 2002. Carnivory, Coevolution, and the Geographic Spread of the Genus *Homo*. *J. Arch. Res.* **10**:1–63. [2]

Stiner, M. C., R. Barkai, and A. Gopher. 2009. Cooperative Hunting and Meat Sharing 400–200 kya at Qesem Cave , Israel. *PNAS* **106**: 13,207–13,212. [2]

Stoakes, H., A. Butcher, J. Fletcher, and M. Tabain. 2011. Long term average speech spectra in Yolngu Matha and Pitjantjatjara speaking females and males. In: Interspeech 2011, pp. 1897–1900. Red Hook, NY: Curran Associates, Inc. [13]

Stout, D. 2002. Skill and Cognition in Stone Tool Production: An Ethnographic Case Study from Irian Jaya. *Curr. Anthropol.* **45**:693–722. [9]

———. 2010. The Evolution of Cognitive Control. *Top. Cogn. Sci.* **2**:614–630. [9]

Stout, D., and T. Chaminade. 2007. The Evolutionary Neuroscience of Tool Making. *Neuropsychologia* **45**:1091–1100. [9]

———. 2012. Stone Tools, Language and the Brain in Human Evolution. *Phil. Trans. R. Soc. B* **367**:75–87. [9]

Stout, D., R. Passingham, C. Frith, J. Apel, and T. Chaminade. 2011. Technology, Expertise and Social Cognition in Human Evolution. *Eur. J. Neurosci.* **33**:1328–1338. [9]

Stout, D., N. Toth, K. D. Schick, and T. Chaminade. 2008. Neural Correlates of Early Stone Age Tool-Making: Technology, Language and Cognition in Human Evolution. *Phil. Trans. R. Soc. B* **363**:1939–1949. [9]

Street, J., and E. Dąbrowska. 2010. More Individual Differences in Language Attainment: How Much Do Adult Native Speakers of English Know About Passives and Quantifiers? *Lingua* **120**:2080–2094. [14, 16]

Strier, K. B. 1987. Activity Budgets of Woolly Spider Monkeys, or Muriquis. *Am. J. Primatol.* **13**:385–395. [2]

———. 1992. Causes and Consequences of Nonaggression in the Woolly Spider Monkey, or Muriqui. In: Aggression and Peacefulness in Humans and Other Primates, ed. J. Silverberg and J. P. Gray, pp. 100–116. New York: Oxford Univ. Press. [2]

Strimling, P., J. Sjostrand, M. Enquist, and K. Eriksson. 2009. Accumulation of Independent Cultural Traits. *Theor. Pop. Biol.* **76**:77–83. [11]

Stromswold, K. 2001. The Heritability of Language: A Review and Metaanalysis of Twin, Adoption, and Linkage Studies. *Language* **77**:647–723. [12, 16]

Sugden, R. 2003. The Logic of Team Reasoning. *Phil. Explor.* **6**:165–181. [2]

Sugie, M., H. Ohba, M. Mizutani, and N. Ohno. 1993. Hard Palate Shape in the Japanese and Indian Children on Moiré Topography with Lateral Movement of a Grating (Article in Japanese). *Kaibogaku zasshi* **68**:522–535. [12]

Sugiyama, L. S. 2004. Illness, Injury, and Disability among Shiwiar Forager-Horticulturalists: Implications of Health-Risk Buffering for the Evolution of Human Life History. *Am. J. Phys. Anthropol.* **123**:371–389. [3]

Sugiyama, L. S., and R. Chacon. 2000. Effects of Illness and Injury on Foraging among the Yora and Shiwiar: Pathology Risk as Adaptive Problem. In: Adaptation and Human Behavior: An Anthropological Perspective, ed. L. Cronk et al., pp. 371–395. New York: Aldine de Gruyter. [2]

Swadesh, M. 1952. Lexicostatistic Dating of Prehistoric Ethnic Contacts. *Proc. Am. Phil. Soc.* **96**:452–463. [15]

Swann, W. B., J. Jensen, A. Gómez, H. Whitehouse, and B. Bastian. 2012. When Group Membership Gets Personal: A Theory of Identity Fusion. *Psychol. Rev.* **119**:441–456. [20]

Swanson, G. E. 1960. The Birth of the Gods: The Origin of Primitive Beliefs. Ann Arbor: Univ. of Michigan Press. [17]

———. 1964. The Birth of the Gods: The Origin of Primitive Beliefs. Ann Arbor: Univ. of Michigan Press. [17, 20]

———. 1966. The Birth of the Gods. Ann Arbor: Univ. of Michigan Press. [19]

Sweetser, E. 1990. From Etymology to Pragmatics: Metaphorical and Cultural Aspects of Semantic Structure. Cambridge: Cambridge Univ. Press. [13]

Tainter, J. A. 1988. The Collapse of Complex Societies. Cambridge: Cambridge Univ. Press. [6]

Tajfel, H., M. G. Billig, R. P. Bundy, and C. Flament. 1971. Social Categorization and Intergroup Behaviour. *Eur. J. Social Psychol.* **1**:149–178. [5]

Tarde, G. 1903. The Laws of Imitation. New York: Holt. [1]

Taves, A. 2009. Religious Experience Reconsidered: A Building Block Approach to the Study of Religion and Other Special Things. Princeton: Princeton Univ. Press. [20]

Tavory, I., E. Jablonka, and S. Ginsburg. 2013. The Reproduction of the Social. In: Scaffolding in Evolution, Culture and Cognition: Vienna Series in Theoretical Biology, ed. L. R. Caporael et al. Cambridge, MA: MIT Press, in press. [18]

Taylor, C. 1989. Sources of the Self: The Makings of Modern Identity. Cambridge, MA: Harvard Univ. Press. [17, 20]

Taylor, P. D. 1992a. Altruism in Viscous Populations: An Inclusive Fitness Model. *Evol. Ecol.* **6**:352–356. [3]

———. 1992b. Inclusive Fitness in a Homogeneous Environment. *Proc. R. Soc. B* **249**:352–356. [3]

Tehrani, J. J., and M. Collard. 2002. Investigating Cultural Evolution through Biological Phylogenetic Analyses of Turkmen Textiles. *J. Anthropol. Arch.* **21**:443–463. [11]

Tehrani, J. J., and M. Collard. 2009. On the Relationship between Interindividual Cultural Transmission and Population-Level Cultural Diversity: A Case Study of Weaving in Iranian Tribal Populations. *Evol. Hum. Behav.* **30**:286–300. [11]

Tennie, C., J. Call, and M. Tomasello. 2009. Ratcheting up the Ratchet: On the Evolution of Cumulative Culture. *Phil. Trans. R. Soc. B* **364**:2405–2415. [5, 6, 11]

Tennie, C., and H. Over. 2012. Cultural Intelligence Is Key to Explaining Human Tool Use. *Behav. Brain Sci.* **35**:242–243. [11]

Tetlock, P. E. 2003. Thinking the Unthinkable: Sacred Values and Taboo Cognitions. *Trends Cogn. Sci.* **7**:320–324. [17, 20]

Thagard, P. 1992. Conceptual Revolutions. Princeton: Princeton Univ. Press. [11]

The Economist. 2011. Killings in Liberia: Nasty Business, Feb. 3. http://www.economist.com/node/18073315. [17]

Thelen, M. H., D. J. Miller, P. A. Fehrenbach, N. M. Frautschi, and M. D. Fishbein. 1980. Imitation During Play as a Means of Social Influence. *Child. Devel.* **51**:918–920. [5]

Thieme, H. 1997. Lower Palaeolithic Hunting Spears from Germany. *Nature* **385**:807–810. [2]

Thomason, S. G., and T. Kaufman. 1988. Language Contact, Creolization, and Genetic Linguistics. Berkeley: Univ. of California Press. [1, 16]

Thurston, W. R. 1987. Processes of Change in the Languages of North-Western New Britain. Canberra: Pacific Linguistics. [15]

———. 1992. Sociolinguistic Typology and Other Factors Affecting Change in North-Western New Britain, Papua New Guinea. In: Culture Change, Language Change: Case Studies from Melanesia, ed. T. Dutton. Canberra: Pacific Linguistics. [13]

Tinbergen, N. 1963. On Aims and Methods of Ethology. *Zt. Tierpsychol.* **20**:410–433. [6]

Tomasello, M. 1994. The Question of Chimpanzee Culture. In: Chimpanzee Cultures, ed. R. Wrangham et al., pp. 301–317. Cambridge, MA: Harvard Univ. Press. [8]

———. 1999. The Cultural Origins of Human Cognition. Cambridge, MA: Harvard Univ. Press. [1, 5, 10, 11, 20]

———. 2003. Constructing a Language: A Usage-Based Approach. Cambridge, MA: Harvard Univ. Press. [14]

———. 2005. Constructing a Language: A Usage-Based Theory of Language Acquisition: Harvard Univ. Press. [1]

———. 2008. Origins of Human Communication. Cambridge, MA: MIT Press. [5, 14, 16]

Tomasello, M., J. Call, J. Warren, et al. 1997. The Ontogeny of Chimpanzee Gestural Signals: A Comparison across Groups and Generations. *Evol. Comm.* **1**:223–259. [5]

Tomasello, M., and M. Carpenter. 2007. Shared Intentionality. *Devel. Sci.* **10**:121–125. [2]

Tomasello, M., M. Carpenter, J. Call, T. Behne, and H. Moll. 2005. Understanding and Sharing Intentions: The Origins of Cultural Cognition. *Behav. Brain Sci.* **28**:675–691. [2, 5, 7]

Tomasello, M., A. C. Kruger, and H. H. Ratner. 1993. Cultural Learning. *Behav. Brain Sci.* **16**:495–511. [5, 11]

Tomblin, J. B., and M. H. Christiansen. 2009. Explaining Developmental Communication Disorders. In: Speech Sound Disorders in Children. In Honor of Lawrence D. Shriberg, ed. R. Paul and P. Flipsen, pp. 35–49. San Diego: Plural Publishing. [12]

Tooby, J., and L. Cosmides. 1989. Evolutionary Psychology and the Generation of Culture: Theoretical Considerations. *Ethol. Sociobiol.* **10**:29–49. [1]

———. 1992. The Psychological Foundations of Culture. In: The Adapted Mind: Evolutionary Psychology and the Generation of Culture, ed. J. Barkow et al., pp. 19–136. New York: Oxford Univ. Press. [1]

———. 2005. Conceptual Foundations of Evolutionary Psychology. In: The Handbook of Evolutionary Psychology, ed. D. M. Buss, pp. 5–67. Hoboken: Wiley. [16]

Tooby, J., and I. DeVore. 1987. The Reconstruction of Hominid Evolution through Strategic Modeling. In: The Evolution of Human Behavior: Primate Models, ed. W. G. Kinzey, pp. 183–237. Albany: SUNY Press. [2]

Townsend, G. C., L. C. Richards, M. Sekikawa, T. Brown, and T. Ozaki. 1990. Variability of Palatal Dimensions in South Australian Twins. *J. Forensic Odonto.* **8**:3–14. [12]

Traugott, E. C. 1980. Meaning-Change in the Development of Grammatical Markers. *Lang. Sci.* **2**:44–61 [1]

Trivers, R. L. 1971. The Evolution of Reciprocal Altruism. *Q. Rev. Biol.* **46**:35–57. [3]

Trudgill, P. 1986. Dialects in Contact. Oxford: Basil Blackwell. [16]

———. 2008. Colonial Dialect Contact in the History of European Languages: on the Irrelevance of Identity to New-Dialect Formation. *Lang. Soc.* **37**:241–280. [16]

———. 2011. Sociolinguistic Typology: Social Determinants of Linguistic Complexity. Oxford: Oxford Univ. Press. [13, 15, 16]

Turchin, P. 2003. Historical Dynamics: Why States Rise and Fall. Princeton: Princeton Univ. Press. [6, 20]

———. 2006. War and Peace and War: The Life Cycles of Imperial Nations. New York: Pi Press. [4, 6, 20]

———. 2008. Arise "Cliodynamics." *Nature* **454**:34–35. [1]

———. 2009. A Theory for Formation of Large Empires. *J. Glob. History* **4**:191–217. [4, 6, 19, 20]

———. 2011. Warfare and the Evolution of Social Complexity: A Multilevel Selection Approach. *Struc. Dynam.* **4**:1–37. [4, 20]

Turchin, P., and S. Gavrilets. 2009. Evolution of Complex Hierarchical Societies. *Soc. History Evol.* **8(2)**:167–198. [4]

Turchin, P., S. Gavrilets, and L. Fortunato. 2012a. Towards a Formal Theory for the Evolution of Human Social Complexity. Nimbios Investigative Workshop. http://www.nimbios.org/workshops/WS_social_complexity. [4]

Turchin, P., and S. A. Nefedov. 2009. Secular Cycles. Princeton: Princeton Univ. Press. [1]

Turchin, P., H. Whitehouse, P. Francois, E. Slingerland, and M. Collard. 2012b. A Historical Database of Cultural Evolution. *Cliodynamics* **3**:271–293. [4]

Turiel, E. 2010. Domain Specificity in Social Interactions, Social Thought, and Social Development. *Child. Devel.* **81**:1467–8624. [20]

Turner, J. C. 1991. Social Influence. Buckingham: Open Univ. Press. [5]

Turner, J. H., and A. Maryanski. 2008. On the Origin of Societies by Natural Selection. Boulder: Paradigm Publishers. [20]

Turner, V. W. 1990. Drama, Fields, and Metaphors: Symbolic Action in Human Society. Ithaca: Cornell Univ. Press. [20]

Tversky, A., and D. Kahneman. 2002. Extensional Versus Intuitive Reasoning: The Conjunction Fallacy in Probability Judgment. In: Heuristics and Biases: The Psychology of Intuitive Judgment, ed. T. Gilovich et al., pp. 19–48. Cambridge: Cambridge Univ. Press. [10]

Tweney, R. 2011. Toward a Cognitive Understanding of Science and Religion. In: Epistemology and Science Education: Understanding the Evolution Vs. Intelligent Design Controversy, ed. R. Taylor and M. Ferrari, pp. 197–212. New York: Routledge. [10]

Tybur, J. M., G. F. Miller, and S. W. Gangestad. 2007. Testing the Controversy. *Hum. Nature* **18**:313–328. [1]

Tylor, E. B. 1871. Primitive Culture: Research into the Development of Mythology, Philosophy, Religion, Art and Custom. London: Murray. [1]

———. 1873. Primitive Culture: Researches into the Development of Mythology, Philosophy, Religion, Language, Art, and Custom (2nd edition). London: John Murray. [20]

Umiltà, M. A., L. Escola, I. Intskirveli, et al. 2008. When Pliers Become Fingers in the Monkey Motor System. *PNAS* **105**:2209–2213. [9]

Urgesi, C., M. A. Salvatore, M. S., and F. Fabbro. 2010. The Spiritual Brain: Selective Cortical Lesions Modulate Human Self-Transcendence. *Neuron* **65**:309–319. [18]

Užgiris, I. C. 1981. Two Functions of Imitation During Infancy. *Intl. J. Behav. Develop.* **4**:1–12. [5]

Valdesolo, P., and D. DeSteno. 2011. Synchrony and the Social Tuning of Compassion. *Emotion* **11**:262–266. [17]

Valdesolo, P., J. Ouyang, and D. DeSteno. 2010. The Rhythm of Joint Action: Synchrony Promotes Cooperative Ability. *J. Exp. Social Psychol.* **46**:693–695. [17]

van Leeuwen, E. J. C., and D. B. M. Haun. 2013. Conformity in Nonhuman Primates: Fad or Fact? *Evol. Hum. Behav.* **34**:1–7. [5]

Van Nierop, O. A., A. C. M. Blankendaal, and C. J. Overbeeke. 1997. The Evolution of the Bicycle: A Dynamic Systems Approach. *J. Design His.* **3**:253–267. [8]

van Schaik, C. P. 1983. Why Are Diurnal Primates Living in Groups? *Behaviour* **87**:120–144. [2]

van Schaik, C. P., and J. M. Burkart. 2010. Mind the Gap: Cooperative Breeding and the Evolution of Our Unique Features. In: Mind the Gap: Tracing the Origins of Human Universals, ed. P. M. Kappeler and J. B. Silk, pp. 477–496. Berlin: Springer-Verlag. [2, 3]

van Schaik, C. P., and P. M. Kappeler. 2006. Cooperation in Primates and Humans. Berlin: Springer-Verlag. [3]

van Schaik, C. P., S. A. Pandit, and E. R. Vogel. 2004a. A Model for Within-Group Coalitionary Aggression among Males. *Behav. Ecol. Sociobiol.* **57**:101–109. [6]

———. 2006. Toward a General Model for Male-Male Coalitions in Primate Groups. In: Cooperation in Primates and Humans: Mechanisms and Evolution, ed. P. M. Kappeler and C. P. van Schaik, pp. 151–171. Berlin: Springer-Verlag. [2]

van Schaik, C. P., G. R. Pradhan, and M. A. van Noordwijk. 2004b. Mating Conflict in Primates: Infanticide, Sexual Harassment and Female Sexuality. In: Sexual Selection in Primates, ed. P. M. Kappeler and C. P. van Schaik, pp. 131–150. Cambridge: Cambridge Univ. Press. [2]

Vansina, J. 1990. Paths in the Rainforests: Toward a History of Political Tradition in Equatorial Africa. London: Currey. [6]

van Veelen, M. J., and J. García. 2010. In and out of Equilibrium: Evolution of Strategies in Repeated Games with Discounting. Tinbergen Institute Discussion paper 10-037/1 http://papers.tinbergen.nl/10037.pdf (accessed June 11, 2013). [3]

van Veelen, M. J., J. García, D. Rand, and M. Nowak. 2012. Direct Reciprocity in Structured Populations. *PNAS* **109**:9929–9934. [3]

Vaughn, K., J. Eerkens, and J. Kantner, eds. 2009. The Evolution of Leadership: Transitions in Decision Making from Small-Scale to Middle-Range Societies. Santa Fe: SAR Press. [6]

Velicer, G. J., and Y. N. Yu. 2003. Evolution of Novel Cooperative Swarming in the Bacterium *Myxococcus Xanthus*. *Nature* **425**:75–78. [3]

Vernes, S. C., D. F. Newbury, B. S. Abrahams, et al. 2008. A Functional Genetic Link between Distinct Developmental Language Disorders. *N. Engl. J. Med.* **359**:2337–2345. [12]

Vigilant, L., M. Hofreiter, H. Siedel, and C. Boesch. 2001. Paternity and Relatedness in Wild Chimpanzee Communities. *PNAS* **98**:12,890–12,895. [2]

Visscher, P. M., W. G. Hill, and N. R. Wray. 2008. Heritability in the Genomics Era: Concepts and Misconceptions. *Nat. Rev. Genet.* **9**:255–266. [12]

Vogeley, K., and A. Roepstorff. 2009. Contextualising Culture and Social Cognition. *Trends Cogn. Sci.* **13**:511–516. [20]

Vrba, E. S. 1995. The Fossil Record of African Antelopes (Mammalia, Bovidae) in Relation to Human Evolution and Paleoclimate. In: Paleoclimate and Evolution with Emphasis on Human Origin, ed. E. S. Vrba et al., pp. 385–424. New Haven: Yale Univ. Press. [2]

Vygotsky, L. S. 1978. Mind in Society: The Development of Higher Psychological Processes. Cambridge, MA: Harvard Univ. Press. [1, 9, 11]

Waddington, C. H. 1957. The Strategy of the Genes. London: Allen & Unwin. [18]

Wade, N. 2009. The Faith Instinct: How Religion Evolved and Why It Endures. New York: Penguin. [20]

Wadley, L. 2010. Were Snares and Traps Used in the Middle Stone Age and Does It Matter? A Review and a Case Study from Sibudu, South Africa. *J. Hum. Evol.* **58**:179–192. [8]

Walker, M. B., and M. G. Andrade. 1996. Conformity in the Asch Task as a Function of Age. *J. Social Psychol.* **136**:367–372. [5]

Walker, R. S., and L. A. Ribeiro. 2011. Bayesian Phylogeography of the Arawak Expansion in Lowland South America. *Proc. R. Soc. B* **278**:2562–2567. [15]

Wang, C.-C., Q.-L. Ding, H. Tao, and H. Li. 2012. Comment on "Phonemic Diversity Supports a Serial Founder Effect Model of Language Expansion from Africa." *Science* **335**:657. [15]

Wason, P. C. 1966. Reasoning. In: New Horizons in Psychology, ed. B. M. Foss, pp. 135–151. Harmondsworth: Penguin. [10]

Weir, A. A. S., J. Chappell, and A. Kacelnik. 2002. Shaping of Hooks in New Caledonian Crows. *Science* **297**:981. [10]

Weisberg, J., M. van Turennout, and A. Martin. 2007. A Neural System for Learning About Object Function. *Cereb. Cortex* **17**:513–521. [9]

Weizman, Z. O., and C. E. Snow. 2001. Lexical Input as Related to Children's Vocabulary Acquisition: Effects of Sophisticated Exposure and Support for Meaning. *Devel. Psychol.* **37**:265–279. [14]

Wells, J., M. H. Christiansen, D. S. Race, D. Acheson, and M. C. MacDonald. 2009. Experience and Sentence Processing: Statistical Learning and Relative Clause Comprehension. *Cogn. Psychol.* **58**:250–271. [14]

West, S. A., A. S. Griffin, A. Gardner, and S. P. Diggle. 2006. Social Evolution Theory for Microorganisms. *Nat. Rev. Microbiol.* **4**:597–607. [3]

West-Eberhard, M. J. 2003. Developmental Plasticity and Evolution. New York: Oxford Univ. Press. [16]

Whallon, R. 1989. The Human Revolution. In: The Human Revolution: Behavioural and Biological Perspectives on the Origins of Modern Humans, ed. P. Mellars and C. Stringer. Princeton: Princeton Univ. Press. [2]

White, L. 1943. Energy and the Evolution of Culture. *Am. Anthropol.* **45**:335–356. [9]

White, N. 1997. Genes, Languages and Landscapes in Australia. In: Archaeology and Linguistics: Global Perspectives on Ancient Australia, ed. P. McConvell and N. Evans, pp. 45–81. Melbourne: Oxford Univ. Press. [16]

Whitehead, H. 2007. Learning, Climate and the Evolution of the Culture Capacity. *J. Theor. Biol.* **245**:341–350. [1]

Whitehouse, H. 1995. Inside the Cult: Religious Innovation and Transmission in Papua New Guinea. Oxford: Oxford Univ. Press. [18]

———. 1996. Rites of Terror: Emotion, Metaphor and Memory in Melanesian Cults. *J. R. Anthropol. Inst.* **2**:703–715. [20]

———. 2000. Arguments and Icons: Divergent Modes of Religiousity. Oxford: Oxford Univ. Press. [18, 20]

———. 2004. Modes of Religiousity: A Cognitive Theory of Religious Transmission. Walnut Creek, CA: Altamira Press. [17, 18, 20]

———. 2008. Cognitive Evolution and Religion: Cognition and Religious Evolution. In: The Evolution of Religion: Studies, Theories and Critiques, ed. J. Bulbulia et al. Santa Margarita, CA: Collins Foundation Press. [20]

———. 2011. The Coexistence Problem in Psychology, Anthropology, and Evolutionary Theory. *Hum. Devel.* **54**:191–199. [20]

Whitehouse, H., and I. Hodder. 2010. Modes of Religiosity at Çatalhöyük. In: Religion in the Emergence of Civilization, ed. I. Hodder. Cambridge: Cambridge Univ. Press. [19, 20]

Whitehouse, H., K. Kahn, M. E. Hochberg, and J. J. Bryson. 2012. The Role for Simulations in Theory Construction for the Social Sciences: Case Studies Concerning Divergent Modes of Religiosity. *Relig. Brain Behav.* **2**:182–201. [18, 20]

Whiten, A., and D. Custance. 1996. Studies of Imitation in Chimpanzees and Children. In: Social Learning in Animals: The Roots of Culture, ed. B. G. Galef and C. M. Heyes, pp. 291–318. Academic Press. [1]

Whiten, A., J. Goodall, W. C. McGrew, et al. 1999. Cultures in Chimpanzees. *Nature* **399**:682–685. [11]

Whiten, A., and R. Ham. 1992. On the Nature and Evolution of Imitation in the Animal Kingdom: Reappraisal of a Century of Research. In: Advances in the Study of Behavior, ed. P. J. B. Slater et al., pp. 239–283. New York: Academic Press. [11]

Whiten, A., R. A. Hinde, K. N. Laland, and C. B. Stringer. 2011. Culture Evolves. *Phil. Trans. R. Soc. B* **366**:938–948. [1]

Whiten, A., N. McGuigan, S. Marshall-Pescini, and L. M. Hopper. 2009. Emulation, Imitation, Overimitation and the Scope of Culture for Child and Chimpanzee. *Phil. Trans. R. Soc. B* **364**:2417–2428. [1, 5, 6]

Whorf, B. L. 1956. Language, Thought and Reality. Cambridge, MA: MIT Press. [13]

Widlok, T. 1999. Living on Mangetti: "Bushmen" Autonomy and Namibian Independence. Oxford: Oxford Univ. Press. [20]

———. 2007. From Individual Act to Social Agency in San Trance Rituals. In: Strength Beyond Structure: Social and Historical Trajectories of Agency in Africa, ed. R. v. Dijk et al. Leiden: Brill. [20]

———. 2010. What Is the Value of Rituals? In: Ritual Dynamics and the Science of Ritual, ed. A. Chaniotis, vol. 2. Wiesbaden: Harrassowitz. [20]

Widule, C. J., V. Foley, and G. Demo. 1978. Dynamics of the Axe Swing. *Ergonomics* **21**:925–930. [7]

Wiessner, P. 1986. !Kung San Networks in a Generational Perspective. In: The Past and Future of !Kung Ethnography, ed. M. Biesele et al., pp. 103–136. Hamburg: Helmut Buske Verlag. [6]

———. 2005. Norm Enforcement among the Ju/'hoansi Bushmen: A Case of Strong Reciprocity? *Hum. Nature* **16**:115–145. [6]

———. 2006. From Spears to M16s: Testing the Imbalance of Power Hypothesis among the Enga. *J. Anthropol. Res.* **62**:165–191. [2]

———. 2009. The Power of One: The Big Man Revisited. In: The Evolution of Leadership: Transitions in Decision Making from Small-Scale to Middle-Range Societies, ed. K. J. Vaugh et al., pp. 195–221. Santa Fe: SAR Press. [2]

Wiessner, P., and A. Tumu. 1998. Historical Vines: Enga Networks of Exchange, Ritual, and Warfare in Papua New Guinea. Washington, DC: Smithsonian Institution Press. [1, 6]

Wilkinson, M. 1991. Djambarrpuyngu. A Yolngu Variety of Northern Australia. Ph.D. thesis, Univ. of Sydney. [13]

Willey, P. 1990. Prehistoric Warfare on the Great Plains. New York: Garland. [3]

Williams, B. J. 1974. A Model of Band Society. Memoir 29: Society for American Archaelogy. [6]

Wilson, D. S. 2002. Darwin's Cathedral: Evolution, Religion, and the Nature of Society. Chicago: Univ. of Chicago Press. [4, 17–20]

———. 2005. Testing Major Evolutionary Hypotheses About Religion with a Random Sample. *Hum. Nature* **16**:419–446. [4, 20]

———. 2007a. Evolution for Everyone. New York: Delacorte Press. [19]

———. 2007b. Group-Level Evolutionary Processes. In: The Oxford Handbook of Evolutionary Psychology, ed. R. Dunbar and L. Barret, pp. 49–55. Oxford: Oxford Univ. Press. [20]

———. 2008. Evolution and Religion: The Transformation of the Obvious. In: The Evolution of Religion: Studies, Theories and Critiques, ed. J. Bulbulia et al., pp. 23–29. Santa Margarita, CA: Collins Foundation Press. [20]

Wilson, D. S., G. B. Pollock, and L. A. Dugatkin. 1992. Can Altruism Evolve in Purely Viscous Populations? *Evol. Ecol.* **6**:331–341. [3]

Wilson, D. S., and E. O. Wilson. 2007. Rethinking the Theoretical Foundations of Sociobiology. *Q. Rev. Biol.* **82**:327–348. [4, 20]

Wilson, E. O. 1975. Sociobiology The New Synthesis. Cambridge, MA: Harvard Univ. Press. [1, 2]

Wilson, M. L., S. M. Kahlenberg, M. Wells, and R. W. Wrangham. 2012. Ecological and Social Factors Affect the Occurrence and Outcomes of Intergroup Encounters in Chimpanzees. *Anim. Behav.* **83**:277–291. [5]

Wiltermuth, S. S., and C. Heath. 2009. Synchrony and Cooperation. *Psychol. Sci.* **20**:1–5. [17, 20]

Wimsatt, W. C. 1986. Developmental Constraints, Generative Entrenchment, and the Innate-Acquired Distinction. In: Integrating Scientific Disciplines, ed. W. Bechtel, pp. 185–208. Dordrecht: Martinus-Nijhoff. [8]

Wimsatt, W. C., and J. R. Griesemer. 2007. Reproducing Entrenchments to Scaffold Culture: The Central Role of Development in Cultural Evolution. In: Integrating Evolution and Development: From Theory to Practice, ed. R. Sansome and R. Brandon, pp. 228–323. Cambridge, MA: MIT Press. [8]

Winata, S., I. N. Arhya, S. Moeljopawiro, et al. 1995. Congenital Non-Syndromal Autosomal Recessive Deafness in Bengkala, an Isolated Balinese Village. *J. Med. Genet.* **32**:336–343. [12]

Winterhalder, B., and E. A. Smith. 1992. Evolutionary Ecology and Human Behavior. New York: Aldine de Gruyter. [2]

Wobst, H. M. 1974. Boundary Conditions for Paleolithic Social Systems: A Simulation Approach. *Am. Antiq.* **39**:147–178. [6]

Wollstonecroft, M. M. 2011. Investigating the Role of Food Processing in Human Evolution: A Niche Construction Approach. *Arch. Anthropol. Sci.* **3**:141–150. [11]

Wolpert, D., K. Doya, and M. Kawato. 2003. A Unifying Computational Framework for Motor Control and Social Interaction. *Phil. Trans. R. Soc. B* **358**:593–602. [9]

Wolpert, L. 1992. The Unnatural Nature of Science. London: Faber and Faber. [10]

Wong, P. C. M., B. Chandrasekaran, and J. Zheng. 2012. The Derived Allele of ASPM Is Associated with Lexical Tone Perception. *PLoS One* **7**:e34243. [12]

Wood, L. A., R. Kendal, and E. Flynn. 2013. Copy You or Copy Me? The Effect of Prior Personally-Acquired, and Alternative Method, Information on Imitation. *Cognition* **127**:203–213. [11]

Woodburn, J. 1970. Hunters and Gatherers: The Material Culture of the Nomadic Hadza. London: British Museum. [20]

———. 1982. Egalitarian Societies. *Man* **17**:431–451. [2]

Woolgar, S. 1982. Laboratory Studies: A Comment on the State of the Art. *Soc. Stud. Sci.* **12**:481–498. [1]

Wrangham, R. W. 1980. An Ecological Model of Female-Bonded Primate Groups. *Behaviour* **75**:262–300. [2]

———. 1999. Evolution of Coalitionary Killing. *Am. J. Phys. Anthropol.* **Suppl 29**:1–30. [5]

———. 2009. Catching Fire: How Cooking Made Us Human. London: Profile. [6]

Wrangham, R. W., and R. Carmody. 2010. Human Adaptation to the Control of Fire. *Evol. Anthropol.* **19**:187–199. [2, 6]

Wrangham, R. W., and D. Peterson. 1996. Demonic Males: Apes and the Origins of Human Violence. New York: Mariner Books. [2]

Wright, R. 2009. The Evolution of God. Boston: Little Brown and Company. [17, 19, 20]

Wright, S. 1932. The Roles of Mutation, Inbreeding, Crossbreeding and Selection in Evolution. *Proc. 6th Intl. Congr. Genet.* **1**:356–366. [11]

Wrong, D. H. 1961. The Oversocialized Conception of Man in Modern Sociology. *Am. Sociol. Rev.* **26**:183–193. [2]

Wu, J., and R. Axelrod. 1995. Coping with Noise in the Iterated Prisoner's Dilemma. *J. Conflict Res.* **39**:183–189. [3]

Xygalatas, D. 2013. Effects of Religious Setting on Cooperative Behavior: A Case Study from Mauritius. *Relig. Brain Behav.* **3**:91–102. [19]

Xygalatas, D., I. Konvalinka, A. Roepstoff, and J. Bulbulia. 2011. Quantifying Collective Effervescence Heart-Rate Dynamics at a Fire-Walking Ritual. *Cogn. Integ. Biol.* **4**:735–738. [20]

Xygalatas, D., P. Mitkidis, R. Fischer, et al. 2013. Extreme Rituals Promote Prosociality. *Psychol. Sci.* doi:10.1177/0956797612472910. [19]

Yeaman, S., R. Bshary, and L. Lehmann. 2011. The Effect of Innovation and Sex-Specific Migration on Neutral Cultural Differentiation. *Anim. Behav.* **82**:101–112. [5]

Yengoyan, A. 1968. Demographic and Ecological Influences on Aboriginal Marriage Sections. In: Man the Hunter, ed. R. B. Lee and I. DeVore. Chicago: Aldine. [6]

Yip, M. 2002. Tone. Cambridge: Cambridge Univ. Press. [12]

Yoffee, N. 1993. Too Many Chiefs? In: Archaeological Theory: Who Sets the Agenda?, ed. N. Yoffee and A. Sherratt, pp. 60–78. Cambridge: Cambridge Univ. Press. [6]

Yopak, K., T. Lisney, S. E. Collin, et al. 2010. Brain Scaling from Sharks to Primates: A Highly Conserved Vertebrate Pattern. *PNAS* **107**:12946–12951. [16]

Zahavi, A. 1975. Mate Selection: A Selection for a Handicap. *J. Theor. Biol.* **53**:205–214. [3]

Zahavi, A., and A. Zahavi. 1997. The Handicap Principle: A Missing Piece of Darwin's Puzzle. New York: Oxford Univ. Press. [20]

Zahn-Waxler, C., M. Radke-Yarrow, and R. King. 1979. Child Rearing and Children's Prosocial Initiations Towards Victims of Distress. *Child. Devel.* **50**:319–330. [14]

Zeshan, U., and C. de Vos, eds. 2012. Sign Languages in Village Communities. Anthropological and Linguistic Insights. Sign Language Typology 4. Berlin: Mouton de Gruyter. [12]

Zeshan, U., and S. Panda. 2011. Reciprocal Constructions in Indo-Pakistani Sign Language. In: Reciprocals and Semantic Typology, ed. N. Evans et al., pp. 91–114. Amsterdam: John Benjamins. [13]

Zhong, C. B., V. K. Bohns, and F. Gino. 2010. A Good Lamp Is the Best Police: Darkness Increases Dishonesty and Self-Interested Behavior. *Psychol. Sci.* **21**:311–314. [19]

Ziman, J., ed. 2000. Technological Innovation as an Evolutionary Process. Cambridge: Cambridge Univ. Press. [11]

Zipf, G. 1935. The Psycho-Biology of Language. Boston: Houghton Mifflin. [13]

Zuckerman, P. 2008. Society without God: What the Least Religious Nations Can Tell Us About Contentment. New York: New York Univ. Press. [20]

Zuidema, W., and B. De Boer. 2009. The Evolution of Combinatorial Phonology. *J. Phonet.* **37**:125–144. [13]

Subject Index

www.ingramcontent.com/pod-product-compliance
Lightning Source LLC
LaVergne TN
LVHW052017200926
848481LV00005B/1520